Persistent Viruses

Academic Press Rapid Manuscript Reproduction

Proceedings of the 1978 ICN—UCLA Symposia on Molecular and Cellular Biology held in Keystone, Colorado, February 1978

ICN–UCLA Symposia on Molecular and Cellular Biology
Volume XI, 1978

PERSISTENT VIRUSES

edited by

JACK G. STEVENS
Department of Microbiology and Immunology
University of California, Los Angeles
Los Angeles, California

GEORGE J. TODARO
National Cancer Institute
National Institutes of Health
Bethesda, Maryland

C. FRED FOX
Department of Microbiology
and Molecular Biology Institute
University of California, Los Angeles
Los Angeles, California

ACADEMIC PRESS NEW YORK SAN FRANCISCO LONDON 1978
A Subsidiary of Harcourt Brace Jovanovich, Publishers

ACADEMIC PRESS, INC.
111 Fifth Avenue, New York, New York 10003

United Kingdom Edition published by
ACADEMIC PRESS, INC. (LONDON) LTD.
24/28 Oval Road, London NW1 7DX

Library of Congress Cataloging in Publication Data

ICN-UCLA Symposium on Persistent Viruses, Keystone, Colo., 1978.
Persistent viruses.

(ICN-UCLA symposia on molecular and cellular biology ; v. 11)
Includes index.
1. Host-virus relationships—Congresses. 2. Virus diseases, Slow—Congresses. 3. Viruses—Congresses. I. Stevens, Jack G. II. Todaro, George J. III. Fox, C. Fred. IV. ICN Pharmaceuticals, Inc. V. California. University. University at Los Angeles. VI. Title. VII. Series: ICN-UCLA symposium on molecular & cellular biology ; v. 11.
QR482.I17 1978 616.01'94 78-12077
ISBN 0-12-668350-6

PRINTED IN THE UNITED STATES OF AMERICA

CONTENTS

D. *Genetic Variability and Expression of Persistent Viruses*

II. PERSISTENT VIRUSES AND DISEASE

A. *Hepatitis Viruses*

C. *Role of Persistent Viruses in Miscellaneous Diseases*

PREFACE

The importance of persistent viruses in human disease is well appreciated, but pathogenetic mechanisms have generally been poorly understood. With the recent advent of improved methods for study, many of these mechanisms have been definitively characterized. This meeting was organized to disseminate current information concerning viral persistence and its role in disease and to promote interactions between workers investigating diverse aspects of these agents. This meeting was the first since the 1974 ICN–UCLA Symposium on Mechanisms of Virus Disease to focus on pathobiological mechanisms in animal virology.

We wish to express our appreciation to the program participants, whose contributions to both the plenary and poster sessions made this meeting a success. We gratefully acknowledge the ongoing financial support from ICN Pharmaceuticals and a generous contract from the National Institutes of Health (National Institute of Neurological and Communicative Disorders and Stroke, National Institute of Allergy and Infectious Diseases, Division of Cancer Research Resources & Centers/National Cancer Institute, National Institute of General Medical Sciences, Fogarty International Center, Viral-Oncology Program/National Cancer Institute), which helped defray the invited speakers' travel expenses. Our special thanks go to Dr. Earl Chamberlayne of the Fogarty International Center for his efficient and helpful service in the capacity of Project Officer for the NIH contract. Finally, we are indebted to a talented collaboration of women, whose clerical and administrative expertise brought this meeting to fruition: Carolyn Hench, Kathy Morris, Bobbi Ottis, Brenda Seidman, Fran Stusser, and Donna Vratari.

ALTERATION OF ESCHERICHIA COLI OUTER MEMBRANE PROTEINS BY PROPHAGES

A MODEL FOR BENEVOLENT VIRUS-CELL INTERACTION

Gordon Edlin[1]

Department of Genetics, University of California, Davis, California 95616

ABSTRACT Lambda lysogens of *E. coli* reproduce more rapidly than isogenic nonlysogenic strains when mixed populations are grown together in glucose-limited chemostats. Lysogens of Phages P1, P2 and Mu also show increased reproductive fitness when grown with the corresponding nonlysogens in glucose-limited chemostats. When the same strains are grown together in batch cultures, in which growth may be limited or not by the glucose concentration, lysogens and nonlysogens show the same reproductive fitness. The outer membrane proteins of bacteria grown in glucose-limited chemostats display a different composition from the same bacteria grown in batch cultures with glucose. In addition the presence of the prophage changes the proportion of the major outer membrane proteins when the lysogens are grown in chemostat cultures. It is suggested that the prophage can modify the membrane proteins of the bacteria by lysogenic conversion and enhance the fitness of the bacteria in certain environments. This seems to be comparable to the situation in animal cells, in which the growth and membrane properties are changed by viral infection or by induction of viral genes. A model is presented suggesting that oncogenes or protoviruses may enhance the growth and development of animal cells under certain conditions. The widespread occurrence of viral genes in animal cells may have resulted from the natural selection of animal cells with increased reproductive fitness.

INTRODUCTION

Lysogeny is the process in which a temperate bacteriophage infects a bacterium without causing lysis and directs the integration of phage DNA into the chromosome of the bacterium. Thereafter, the phage DNA (the prophage) is repli-

[1]Department of Genetics, University of California, Davis, California 95616

ISBN 0-12-668350-6

cated along with the bacterial DNA. The growth and physiology of the bacterial lysogen are indistinguishable from that of the non-lysogen except that the lysogen is immune to infection by homologous phage. This immunity has generally been used to explain the evolutionary selection of lysogenic bacteria (1).

Bacteriophages such as lambda, P2 and Mu all integrate their DNA into the *E. coli* chromosome. Lambda has one specific integration site and can use either phage or bacterial recombination enzymes to integrate the prophage DNA. P2 integrates at several chromosomal locations and is unusual in that the prophage is not inducible by ultraviolet light (2). Mu can integrate at many sites within any gene of the *E. coli* chromosome by "illegitimate recombination" that does not require the usual recombination enzymes. It has been pointed out that Mu lysogens have properties that are similar to eucaryotic cells infected by SV40 virus (3).

The stable association of phage and bacteria in lysogeny appears to be widespread in nature. Most bacteria isolated in nature possess several prophages as well as other extrachromosomal elements (4). In some cases it can be shown that the prophages modify the surface properties of lysogenic bacteria as well as confering immunity to infection by phage of the same type (5). The survival and reproduction of bacteria in nature depend on their ability to attach to specific surfaces or tissues. In other words, a bacterial colony must become established in a favorable ecological niche. It might be the mucosal lining of the intestine, a root on a legume, or a habitable rock in a river bed (6, 7, 8). The ability of bacteria to attach to surfaces by formation of a "glycocalyx" is often correlated with lysogeny (9). Moreover, there are specific examples of lysogenic conversion in which the prophage genome alters biochemical properties of the bacteria. The O-antigens in *Salmonella anatum* are changed by the phages ε^{15} and ε^{34}. Also harmless strains of *Corynebacterium diphtheriae* became toxin producers in the presence of phage β (5).

Most of the lysogenic bacteria survive in nature in harmony with their environments and prophage. In fact, the human body contains more bacteria in the intestinal tract than all the rest of the bodys' cells put together, and presumably we could not survive without these bacteria (10). To cause a disease a pathogen must multiply in host tissues, resist the host defenses and damage the host in some way (8). Bacteriophage may play a special role in the reproductive fitness of bacteria in particular environments by modifying bacterial colonization or enhance growth. According to this view the symbiotic association of prophages and bacteria is a consequence of natural selection and is clearly of evolutionary advantage to phage and bacterium. Not only do lysogenic bacteria acquire immunity to infection by phage of a similar type but these lysogens may be reproductively more fit. Fitness, in this sense, refers to the contribution that lysogenic bacteria make to the gene pool of successive generations (11). The prevalence of lysogenic bacteria found in nature is in accord with these views.

We have shown that lambda lysogens of E. coli are more fit than non-lysogens when both strains are grown together in glucose-limited chemostats. In batch cultures at any glucose concentration both strains reproduce at identical rates (12, 13). Lysogens carrying the prophages P1, P2 or Mu are also reproductively more fit than non-lysogens in chemostats (14). The increased fitness of the lysogenic strains correlates with changes in the composition of the major outer membrane proteins.

These observations are similar to changes observed in the growth rates and membranes of animal cells transformed by viruses. A basis for these changes is found in the protovirus and oncogene hypotheses (15, 16, 17). The association of viral genes with animal cells has a long evolutionary history (18, 19) and there is evidence that viral gene expression alters membranes (20, 21).

A model is presented suggesting that animal cells carrying viral genes or genomes result from the natural selection of cells with increased reproductive fitness in certain environments. In this view viral genes are as essential to cell survival as are cellular genes. Transformation and unregulated growth of cancer cells is an anomaly. It is perhaps analogous to the detrimental hyperimmune response of an animal to a harmless viral or bacterial infection (22).

RESULTS

Physiological Experiments. In order to test for the reproductive fitness of one organism versus another, the frequency of each organism in a mixed population must be measured. In the case of bacteria growing together in mixed populations, each genotype must be distinguishable from the others. We have devised techniques for comparing the relative fitness of lysogenic and non-lysogenic bacteria growing together under a number of different growth conditions.

Prophage are maintained in the integrated state in lysogenic bacteria by a repressor protein that binds to an operator site on the DNA and prevents expression of all prophage genes that would be required to synthesize phage particles. The maintenance of the prophage in the bacterial DNA requires the continuous presence of functional repressor proteins. Phage mutants can be isolated which produce a repressor protein that is thermolabile. Bacteria which are lysogenic for this class of mutant prophages are stable lysogens at low temperatures (25°C-30°C) but are induced to produce phage and are lysed at high temperatures (38°C-42°C). Mutants producing thermolabile repressor proteins have been isolated for the bacteriophages lambda, P1, P2 and Mu (13, 14).

It is a simple matter to determine if bacteria are lysogenic or not using these mutant phages. A lysogenic bacterium will produce a visible colony when plated on nutrient agar and incubated at 30°C, but will fail to produce a visible colony at 42°C since the bacterium will lyse. In a mixed population of lysogenic and non-lysogenic bacteria, the frequency of each type is easily determined. Bacteria are plated on nutrient agar plates and incubated at 30°C and 42°C. The number of colonies appearing at 30°C are a measure of the total number of bacteria in the population whereas the colonies appearing at 42°C measures the number of non-lysogenic bacteria. The number of lysogenic bacteria is the difference between these two numbers.

This method has been used to examine the relative reproductive fitness of isogenic lambda lysogens and non-lysogens. In these experiments an important control is to ensure that the phage are not transmitted from the lysogenic to the non-lysogenic bacteria. This is accomplished in the lambda lysogens by using lambda mutants that cannot be induced from the prophage state (*ind*$^-$). In addition, these lambda phage carry nonsense mutations in another gene essential for phage development (*sus* J), so that, in the rare instance that a prophage is excised, it cannot develop and produce viable phage, which can subsequently infect the non-lysogenic bacteria.

Figure 1 shows the results of an experiment in which a mixture of *E. coli* lambda lysogens and non-lysogens are grown together in either batch or chemostat cultures with limiting glucose as the energy source. In batch cultures both the lambda lysogen and non-lysogen reproduce at identical rates. Except for daily fluctuations the frequency of each strain in the population remains constant for at least forty generations. The same result is obtained in batch cultures with either limiting or excess glucose or with other carbon sources such as lactose or glycerol.

A very different result is observed when the lambda lysogen and non-lysogen are grown together in a glucose-limited chemostat. The lambda lysogen is reproductively more fit and rapidly becomes the most frequent strain of the population, reaching a frequency of 99.9% within 30-40 generations. It is interesting that in all the chemostat experiments that have been done, the non-lysogenic bacteria are never completely eliminated from the population although their frequency may become as low as 0.01%. The increased reproductive fitness of

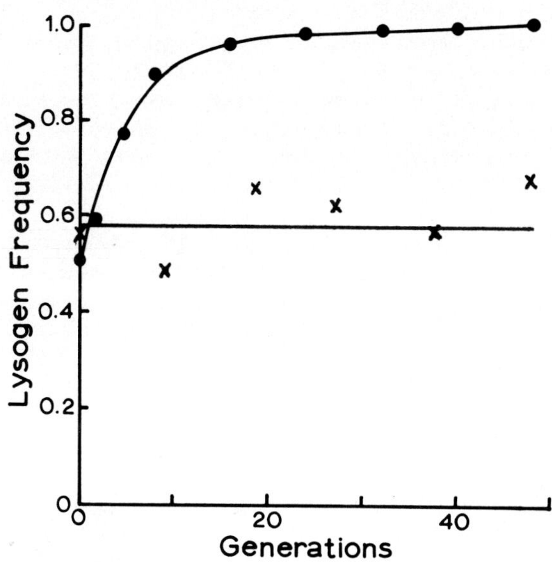

FIGURE 1. Reproductive fitness of the λ lysogen in batch and chemostat cultures. E. coli strain (AB) and the lysogen (ABλ) were grown together in either batch or chemostat cultures. Bacteria were grown in M9 minimal medium plus 0.01% glucose and 20 μg/ml methionine. The batch cultures were diluted 10^6 daily into fresh medium. The chemostat cultures were grown continuously with a generation time of 4 hours. Samples were plated daily on nutrient agar and incubated at 30°C and 42°C. Only non-lysogens give colonies at 42°C. x - - - x, frequency of ABλ in a mixed population of AB and ABλ growing in a batch culture; ● - - - ●, frequency of ABλ in a mixed population of AB and ABλ growing in a chemostat culture.

lambda lysogens in chemostat cultures also is observed for other carbon sources. The fitness does not depend on the generation time of the bacteria in the chemostat between generation times of 3-8 hours.

To fully appreciate these results it is important to understand the essential difference between growth in glucose-limited batch cultures and glucose limited chemostat cultures. In a glucose-limited batch culture (0.01% glucose) bacteria grow exponentially at the maximum generation time for the particular medium until the glucose is exhausted. Thus, for glucose concentrations between 0.01% and 0.4% bacteria grow with the same generation time - only the final yield of cells varies. In batch cultures the bacteria are not starved until the carbon source is actually exhausted, at which point growth stops. In chemostat cultures, on the other hand, bacteria are constantly starved for glucose. The generation time in chemostat cultures in determined by the rate at which fresh medium flows into the culture. In the culture vessel itself the concentration of glucose is essentially zero. Each drop of medium that flows in is immediately utilized.

Although most bacteria are grown in batch cultures in the laboratory for convenience, chemostat cultures probably more closely resemble the conditions that exist in nature. Smith has pointed out that the generation time for bacteria growing in the gut of mice is estimated to be 6 hours (7). Woods and Foster have also discussed continuous versus batch cultures as they relate to actual *in vivo* growth conditions bacteria encounter (23). A portion of their discussion appears below.

> "When the host is complex, and has differentiated tissues, it will have some form of circulatory or translocatory system. The bacterium will therefore be in an environment which is constantly renewed in the sense that there will be, due to the activities of the host, both a continual replenishment of growth metabolites and a continual removal of waste products of metabolism. The situation is essentially similar for organisms living in an intestinal tract. The system is therefore analogous in many ways to continuous culture and less like the batch cultures that are the source of so much of the bacterial material used for metabolic studies *in vitro*."

Having shown that lambda lysogens of *E. coli* are reproductively more fit in continuous chemostat cultures, we tested the generality of the phenomenon. *E. coli* strains lysogenic for the phages P1, P2 and Mu were grown together with the isogenic non-lysogenic strains. In each case it was found that in batch cultures the reproductive rates were identical whereas in chemostat cultures the lysogens reproduced more rapidly (14).

Further conformation of the fitness of lysogenic bacteria in chemostat cultures was obtained by finding a condition in which the relative fitness could be reversed. Usually all cultures are grown aerobically. However, when the chemostat cultures are maintained under anaerobic conditions, the non-lysogenic bacteria reproduce more rapidly than the lambda lysogens. Figure 2 shows the result of such an experiment. A mixed population of the lambda lysogen and non-lysogen is

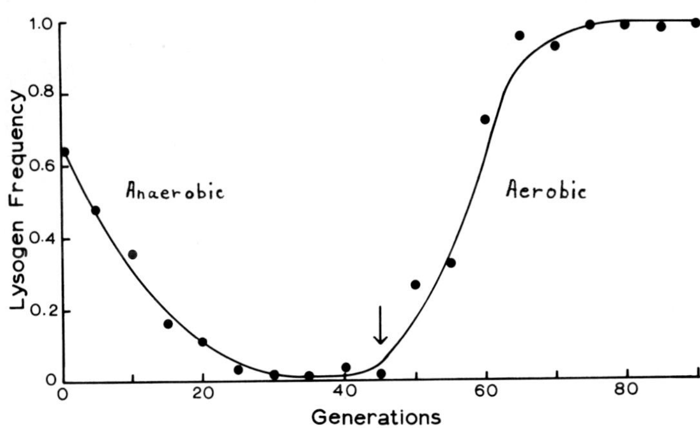

FIGURE 2. Reproductive fitness under aerobic and anaerobic growth. E. coli strain (AB) and the lysogen (ABλ) were grown together in a glucose limited chemostat (0.01%) under anaerobic conditions. A mixture of N_2-CO_2 (95%-5%) was bubbled through the culture. After several days of anaerobic growth, the culture was switched to aerobic growth as indicated by the arrow. The frequency of the lysogen and non-lysogen in the population was measured by plating samples on nutrient agar and incubating the plates at 30°C and 42°C. Only the non-lysogens give colonies at 42°C.

grown anaerobically in a chemostat culture. When the frequency of the lambda lysogen has fallen to about 1%, the chemostat is switched to aerobic growth. The lysogenic bacteria reproduce more rapidly and quickly predominate in the population.

Biochemical Experiments. What is the biochemical basis for the increased reproductive fitness of lysogenic bacteria which are growing aerobically in glucose-limited chemostat cultures? Since the same concentration of glucose is presumably equally available to lysogenic and non-lysogenic bacteria, a difference in growth rate must reflect increased uptake of glucose or more efficient utilization.

The outer membrane of E. coli and S. typhimurium has been shown to regulate the permeability of the bacterium to molecules of particular sizes. The proteins that are involved in this process are the most abundant proteins in the outer membrane and have been termed "porins" (24, 25, 26). Our hypothesis is that the increased fitness of the lysogenic bacteria in chemostat cultures is due to more efficient uptake of glucose from the medium. As described earlier the concentration of glucose in a chemostat culture is essentially zero. Each drop of fresh medium is immediately used up by the bacteria. If the prophages were able to modify the outer membrane proteins, the permeability of glucose might be changed. Our initial approach was to examine the total membrane proteins of lysogenic and non-lysogenic strains of E. coli grown in batch and chemostat cultures. Preliminary experiments indicated that most of the differences were confined to the outer membrane proteins so further experiments were designed to examine the outer membrane proteins under the conditions that affect reproductive fitness.

Figure 3 shows the outer membrane proteins from E. coli strain ABλ which have been separated by electrophoresis on a SDS-polyacrylamide gel. The outer membrane proteins displayed in column A are from lambda lysogens growing exponentially in batch culture with excess glucose. The same pattern of protein bands is obtained when the bacteria are grown in batch culture with limiting glucose (unpublished results). The non-lysogen also gives the same pattern when grown in batch cultures. That is, both lysogenic and non-lysogenic bacteria have identical outer membrane protein compositions if grown in batch cultures.

Column B shows the outer membrane protein from the lambda lysogen grown in a glucose limited chemostat. The relative amounts of the various outer membrane proteins is quite different when the lysogen is grown in batch or chemostat cultures. Precisely the same differences are observed if the non-lysogen is grown in batch or chemostat cultures with one significant exception. The most abundant outer membrane proteins are labeled I and II in figure 3. Actually band I consists of 2 proteins which are poorly separated. When the lambda lysogen is grown in a glucose-limited chemostat, protein II is reduced in amount by 40%-50% as compared to the amount found when the non-lysogen is grown in the chemostat (column B, figure 4; and unpublished results). In general, when the non-lysogen is grown in the chemostat or in batch culture, the amounts of protein I and II are identical. If the lambda lysogen is grown in the chemostat, the amount of protein II is always markedly reduced (column B, figure 3). This change in the relative abundance of the major outer membrane proteins correlates with the increased fitness that is observed for the

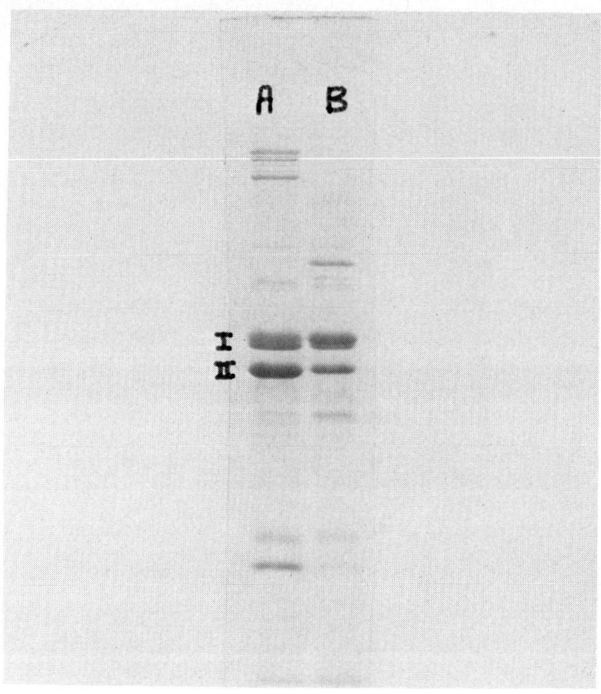

FIGURE 3. Outer membrane proteins from E. coli strain AB257λ. Lysogenic bacteria were grown in M9 minimal medium plus 0.01% glucose and 20 μg/ml methionine. 500 ml of culture was collected by centrifugation and disrupted by sonication. Membranes were pelleted by centrifugation and the outer membrane was prepared by solubilizing the inner membrane in 2.5% sarkosyl. After sarkosyl treatment the outer membrane fraction was pelleted by centrifugation and resuspended in T buffer (Tris, 10 mM. EDTA, 5 mM; betamercaptoethanol, 5 mM; pH 7.8). 60 μg of protein was denatured by boiling in SDS for 3 minutes and applied to a 14% SDS-polyacrylamide gel according to the method of Ames (27).

A. AB257λ grown in batch culture.
B. AB257λ grown in chemostat culture.

lambda lysogen grown in a chemostat culture (figure 1).

The results suggest that the reproductive fitness of E. coli bacteria is determined by the relative amounts of the major outer membrane proteins. Figure 2 shows that the fitness of the non-lysogen and lysogen is different in aerobic and anaerobic chemostat cultures. What happens to the outer membrane proteins under these conditions? Figure 4 shows the outer membrane proteins for the non-lysogenic strain under

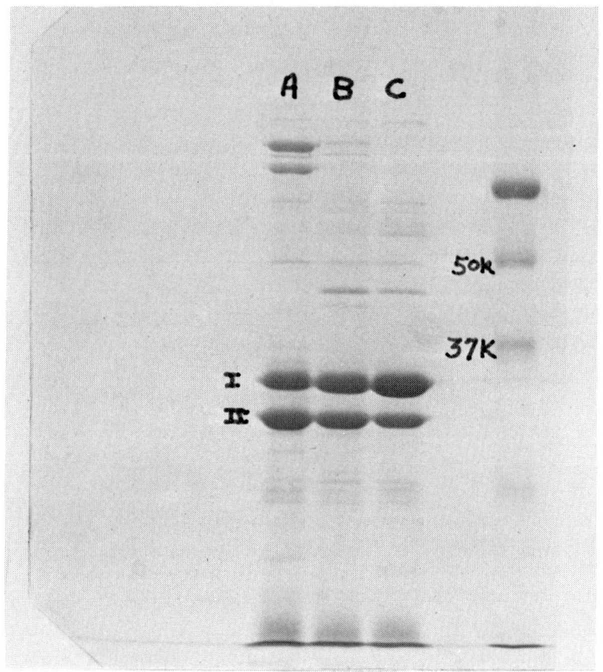

FIGURE 4. Outer membrane proteins from E. coli strain AB257. Bacteria were grown and membranes were prepared as described in figure 3. 60 μg of protein was applied to a 10% SDS-polyacrylamide gel.

A. Strain AB257 grown in batch culture.
B. Strain AB257 grown in an aerobic chemostat culture.
C. Strain AB257 grown in an anaerobic chemostat culture.

three different growth conditions. Column A of figure 4 shows the pattern for the non-lysogen grown in batch culture which is the same as that observed for the lysogen grown in batch culture (column A, figure 3). Column B and C of figure 4 show the outer membrane proteins of the non-lysogen grown in a chemostat either aerobically (B) or anaerobically (C). Under anaerobic conditions the relative amount of protein II has decreased significantly as compared to the amount seen in the aerobic chemostat culture. Figure 2 shows a reversal of fitness of the lambda lysogen and non-lysogen under aerobic and anaerobic conditions. Figure 4 shows that the non-lysogen undergoes a corresponding change in the relative amounts of the outer membrane proteins I and II under aerobic and anaerobic growth. The change in protein II in the non-lysogen grown anaerobically in the chemostat (column C, figure 4) appears to be the same as that caused by the lambda prophage when the lysogen is grown aerobically in the chemostat (column B, figure 3). Thus, there appears to be a strong correlation between the reproductive fitness of bacteria and the relative proportions of the most abundant outer membrane proteins. These outer membrane proteins are thought to mediate the transport of nutrients from the environment (24, 25).

DISCUSSION

The studies described above and in previous reports (13, 14) show that prophages can enhance the reproductive fitness of their host bacteria in certain environments. In batch culture conditions lysogens and non-lysogens grow equally well. In chemostat cultures lysogens have a reproductive advantage as long as oxygen is supplied, but are at a growth disadvantage under anaerobic conditions. The mechanism underlying these various growth changes seems to reside in the distribution and amounts of various outer membrane proteins. In particular, the most abundant outer membrane proteins (porins) change quantitatively and in a predictable manner under conditions where bacterial fitness changes.

These observations are understandable within an evolutionary framework. Natural selection operates on reproduction and fitness. Organisms that reproduce more effectively become the most prevalent members of a population. Bacteria and prophage mutually benefit from increased reproductive fitness. It seems reasonable that phage genes would have evolved to assist bacterial growth under certain environmental conditions. Since viruses readily disseminate genetic information within a population of bacteria (4), bacteria would eventually benefit by incorporation of a set of "fitness genes". In retrospect, adaptation of the outer membrane of the bacteria cell to optimize growth in environments of limited or depleted nutrient supply would be of evolutionary advantage. Several different temperate bacteriophages are able to increase the fitness of *E. coli* in glucose limited chemostats (13, 14) and in each instance a quantitative change is observed in the major outer membrane proteins (unpublished experiments).

The prophage-bacteria experiments outlined here have obvious similarities to the oncogene-cell model in animals (17). The analogy is shown below.

Similarities	Phage-Bacteria	Oncogene-Cell
viral genes in DNA	yes	yes
increased growth rate	yes	yes
membrane changes	yes	yes
increased glucose uptake	probable	yes

A Model of Benevolent Virus-Cell Association

I would like to expand briefly on these observations and present a model that focuses on the beneficial aspects of virus-cell interactions. Oncogenes have a long evolutionary history of association with cells (18, 19). I propose that this association has resulted from the natural selection of cells with increased reproductive fitness or function within the animal. Consider an animal which lives in a state of semi-starvation or of limited nourishment. Cells that can reproduce in nutritionally impoverished environments or that can take up nutrients more effectively are more fit, since their genetic information will be passed on to subsequent generations. This may have led to an early symbiotic association of viruses and cells.

These ideas have additional implications for human disease and cancer. Consider the following quotations from several recent articles.

> "Today, cancer viruses rarely win the headlines they rated when the 'War on Cancer' was declared some 6 years ago." (28).

> "Cancer in both animals and humans is acknowledged to be a multifactorial disease in which genetic background, host defenses, and environmental influences play critical roles." (29).

> "The factors known to affect the susceptibility of animals and cells to virus infection include the effects of age, sex, genetic constitution, nutrition, body temperature, metabolic activity, season of the year, stress, trauma and intercurrent infection." (30).

> "There is increasing epidemiological evidence that nutrition plays a dominant role in the pathogenesis of several types of human cancers." (31).

It should be apparent from these few quotes that singling out viruses as *the causative* agents in infectious disease and cancer is a great oversimplification. Viruses and oncogenes may contribute genetic information to the growth and development of the cell which has been and may still be essential. Perhaps future insights will provide an understanding of the factors that have contributed to the evolutionary association of viruses and cells and to the factors that can upset this otherwise harmonious and beneficial association of virus and cell. As has been true in the past, an understanding of the prophage-bacteria association may provide the basis for an animal cell model.

ACKNOWLEDGMENTS

I want to thank Crystal DiModica who managed to type between the lines to the very end. I am especially indebted to John Stanovich for sticking to the gels through thick and thin.

REFERENCES

1. Campbell, A. (1961). Evolution 15, 153.
2. Bertoni, L.E., and Bertoni, G. (1971). Adv. Genet. 16, 199.
3. Razzaki, T., and Bukhari, A. I. (1975). J. Bacteriol. 122, 437.
4. Reanney, D. C. (1976). Bacteriol. Rev. 40, 552.
5. Barksdale, L., and Arden, S. B. (1974). Ann. Rev. Microbiol. 28, 265.
6. Savage, D. C. (1972). Symp. Soc. Gen. Microbiol. 24, 25.
7. Smith, H. (1976). Symp. Soc. Gen. Microbiol. 26, 299.
8. Smith, H. (1977). Bacteriol. Rev. 41, 475.
9. Costerton, J. W., Geesey, G. G. and Cheng, K-J. (1978). Sci. Amer. 238, 86.
10. Keusch, G. T. (1974). Natural History 83, 74.
11. Demetrius, L. (1977). Amer. Natur. 111, 1163.
12. Edlin, G., Lin, L., and Kudrna, R. (1975). Nature 255, 735.
13. Lin, L., Bitner, R., and Edlin, G. (1977). J. Virology 131, 554.
14. Edlin, G., Lin, L., and Bitner, R. (1977). J. Virology 131, 560.
15. Temin, H. (1971). Ann. Rev. Microbiol. 25, 609.
16. Temin, H. (1974). Ann. Rev. Genet. 8, 155.
17. Heubner, R. J., and Todaro, G. J. (1969). Proc. Natl. Acad. Sci. USA 64, 1087.
18. Benveniste, R. E., and Todaro, G. J. (1977). Proc. Natl. Acad. Sci. USA 74, 4557.
19. Gillespie, D., and Gallo, R. C. (1974). Science 188, 802.
20. Ito, Y. Brocklehurst, J. R., and Dulbecco, R. (1977). Proc. Natl. Acad. Sci. USA 74, 4666.
21. Thomas, L. (January 2, 1978) in "The New Yorker", pp. 40-41.
22. Woods, D. D., and Foster, M. A. (1964). Symp. Soc. Gen. Microbiol. 14, 30.

23. Nakae, T. (1976). J. Biol. Chem. 251, 2176.
24. Nakae, T. (1976). Biochem. Biophys. Res. Commun. 71, 877.
25. Bavail, P., Nakaido, H., and von Meyenburg, K. (1977). Molec. gen. Genet. 158, 23.
26. Ames, G. F. (1974). J. Biol. Chem. 249, 634.
27. Marx, J. L. (1978). Science 199, 161.
28. Klein, P. A., and Smith, R. T. (1977). Ann. Rev. Med. 28, 311.
29. Burrows, R. (1972). Symp. Soc. Gen. Microbiol. 22, 303.
30. Wynder, E. L. (1976). Fed. Prod. 35, 1309.

THE LUCKÉ TUMOR: A MODEL FOR PERSISTENT VIRUS[1] INFECTION AND ONCOGENESIS

Allan Granoff and Robert F. Naegele

Division of Virology, St. Jude Children's Research Hospital, P. O. Box 318, Memphis, Tennessee 38101

ABSTRACT

The frog renal adenocarcinoma (Lucké tumor) and its etiological agent, the Lucké herpesvirus (LHV), offer a unique model for the study of the relationship of persistent virus infection to cancer etiology. A salient feature of the Lucké tumor-virus relationship is the effect of temperature on the presence or absence of virus particles. A variable number of tumor cells *in vitro* or *in vivo* maintained at low temperature ($<12^{o}C$) contain virus particles. In contrast, they are not present in tumor cells *in vitro* or *in vivo* at temperatures above 12°C. Although virions cannot be detected in tumor cells at the higher temperature, virus-specific RNA and membrane antigens are present in such cells.

LHV extracted from naturally occurring Lucké tumors induces virion-free Lucké tumors in developing frog embryos. When virus-free tumor fragments from an induced tumor are incubated at low temperature *in vitro*, herpesvirus can be recovered; the virus is oncogenic when injected into developing embryos.

Preliminary experiments indicate that the LHV genome has a molecular weight of about 66×10^6 suggesting that it may carry only the minimal genetic sequences required for specifying a herpesvirus.

These results will be discussed in the framework of persistent virus infection.

INTRODUCTION

The renal adenocarcinoma of the leopard frog, *Rana pipiens*, commonly called the Lucké tumor, is one of the

[1]This work was supported by USPHS Research Grant CA 07055, Cancer Center Support (Core) Grant CA 21765, Childhood Cancer Research Center Grant CA 08480 from the National Cancer Institute, and by ALSAC.

ISBN 0-12-668350-6

oldest tumors of animals to be assigned a virus etiology (1,2). However, it took some 20 years from the time of its discovery to establish an association of a particular virus, a herpesvirus, with the tumor (3,4). Another 10-17 years passed before there was sufficient evidence to establish, unequivocally, an etiologic role of the herpesvirus in tumor formation (5,6,7). Over the years a body of evidence has accumulated from this system which strengthens the notion that it is an excellent model for the study of persistent virus infection, as well as for herpesvirus-induced oncogenesis (for review see Ref. 8). This paper will briefly review some of the salient features of the Lucké tumor and the causative Lucké tumor herpesvirus (LHV), with emphasis on those features that reflect on persistent virus infection, the subject of this symposium.

GENERAL FEATURES OF THE LUCKÉ TUMOR AND THE LUCKÉ HERPESVIRUS

The major geographical areas where Lucké tumor-bearing frogs are found include the north central and northeastern parts of the United States and adjacent southern Canada. The tumors are usually highly differentiated renal adenocarcinomas. In the past years the frequency of tumor-bearing frogs taken directly from nature varied from 1-9%. However, for reasons as yet unknown, the population of northern *R. pipiens* has been drastically reduced in recent years, affecting the availability of tumor-bearing frogs.

A salient feature of the Lucké tumor, which bears on persistent virus infection, is the effect of temperature on the presence or absence of acidophilic intranuclear inclusion bodies and virus. Tumor cells of frogs, either in hibernation in nature or maintained at low temperature ($<12^{\circ}C$) in the laboratory, contain intranuclear inclusions and herpesvirus particles. The virus can be physically extracted from such tumors for biological and biochemical studies. In contrast, neither inclusions nor virus is found in tumor cells of frogs captured in the spring or summer, or maintained in a laboratory at temperatures above $12^{\circ}C$. Virus particles can be made to disappear from virus-containing tumors by placing the tumor-bearing animal or *in vitro* cultured tumor tissue from such an animal at elevated ($>12^{\circ}C$) temperature. The cells containing virus particles die and those that appear not to contain virus survive and multiply. The opposite effect can be demonstrated by placing frogs with virion-free tumors, or *in vitro* cultured tumor fragments from them, at low temperature

(<12°C). This procedure induces virus replication in a variable number of tumor cells and these cells die, also. It is clear that long-term cell survival is incompatible with virus replication. Table 1 summarizes experiments upon which these conclusions are based.

Since virus replication is not induced in every cell at low temperature (8), it is possible that some cells do not contain a complete virus genome, or, if they do, some cellular or virus factor(s) controls full virus gene expression in these cells. It is also possible that with extended time all cells would eventually produce virus. Whatever the reason, virion-free tumor cells at low temperature must contribute to maintainence of persistent virus infection. It should be emphasized that although the presence or absence of virus particles is temperature-dependent, the tumor is not, *i.e.*, it may be present in an animal regardless of the temperature at which it is held.

LHV is identical in size, morphology, and structure to other herpesviruses of both higher vertebrates (*e.g.*, humans) and lower vertebrates (*e.g.*, fish) and shares with them a number of biological and biochemical features (8). The developmental sequence of LHV in tumor cells also appears similar to other herpesviruses. However, it has only been determined by reconstruction from static images of virus seen by electron microscopy in thin sections of virus-containing tumor cells (17). The virus has not been grown *in vitro* under circumstances where a time sequence can be used to study the kinetics and molecular and biological events of its replicative cycle (18). The latent persistence of LHV in host tumor cells following primary infection is also compatible with the properties of other herpesviruses. The LHV genome persists in infected cells that are not producing virus in a state yet to be discovered and, as will be discussed shortly, the virus genome is only partially expressed under these circumstances.

The effect of temperature on LHV replication and the relationship of this property to establishment of a persistent infection have some counterparts in experimental situations with other mammalian herpesviruses, such as herpes simplex virus and its temperature-sensitive mutants (19,20,21,22,23).

THE LUCKÉ HERPESVIRUS DNA

The DNA of LHV is a linear, double-stranded molecule with a base composition of 45-47% guanine + cytosine (24, 25). We have recently been analyzing, by contour length measurements, the molecular weights of the DNAs of a number

TABLE 1
EVIDENCE FOR THE EFFECT OF TEMPERATURE ON LUCKÉ TUMOR HERPESVIRUS

Nuclear inclusions more frequent in winter and spring than in summer and fall (9).

Frogs held at 20-24°C; no inclusions. Frogs held at 5°C; inclusions (10).

Biopsy of 20-24°C frog tumor; no inclusions. When frogs moved to 4°C, inclusions present after 5-7 months (11).

Fragments of inclusion-positive tumor transplanted to anterior eye; frogs placed at 21-24°C; no virus present after 15 days (12).

Tumor-bearing frogs held at 4°C; when moved to 20-22°C much debris and virus in lumina; no inclusions after 7 days (13).

Fragments of inclusion-negative tumor transplanted to anterior eye; frogs placed at 7-8°C; virus present after 4-7 months (14).

No virus in urine from tumor-bearing frogs held at 25°C; urine from tumor-bearing frogs held at 4°C contains virus (15).

Explants of inclusion-negative tumors held *in vitro* at 7.5^{o}C; virus present after 3 months (16).

---

of herpesviruses from lower vertebrates. A summary of preliminary data is presented in Table 2. As can be seen, the LHV DNA has a molecular weight significantly lower than the DNAs of other herpesviruses from lower vertebrates and much lower than the DNA of human HSV type 1. Our result with channel catfish virus corroborates the value reported by Sheldrick (26). The molecular weight of LHV

TABLE 2
MOLECULAR WEIGHTS OF HERPESVIRUS DNAs FROM A HUMAN AND LOWER VERTEBRATES[a]

Herpesvirus	Genome Molecular Weight ($X10^6$)	
	Range	Average
HSV-Type 1	89.1 - 107.5	94.8
Channel catfish	66.6 - 85.8	79.3
Trout	69.9 - 83.8	78.4
Frog virus 4	73.9 - 79.2	77.2
Lucké tumor	62.5 - 73.4	65.5

[a]Molecular weights were determined by measurements from electron micrographs of contour lengths of each viral genome obtained from purified virions. PM_2 was used as a reference for obtaining the values given.

roughly corresponds to the long DNA segment of unique base sequence, designated "L", present in other herpesvirus genomes (27). This correspondence in size may be fortuitous and of no biological consequence. For example, the size of the LHV "L" segment may be less than that of other herpesviruses and have as well, a short ("S") region (27), the total size being equal to that of the "L" region of other members of this virus group. On the other hand, it may represent minimal but fundamental genetic information required for specification of a herpesvirus, for oncogenic activity of the virus, and perhaps for its ability to establish persistent virus infection.

TRANSMISSION EXPERIMENTS

The first unequivocal evidence that a virus was the etiological agent of the frog renal tumor was provided by Tweedell (5) who demonstrated that *R. pipiens* embryos or larvae injected with herpesvirus-containing cytoplasmic fractions of tumor extracts from cold tumors developed typical Lucké tumors. The animals were held at 20-25^oC and the tumors that developed were virus-free. As will be discussed later, virus can be induced from these virus-free tumors by low temperature treatment. Comparable extracts of normal, adult frog kidneys or of virus-free tumors did not induce tumors. Enveloped LHV, purified by rate zonal centrifugation of cytoplasmic fractions of Lucké tumor cells, has also been shown to be oncogenic (6) as has LHV-containing ascitic fluid from a frog bearing a cold tumor (28).

PERSISTENT VIRUS INFECTION AND THE LUCKÉ TUMOR

We may define a persistent infection as one in which the agent persists in the host for long periods of time with or without some overt manifestation of infection. The agent may or may not be recoverable by usual cultivation methods. According to this definition LHV and the Lucké tumor must be considered an example of a virus -host system with a natural capacity to exhibit persistent infection; neoplastic transformation is an accompaniment of the infection. We will now examine some properties of the Lucké herpesvirus-Lucké tumor system that fulfill the criteria of a persistent virus infection. As discussed elsewhere in this monograph, tumor viruses other than herpesviruses, such as SV40, polyoma virus, and the RNA ones, may also be considered examples of viruses that establish persistent infection usually with accompanying cell transformation. Some analogies between LHV and these viruses, as related to persistent infection, will become apparent in the following discussion.

Initial infection of the proper target cell *in vivo* (most likely pro- or mesonephros) is followed by malignant transformation in the absence of virus replication at temperatures above 12°C. Since virus replication can be induced in a variable number of cells at lower temperatures, all of the LHV genetic information must be present in these cells at the higher temperatures. Several years ago we (7) devised an experiment to fulfill, as nearly as possible, Koch-Henle's postulates for identifying the causal agent of the Lucké tumor. This experiment also exemplifies persistent virus infection by LHV. Briefly, LHV extracted from a naturally occurring cold tumor was injected into frog embryos which were then grown at 20-22°C. These animals developed typical virus-free renal adenocarcinomas between 3 and 8 months after inoculation. No tumors developed in surviving sham-infected controls. Fragment tissue cultures of an induced tumor were made and placed at either 22°C or 7.5°C and examined periodically by light and electron microscopy for the presence of intranuclear inclusions and herpesvirus. By 70 days 39% of the tumor cells maintained at the low temperature contained virus, while those maintained at the higher temperature were virus-free.

This part of the experiment demonstrated that the LHV genome had persisted in the tumor and was induced by low temperature treatment. That the virus observed was the same as the one inoculated was established by isolating virus from the virus-containing low temperature tissue culture tumor fragments and inoculating it into frog embryos: 65% of the embryos developed tumors in 4-5 months. Controls inoculated with homogenates of the virus-free high temperature tumor fragments incubated at 22°C did not develop tumors.

It is clear from the above experiment and others on induction of virus from virion-free tumor fragments *in vitro* (Table 1), that the intact host does not control virus replication *in vivo* (*e.g.*, by some humoral or cellular immune mechanism) and that temperature alone is the inducing agent. However, it is possible that at the cellular level, a cellular temperature-dependent factor may be operative in regulating the expression of LHV genetic information. Therefore, both cell and virus may have temperature-sensitive functions, each of which contributes to the observed temperature effect and to the establishment of persistent virus infection. As mentioned earlier, a cell function may be particularly important at low temperature in modulating the number of cells that become productively infected and die and those that survive because virus replication is not induced.

The question of whether there is some expression of LHV information in virion-free cells at temperatures >12°C has been answered by two types of experiments. Evidence for transcription of LHV information in virus-free Lucké tumor cells has been obtained from molecular hybridization experiments (29). Normal frog kidney RNA did not bind to viral DNA and purified virus-specific RNA from tumors did not hybridize to cellular DNA. However, labeled RNA extracted from tumors hybridized with DNA extracted from purified LHV. Thus, virus-free tumor cells contain viral transcripts and we have recently found evidence for a viral translation product in them (30). By indirect immunofluorescence we were able to demonstrate LHV-specific antigen in the membranes of virus-free tumor cells regardless of whether the tumor also contained some cells with virus, *i.e.*, both low and high temperature virus-free tumor cells were membrane antigen positive. As expected, virus membrane antigen could also be detected in cells containing virus particles. A variety of normal *R. pipiens* cells were negative for LHV antigen.

DISCUSSION

Table 3 summarizes characteristics of the Lucké herpesvirus that make it a candidate for further studies, not only of viral oncogenesis, but of persistent virus infection as well. It is quite clear that at temperatures above 12°C, tumor cells are nonpermissive for virus replication but permissive for partial LHV gene expression and for malignant transformation. At lower temperature, the transformed cell becomes permissive for virus replication and cell death follows. Since low temperature does not induce virus replication in every cell (8), although some LHV genetic information is expressed in the nonproducing cells (30), the persistent infection is maintained and viral genetic information is kept intact under both conditions. As already discussed there are a number of possibilities for explaining the absence of virus replication in all cells at low temperature; cellular control, subgenomic viral information, or asynchrony of the induction mechanism and need for extended low temperature treatment.

There is little doubt that the Lucké herpesvirus and the tumor induced by it constitute a profitable model system for exploitation of viral oncogenesis and persistent virus infection. It must be emphasized that this is a naturally occurring tumor and persistent virus infection, unlike many of the laboratory created experimental systems, several of which have been discussed during this symposium. Whether the establishment of persistent infection is requisite for tumor formation - or vice versa - is unknown. However, only tumor cells appear to be infected and there is no evidence for sustained infection of normal kidney cells or of cells of other organs (R. Naegele, unpublished observation). This is unlike other tumor-virus-cell systems where normal cells may be persistently infected (*e.g.*, RNA tumor viruses) or where the phenotype of the transformed cell can revert to a normal one by experimental manipulation of temperature (*e.g.*, cells transformed by ts mutants of herpes simplex virus, adenovirus, SV40, polyoma). Induction of virus replication in other virus-cell systems frequently requires the use of antimetabolites (*e.g.*, BrdU) unlike the Lucké tumor where induction occurs in nature when tumor-bearing frogs go into hibernation.

Further research on this potentially useful model system will be delayed for several reasons. (1) The *R. pipiens* population where the tumor occurs has been drastically reduced in recent years affecting the resource of tumor-bearing frogs. (2) There is no available cell culture system that would provide adequate amounts of virus

TABLE 3
CHARACTERISTICS OF LUCKÉ HERPESVIRUS PERSISTENT INFECTION

Natural host	*R. pipiens*
Tissue and cells in which virus persists	Kidney; adenocarcinoma; epithelial cells
Effect of temperature on virus gene expression	
$<12^{\circ}C$	Permissive; virus replication; cell death
$>12^{\circ}C$	Nonpermissive; virus-specific RNA and membrane antigen(s) present; cell proliferation
Physical state and relationship of virus genome to cell genome	Unknown

and which would allow the various parameters of virus-cell interaction to be studied under controlled conditions. There has been a report that LHV can be grown *in vitro* in a frog embryo pronephros cell line and that cells appear to be persistently infected for a limited number of cell passages (18). Unfortunately, we have been unable to confirm this using the same cell line and numerous preparations of LHV.

Significant advances await the isolation and propagation of LHV in tissue culture. With this accomplished, the Lucké tumor and the Lucké herpesvirus experimental model may provide new and relevant information on both viral oncogenesis and persistent virus infections.

ACKNOWLEDGMENTS

We are indebted to Dr. Gopal Murti, of our department, for contour length measurements of herpesvirus DNAs.

REFERENCES

1. Lucké, B. (1934). Amer. J. Cancer 20, 352.
2. Lucké, B. (1938). J. Exp. Med. 68, 457.
3. Fawcett, D. W. (1956). J. Biophys. Biochem. Cytol. 2, 725.
4. Lunger, P. D. (1964). Virology 24, 138.
5. Tweedell, K. S. (1967). Cancer Res. 27, 2042.
6. Mizell, M., Toplin, I., and Isaacs, J. J. (1969). Science 165, 1134.
7. Naegele, R. F., Darlington, R. W., and Granoff, A. (1974). Proc. Natl. Acad. Sci. 71, 830.
8. Granoff, A. (1973). In "The Herpesviruses" (A. Kaplan, ed.), pp. 627-640. Academic Press, New York.
9. Lucké, B. (1952). Ann. N. Y. Acad. Sci. 54, 1093.
10. Roberts, M. E. (1963). Cancer Res. 23, 1709.
11. Rafferty, K. A., Jr. (1965). Ann. N. Y. Acad. Sci. 126, 3.
12. Zambernard, J., and Mizell, M. (1965). Ann. N. Y. Acad. Sci. 126, 127.
13. Zambernard, J., and Vatter, A. E. (1966). Cancer Res. 26, 2148.
14. Mizell, M., Stackpole, C. W., and Halperen, S. (1968). Proc. Soc. Exp. Biol. Med. 127, 808.
15. Granoff, A., and Darlington, R. W. (1969). Virology 38, 197.
16. Breidenbach, G.P., Skinner, M.S., Wallace, J.H., and Mizell, M. (1971). J. Virol. 7, 679.
17. Lunger, P.D., Darlington, R.W., and Granoff, A. (1965). Ann. N. Y. Acad. Sci. 126, 289.
18. Tweedell, K., and Wong, W. Y. (1974). J. Natl. Cancer Inst. 52, 621.
19. Macnab, J. C. (1974). J. Gen. Virol. 24, 143.
20. Takahashi, M., and Yamanishi, K. (1974). Virology 61, 306.
21. Darai, D., and Munk, K. (1976). Intl. J. Cancer 18, 469.
22. Braun, R., Flugel, R. M., and Munk, K. (1977). Nature 265, 744.
23. O'Neill, F. J. (1977). J. Virol. 24, 41.
24. Wagner, E. K., Roizman, B., Savage, T., Spear, P. G., Mizell, M., Durr, F. E., and Sypowicz, D. (1970). Virology 42, 257.
25. Gravell, M. (1971). Virology 43, 730.
26. Sheldrick, P. (1978). In "Oncogenesis and Herpesviruses" (3rd International Symposium, Cambridge, Mass.), in press.

27. Honess, R. W., and Watson, D. H. (1977). J. Gen. Virol. 37, 15.
28. Naegele, R. F., and Granoff, A. (1972). J. Nat. Cancer Inst. 49, 299.
29. Collard, W., Thornton, H., Mizell, M., and Green, M. (1973). Science 181, 448.
30. Naegele, R. F., and Granoff, A. (1977). Intl. J. Cancer 19, 414.

EBV-PERSISTENCE IN HUMAN LYMPHOID AND CARCINOMA CELLS

George Klein

Department of Tumor Biology
Karolinska Institutet
S 104 01 Stockholm 60
Sweden

ABSTRACT This is a brief summary of available evidence concerned with i) the interaction of Epstein-Barr virus (EBV) with human B lymphocytes, ii) virally induced phenotypic changes and iii) the nature of the association between EBV and three EBV related diseases, infectious mononucleosis (IM), Burkitt's lymphoma (BL) and nasopharyngeal carcinoma (NPC).

INTRODUCTION

Epstein-Barr virus (EBV) is a ubiquitous virus in man (for reviews see 1, 2, 3, 4, 5). Closely related lymphotropic herpesviruses exist in the large apes, most of the African monkeys studied, but not New World monkeys that have other, unrelated lymphotropic herpesviruses. The human EBV has a natural tendency to establish latent infection in the B-lymphocytes. Latency is by far the most usual consequence of infection. It can be studied in "transformation experiments" in vitro. Entry into the viral cycle is the rare exception. Recent UV-inactivation experiments (6, 7) indicate that most and perhaps all of the viral genome is required for the successful establishment of latency. This may be related to the fact that the virus has evolved a natural capacity to establish its latency in the B-lymphocyte. If so, transformation is probably a highly regulated process. Numerous sequences may participate in the regulation of this process, distributed over the large viral genome.

The "natural" transforming, i.e. latency-establishing role of the virus is confirmed by the fact that the wild type virus, isolated from the throat washings of infectious mononucleosis patients, is trans-

ISBN 0-12-668350-6

forming, rather than lytic, just like all known laboratory strains, with one exception (discussed below). This situation is very different from Herpes simplex virus that is normally lytic and can transform only after partial inactivation of the viral genome? It cannot be excluded, however, that the interaction of H.simplex with the ganglion cell, i.e. the host cell where the virus normally becomes latent (8), is based on similar principles as the interaction of EBV with its lymphocyte host.

The following discussion is concerned with three topics: i) the primary transforming (immortalizing) interaction of EBV with the human B-lymphocyte in vitro; ii) changes in cellular phenotype associated with EBV-transformation and iii) discussion of the three EBV-related diseases, mononucleosis (IM), Burkitt's lymphoma (BL) and nasopharyngeal carcinoma (NPC).

i) EBV-TRANSFORMATION OF THE HUMAN B-LYMPHOCYTE

Exposure of human lymphocytes to transforming strains of EBV leads to a series of events that have been only partly explored. The first step is adsorption of enveloped virus to EBV-receptors that are present on all B-lymphocytes, but not on T-lymphocytes (9, 10, 11). There is a close steric relationship, if not identity, between the C3 receptor of the B-lymphocyte and its EBV-receptor (12, 13, 14, 15). Virus penetration is followed by the appearance of the nuclear antigen, EBNA (16, 17), activation of polyclonal immunoglobulin synthesis (18) and a cellular S-phase (17, 19), followed by mitosis. This sequence of events occurs in B-cells only. Some O-cells (cells without B and T lymphocyte markers) can also absorb EBV, but do not respond with either EBNA or DNA synthesis, nor do they transform, suggesting some intracellular restriction. T-cells do not adsorb virus at all (20).

In EBV-infected B-cell cultures, EBNA positive blasts aggregate in characteristic large clumps, readily recognized at low power. A weak or two later permanent ("immortal") lines emerge that carry multiple viral genome copies and express the nuclear antigen in 100% of the cells.

The state of the viral genome in EBV-transformed cells is a point of special interest, in view of its unique features. All primary EBV-transformed lines that have been tested, carried multiple copies of the viral genome although some lines that do not fall into the category of primary transformants had only one genome copy. The main EBV-genome number per cell in a variety of lines was found to be 38 (5, 21). Interestingly, exactly the same figure was found for a collection of EBV-carrying Burkitt lymphoma biopsies (22). There was a wide variation, however, ranging from 10 to 70 copies per cell.

A small and not precisely defined part of the multiple viral genomes are covalently integrated with the host genome, whereas the major part exists as free, covalently closed circles (plasmids) (23). The precise relationship between the integrated and the free forms is not known. It is interesting that the characteristic EBV genome number of each line is usually maintained at a relatively constant level from year to year.

Even if normal lymphocytes are EBV-infected at a low multiplicity, i.e. only one infectious particle per transformed cell, lines emerge with multiple EBV-genome copies (24). In view of the fact that the virus induces DNA synthesis as an obligatory step in the transformation process, it is conceivable that multiple viral genomes must be produced before transformation can be achieved. This could be required e.g. for the integration of one or two crucially important copies in strategical positions. Alternatively, the amplification of the incoming viral genome may be an epiphenomenon, perhaps merely a secondary consequence of the virally induced DNA synthesis.

In this contect, the existence of "secondarily" derived lines with only one EBV-genome copy is of interest. Two such lines have been reported. One is an EBV-converted subline of the American, EBV-negative Burkitt lymphoma, Ramos, designated AW-Ramos. It carries one integrated EBV genome((25). Perhaps viral genome amplification was not necessary in this case since the line was already immortal prior to EBV-conversion. It is also noteworthy that AW-Ramos was converted by exposing Ramos cells to the P3HR-1 virus variant (26) which is non-transforming for resting B-lymphocytes (27). This does not account for the presence of one viral genome by itself, however, because other P3HR-1 virus converted cell lines of Ramos and other established lymphoma cells carry multiple EBV-genome copies (28, 29).

Another line that was reported to carry one EBV-genome copy only is a clone of P3HR-1 itself. It was isolated after chronic exposure to cycloheximide (30). It is not clear if and how cycloheximide brought about this change. It cannot be excluded that it merely selected a non-inducible clone that has appeared in the P3HR-1 population by spontaneous variation. Cycloheximide is known to induce the viral cycle in a proportion of the cells (30). Since induction is always suicidal, it is conceivable that chronic inducing treatment selects for non-inducibles. They have usually lower EBV-genome numbers than the inducible lines. However this may be, the existence of a P3HR-1 subline with only one viral genome does not conclusively prove that one genome is sufficient to achieve transformation, because the line is tumor derived and aneuploid and could well have been "superimmortalized" by cytogenetic changes. If so, it may no longer require the presence of the viral genome for the maintenance of immortality (cp. the discussion below in relation to the pathogenesis of Burkitt's lymphoma).

The hope was often expressed that a better understanding of the virus-cell interaction may reveal disease-related differences between the three EBV-associated pathological conditions, IM, BL and NPC. So far no systematic differences were found, however. Lines derived from the three conditions showed essentially the same range of variation, with the reservation that NPC-derived epithelial cell lines have not yet been established. NPC-related EBV genome studies were either made on non-representative, EBV-carrying lymphoid lines or on representative but not readily available NPC biopsies or nude mouse passaged solid tumors.

A difference in circle size (31) has been found between certain IM-derived and other (BL or normal derived) lines. At first, this was thought as possibly relatable to an IM-associated distinct viral substrain. Later it turned out, however, that cell lines with the exceptional, small viral genome could all be related to the same viral substrain, originating from a single mononucleosis derived line (833 L), rather than to the disease condition.

In EBV transformed lymphoid lines, the productive virus cycle is always very severely restricted. The precise degree of the restriction varies from line to line. Some regularities can be discerned. EBV-transformed cord blood cell lines are mostly non-producers. In vivo transformed or spontaneously established lines from adults tend to be low level producers in about half of the cases and non-producers in the other half. EBV-transformed lines of non-human primates, marmosets or squirrel monkeys tend to be regular and stable producers at a low to moderate level. These differences are due to cellular, rather than viral factors, since they are also seen in comparisons between lines transformed by the same viral substrain (e.g. the B95-8 virus). They may be due to subtle differences in B-lymphocyte differentiation. This is also supported by the fact that hybrids derived from the fusion of two different EBV-carrying B-cell lines show a dominance of producer status over non-producer status (32, 33). In contrast, fusion between producer lines and cells of other differentiation types, e.g. fibroblasts or epithelial cells regularly lead to the extinction of spontaneous productivity (34). A similar pattern has been seen for other differentiated B-cell characteristics, viz. extinction of the phenotypic property after fusion with non-B cells, in contrast to the good maintenance and autonomous expression of the same properties after the fusion of two B-cell lines (35).

Different producer lines differ from each other with regard to the "blocks" or "bottlenecks" of the virus cycle. Some (prototype Raji) make a small number of early antigen (EA) positive cells but go normally no further (36). Other lines (prototype Daudi, Jijoye) proceed regularly from EA to viral DNA synthesis, tightly followed by late antigen (VCA) expression (36). In hybrids between these two cell types (prototype Raji/ Daudi) the block to viral DNA synthesis is dominant over the absence of such a block, suggesting negative control (32; Moar, M., and Klein, G., to be published).

It is likely that the switch-on of EA is regulated by a "flip-flop" type control, with a certain fixed probability of entry into the viral cycle per division and per cell. This control of spontaneous EA production is dominant in hybrids, as already mentioned, suggesting a positive control. Although EA is a prerequisite for viral DNA replication and VCA production, it is not sufficient for the latter, in view of the additional, negative control that regulates this later step.

The controls that regulate the viral cycle have been also studied after experimental induction. Induction can be achieved in two major ways: with halogenated pyrimidines, like BUDR and IUDR (37, 38, 39) or after superinfection with the non-transforming (P3HR-1) virus substrain (40, 41).

The mechanism of viral induction by these agents is not known. It is noteworthy, however, that the same halogenated pyrimidines induce C-type viruses, carried in the form of preexisting proviral DNA copies in the cell

DNA replication and cell division, thereby increasing cytogenetic variation. Among the variants generated, some may have the cytogenetic anomaly that liberates the cell from growth controlling feedbacks. In the case of the B-cell, changes involving the distal part of a chromosome 14 appear to be particularly relevant.

In this context it is important that both Burkitt or Burkitt-like and non-Burkitt lymphomas were found without EBV but with the 14q+ marker (67). They were all B-cell lymphomas. It is conceivable that these EBV-negative B-cell lymphomas are initiated by other, viral or non viral agents but undergo later the same "convergent" chromosomal evolution before they reach autonomy.

Nasopharyngeal carcinoma, the second EBV-carrying malignancy, shows a remarkably sharp association with the viral genome. Multiple viral genomes are found in the epithelial cells, not in the infiltrating lymphocytes (72, 73). In a correlated histological and EBV-DNA study (74) we found that 100% of the low differentiated or anaplastic NPC carry the viral genome. The cells regularly express the EBNA-antigen. In this case, we again face the dilemma of reconciling a ubiquitous virus with a relatively rare tumor. The sharp tumor/virus association, limited to one histological type, and together with the absence of the viral genome from other tumors of this and other anatomical regions (except Burkitt lymphoma) suggest more than a casual association. It is possible that NPC is associated with a specific subvariant of EBV, but there is no evidence to support this at present. Another alternative is that host co-factors contribute to tumor development. Whereas in Burkitt's lymphoma the epidemiological evidence (time space clustering) speaks for environmental co-factors, the strong ethnic predominance of NPC (75) suggests genetic, rather than environmental co-factors.

ACKNOWLEDGMENTS

This work was supported by Contract No. N01 CP 33316 within the Virus Cancer Program of the National Cancer Institute.

REFERENCES

1. Klein, G. (1975). New Engl. J. Med. 293, 1353.
2. Klein, G. (1973). The Epstein-Barr Virus. In: The Herpesviruses. Ed. A.S. Kaplan, pp. 521-555, Academic Press, New York.
3. Epstein, M.A. (1970). Adv. Cancer Res. 13, 384.
4. Miller, G. (1971). Yale J. Biol. Med. 43, 358.
5. zur Hausen, H. (1975). Biochim. Biophys. Acta 417, 25.
6. Henderson, E., Heston, L., Grogan, E., and Miller, G. (1978). J. Virol. 25, 51.
7. Dalens, M., and Adams, A. (1977). Virol. 83, 305.
8. Stevens, J.G. (1978). Adv. Cancer Res., in press.
9. Jondal, M., and Klein, G. (1973). J. Exp. Med. 138, 1365.
10. Einhorn, L., Steinitz, M., Yefenof, E., Ernberg, I., Bakacs, T., and Klein, G. (1978). Cell. Immunol. 35, 43.
11. Greaves, M.F., Brown, G., and Rickinson, A.B. (1975). Clin. Immunol. Immunopathol. 3, 514.
12. Jondal, M., Klein, G., Oldstone, M.B.A., Bokish, V., and Yefenof, E. (1976). Scand. J. Immunol. 5, 401.
13. Yefenof, E., Klein, G., Jondal, M., and Oldstone, M. B.A. (1976). Int. J. Cancer 17, 693.
14. Yefenof, E., and Klein, G. (1977). Int. J. Cancer 20, 347.
15. Klein, G., Yefenof, E., Falk, K., and Westman, A. (1978). Submitted for publ.
16. Reedman, B.M., and Klein, G. (1973). Int. J. Cancer 11, 499.
17. Einhorn, L., and Ernberg, I. (1978). Int. J. Cancer 21, 157.
18. Rosén, A., Gergely, P., Jondal, M., Klein, G., and Britton, S. (1977). Nature 267, 52.
19. Robinson, J., and Miller, G. (1975). J. Virol. 15, 1065.
20. Einhorn, L., Steinitz, M., Yefenof, E., Ernberg, I., Bakacs, T., and Klein, G. (1978). Cell. Immunol. 35, 43.
21. Lindahl, T., Klein, G., Reedman, B.M., Johansson, B., and Singh, S. (1974). Int. J. Cancer 13, 764.
22. Klein, G. (1975). Cold Spring Harbor Symp. on Quant. Biol. 39, 783.
23. Adams. A. (1978). In: The Epstein-Barr Virus, A.M. Epstein and R. Achong, eds., in press.
24. Sugden, B., and Mark, W. (1977). J. Virol. 23, 503.

25. Andersson, M., and Lindahl, T. (1976). Virol. 73, 96.
26. Klein, G., Giovanella, B., Westman, A., Stehlin, J., and Mumford, D. (1976). Intervirol. 5, 319.
27. Menezes, J., Leibold, W., and Klein, G. (1975). Exp. Cell Res. 92, 478.
28. Klein, G., Zeuthen, J., Terasaki, P., Honig, R., Billing, R., Jondal, M., Westman, A., and Clements, G. (1976). Int. J. Cancer 18, 639.
30. Tanaka, A., Nonoyama, M., and Hampar, B. (1976). Virol. 70, 164.
31. Adams, A., Bjursell, G., Kaschka-Dierich, Ch., and Lindahl, T. (1977). J. Virol. 22, 373.
32. Klein, G., Clements, G., Zeuthen, J., and Westman, A. (1976). Int. J. Cancer 17, 715.
33. Ber, R., Klein, G., Moar, M., Povey, S., Rosén, A., Westman, A., Yefenof, E., and Zeuthen, J. (1978). Int. J. Cancer, in press.
34. Glaser, R., and Rapp, F. (1975). Progr. Med. Virol. 21, 43.
35. Klein, G., Terasaki, P., Billing, R., Honig, R., Jondal, M., Rosén, A., Zeuthen, J., and Clements, G. (1977). Int. J. Cancer 19, 66.
36. Moar, M., Siegert, W., and Klein, G. (1977). In: Haematology and Blood Transfusion, vol. 20, eds. Tierfelder, Rodt and Thiel, pp. 283.
37. Hampar, B., Derge, J.G., Martos, L.M., and Walterk, J.L. (1972). Proc. Nat. Acad. Sci. 69, 78.
38. Gerber, P. (1972). Proc. Nat. Acad. Sci. 69, 83.
39. Klein, G., and Dombos, L. (1973). Int. J. Cancer 11, 327.
40. Klein, G., Dombos, L., and Gothoskar, B. (1972). Int. J. Cancer 10, 44.
41. Henle, W., Henle, G., Zajac, B., Pearson, G., Waubke, R., and Scriba, M. (1970). Science 169, 188.
42. Nyormoi, O., Klein, G., Adams, A., and Dombos, L. (1973). Int. J. Cancer 12, 396.
43. Klein, G., Clements, G., Zeuthen, J., and Westman, A. (1977). Cancer Letters 3, 91.
44. Shapiro, I., Andersson-Anvret, M., and Klein, G. (1978). Intervirol., in press.
45. Klein, G., Lindahl, T., Jondal, M., Leibold, W., Menezes, J., Nilsson, K., and Sundström, Ch. (1974). Proc. Nat. Acad. Sci. 71, 3283.
46. Menezes, J., Leibold, W., Klein, G., and Clements, G. (1975). Biomed. 22, 276.
47. Steinitz, M., and Klein, G. (1975). Proc. Nat. Acad. Sci. 72, 3518.

48. Steinitz, M., and Klein, G. (1976). Virol. 70, 570.
49. Steinitz, M., and Klein, G. (1977). Europ. J. Cancer 13, 1269.
50. Yefenof, E., and Klein, G. (1976). Exp. Cell Res. 99, 175.
51. Yefenof, E., Klein, G., Ben-Bassat, H., and Lundin, L. (1977). Exp. Cell Res. 108, 185.
52. McConnell, I., and Lachmann, P.J. (1976). Transpl. Rev. 32, 72.
53. Montaigner, L., personal communication.
54. Henle, G., Henle, W., and Diehl, V. (1968). Proc. Nat. Acad. Sci. 59, 94.
55. Evans, A.S., Niederman, J.C., and McCollum, R.W. (1968). New Engl. J. Med. 279, 1121.
56. Golden, H.O., Chang, R.S., Lou, J.J., and Cooper, T. Y. (1971). J. Infec. Dis. 124, 422.
57. Pereira, M.S., Blake, J.M., and Macrac, A.D. (1969). Brit. Med. J. 4, 526.
58. Pereira, M.S., Field, A.M., Blake, J.M., Rodgers, F. G., Bailey, L.A., and Davies, J.R. (1972). Lancet 1, 710.
59. Klein, G., Svedmyr, E., Jondal, M., and Persson, P.O. (1976). Int. J. Cancer 17, 21.
60. Svedmyr, E., and Jondal, M. (1975). Proc. Nat. Acad. Sci. 72, 1622.
61. Bakacs, T., Svedmyr, E., Klein, E., Rombo, L., and Weiland, D. (1978). Cancer Letters, in press.
62. Britton, S., Andersson-Anvret, M., Gergely, P., Henle, W., Jondal, M., Klein, G., Sandstedt, B., and Svedmyr, E. (1978). New Engl. J. Med. 298, 89.
63. Andersson, M., Klein, G., Ziegler, J., and Henle, W. (1976). Nature 260, 357.
64. Nilsson, K., and Pontén, J. (1975). Int. J. Cancer 15, 321.
65. Manolov, G., and Manolova, Y. (1972). Nature 237, 33.
66. Jarvis, J.E., Ball, G., Rickinson, A.B., and Spstein, M.A. (1974). Int. J. Cancer 14, 716.
67. Zech, L., Haglund, U., Nilsson, K., and Klein, G. (1976). Int. J. Cancer 17, 47.
68. Nilsson, K., Giovanella, B.C., Stehlin, J.S., and Klein, G. (1977). Int. J. Cancer 19, 337.
69. Burkitt, D. (1967). In: Treatment of Burkitt's Tumours, (J.H. Burchenal and D.P. Burkitt, eds.), UICC Monogr. pp. 94-101. Springer-Verlag, Berlin and New York.
70. de Thé, T., and Geser, A. (1976). C.R. Acad. Sc. Paris 282, 1387.

71. Burkitt, D. (1969). J. Nat. Cancer Inst. 42, 19.
72. Klein, G., Giovanella, B.C., Lindahl, T., Fialkow, P.J., Singh, S., and Stehlin, J.S. (1974). Proc. Nat. Acad. Sci. 71, 4737.
73. Wolff, H., zur Hausen, H., and Becker, V. (1973). Nature New Biol. 244, 245.
74. Andersson-Anvret, M., Forsby, N., Klein, G., and Henle, W. (1977). Int. J. Cancer 20, 486.
75. Ho, J.H.C. (1972). Adv. Cancer Res. 15, 57.

IMMUNOBIOLOGY OF PERSISTENT VIRUS INFECTION[1]

Michael B. A. Oldstone, Michael J. Buchmeier, and Michael V. Doyle

Research Institute of Scripps Clinic and Research Foundation, La Jolla, California 92037

INTRODUCTION

While vesicular stomatitis virus has proved to be a panacea in the study of the molecular biology of viruses, in a similar vein, studies of lymphocytic choriomeningitis virus (LCMV) infection have provided the rosetta stone for beginning to understand viral immunobiology and also cell-cell recognition. Observations with LCMV have yielded several fundamental concepts in modern biology. For example, the pioneering work by Wally Rowe (1) in the mid 1950s, who used various modalities of immunosuppression while studying acute LCMV infection in mice, opened the field of viral immunopathology. Later work by Cole, Nathanson, Gilden and associates provided the first evidence that a subpopulation of lymphocytes bearing markers for theta isoantigen (thymus derived T cells) killed virus infected targets _in vitro_ and _in vivo_ (2, 3). Unique experimental results by the Canberra group of Zinkernagel, Doherty and Blanden (4, 5) showed that cell-cell recognition between specifically primed cytotoxic T lymphocytes and virus infected targets required not only recognition of specific viral antigens on the cell surface but also a fit within the major histocompatibility (H-2) complex at the D or K region.

Still other major advances in biomedical concepts based on early findings by Traub (6) were to come. Eric Traub (6) noted the natural occurrence of persistent LCMV infection in mice whose tissues retained infectious virus throughout the animal's life. The results of Traub's observations, coupled with the persistence of virus during an animal's lifetime in the absence of free circulating antibody, led Burnet to postulate the concept of immunologic tolerance (7). This concept that a host could not respond immunologically to infecting virus after _in utero_ or at birth inoculation remained unchallenged until the

[1]This work was supported by U.S.P.H.S. grants AI-09484 and NS-12428, N.I.H. Postdoctoral Fellowship No. 1 F32 AI 05380 (M.B.), and Arthritis Foundation Postdoctoral Fellowship (M.D.).

ISBN 0-12-668350-6

late 1960s when Oldstone and Dixon provided the first evidence that such persistently infected mice do, in fact, mount humoral immune responses against their infecting virus. The antibody was there but undetected previously owing to its binding to virus and forming virus-antibody immune complexes (8, 9). Several investigators (reviewed in 10) used this information to show that in most, if not all, virus infections a host immune response is elicited and results in the formation of immune complexes. The biomedical importance of the initial observations of virus-antibody immune complexes in LCMV infection lay in defining the events involved, which made possible later studies of virus-induced immune complex disease in a variety of models (10). A most recent novel biological finding is that persistent LCMV infection of neuroblastoma cells significantly impairs their ability to make a differentiated or luxury product, such as the enzymes that make or degrade acetylcholine, the major neurotransmitter (11). Although infected neuroblastoma cells are deficient in making the neurotransmitter or a synapse, their vital functions of normal growth rates, cloning efficiency, protein, RNA and DNA metabolism are not disturbed. This finding indicates a new mechanism whereby viruses can cause cellular injury, i.e., alteration of differentiated function of specialized cells of the host without their destruction.

Despite much well documented and important biological knowledge gleaned from the study of LCMV natural or experimentally induced infection, neither the underlying biochemical nor molecular events associated with these phenomena are clear. Therefore our focus and goals of the past few years have been to purify, identify and quantify the polypeptides of LCMV. With this knowledge concerning specific viral polypeptides and the raising of monospecific antisera to these antigens we will be able to study how LCMV is assembled and expressed on a cell surface. More important we will be able to determine the specific antigen(s) that initiates T cell and/or B cell proliferation, T cell recognition and resultant T cell killing of virus infected cells and the specific viral antigens and antiviral antibodies formed in complexes in the circulation and deposited in various tissues to induce immune complex disease.

Although it has been clear since the late 1960s that adult mice persistently infected by LCMV either at birth or <u>in utero</u> produce antiviral antibodies (8, 9), it is not known against which specific viral antigen(s) the response is directed. The question is, do persistently infected mice make a humoral immune response against all virion structural polypeptides or a selected few? To answer this question, to understand the induction of cytotoxic T cell responses and to determine the viral antigen(s) associated with the H-2 complex recognized by sensitized T cells, we initiated studies of the structual proteins of LCMV (12-14).

Structural Proteins of LCMV. After one infects various tissue culture cells with LCMV, a defective interfering virus (DIV) is generated rapidly. With stocks of LCMV that are relatively free of DIV, and knowing the kinetics of DIV production (15), one can label LCMV with ^{35}S methionine and ^{3}H glucosamine, purify extracellular virions and identify viral polypeptides in the cytoplasms of infected cells (13, 14). The purified LCM virions contained three major ^{35}S methionine labeled polypeptides (Figure 1). The largest of these was a 63,000 dalton nonglycosylated polypeptide (NP) found associated with viral RNA in a ribonucleoprotein complex. The other two were glycopeptides, termed GP-1 and GP-2, with molecular weights of approximately 54,000 and 35,000 daltons, respectively. These two glycopeptides were shown by proteolytic digestion to be located on the virion surface. Using monospecific antiserum to LCMV to form immunoprecipitates, we also examined the polypeptide profiles by SDS-polyacrylamide gel electrophoresis on 10 and 12.5% gels and found similar results.

Two dimensional tryptic peptide maps of each of the major viral polypeptides were prepared to determine whether each was a distinct product of viral translation. Maps were prepared from polypeptides labeled internally with ^{35}S methionine and externally with ^{125}Iodine (16). Figure 2 shows representative ^{35}S methionine peptide maps and demonstrates that each polypeptide is a unique product by this criterion.

When LCMV infected cells were examined for these products, NP was present inside the cell but not expressed on the plasma membrane, whereas the glycopeptides were expressed on the cell surface. In terms of biological activity attributable to the viral polypeptides, when antibodies to the nonglycosylated NP were used, specifically sensitized cytotoxic T cells, which recognize LCMV infected targets, were not blocked in their killing. In contrast, $F(ab')_2$ antibodies to GP-1 and GP-2 abrogated cytotoxic T cell killing in a dose dependent response. These experiments indicated that primed cytotoxic T cells recognized either the GP-1 or GP-2 or both antigens on the cell surface. Currently, monospecific antisera to GP-1 and to GP-2 are being raised so that we can determine whether the cytotoxic T cell recognizes both or only one of the glycopolypeptides. In addition, the use of these antisera, as well as monospecific anti-H2 reagents in biochemical studies with membrane purification and in cell topography mapping studies using co-capping and immune electron microscopy, should enable us to determine whether there is a physical association between the H2 complex and the viral glycopeptides.

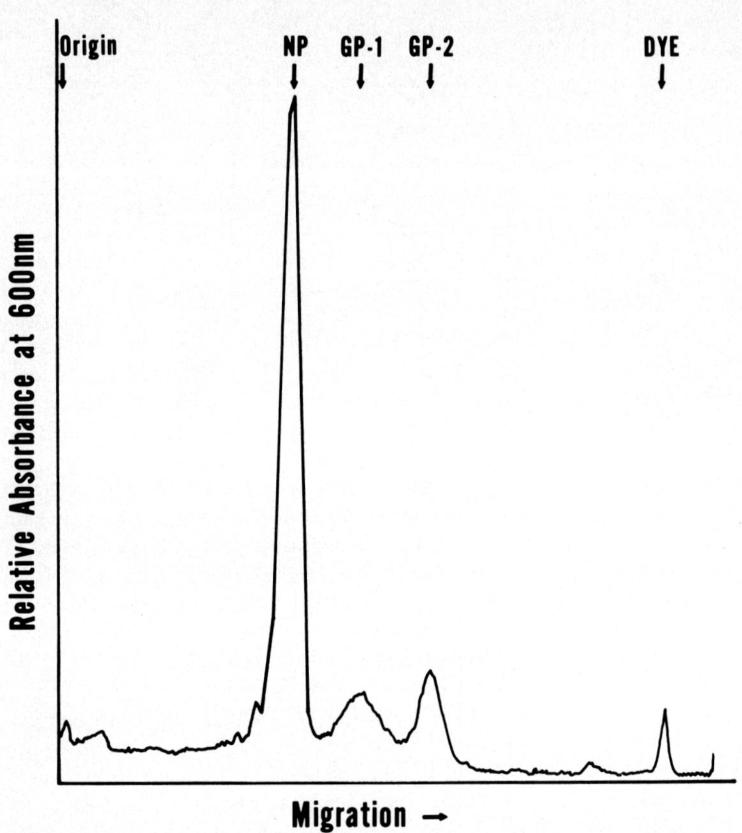

FIGURE 1. Polypeptides of LCM virus. A preparation of purified LCMV, radiolabeled with ^{35}S methionine, was subjected to SDS polyacrylamide gel electrophoresis under reducing conditions. The gel was dried, and an autoradiogram prepared. The figure is a densitometer tracing of the autoradiogram. Three major polypeptide species were observed, and are indicated by NP, GP-1 and GP-2 at the top of the figure.

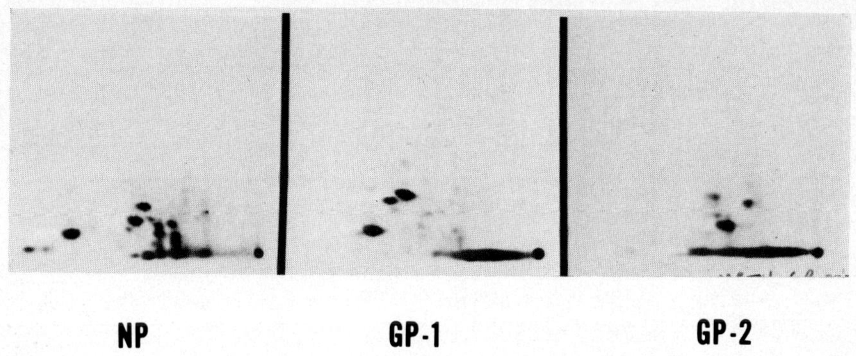

FIGURE 2. Tryptic peptide maps of the major polypeptides of LCMV. A preparation of ^{35}S methionine labeled LCMV was separated into its polypeptides as described in Figure 1. The individual polypeptides were localized by staining and autoradiography, then excised from the gel, and digested with TPCK trypsin. The soluble polypeptides were mapped by high voltage electrophoresis from right to left, and thin layer chromatography from bottom to top. Autoradiograms of the plates are shown.

Immunoglobulins Raised Against Specific Structural LCMV Polypeptides During Persistent LCMV Infection. With the identification of the structural polypeptides of LCMV, we can learn whether the infected host makes antibodies against all these polypeptides or not. Since mice continuously produce immunoglobulins (presumably antiviral antibodies) during persistent virus infection and these complex to viral antigens, which are in excess, it is necessary to isolate the virus-immunoglobulin complex. Such virus-immunoglobulin complexes deposit in several tissues of persistently infected mice, predominantly the glomeruli of the kidney. We recovered the immunoglobulin deposited in the renal glomeruli by mincing these tissues and using a low molarity, low pH buffer to dissociate the virus antigen from antibody. Then we passed the dissociated mixture through an Amicon filter, which retained immunoglobulins but passed LCMV polypeptides. The immunoglobulins were removed from the filter and purified on sucrose gradients. Immunoglobulin was so obtained from separately pooled kidneys of 3 different strains of mice persistently infected with LCMV, and as a control from non-LCMV infected AKR/J mice (Table 1). AKR/J mice make and deposit immune complexes with specificity for murine leukemia viral antigen (17).

TABLE I
YIELDS OF IMMUNOGLOBULIN ELUTED FROM KIDNEYS OF LCMV-CARRIER AND UNINFECTED MICE[a]

LCMV carrier	Strain of mouse	Age mo.	No. kidneys	μg Ig/ kidney	Total Ig recovered (μg)
Yes	BALB/WEHI Nu/+	12-18	112	5.7	640
Yes	CBA/WEHI Nu/+	12-18	56	6.3	352
Yes	SWR/J	4-6	110	15.	1644
No	AKR/J	6-10	58	2.9	168

[a] All mice had immune complex disease. The AKR/J kidney pool had been analyzed previously and was shown to have antibodies against the major oncornavirus polypeptides.

The specificity of the eluted immunoglobulin was tested using an immune precipitation assay (13). For the assay, radiolabeled virus was disrupted by incubation with non-ionic detergent and ribonuclease, then reacted with eluted immunoglobulin. The resultant immune complexes were precipitated with a second antibody directed against mouse gamma globulins. To establish the specificity of the reaction, the immunoglobulins recovered from the glomeruli were also reacted against radiolabeled Pichinde virus (PV), a serologically unrelated, but structurally similar member of the arenavirus group (18-20). Both the LCMV and PV had been doubly radiolabeled with ^{35}S methionine and ^{3}H glucosamine to mark polypeptides and glycopeptides, respectively. Two aspects of the specificity of the immune reaction were examined. First, the quantity of each radiolabeled precipitate (corrected for isotope crossover) was expressed as a percent of the offered CPM (Table 2). Second, the specific viral polypeptides precipitated were identified by analysis on SDS gels (Figure 3). As shown in Table 2, each of the eluates prepared from LCMV carrier mice specifically precipitated LCMV protein, but did not precipitate PV protein. Values for ^{35}S methionine (indicative of NP) ranged from 31 to 48% of the offered CPM compared with background levels of 5 to 7% of ^{35}S methionine labeled PV. Specificity was also seen for viral glycopeptides. The LCMV eluates precipitated 14 to 30% of the ^{3}H glucosamine label while only 3 to 5% of PV specific ^{3}H label was precipitated. The AKR/J controls precipitated only background levels of both viruses.

The specificity of viral polypeptides precipitated by the LCMV carrier Ig preparation was determined by SDS-polyacrylamide gel analysis of the immune precipitates. Figure 3 shows that precipitates formed with each of the three LCMV specific eluates contained the NP, GP-1 and GP-2 polypeptides. Although all of the eluates precipitated similar quantities of NP, their content of GP-1 and GP-2 varied. The precipitate formed with immunoglobulin eluted from CBA mice glomeruli contained a proportionately larger amount of both GP-1 and GP-2 than did those of the SWR/J or BALB/WEHI mice. The AKR/J eluate precipitated only background levels of LCMV polypeptides identifiable on gels.

Hence, it is clear from these data that mice persistently infected with LCMV produce antibodies directed against the full spectrum of LCMV known structural antigens.

TABLE II
IMMUNE PRECIPITATION OF LCM VIRUS (LCMV) AND PICHINDE VIRUS (PV) PROTEINS BY Ig ELUTED FROM LCMV CARRIER AND CONTROL MOUSE KIDNEYS

Ig eluted from persistently infected mice (strain)	μg Ig reaction	Viral antigen	Percent of input CPM precipitated[a]	
			^{3}H glucosamine	^{35}S methionine
BALB/WEHI Nu/+ (LCMV infected)	4.2	LCMV	16.1	31.0
		PV	4.1	4.9
CBA/WEHI Nu/+ (LCMV infected)	2.8	LCMV	30.5	34.2
		PV	4.1	6.4
SWR/J (LCMV infected)	6.5	LCMV	14.2	48.3
		PV	3.0	4.9
AKR/J (MuLV infected)	3.2	LCMV	4.8	4.9
		PV	5.3	7.2

[a]Mean input CPM: LCMV, ^{3}H 17,000; ^{35}S 9600; PV, ^{3}H 17,000; ^{35}S 8300.

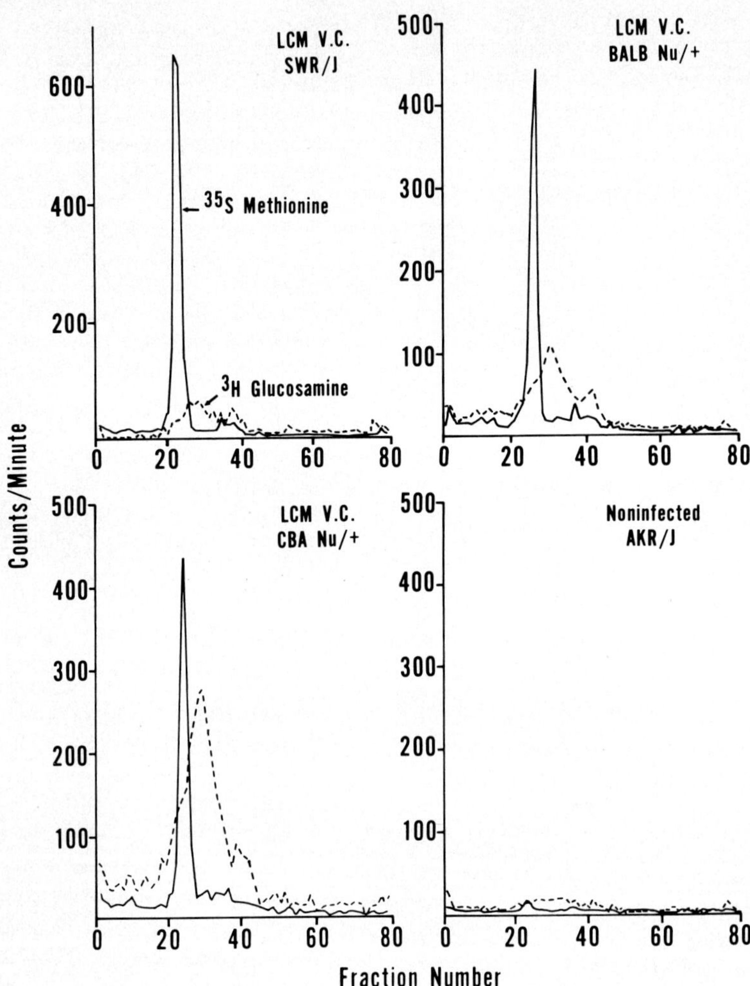

FIGURE 3. Immune precipitation of the polypeptides of LCMV by Ig eluted from the kidneys of virus carrier (LCM V.C.) mice of 3 strains and control mice. Viral antigen was disrupted by incubation with NP-40 and RNase, then reacted with the following quantities of eluted immunoglobulin: SWR/J, 6.5 μg; CBA Nu/+ 2.8 μg; BALB Nu/+ 4.2 μg; and AKR/J 3.2 μg. The resultant immune precipitates were collected by precipitation with antibody to mouse Ig, washed, and analyzed by SDS-PAGE on 12.5% gels.

Unresolved Dilemma: Failure to Clear LCMV in Persistently Infected Mice. Although, as just shown, murine B lymphocytes (plasma cells) respond to LCMV antigens by producing humoral antibody responses against all the known structural polypeptides of LCMV during the entire life of an infected animal, yet LCMV is not eliminated. The question is, why? To address this question, we have begun studies aimed first at determining the competency of T cells specifically committed against LCMV and second of T and B antigen binding cells in persistently infected mice. These parameters are being compared to those in noninfected, immune and acutely infected age and sex matched controls. Our initial studies, recently completed, determined whether infectious virus persists in T or B lymphocytes of adult mice persistently infected with LCMV since birth. Replicating virus was found in subpopulations of circulating blood mononuclear cells harvested from BALB/WEHI and SWR/J mice persistently infected with LCMV. Infectious virus was demonstrated using an infectious center assay on Vero cell monolayers (21, 22). Lymphoid cells isolated from both primary and secondary lymphoid organs of such virus carrier mice also carried infectious virus. The number of infectious centers formed varied among the persistently infected animals from 1 to 10,000 per 10^6 viable mononuclear cells. However, within any individual persistently infected mouse sampled weekly, the number of infected lymphoid cells within the peripheral blood was constant over a 4 to 6 week period. Virus was not present in a fully infectious form either on the cell surface or within the cell. Mononuclear cells replicating LCMV (Figure 4) were found in the bone marrow derived, thymus derived and adherent cell populations (Tables 3, 4). Similar results have recently been found by Popescu and his associates (23, 24).

Infectious centers were also found in peripheral blood and splenic mononuclear cells of adult mice acutely infected with LCMV. Here, the levels of such cells rose to a peak of 100 per 10^6 mononuclear cells by day 3 to 4 after intraperitoneal inoculation with virus and then decreased to below detectable levels (less than 1 per 10^6 viable cells) by day 8 or 9.

Thus, in numerous persistently or acutely infected adult mice studied, 0.5 to 3% of the total peripheral blood mononuclear cell population contained infectious virus which scored as infectious centers. Tests involving adherence to glass or plastic surfaces, rabbit antibody to mouse immunoglobulin, anti-theta serum and complement and nude mice in a variety of experiments indicated that adherent cells, B cells and T cells all scored as infectious centers (Tables III and IV). Experiments are currently aimed at determining whether the function of these various subpopulations of lymphocytes are impaired owing to their harboring of LCMV. The presence of viral

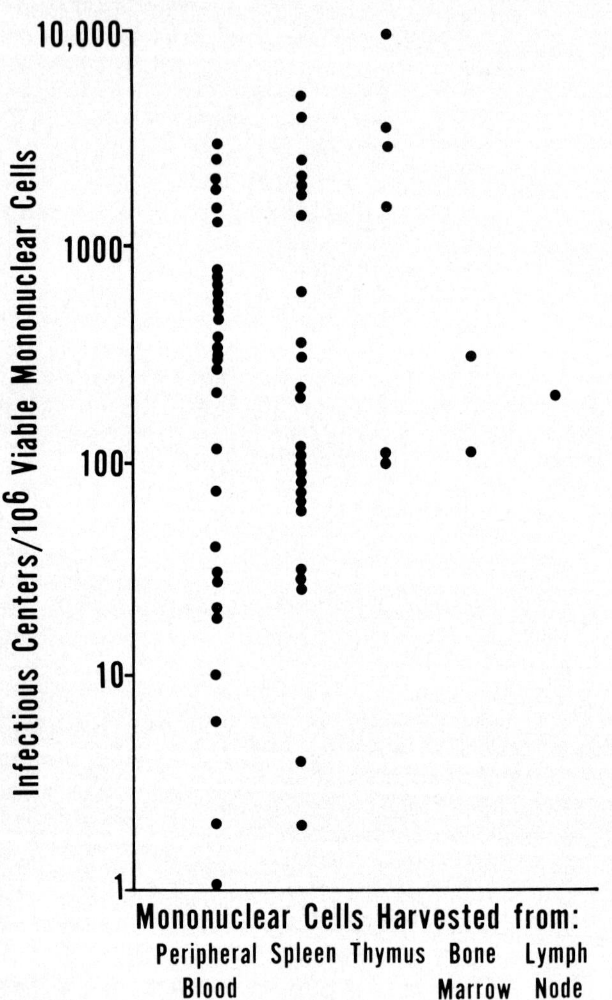

FIGURE 4. Mononuclear cells scoring as infectious centers and harvested from lymphoid tissues of adult BALB/WEHI mice which had been infected at birth with LCMV. Ficoll-Hypaque purified mononuclear cells from a variety of lymphoid tissues were added to 60 x 15 mm Petri dishes containing confluent Vero target cell monolayers. Plates were overlaid with 0.5% agarose in BME and incubated for 7 days. The plaques were then

TABLE III
THYMUS-DERIVED (T) LYMPHOCYTES HARVESTED FROM ADULT BALB/WEHI MICE WHICH HAD BEEN INFECTED AT BIRTH WITH LCMV SCORE AS LCMV INFECTIOUS CENTERS[a]

Source of mononuclear cells	Treatment	Infectious centers per 10^6 viable cells
Peripheral blood	None	2600
	Anti-thymocyte serum	2400
	Guinea pig complement	2200
	Anti-thymocyte serum plus complement	700
Spleen	None	2400
	Anti-thymocyte serum	2500
	Guinea pig complement	2300
	Anti-thymocyte serum plus complement	600
Thymus	None	3500
	Anti-thymocyte serum	3700
	Guinea pig complement	3400
	Anti-thymocyte serum plus complement	100

[a]Specificity of the antiserum to murine thymocyte was shown both by specific killing of thymocytes and by the inhibition of T lymphocyte mitogenic response to concanavalin A. In contrast the anti thymocyte serum did not abrogate the expected B lymphocyte response to lipopolysaccharide, a B cell mitogen. Viral specificity was shown by the ability of monospecific antibodies to LCMV and complement to abrogate infectious center formation. Other tests showed that infectious LCMV was carried inside the T lymphocyte and not on its outer plasma membrane. Further, the virus was not present in a mature form as freezing and thawing inhibited infectious centers.

---

developed by overlaying with 0.5% agarose in BME containing 0.01% neutral red. Results are expressed as the number of infectious centers per 10^6 viable mononuclear cells. Each point represents data from a single experiment.

TABLE IV
BONE MARROW (B) DERIVED LYMPHOCYTES, ADHERENT CELLS AND MACROPHAGES HARVESTED FROM ADULT BALB/WEHI THYMUSLESS (Nu/Nu) MICE WHICH HAD BEEN INFECTED AT BIRTH WITH LCMV SCORE AS LCMV INFECTIOUS CENTERS[a]

Exp	Source of mononuclear cells	Treatment	Infectious centers per 10^6 viable cells
1.	Peripheral blood	None	1100
		Anti-thymocyte serum plus complement	1100
	Spleen	None	450
		Anti-thymocyte serum plus complement	400
2.	Peripheral blood	None	500
		Depletion of adherent cells	350
	Spleen	None	180
		Depletion of adherent cells	100

[a]Specificity for B lymphocytes was further shown by experiments using a monospecific antibody to murine Ig and complement to deplete B cells. Among the adherent cell population, macrophages were identified on the basis of ingestion of zymosan particles. Adherent cells were overlaid with Vero cells to test for infectious centers. As with the results shown in Table 3 for T lymphocytes (see legend), specificity for LCMV was determined using an antiviral antibody. Virus was found harbored inside (not on the surface) of B lymphocytes and was not in a fully infectious form. Similar results were obtained with thymus sufficient, non nude mice.

antigens on the surface of LCMV infected lymphocytes, coupled with the use of monospecific antiviral antibodies and antiviral electronic cell sorting techniques should permit the enrichment of subpopulations of lymphocytes scoring as infectious centers. We can then do defined studies of their differentiated functions against LCMV and nonviral antigens.

Conclusions. It is quite evident that observations of major importance in virology, immunology and cell biology have been made in the study of both acute and persistent LCMV infection. Now, with the availability of purified LCMV structural polypeptides, monospecific antibodies directed against these polypeptides, genetically well defined inbred mouse strains and the recent technical developments in eliciting secondary T cell priming *in vitro* (25) it should be possible in the future to define these biological observations in molecular and biochemical terms.

ACKNOWLEDGMENTS

This is Publication No. 1493 from the Department of Immunopathology, Research Institute of Scripps Clinic and Research Foundation, La Jolla, California 92037.

REFERENCES

1. Rowe, W. P. (1954). Res. Report Naval Medical Research Institute 167, 167.
2. Cole *et al*. (1972). Nature 238, 335.
3. Gilden, R. *et al*. (1972). J. Exp. Med. 135, 860.
4. Zinkernagel, R. *et al*. (1974). Nature 248, 701.
5. Doherty, P. *et al*. (1975). J. Exp. Med. 141, 502.
6. Traub, E. (1936). J. Exp. Med. 63, 847.
7. Burnet, F. M., and Fenner, F. (1949). In "The Production of Antibodies", 2nd ed., p. 104. McMillan and Co., Ltd., Melbourne.
8. Oldstone, M. B. A., and Dixon, F. (1967). Science 158, 1193.
9. Oldstone, M. B. A., and Dixon, F. (1969). J. Exp. Med. 129, 483.
10. Oldstone, M. B. A. (1974). In "Progress in Medical Virology", vol. 19, (J. L. Melnick, ed.), p. 84. S. Karger, Basel.
11. Oldstone, M. B. A. *et al*. (1977). J. Cellular Physiology 91, 459.
12. Welsh, R. M. *et al*. (1976). Virology 73, 59.
13. Buchmeier, M. J., and Oldstone, M. B. A. (1978). J. Immunol., in press.
14. Buchmeier, *et al*. (1978). Virology, in press.
15. Welsh, R. M., and Oldstone, M. B. A. (1977). J. Exp. Med. 145, 1459.
16. Elder, J. H. *et al*. (1977). J. Biol. Chem. 252, 6510.
17. Oldstone, M. B. A. *et al*. (1976). J. Virol. 18, 176.
18. Rowe, W. P. *et al*. (1970). J. Virol. 5, 289.
19. Murphy, F. A. *et al*. (1970). J. Virol. 6, 507.

20. Vezza, A. C. *et al*. (1977). J. Virol. 23, 776.
21. Doyle, M., and Oldstone, M. B. A. (1978). ICN Meeting on Persistent Virus Infections, Keystone (abstr).
22. Doyle, M., and Oldstone, M. B. A. (1978). J. Immunol., submitted.
23. Popescu, M. *et al*. (1978). Zeitschrift fur Naturforschung, in press.
24. Popescu, M. *et al*. (1978). J. Exp. Med., submitted.
25. Dunlop, M., and Blanden, R. (1976). Immunology 31, 171.

ROLE OF DI, VIRUS MUTATION, AND HOST RESPONSE IN PERSISTENT INFECTIONS BY ENVELOPE RNA VIRUSES[1]

John J. Holland, Bert L. Semler, Charlotte Jones, Jacques Perrault, Lola Reid, and Laurent Roux

Department of Biology
University of California, San Diego
La Jolla, California 92093

ABSTRACT

We report here the chemical and biological characterization of VSV infectious virus and DI released from BHK_{21} carrier cells after more than four years of persistence. The infectious virus is a very slow-growing, small plaque mutant with slight temperature sensitivity. Temperature sensitivity does not explain the lack of cytopathology in carrier cells since these cells survive incubation for 7 days at 25°C or 32°C. Virus mutations alone do not explain the lack of cytopathology since cloned virus free of DI destroys 100% of BHK_{21} cells exposed at any multiplicity, whereas this infectious virus mixed with DI from the carrier cells does not destroy all exposed BHK_{21} cells, and a new carrier state is readily re-established. Oligonucleotide maps of RNA of virus from long term carriers show that these have numerous mutations (by comparison with the original input virus), and that many DI from carrier cells are unique in their oligonucleotide map patterns as compared to a variety of VSV DI from other sources (including the DI originally used to establish the persistent infection), but that they are otherwise similar to the original input virus. Using sequencing and annealing, we have found that common DI from acute infections and certain DI from persistent infections conserve the 5' end of standard virus RNA and not the 3' OH terminus of the virion RNA. We present data showing that tumorigenic BHK_{21} cells or Hela cells lose their tumorigenicity (as assayed in nude mice) when they are persistently infected with VSV and other enveloped RNA viruses. This appears to be due to recognition of viral envelope proteins at the cell surface, and some radiation-sensitive cell(s) of the nude mice are involved. Although carrier cells do not usually form tumors in nude mice, they do form a benign nodule from which carrier cells, carrier virus and DI can be reisolated after months or years in an *in vivo* environment.

ISBN 0-12-668350-6

These carrier cells shed virus and DI in vivo and in vitro, and one selected carrier cell population is able to form rapidly growing tumors which also shed virus and DI in vivo for many months despite low levels of neutralizing antibody production. We have kept six different long term carrier cultures in a slowly-growing (and/or non-growing) state for four months and found that one of these has ceased shedding mature infectious virus or DI but that both can be "rescued" by cocultivation or by stimulation of renewed growth of the carrier, or by wild type virus challenge. Finally, we describe preliminary evidence that DI may also be involved in long term persistent infection by paramyxoviruses, since persistent Sendai virus infections generally require DI for establishment and produce both viral size and DI size intracellular nucleocapsid throughout the course of persistence.

INTRODUCTION

Huang and Baltimore[1] were the first to provide a clear, unifying definition of DI, and to emphasize their ubiquity among animal viruses of nearly all classes. They also speculated that DI might play an important role in nature, in recovery from acute virus infections, and in persistent virus infections. Since that time our laboratory has been testing that hypothesis and studying the biology and biochemistry of DI in some detail. We have found that DI can provide prophylactic protection against acute virus infection[2], that they can both be generated and replicate true in the central nervous system of mice[3], and that they are required for, and remain present throughout, long term persistent infections of BHK_{21} cells in culture by otherwise 100% cytocidal vesicular stomatitis virus (VSV) infection[4]. Since details of this work and the relevant literature in the field have been presented in two recent volumes of ICN-UCLA symposia[5,6] and in recent manuscripts[7,8], they need not be reviewed here. Since that time an extensive review of DI has also appeared[9], and a number of other laboratories have also demonstrated a cell protective role of DI and/or their association with persistence of a variety of RNA viruses. These include lymphocytic choriomeningitis virus in vivo[10] and in vitro[11], Japanese encephalitis virus[12], rabies virus[13,14], Sindbis virus (in insect cells)[15], and reovirus in mouse L cells[16].

Other studies have implicated temperature sensitive mutants as a major factor in RNA virus persistence[17,18,19,20]. In reference 19, DI and interferon roles in persistence were also proposed. Others have also found interferon in persistent VSV infection of L cells for periods of 1 to 2 months[20]. We summarize here the results of most recent studies on DI and long term persistent infections mediated by them in cells in culture and describe evidence for mutations occuring during persistent infection and for virus alteration of *in vivo* behavior of persistently infected cells. Since a number of different approaches are outlined here, most of them will be summarized in tabular form and the detailed studies of each will be reported elsewhere.

RESULTS

DI Prophylaxis of CNS Infections of Mice. We showed earlier[2] that VSV DI can attenuate the severity of virulent VSV infections. Large numbers of DI infected intracerebrally along with small lethal doses of standard virus completely protected most mice against death (or changed the nature of the disease and prolonged time of death when used with high challenge doses of standard virus). Crick and Brown[21] obtained evidence which suggested that this DI prophylaxis might be due to large amounts of antigen in the DI conferring immunological protection along with some induction of nonspecific antiviral resistance mechanism. However, Jones and Holland (data to be published) have found that this DI prophylaxis is mainly due to true homologous autointerference since UV-inactivated DI (or UV-inactivated virus) did not protect mice against death when given together with any lethal dose of virus. This data is summarized in Table I.

TABLE 1

CHARACTERISTICS OF *IN VIVO* PROTECTION OF MICE BY VSV DI PARTICLES[a]

1. Prophylactic protection requires viable DI. UV-inactivated DI or UV-inactivated standard virus did not provide prophylaxis.
2. UV-inactivated standard virus or UV-inactivated DI do immunize against later challenge.
3. UV-inactivated DI of some kinds (but not UV-inactivated standard virus or UV-inactivated DI of other kinds) will lengthen time until death of mice challenged at the same time.
4. Approximately $3x10^8$ physical particles of active DI is the minimum number able to protect against simultaneous challenge with low dosages of standard virus.

[a]Charlotte Jones and John Holland (1978) data to be published.

We did observe a prolongation of time until death with one class of DI (possibly one with some double-stranded RNA of the "snapback" type[22,23]) but not with another class at the same dosage. We also showed that we could protect with lower dosages of DI (approximately $3x10^8$ per brain) than were employed previously. This is probably the minimum number required to assure that an active DI coinfects every neuron infected by incoming challenge virus during the critical first cycle of replication in the brain. In any case, DI prophylaxis requires true homologous interference by biologically active DI.

Long Term Persistent Infection of BHK_{21} Cell Culture by VSV Virus Plus DI. We have described in detail elsewhere[5-8] the major role played by DI in establishment and maintenance of long term persistent infections of highly susceptible BHK_{21} cells in culture. We have also observed mutation of the virus to a slightly temperature sensitive mutant which exhibits a small plaque characteristic and very slow growth at all temperatures. Table II summarizes our experiences with this long term carrier over 4 1/2 years of persistence.

TABLE II

CHARACTERISTICS OF LONG TERM (OVER 4 YEARS) PERSISTENT $VSV_{IND.}$ INFECTION OF BHK_{21} CELLS (CAR_4)

1. DI are required to establish persistence (along with standard virus).
2. All or nearly all cells remain infected indefinitely (over 4 years).
3. Carrier cells grow at near normal rates (except for periods of "crisis").
4. Standard virus is shed continuously at very low levels, but with cyclic fluctuations in amount.
5. DI are shed continuously in fluctuating amounts.
6. Cured clones of cells can be isolated under antiserum and these are 100% destroyed by all m.o.i. of DI-free virus isolated from carrier cells.
7. Carrier virus plus carrier DI readily reestablish long term persistent infection in BHK cells.
8. Long term carriers shed very slowly-growing, temperature sensitive, small plaque mutants with considerably altered oligonucleotide maps.
9. Small plaque mutant virus from long term carriers do not interfere with wild type virus, but carrier DI (or carrier cell DI RNP) interfere strongly.
10. Carrier cells are stable at 25°C, 32°C, and 37°C.
11. Carrier cells shed a wide variety of different DI at different times (with quite different sizes and oligonucleotide maps) and all that have been examined lack transcriptase activity.

Very little that is new has occurred with this long term carrier (CAR_4 during the 2 years since we last reported its behavior[5-8]. It is still showing virus antigen in all or nearly all cells, and still producing only small amounts of mature infectious virus and mature DI. The amounts of virus and DI released are still varying in a cyclic manner akin to the cyclic DI effects reported by Palma and Huang in a continuously passaged acute infection[24]. Recently we have characterized the DI released from these long term carriers and we find a variety of different kinds released at different times with a variety of different oligonucleotide map patterns. Some of these are quite unusual in comparison with most common DI but none have yet been found to be capable of transcribing genetic information. (Holland, J., B. Semler, and C. Jones, in preparation,1978). We have also characterized the RNA genome of the small plaque, ts virus being released after many years of persistence and we find a remarkably changed oligonucleotide map pattern (Holland, J., B. Semler, and C. Jones in preparation, 1978). The results indicate that there are probably over a hundred mutations accumulated by 4 years of persistence. This is more remarkable in view of the stability of the map patterns of serially passaged virus (reference 25, and Holland, Semler, and Jones, in preparation, 1978). Many of these multiple mutations may of course play no role in persistence, but some almost certainly must, and evolution of virus toward an enfeebled small plaque mutant with ts character must also be a result of many of these mutations. We are presently carrying out peptide mapping to determine the proteins most affected by these mutations. In any case, these virus populations evolving in long term persistence over many years can hardly be viewed as simple point mutants causing a single ts change in one protein or a single small plaque expression of one protein.

Characteristics of DI Recovered from Long Term Persistent Infections. In addition to oligonucleotide mapping we have begun characterizing RNA of DI recovered from persistent infections by a variety of other techniques. Perrault and Leavitt[26,27] have recently shown that most VSV DI have inverted complementary terminal sequences not found in infectious virus RNA. Perrault *et al.*[28] have shown that one terminus is the same as the 5' end of the infectious virion RNA (and its complement forms the 3' end of the subgenomic recombinant DI RNA molecule). We have also shown[28] that this new 3' OH end acts as the template for the synthesis of the short (45 nucleotide)polymerase product synthesized by DI *in vitro*[29,30]. This small RNA has now been sequenced

(Semler, B., Perrault, J., Abelson, J., and Holland, J., manuscript in preparation, 1978) and thereby provides the sequence of the 5' end of infectious VSV virions and of both the complementary 5' and 3' ends of DI. Figure 1 shows this comparative structural difference between infectious virus and DI, and the origin and sequence of the DI polymerase product RNA templated by the minus strand of DI RNA.

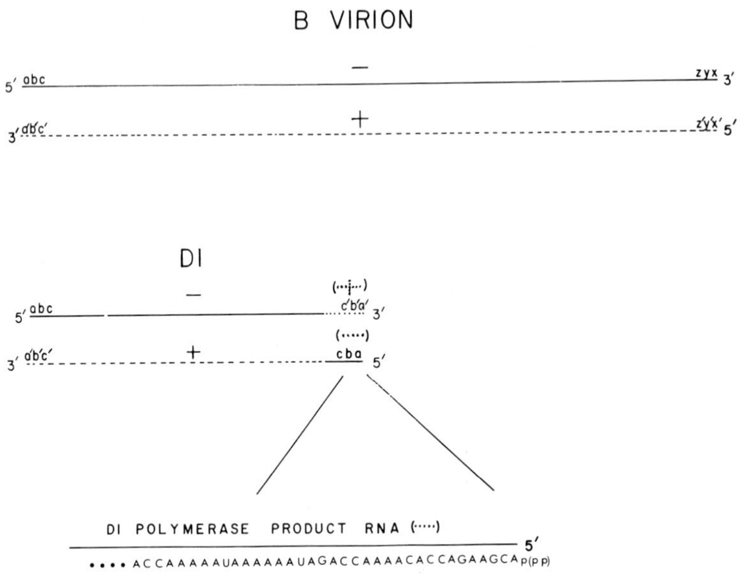

FIGURE 1. Schematic diagram of the structure of VSV viral RNA and of the RNA of the most common class of DI (DI derived from the 5' half of virion RNA). A rare class of DI from the 3' half of virion RNA (HR long T) retains both the 3' and 5' termini of virion RNA (Perrault, Semler, and Holland, manuscript in preparation, 1978).

Sendai virus DI RNA show close similarity to the major class of VSV DI RNA [43].

It can be seen that the product RNA is extremely rich in adenosine residues and that the three 5' terminal nucleotides (ACG) are identical to those reported for the 5' end of virion RNA and for the virion transcriptase leader RNA by Banerjee _et al._[31], and complementary to the ....Py GU 3' OH terminus of virion RNA[31,32]. Although these last 3 residues at each end of the virion show that terminal complementarity might occur in the virus as well as in the DI, this is not the case because the DI polymerase product sequence shown in Figure 1 does not anneal to virion RNA as it does to DI RNA[28], and no other evidence for extensive self-complementarity of virion RNA termini could be obtained[26,28].

We have only recently begun to characterize the RNA termini and polymerase product RNAs of DI from long term persistent infections, but our evidence thus far is summarized in Table III. This is very preliminary data on only a few out of numerous carrier DI but so far none of them are unusual as compared to common VSV DI (except for some with very unusual oligonucleotide maps, which have not yet been characterized for other properties). At present it seems that the nature of the important RNA termini is not greatly altered as the virus and DI evolve over years of persistence in cell culture.

TABLE III

CHARACTERISTICS OF COMMON VSV DI WHICH ARE SHARED BY DI ISOLATED FROM CARRIER CELLS

1. Shortened (subgenomic) RNA with oligonucleotide maps showing only a subset of total viral information.
2. Non-transcribing (but synthesize a 46 nucleotide product complementary to the 3' OH end of the DI RNA).
3. Strongly interfering.
4. Contain inverted terminal complementary sequences not found in infectious virion RNA.
5. Contain a new 3' OH end not found in the infectious virion RNA (and this end is complementary to the 5' end of both DI RNA and virion RNA).
6. Contain RNA sequences deriving mainly from the 5' half of the virion genome RNA.
7. Some carrier DI contain only minus strand RNA, some have plus and minus, and some contain "snapback" RNA[26,27].

VSV Carrier State in Other Cell Lines. Youngner and his colleagues[18] have reported that L cells persistently infected following initial inoculation with VSV plus DI exhibit a different cell-virus relationship from that which we observe with BHK_{21} cells. They conclude that DI are not necessary for persistence in L cells. Rather, their evidence favors a major role of ts mutations such that the mutated virus does not kill the carrier cells at 37°C. Furthermore, they find that ts virus presumably freed of DI by cloning is able to interfere with replication of wild type VSV. Since this is clearly not happening with our BHK_{21} carriers we have explored other VSV-cell carrier systems.

Our L cells unfortunately do not allow establishment of long term persistent infection with VSV. When infected with virus alone all cells are killed, and when infected with virus plus DI only short term persistence ensues with shedding of virus and DI for periods of 1 to several months followed by curing (i.e., total elimination of all traces of virus)[7]. However, our HeLa cells offer an unusual host type which fails to replicate VSV DI in these cells. This can be done only with difficulty since all cells are destroyed or recover and cure even in the presence of DI. However, we have established persistent VSV infection of these HeLa cells on 2 out of 9 attempts and these lines are designated HeLa Car_{44} and HeLa Car_{49}. The former HeLa carrier has already been described[7]. It apparently resulted from surviving cells selected for their ability to produce DI, since these were produced during the 10 months of persistence before this carrier "cured". Nearly 2 years later we established another VSV carrier state in this HeLa cell line and it showed different properties. The properties of these two carriers are outlined in Table IV. The second HeLa cell carrier (HeLa Car_{49}) differs significantly from the first (HeLa Car_{44}) in not producing detectable numbers of DI. It also differs from the BHK_{21} cell carrier (Car_4) in that most cells lack virus antigen and are fully susceptible to wild type virus challenge when the monolayer is healthy between frequent periods of crisis.

TABLE IV

CHARACTERISTICS OF TWO DIFFERENT VSV_{IND} CARRIER CULTURES ESTABLISHED FROM A HELA CELL STRAIN WHICH DOES NOT REPLICATE DI

HELA CAR_{44}

1. Established with difficulty using virus plus DI (3 other attempts to establish persistence failed).
2. All or nearly all cells are infected and resistant to homologous virus challenge.
3. Sheds mature infectious ts small plaque virus and mature DI (i.e., a DI producing cell mutant was selected).
4. Frequent severe crises of cytopathology occurred.
5. Carrier spontaneously "cured" after 10 months of persistence, with no trace of virus remaining, and no virus RNA (or DNA) present by annealing.

HELA CAR_{49}

1. Established with difficulty in the same cell line (2 years after CAR_{44}) using virus plus DI (4 other attempts failed).
2. Cells in severe cytopathic crisis for most of the first 4 months.
3. Strong tendency to "curing" (one of several matched subcultures frequently ceased shedding virus.
4. When carrier is not in crisis most cells lack virus antigen and are as sensitive as normal HeLa cells to wild type virus challenge (100 p.f.u. per cell produced by carrier cells and controls).
5. Mature ts, small plaque infectious virus is shed into the medium.
6. No mature DI or immature DI RNP are detectable by amplification assay.
7. Small "comet" plaques frequently appear on carrier cell monolayers preceding cytopathic crisis.
8. Shed virus rapidly destroys all BHK_{21} cells at all m.o.i., and rapidly kills all HeLa cells at high m.o.i. but only slowly at low m.o.i. and sometimes forms a carrier (without DI) in HeLa cells at low m.o.i.
9. Shed virus does not interfere with wild type virus.

This type of carrier seems to be a precarious balance between infected cells shedding virus and normal fully susceptible cells able to replicate and provide new hosts when the others die. This explains the frequent crises and strong tendancy to curing. This type of carrier represents virus persistence at the population level but probably not at the level of individual cells (which are constantly dying of infection). It should be noted that these HeLa cells are relatively poor hosts for wild type VSV, yielding only about 100 p.f.u./cell (as compared to 20,000 p.f.u/cell for BHK_{21} cells). Other cell types which are poor hosts for VSV (RK_{13} rabbit cells, MDBK bovine cells, MDCK canine cells and PK_{15} pig cells) showed rapid curing after initial shedding of virus for days or weeks. These cells are also low interference cell lines[33,34,1,11].

Another interesting cell type is found in insect cells which readily become persistently infected by VSV[35] but which do not shed levels of DI detectable on BHK_{21} cells[7]. Perhaps DI are not formed (or needed for persistence) in these cells, or else they are unusual DI which do not mature beyond the RNP level, or which cannot replicate in BHK_{21} cells, or which require very low temperatures.[36]

Transplantation of Carrier Cells into Nude Mice for Prolonged Periods. During the past 2 years we have begun an investigation of the behavior of these *in vitro*-established carrier cells in an *in vivo* environment. Surprisingly, we found that all of our BHK_{21} and HeLa cells persistently infected with any of a variety of enveloped RNA viruses (VSV, rabies, measles, mumps or influenza virus) lost their ability to form rapidly growing tumors upon subcutaneous injection into athymic nude mice (Reid, L., Jones, C., and Holland, J., manuscript submitted). These results are briefly summarized in Table V. Presumably these findings are due to surface expression of virus envelope glycoproteins at the plasma membrane of persistently infected cells. The nature, and mechanism of action of the effector "immunocytes" remains to be determined. It will be interesting to see whether the antiviral antibody response mounted (albeit weakly) in mice carrying these persistently infected cells ultimately leads to virus populations with altered surface antigens.

TABLE V

CHARACTERISTICS OF LONG TERM CARRIER CELLS (CAR_4) TRANSPLANTED INTO NUDE (ATHYMIC) MICE

1. Tumorigenicity of persistently infected cells is greatly suppressed as compared to uninfected control BHK_{21} cells (this was true also for BHK_{21} cells persistently infected with rabies virus, mumps virus or influenza type A virus, and for HeLa cells persistently infected with VSV or measles virus).
2. Small subcutaneous nodules are formed which are well encapsulated with a strong fibrous reaction and a marked lymphocytic infiltration (which are both absent in normal cell tumors). (Occasionally in some mice, tumors will form but much later than with BHK_{21}.)
3. Carrier cells reisolated from these small nodules after many months quickly re-establish in cell culture and are shedding mature virus and DI at low levels, and are resistant to homologous superinfection.
4. Irradiation of nude mice eliminates some immunocyte responsible for suppressing growth of persistently infected cells, and these carrier cells then form large rapidly-growing tumors which shed virus and DI.
5. Long after the BHK_{21} cells have formed tumors, some subcutaneous nodules of carrier cells evolve into tumor producing cell populations. Many of these are merely "cured" carrier cells expressing the tumorigenic capacity of BHK_{21} cells, but some are shedding virus and DI. One of the latter virus-shedding tumors has been serially passaged in nude mice for more than one year and its characteristics are described below:
   a) Cells isolated from this virus-shedding tumor are karyotypically BHK_{21} cells, and they form rapidly-growing, soft, diffuse tumors upon reimplantation into other nude mice (even mice preimmunized with whole virion antigens).
   b) All, or nearly all, of these tumor cells are persistently infected with VSV (i.e., are antigen positive and resistent to homologous virus challenge).
   c) Both *in vitro* and *in vivo* these tumor cells continuously shed large amounts of small plaque, ts, infectious virus, and DI of a variety of sizes and genome subsets.
   d) Re-establishment of new carriers using this shed virus plus DI in normal BHK_{21} cells produces a carrier cell population able to form slowly-growing tumors rather than nodules.

A striking finding in animals carrying the virus-shedding soft tumors is the large amount of mature virus (and DI) replicated continuously in the tumors and the relatively low levels of virus (or lack of virus) in mouse organs, such as spleen, liver, kidney, brain, etc. Further work is needed to determine whether any of these organs is persistently infected and shedding low levels of virus, or whether they are continuously eliminating virus circulated to them from the tumors. This system should lend itself to quantitative cellular, biochemical and immunological studies of large numbers of persistently-infected cells in an *in vivo* environment. This should be applicable not only to persistently infected tumor xenografts, as reported here, but to syngeneic and isogeneic grafts of non-tumorigenic mouse cells as well.

A Very Slowly-Growing Latent VSV Carrier Culture (CAR_6) of BHK_{21} Cells Producing No Mature Infectious Virus. Since VSV is amenable to biochemical and genetic analysis it would be desirable to have available a carrier culture which resembled SSPE brain cells in producing no mature infectious virus except after cell cocultivation with susceptible normal cells[36,37]. However, even after 4 1/2 years of persistence our BHK_{21}-VSV carrier (CAR_4) is still producing low levels of infectious virus, as it was 2 years earlier[7]. Recently we established 7 new BHK_{21}-VSV carrier cell cultures using virus plus DI isolated from the original CAR_4 carrier after 4 years of persistence *in vitro*. However, unlike all previous carrier cell populations we did not passage these 7 new carrier cultures when the persistently-infected cells became confluent. Instead, we deliberately allowed them to overgrow, to undergo density dependent inhibition, to fuse, to detach from the monolayer, to round up, etc., while remaining in the original culture flask for 4 months. We merely replaced spent medium with fresh medium each week (or when acid production required replacement). After 4 months of such "neglect" the cells in all of these 7 carrier cultures were very slow growing and sickly in appearance (many giant cells, loosely attached and rounded cells, extremely granular cytoplasm, etc. When these cultures were examined for the presence of virus or virus antigen four months after establishment, 3 of them were found to be typical BHK-VSV carriers similar to CAR_4 described in Table II, 2 had "cured", one was lost, and one, which is designated CAR_6, was found to have unique properties in that it sheds no mature virus although all cells contain viral antigen, resist wild-type challenge and retain the capacity to produce mature virus only after resumption of growth for several weeks or after cocultivation with growing susceptible normal BHK_{21}

The characteristics of this CAR_6 persistently infected culture are outlined in Table VI.

TABLE IV

CHARACTERISTICS OF BHK_{21}-VSV_{IND} CAR_6

1. Established in BHK_{21} cells, using CAR_4 virus plus DI after 4 years of persistence. (Only one of 7 such carriers.)
2. Cell growth was discouraged by cell crowding in the original flasks for 4 months after persistence was established.
3. All cells are persistently infected.
4. No mature virus shedding is detectable in medium.
5. No mature DI are detectable in medium.
6. Cocultivation with normal BHK_{21} cells caused slowly-evolving c.p.e. and eventual release of tiny amounts of infectious small plaque virus and DI (after several days to more than one week).
7. Amplification assays detect DI RNP in cytoplasmic extracts (2 stage amplification required).
8. Superinfection with wild type VSV "rescues" some mature DI despite over 99.9% interference with virus yield.
9. CAR_6 cells are very sickly, and only very slow resumption of growth was achieved.
10. Twenty days following resumption of slow growth (after trypsin dispersions and cell passage), small amounts of mature infectious small plaque virus resumed being shed into the medium.

These slowly growing carrier cells seem to bear at least superficial resemblance to SSPE[37,38] in producing no mature virus until after cocultivation. The situation may also be analogous in that brain cells are mostly non-growing (neurons) or slowly-growing (glial and other cell types). It seems clear from Table VI that DI RNP (but no mature DI or virus) are present in CAR_6 cells, and this can explain their resistance to homologous challenge. A number of possible explanations can be given for the total failure of CAR_6 to produce mature infectious virus. The simplest explanation is that mutations have accumulated which prevent virus maturation but allow RNP assembly and replication in the cytoplasm. Such debilitating virus mutations might be selected against in the usual type of carrier with rapidly growing cells (which would outgrow infected cells after curing occurs unless mature virus is produced to reinfect them). In very slowly-growing or static cell populations, maturation of infectious virus would not be required for maintenance of viral genome replication in all cells <u>so long as none of them</u>

cured. If curing occurs the population of cured cells will either outgrow the sickly infected cells to give a cured culture, or some of the growing population of cured cells will more likely fuse with remaining infected cells to yield mature virus able to reinfect all cured cells (in effect this is a form of cocultivation rescue of virus maturation potential). It is not known if curing requires rapid cell growth but that is quite likely. Curing occurs at a high level when infected cell populations are cloned (where cell-to-cell spread of virus is minimized initially by enormous medium volume to cell volume ratios), or whenever persistently-infected cells are grown as very light monolayers under antiserum to prevent mature virus spread[7]. If this is the case, then rapidly growing carrier cultures, such as CAR_4, might often contain cells in which the virus has mutated to prevent maturation, but the rapid cell growth conditions would tend to mask this condition (which might in fact commonly occur *in vivo* where many cells are non-growing or slowly-growing), and where antibody responses minimize virus spread.

There are of course many more trivial explanations than that of viral envelope protein mutation for lack of virus shedding (i.e., the combination of DI plus sickly, non-growing cells might result in very unstable mature virus; or produce such low yields of cell virus components that the probability of maturation of infectious virus is extremely low, etc.). If we can maintain these CAR_6 cells in their present non-producer condition and slowly grow enough of them for biochemical studies without inducing curing and/or virus maturation, they should prove useful for study of enveloped RNA virus latency.

DI Particles in Acute and Persistent Sendai Virus Infections. Although measles and other paramyxoviruses are among the most studied of enveloped RNA viruses with regard to persistence[37-40,17], little is known regarding the role of DI in paramyxovirus persistence. Because Sendai virus DI can be studied reasonably well[41-43] as compared to viruses such as measles, we have followed the role of Sendai DI particles by analysing the nucleocapsid RNAs isolated from BHK infected cells. In acute infections it was found that DI particle pools (which have a relatively low infectivity of around 5×10^8 PFU/ml) and a high hemagglutinating activity [about 10^3 HAU/ml]) could inhibit the replication of viral 50S RNA when added along with standard virus stock ($>10^9$ PFU/ml, <500 HAU/ml). Moreover, the cells infected with standard virus plus DI virus survived and established long term persistent infections, whereas standard virus alone usually destroyed 100% of the cells.

Three separate persistently infected BHK cultures were obtained. They are resistant to homologous virus challenge although they are completely sensitive to VSV. When stained with fluorescein-labelled anti-Sendai antibody, they show cytoplasmic antigen in 100% or nearly 100% of the cells. When they are labelled with ^{3}H-uridine in the presence of actinomycin D, viral nucleocapsids can be isolated which contain both 50S full length and smaller DI RNAs. The relative amounts of standard virus to DI RNAs varies from 1/4 to 1/100 in different experiments, suggesting that there is a dynamic balance (DI cycling?) involved in the carrier state. Finally, these carriers are shedding standard and DI physical particles but these particles become infectious only when they are treated with trypsin after release into the culture medium.

Further characterization of these DI RNAs is in progress to determine the role they may play in the maintenance of the carrier state caused by this paramyxovirus.

DISCUSSION

It seems clear from all of the above that there are many possible kinds of virus cell interactions leading to enveloped RNA virus persistence. Persistence may exist only at the cell population level with smouldering extracellular spread of virus to growing, susceptible, low yielding cells (as with HeLa CAR_{49}), or it may be confined to individual cells in a slow growing or non-growing state, which produce no mature virus (as with CAR_6), or most cells may be infected and growing and yielding virus and DI (as with CAR_4). Other types, such as insect cell carriers, remain largely unexplored at the biochemical level. It also seems clear that there is no single explanation for persistence. Certainly DI alone cannot explain all persistence, nor can ts mutations of a certain kind (the evidence here indicates long term accumulation of massive numbers of mutations - all of which, or only part of which, may contribute to persistence). Clearly the cell type and the state of cell growth is an important factor, and the nude mouse studies show that the host and its immune responses are critical determinants of persistence. Particularly intriguing is the enormous virus production in the rapidly-growing carrier cell tumors resulting in seeding of virus throughout the body of these T cell deficient mice without <u>apparent</u> virus persistence in mouse organs. This requires further exploration to determine whether foci of persistence are, or are not, established

perpetuating in populations, since there is no fossil record of extinct viruses. Variola may be the first candidate in modern times.

The ability of a virus to persist in an individual host may, however, be critical for its perpetuation in a population, as discussed below.

Eradication is the converse of perpetuation and represents the ultimate method for control of an infectious disease. To determine the potential for eradication, it is necessary first to understand the requirements for perpetuation. Thus the subject is highly relevant to practical goals in public health and preventive medicine.

While we emphasize viral infections here, the principles governing perpetuation also apply to other microbial diseases, such as gonorrhea (1). Furthermore, eradication is a potential goal for agents such as tuberculosis, diphtheria, and pertussis (2).

Viruses in Human Populations: Major Ecological Patterns. The principal ecological patterns of viruses in host populations are set forth in TABLE 1, with particular reference to human infections. In the majority of instances, a virus persists in only a single species and is transmitted directly from host to host. An infrequent variant of this pattern involves an intermediate insect vector, as with dengue. The other major pattern is that of the zoonotic viral infections, where the agent is maintained in another species, but may be transmitted to humans. In most such instances (ie, arboviruses), man is a "dead end" host, although occasionally the agent may be subsequently maintained by human-to-human passage (urban yellow fever, perhaps influenza).

In this brief review, we will limit attention to the maintenance of a virus within a single naturally infected species.

Parameters Which Determine Perpetuation. Descriptive epidemiologic studies, reviewed by Matumoto (3), have indicated a number of parameters which are important determinants of virus perpetuation. These parameters fall into two rubrics, those which concern the population and those specific for the agent.

Population parameters include: (i) The size of the population. At a given incidence the absolute number of infections is proportional to population size. As incidence drops, the chain of infection is more likely to be interrupted in small than in large populations. (ii) Population turnover. The rate of replacement of older immune individuals by young susceptibles influences the proportion of the population which

TABLE 1
ECOLOGICAL PATTERNS OF THE INTERACTION OF VIRUSES AND HOST POPULATIONS: human examples

1. INFECTION CONFINED TO A SINGLE VERTEBRATE HOST: most human viruses

2. VECTOR-BORNE WITH A SINGLE VERTEBRATE HOST: dengue, urban Yellow Fever

3. MULTIPLE VERTEBRATE HOSTS (with or without a vector): ?influenza, jungle Yellow Fever

4. TARGET POPULATION NOT INVOLVED IN VIRUS PERPETUATION: arboviruses, vesicular stomatitis, rabies, arenaviruses

TABLE 2
REPORTED CASES OF MEASLES IN CITIES OF NORTH AMERICA, 1921-1940. After Bartlett (6).

City	Population	Average Annual Reports	Years with a Negative Month	Estimate Per Cent Reported
New York	7,500,000	21,000	0	17%
Chicago	3,400,000	11,000	0	19%
Philadelphia	1,900,000	8,000	0	25%
Detroit	1,600,000	8,000	0	20%
Los Angeles	1,500,000	3,500	0	14%
Montreal	1,000,000	3,500	0	21%
Cleveland	900,000	4,000	1	33%
Baltimore	900,000	5,500	0	39%
Boston	800,000	4,000	0	37%
Toronto	700,000	3,900	0	36%
Washington	700,000	2,300	0	22%
Pittsburgh	700,000	3,800	0	35%
Milwaukee	600,000	5,300	0	56%
Buffalo	600,000	2,200	0	24%
Minneapolis	500,000	2,800	0	35%
Vancouver	300,000	900	20	16%
Rochester	300,000	1,200	3	24%
Dallas	300,000	1,400	18	29%
Akron	200,000	900	8	24%
Winnipeg	200,000	2,100	7	55%

is susceptible, and this may be critical for perpetuation. (iii) Density of population. Density influences contact rate and therefore the probability of transmission. (iv) Immunity level. If immunes cannot serve as links in a chain of infection, the density of susceptibles is a determinant of the probability of transmission.

Agent-Specified parameters include: (i) Infectiousness or transmissibility. Variation between viruses in the probability that an infected individual transmits to contact is well established. Although a difficult parameter to quantify, transmissibility may be measured indirectly in household studies (4). (ii) Duration of infection. The time during which an infected individual can transmit may be critical to the perpetuation of a virus, particularly in small populations. (iii) Generation time. The interval between initiation of successive infections in a series, which may be approximated by the latent period plus half the period of infectivity. (iv) The prevalence of infections at a specific time will also influence the probability of subsequent perpetuation or disappearance.

The 3 variables, transmissibility, duration of infection, and generation time, determine the rate of spread of an agent through a population (for a given set of population parameters). Paradoxically, an agent which spreads rapidly may exhaust susceptibles and disappear more rapidly from a small population than a virus which moves indolently through the same population. For small populations, a crude prediction of perpetuation may be made by determining the number of susceptibles entering the population per generation period. If this number is less than one, persistence will almost certainly not occur.

DESCRIPTIVE EPIDEMIOLOGIC DATA

Observations on the perpetuation of viruses in populations are scattered through the epidemiologic literature (3). No attempt has been made to compile these systematically for this review. Instead, a few selected examples will be considered, to illustrate the impact of some of the parameters noted above. Particular attention will be devoted to population size, population turnover, immunity levels, seasonal cycles, and to a comparison of viruses with differing patterns of pathogenesis.

Large Human Populations: Measles. To follow the pattern of virus infection in large populations, it is necessary to rely on reportable diseases which are clinically distinctive, and have a high case:infection ratio. Measles is one of the few virus diseases which meets these criteria and its behavior in populations has been a recurrent subject of interest to epidemiologists (5-9). Even measles has its limitations as a subject for study, since only 10-55% of cases are reported.

Although measles is an ubiquitous and highly infectious virus, its ability to perpetuate in human populations is surprisingly fragile. Bartlett, in his classical studies (5, 6), found that in urban populations of less than 500,000, measles would periodically disappear. As TABLE 2 shows, in 5 cities of 200-300,000, the disease faded out for at least one month in 56 of the 100 years recorded. Black (7) compiled data on measles in island populations, which led to a similar estimate of critical population size.

Since a population of 300,000 might be expected to experience about 5,000 cases of measles in an average year, it is at first surprising that this would be insufficient to maintain the infection. The explanation becomes more apparent when the seasonal cycle of measles is examined (FIGURE 1). Data from a large U.S. city (Baltimore, Maryland) show that the seasonal cycle is very marked so that only about 1 per cent of the annual total occurs in the 5 low months (August-December). In the lowest month (August) less than 1/1250 of the annual total occurs, per 12-day generation period. This is about 4 cases in a population of 300,000, and there is considerable statistical variation around this average. From this viewpoint, it seems entirely plausible that periodic fadeouts would occur in populations below 500,000.

This analysis suggests that persistence of a short cycle infection in a large population is dependent upon the relationship between incidence during the seasonal trough and the generation time. Influenza, with a generation time of 1-3 days, might be unable to persist in any human population in the absence of antigenic variation.

It has been assumed in the foregoing that measles infects close to 100 per cent of the population, or that the average annual number of infections is equal to annual births. FIGURE 2 indicates that this is essentially correct, and that in one urban area (Baltimore, Maryland, 1900-1931) the average age of infection was 5.2 years. This curve, which defines the susceptible segment of the population (about 9 per cent) has several implications. First, the seasonal fluctuation in number of susceptibles is relatively slight (estimate ±15 per cent in a year of high incidence and considerably less in a year of low incidence). Thus, seasonal flux cannot be explained by exhaustion of susceptibles, although this plays a role in epidemic years. Second, it is possible to estimate the threshold number of suscep-

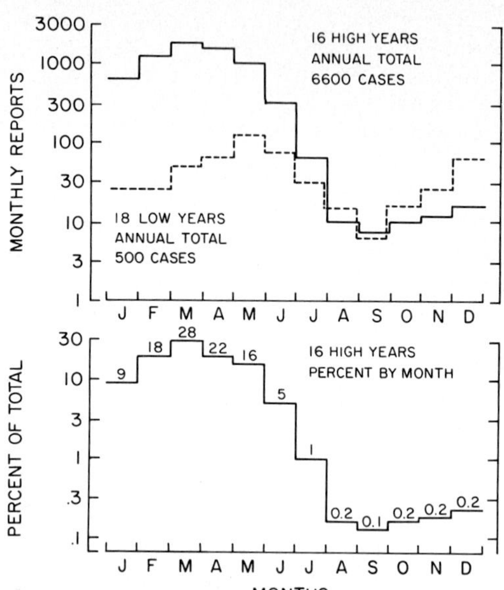

FIGURE 1. Reported measles by month for Baltimore, Maryland, 1928-1961. The 16 high years ranged from 4,400-18,600, and the low years from 100-4,300. It is estimated that about 35% of cases were reported. After Yorke (8).

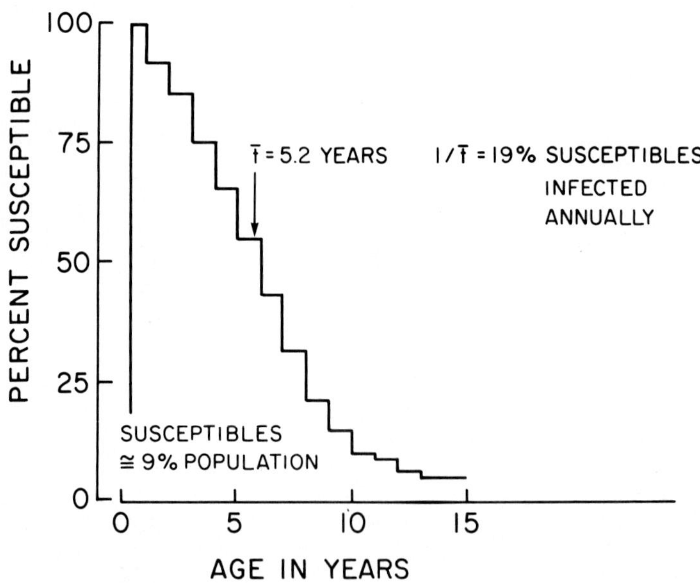

FIGURE 2. Age-specific susceptibility to measles, Baltimore, Maryland, 1900-1931. After Hedrick (9).

tibles required to perpetuate measles and also to project the maximum reduction in susceptibles which might be achieved by immunization. This approach is further developed in the concluding section.

Isolated Human Populations. Primitive isolated human populations, which have minimal contact with the outside world, are occasionally available for serological study. These populations provide a unique opportunity to determine the ability of viruses to perpetuate in their natural host. The studies of Black (10, 11) and others (4, 13, 14) have delineated two principal patterns (TABLE 3) for agents which have no extra-human reservoir. Viruses which are eliminated following primary infection tend to cause abrupt outbreaks after introduction into the population and then die out until reintroduced. Viruses which are capable of persisting in individual hosts are able to perpetuate in small isolated groups. Epidemiologically, a distinction must be made between two kinds of persistent infection: those, like the herpesviruses, which are accompanied by longterm (if intermittent) shedding, and those where persistently infected individuals are not infectious, such as in measles or rubella panencephalitis.

FIGURE 3 shows the age distribution of neutralizing antibodies to poliovirus types 1, 2, and 3 in the isolated Eskimo village of Narssak (population 450) in southern Greenland, from a study by Paffenbarger (15). The profiles are stiking and show that type 1 virus had infected essentially the entire population 25 years previously, while type 2 virus had caused a similar wave of infections 15 years previously and type 3 had been present only one year prior to the collection of sera in 1952. Also, it is clear that

TABLE 3
HUMAN VIRUSES WHICH PERSIST IN
SMALL POPULATIONS OR IN INDIVIDUALS

Individuals	Persist in Small Populations	Agents
+	+	HSV-1, HSV-2, VZV, CMV, EBV, HBV Kuru ?Adeno, ?BKV, ?JCV
+	-	Measles, Rubella Creutzfeldt Jacob

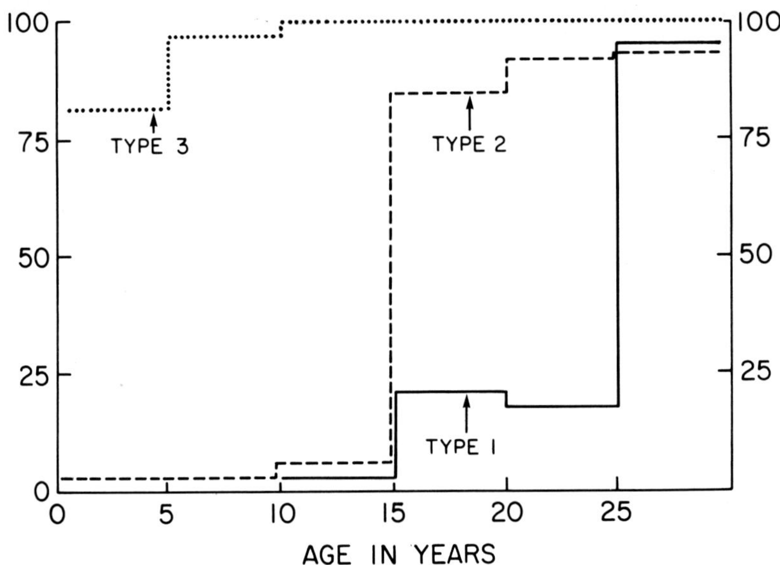

FIGURE 3. Age distribution of polio antibodies in an isolated Eskimo Village, Marssak, Greenland. After Paffenbarger (15).

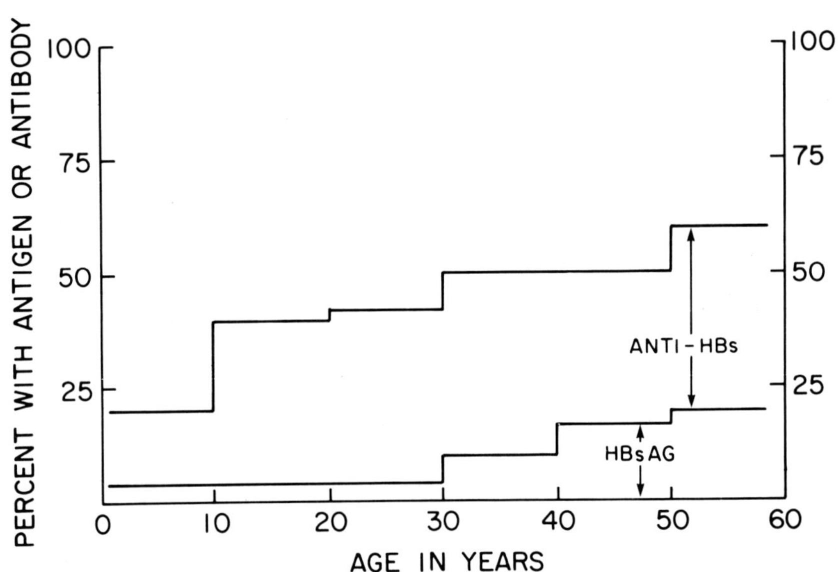

FIGURE 4. Age distribution of HBsAg and Anti-HBs in eskimos of southwest Greenland. After Skinhoj (12).

following its intrusion and exhaustion of susceptibles, each virus had essentially disappeared from the population. (The type 1 antibodies at 15-24 years undoubtedly represent cross-reactions due to infection with type 2 virus).

In contrast, hepatitis B (HB) produces a considerable proportion of persistent infections, particularly in primitive populations. FIGURE 4, from a study by Skinhoj (9) of Eskimos in southwest Greenland, shows that infections have occurred at all ages, with a gradual rise in cumulative incidence to 60 per cent by age 60. The presence of HBsAg positives documents the mechanism of perpetuation.

Small Animal Populations. Animal populations differ radically from human populations in their relatively rapid rate of turnover, and this is a critical variable in virus perpetuation. Depending upon the circumstances, mean life expectancy in many animal populations ranges from 6-12 months, in contrast to a range of 30-70 years for humans. In contrast to these enormous differences, the kinetics of infection are similar in an individual human or animal host. For instance, the median incubation period (to rash) is about 14 days in smallpox and 10 days in mousepox.

Differences in population turnover undoubtedly play an important role in the perpetuation of certain acute viral infections in wild animal populations. Thus, rabies can be maintained in the rather sparse fox population by moving back and forth between adjacent geographic areas. Infections with long incubation periods (such as progressive pneumonia and scrapie of sheep) may persist unrecognized because most animals are slaughtered prior to the onset of frank clinical signs.

Laboratory animal colonies have provided an opportunity to document the quantitative impact of population turnover. Observations in such colonies are of more than academic interest, since unwanted virus infections present an important practical problem which frequently confounds biomedical research (16).

Ectromelia. Ectromelia or mousepox (17) closely resembles variola in man. Infection is initiated either by entry through the skin or by inhalation, and in the systemic infection which follows the liver and spleen are critically involved target organs, and there is a generalized pox-like rash of the skin from which aerosolized virus is transmitted. As in smallpox, infections can be fatal (depending upon many variables, including virus strain, and age and genotype of mouse), but recovered mice are solidly immune and have cleared virus from their tissues. Virus persistence, if it occurs, is at most a rare pheonomenon, inadequate to

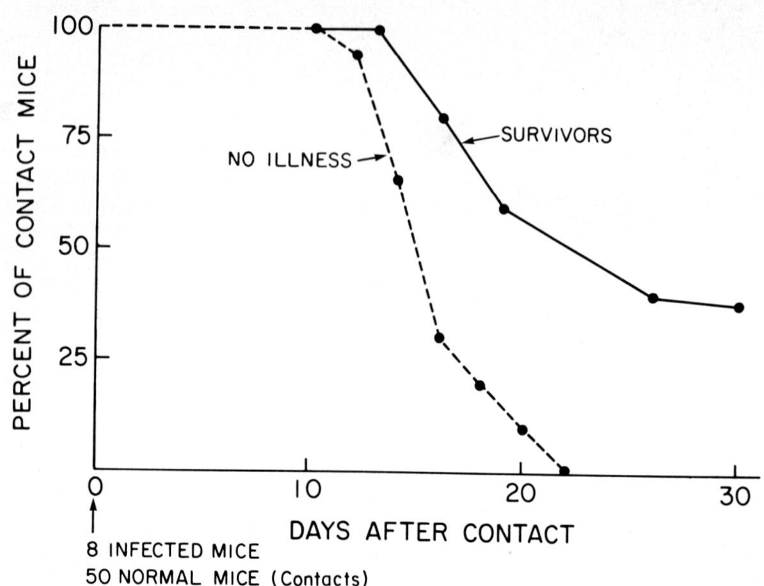

FIGURE 5. An acute "outbreak" of ectromelia in a closed mouse colony to show survival curves. After Fenner (20).

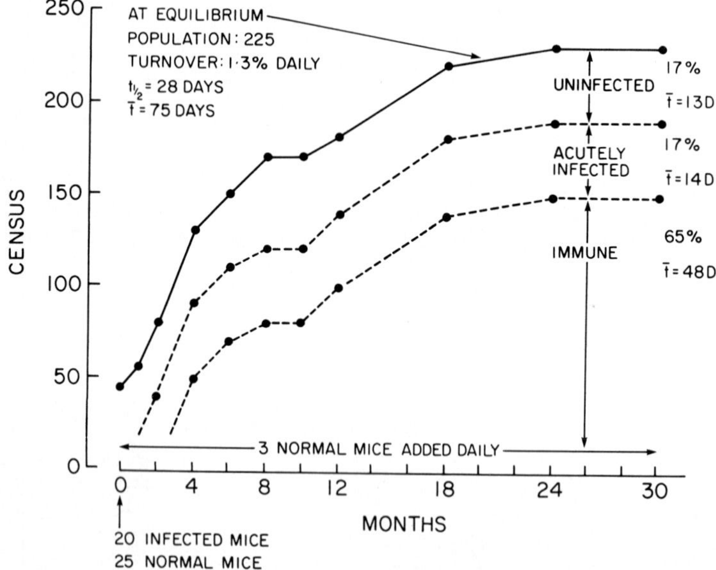

FIGURE 6. Persistent ectromelia infection in a small mouse colony to show population dynamics. Reconstruction from data of Greenwood et al (21), interpreted in light of the findings of Fenner (17, 18, 20, 22, 23).

perpetuate infection in a population (18). Ectromelia is a relatively infectious agent and a notorious cause of acute and devastating epizootics (19). When a few infected mice are placed in a larger group of susceptible animals, the virus spreads rapidly, and all mice are infected within two incubation periods (FIGURE 5). Many mice die, and all the survivors are immune.

In these circumstances, it would not be expected that ectromelia could be perpetuated in a small mouse population. However, a series of classical studies by Greenwood and his colleagues (21) and subsequently by Fenner (17, 18, 20, 22, 23), elegantly demonstrated that this virus was readily perpetuated in mouse populations of 100-200 animals.

FIGURE 6 summarizes one experiment, based on Greenwood's data, which have been interpreted and consolidated in light of Fenner's subsequent publications (17, 18, 20, 22, 21). A mouse colony was established with 25 normal and 20 infected mice, and 3 uninfected mice were added daily and animals were only removed if they died. During the 30-month observation period shown, the total colony size gradually rose to a constant level of about 230 animals. At equilibrium, about 40 per cent of this population died and was replaced each month and the mean lifetime of animals entering the colony was about 75 days. The infection was maintained throughout the existence of the colony and at equilibrium about 1/5 the mice were not yet infected, 1/5 were actively infected, and 3/5 were immune survivors. The mortality produced by acute infection was 50-60%, typical for the Hampstead mouse passage strain. Although the minimal population size in which ectromelia could perpetuate was not determined, Fenner (18) observed one persistently infected population which stabilized at 70 animals.

Rat Virus. Rat virus is a parvovirus of rats which is mainly transmitted as an enteric infection. This agent was unwittingly maintained (24) in a juvenile colony of about 1000 animals of which 250 5-month-old animals were replaced monthly with 250 susceptible weanlings. Rat virus spread at a rate such that 50 per cent of animals were infected by age 5 months and the overall proportion of immunes averaged about 30 per cent. If the population had been permitted to remain for a normal life expectancy, it is likely that all animals would have converted to immunes; under such conditions the agent might have failed to perpetuate.

Measles Virus. An interesting study (25) of the ecology of measles virus in rhesus monkeys indicated that animals in the wild were free of infection. Following capture, monkeys rapidly acquired infection during holding and shipment to the U.S., so that all animals had been infected within 8 weeks. Obviously, measles infection could not be maintained in a stable rhesus colony. However, the rapid turnover in the exporters' compounds in India, where monkeys were held for only a few days to a month, constantly introduced enough susceptibles to perpetuate the virus.

APPLICATION OF COMPUTER SIMULATION TO VIRUS PERPETUATION

Mathematical models of infectious diseases have been applied mainly to analysis of epidemics (26-29). A deterministic model was described by one of us (8, 30) for the simulation of seasonal variation in infections over a period of years. Since computer simulations based on this formulation described the seasonal trough, the model provided some insight into the problem of virus perpetuation.

We now have developed a modified and simplified analysis. Application of this analysis to measles illustrates the method and shows how the model may be used to examine the role of various parameters in perpetuation of an infectious agent.

It has been suggested above that the requirement for large populations for the perpetuation of measles is due to the marked seasonal fluctuation in measles. The following analysis attempts to quantify the parameters which determine the seasonal cycle and then examine the relationship of seasonality and persistence. This question is attacked in steps.

The Model. (i) Using data from New York City (8, 30), an average two-year cycle of measles incidence is constructed which reflects the natural biennial alternation between years of high and low incidence. Monthly cases are spread by individual days to give a smooth curve (FIGURE 7).

(ii) The number of susceptibles is tabulated on a daily basis, by a bookkeeping procedure which accounts for both losses and gains. Losses on a particular day are equivalent to the cases with onset one incubation period (taken as 12 days) later. Gains are the number of new susceptibles entering the population. If all persons eventually acquire measles, then the average daily number of cases (33,600 in two years or 46 daily) is equal to the daily gain in susceptibles.

This procedure generates a periodic curve of the fluctuation in susceptibles around an average level. This average is independently estimated as described subsequently. The number of susceptibles may be expressed as the mean age of acquisition of infection ($\bar{t}$), by dividing the absolute

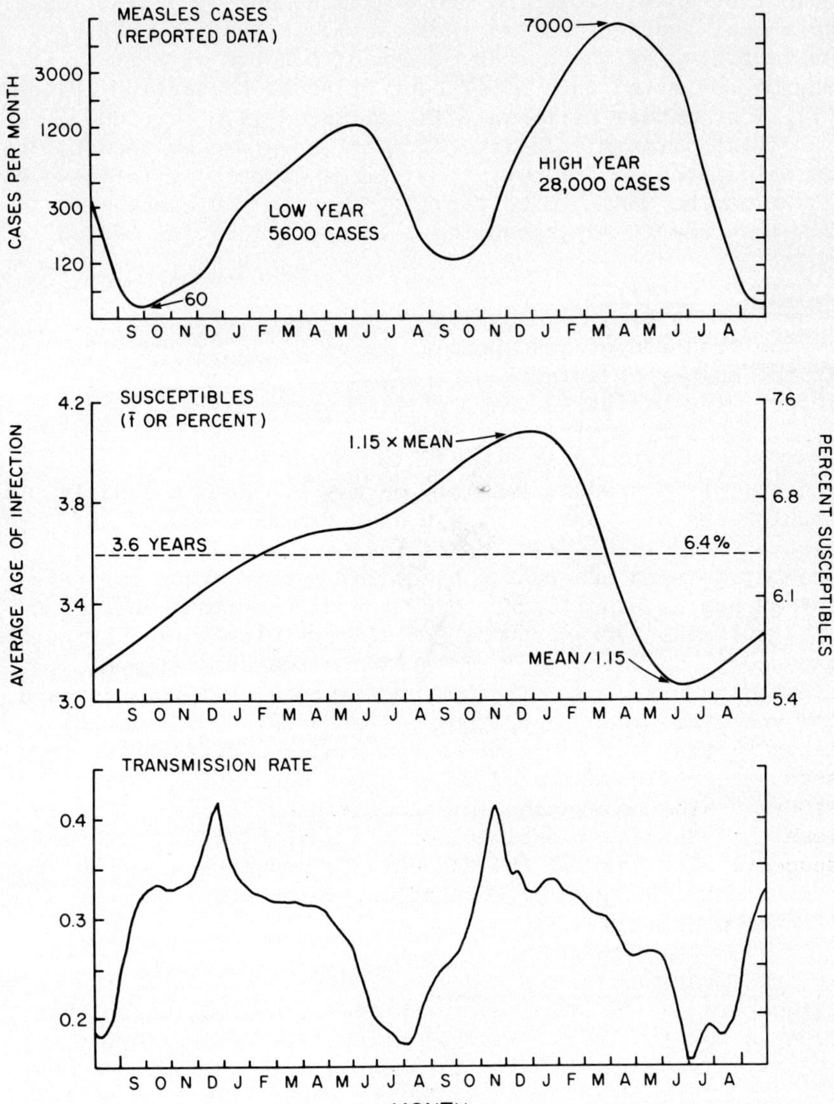

FIGURE 7. Computer simulation of seasonal fluctuation in measles. The daily number of cases (top panel) has been used to generate the curves showing seasonal variation in number of susceptibles and in transmission rate. The high and low years are based on reported cases for New York City (8), and it is assumed that 16,800 susceptibles enter the population annually, equivalent to a birthrate of 1.8 per cent in a population of 930,000. See text for details.

number of susceptibles by the annual number of births (equal to annual average measles incidence). Since $\bar{t}$ expresses susceptibles as the number of annual cohorts of births, it may be converted into a per cent of total population (FIGURE 7), if an annual birth rate is estimated (1.8 per 100).

(iii) Transmissibility of measles varies by season, and is calculated as follows. The number of persons infected on day t can be taken to be C(t+12), that is the cases with onset on day (t+12), assuming a 12-day incubation period. Then

$$\frac{C(t+12)}{S(t)}$$

is the fraction of susceptibles infected on day t. If C(t) is the number of infectious individuals on day t, assuming infectivity begins at onset of illness, then

$$T(t) = \frac{C(t+12)}{S(t)} \frac{1}{C(t)}$$

where T(t) is transmissibility on day t. From the daily tabulations of C and S, values of T can be computed and plotted (FIGURE 7). (Transmissions are actually spread around the mean generation time, but a correction for this spread has no significant influence on the plots in FIGURE 7).

(iv) The average number of susceptibles is set by trial-and-error, in which the simulation is run at different average values of S. The seasonal curves of T for high and low years are then superimposed, and a value of S (mean) taken as that which minimizes the high-low differences in seasonal configuration of T. (In the simulation shown in FIGURE 7, the mean number of susceptibles ($\bar{t}$) is 3.6 years, less than Hedrick's estimate of 5.2 years (FIGURE 2). This suggests that part of the susceptible population may not be mixing with the general population, which might be true of infants under age 2.)

<u>Observations on the Model</u>. FIGURE 7 sets forth the major parameters in the model over a biennial period, which then repeats indefinitely. Several points are noteworthy.

(a) Transmissibility (T) shows a rather small seasonal variation around the mean (± 20 per cent), with somewhat greater variation during transitory seasonal extremes.

(b) Transmissibility is on the whole higher (roughly 3%) in the low than the high year, which is an unexpected finding. Thus, the larger number of cases in high incidence years is due to the accumulation of susceptibles at the beginning of the seasonal rise. On the other hand, the seasonal drop in low years occurs without any decrease in susceptibles. Thus, cyclic changes in both S and T interplay in producing the familiar biennial cycle of measles.

(c) An outbreak of measles can be reconstructed. At first, susceptibles are numerous and the number of cases is low but increasing. It may take a number of generations of growth in the case rate before susceptibles are affected substantially. Then susceptibles drop to some critical level where the case rate peaks. Temporarily, each infective is only replacing himself. The continuing drop in the level of susceptibles makes it less and less likely that an infective will transmit the infection so the level of cases drops more and more steeply.

If the initial level of susceptibles was far above the critical level needed for one infective to infect one susceptible, then at the end of the outbreak the number of susceptibles will be far below that level. After a considerable interval, susceptibles will climb back to the critical level and will continue to climb since it will now take a long time before the level of infectives grows high enough to initiate a decline in susceptibles.

This picture ignores seasonal variation. Soper (31) incorrectly believed that natural oscillations were sufficient to account for fluctuation in incidence. Without a computer he could not tell what happens over a long time. Had he carried out his calculations over several cycles, he could have found the amplitude of oscillation slowly damping down to a constant endemic level (8, 30). We may think of the susceptible level as similar to a pendulum swinging back and forth past equilibrium. Seasonal variation gives the pendulum a shove every year and these regular shoves can have a major cumulative effect.

Level of Susceptibles and Perpetuation. Isolated large populations often have a high level of susceptibles, and when infection is introduced a roaring epidemic may occur, which reduces susceptibles to a very low level. During the extended period before susceptibles can accumulate, a fadeout is possible even in populations larger than Bartlett's critical 300,000. Hence a situation of isolation is conducive to fadeout of an infectious agent.

The simulation model may be manipulated to examine the influence of different variables on perpetuation. An example is shown in FIGURE 8, in which a single infection is introduced during the period of peak transmissibility (mid-January). If susceptibles are at a level of 6 per cent or higher, the virus persists and begins to move into the biennial cycle. With a relatively small reduction in susceptibles (to about 5 per cent), fadeout occurs during the first seasonal trough. This simulation is conducted with a number of susceptibles roughly equivalent to those seen in a population of 900,000, and is, therefore, relevant to what

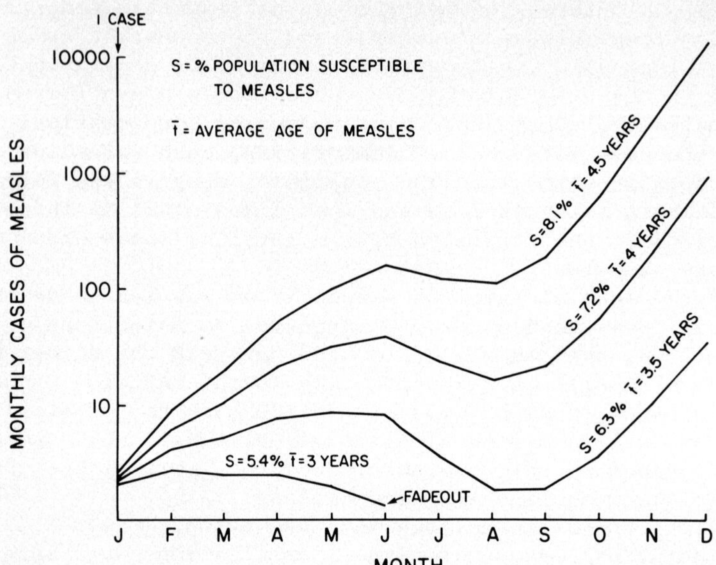

FIGURE 8. Computer simulation of measles following introduction of one case at seasonal peak (January) to show effect of variation in the per cent of the population which is susceptible to measles. The simulation is based on a population of 930,000 with an annual birthrate of 1.8 per cent, and uses the seasonally variable transmission rates shown in the prior figure. See text for details.

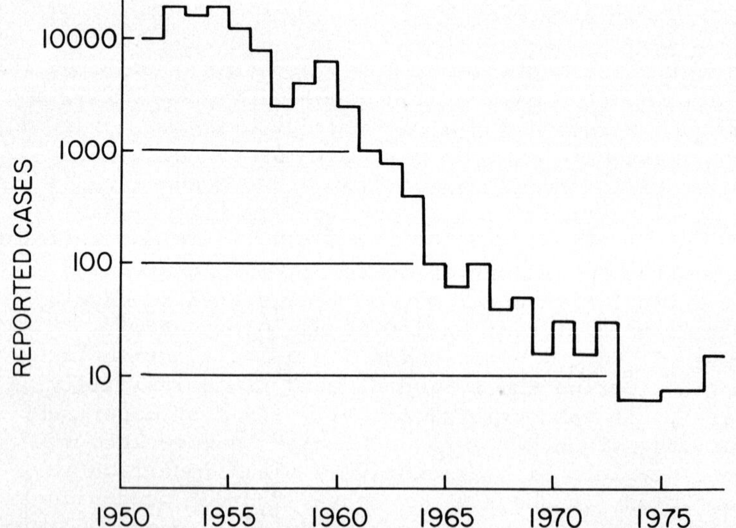

FIGURE 9. Reported cases of paralytic poliomyelitis, USA, 1951-1977. After Center for Disease Control (32)

might be expected in a city of moderate size, which has experienced a fadeout followed by reintroduction of measles. It indicates the difficulty in preventing measles above a critical level of susceptibles and the potential for control if immunization has reduced susceptibles below that level.

ERADICATION: PUBLIC HEALTH IMPLICATIONS OF VIRUS PERPETUATION

The recent dramatic successes in global smallpox control have demonstrated that eradication is an attainable objective for selected viral infections of man. The salient epidemiological features of smallpox which made eradication possible were (2): (i) The relatively long incubation period (about 14 days) and low infectiousness (requiring close direct contact) cause variola to spread slowly within or between rural communities. This made it practical to abandon mass immunization in favor of a search-and-containment strategy, in which local outbreaks were identified and aborted by intensive immunization around each focus. (ii) The marked seasonality of smallpox meant that, even in epidemic areas, the number of cases was relatively low during the trough period. Exploitation of this phenomenon played a key role in achieving eradication in India and Bangladesh. (iii) The high ratio of cases to infections (approaching unity) and the ready recognition of smallpox were also of great importance.

In light of this experience, it is of interest to examine the current status of several important viral infections of man. Consideration is limited to the U.S., where copious data are available.

Poliomyelitis. The annual incidence of poliomyelitis in the U.S. is shown in FIGURE 9, for 1951-1977. Paralytic cases have been reduced from a level of 10,000-20,000 to 10-20, or about 1,000-fold. The epidemiologic classification of cases for 1969-1977 is set forth in FIGURE 10. Putting aside vaccine-associated cases and imported cases, it can be seen that the last outbreak of poliomyelitis occurred in 1972. During the 5 years 1973-1977, there have been a total of only 7 endemic cases with no vaccine association, and the scattering of these 7 in space and time makes it possible that they were related to the vaccine virus. Clearly, the U.S. is on the threshold of eradication of indigenous natural poliomyelitis.

In searching for an explanation of the incipient fadeout of wild poliovirus, an attempt has been made to estimate the reduction in the pool of susceptibles attained by oral poliovirus vaccination. FIGURE 11 compares the age-specific distribution of susceptibles in 1955, at the threshold of

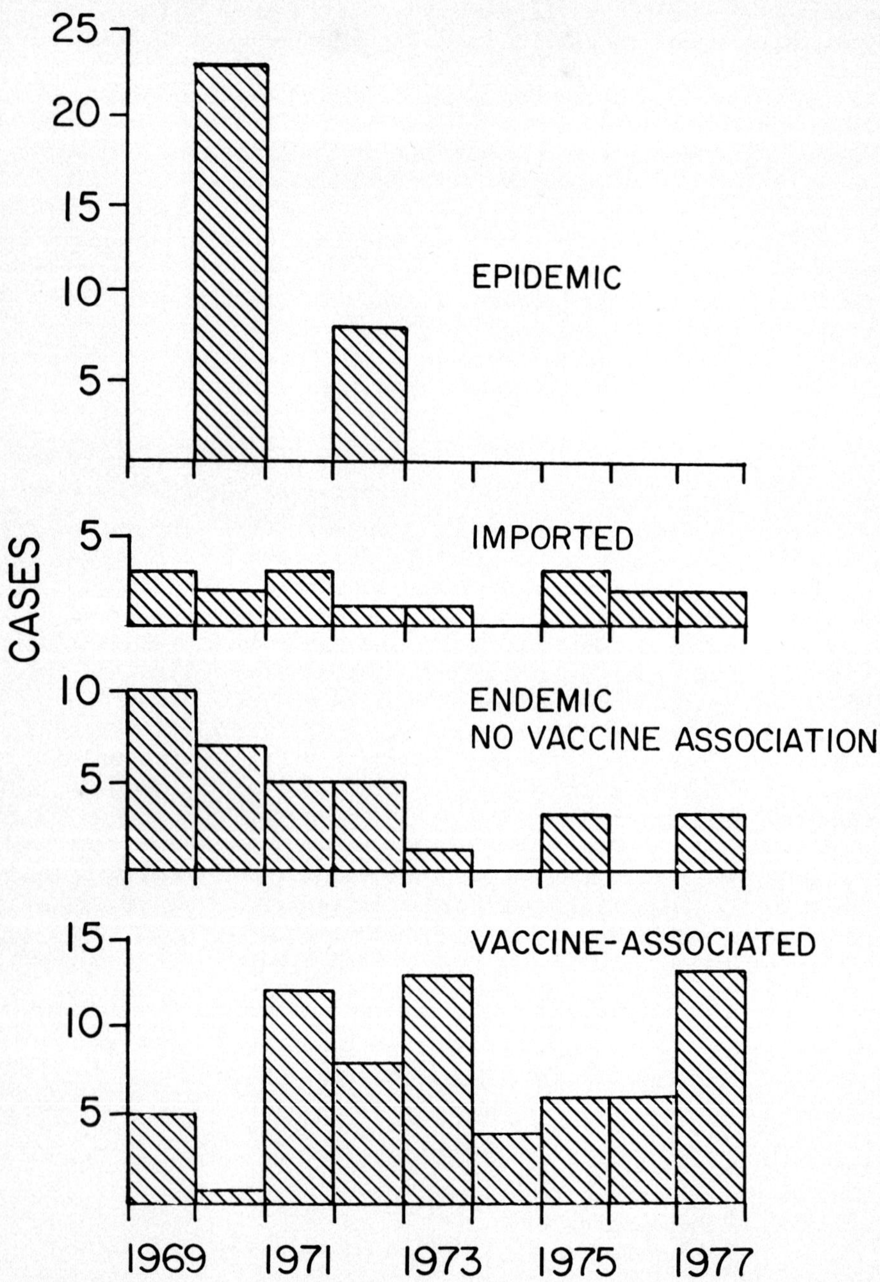

FIGURE 10. Paralytic poliomyelitis by epidemiologic category USA, 1969-1977. After Communicable Disease Center (32,33).

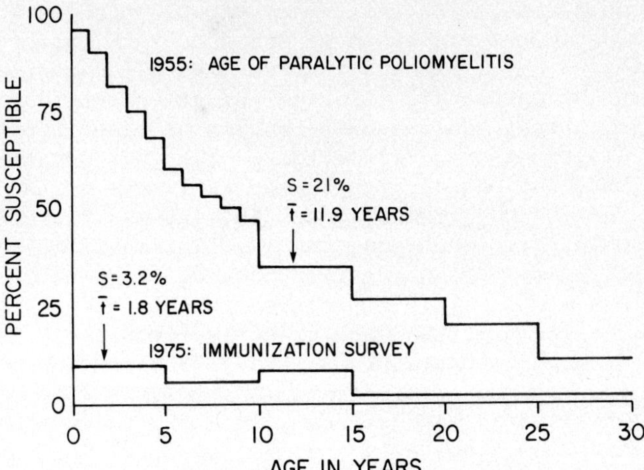

FIGURE 11. Age distribution of poliomyelitis susceptibles in the U.S., for 1955 and 1975. See text for derivation of curves. Based on data of Hall et al (34) and CDC (32).

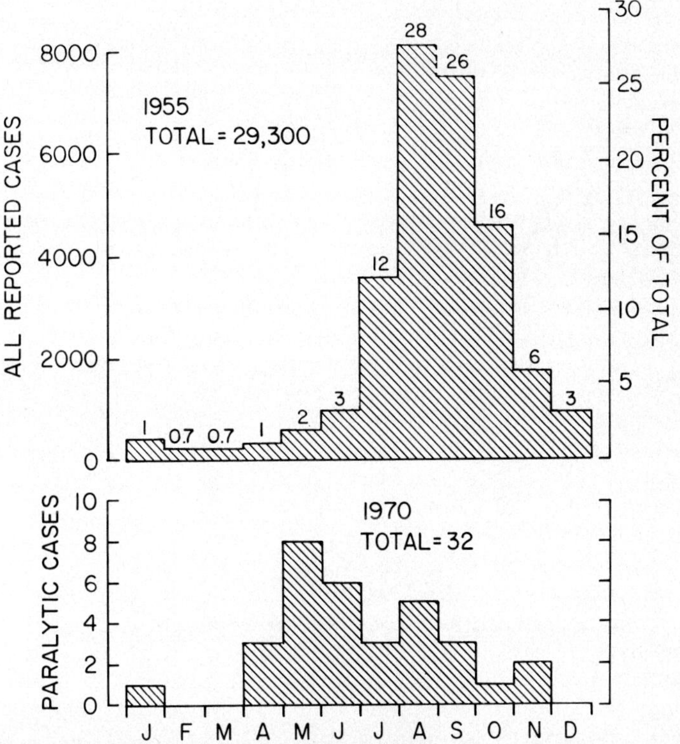

FIGURE 12. Seasonal distribution of poliomyelitis, U.S., 1955 and 1970. After CDC (35, 36).

mass immunization, with the number of susceptibles in 1975. The 1955 estimates are based on the age distribution of paralytic poliomyelitis and involve several assumptions: (i) that the cumulative lifetime infection rate was 100 per cent; (ii) that the age distribution of cases reflects the age distribution of infections. The 1975 data are then based on the National Health Survey estimates of the proportion of persons not immunized and involve a different set of assumptions: (i) that essentially all immunizing infections are conferred by vaccine; (ii) that all vaccinated individuals are immune; (iii) no correction has been made for unimmunized children who were silently immunized by contact with vaccine recipients. Thus, the 1975 estimates must be interpreted with caution.

These estimates suggest that, in 1955, the average age of infection was about 12 years, *ie*, about 21 per cent of the population was susceptible. This compares with 1975 estimates that place the average age of immunization at under 2 years, equivalent to a susceptible pool of about 3 per cent of the population. Thus, there has been an estimated 85 per cent reduction in the number of susceptibles.

An apparent 7-fold reduction in susceptibles has led to a reduction in poliomyelitis of 1000-fold and brought the U.S. to the threshold of eradication. This remarkable achievement suggests that the pool of susceptibles has dropped below the level required for perpetuation of wild poliovirus. This unpredicted development requires explanation. An examination of the seasonal cycle of poliomyelitis (FIGURE 12) suggests a possible mechanism. In the era prior to immunization (1955 is shown) poliomyelitis showed a marked winter trough, with no more than one per cent of the annual total occurring in each of the 4 months January-April. The seasonal cycle was maintained during the period of vaccine-induced declining incidence, as shown for 1970 (the last year with enough cases to demonstrate seasonality). In 1970, there was only one case during the period January-March, and again incidence dropped to zero in December. Since paralytic cases represent only one per cent of all infections, the 1970 data do not prove that the virus was unable to overwinter. However, when total infections for the entire U.S. drop to a level of the order of 100 per month, it is plausible that the chain of infection might be entirely interrupted in major sections of the country.

The hypothesis that wild poliovirus is on the threshold of eradication in the U.S. has several practical implications which deserve brief mention: (i) Importation becomes an important potential source of future poliomyelitis, and a requirement for adequate immunization of all entrants into the continental U.S. seems logical. (ii) It is important to maintain immunization levels to minimize the possibility that wild poliovirus could be re-established following the foreign introductions which are now occurring (FIGURE 10).

Measles. Measles represents an infection which has been markedly reduced in the U.S. by immunization, but which is far from eradicated. FIGURE 13 shows that prior to live measles virus vaccine, about 450,000 cases were reported annually or 15 per cent of an estimated annual 3,000,000. Reported cases have decreased by immunization to a level which is 5-10 per cent of the pre-vaccine era. However, during the last 10 years the number of cases reported annually has remained at a plateau level, and the absence of a downward trend has markedly damped earlier enthusiastic predictions of eventual eradication.

It is interesting to examine the possibility of measles eradication from the viewpoint of virus perpetuation. As a first step, it is important to estimate the size of the susceptible population prior to and after mass measles immunization. FIGURE 14 compares the age-specific distribution of susceptibles for a typical pre-vaccine period (Baltimore, Maryland, 1900-1931) with estimates for the U.S. for 1975. The susceptible population is estimated at about 9 per cent prior to immunization compared to about 4 per cent currently.

If these estimates are applied to the population size requirements for measles perpetuation (TABLE 4), it can be estimated that close to 50,000 susceptibles are required to maintain measles in a city in the U.S. If the size of the susceptible pool is reduced by 55 per cent through immunization, the absolute number of susceptibles in a large city, such as New York, remains well above the threshold for perpetuation. However, a city of moderate size, such as Baltimore, drops below the estimated threshold requirement for measles perpetuation. It is interesting that measles reports (TABLE 5) indicate that the disease continues to be constantly present in New York, as it was in Baltimore before mass measles immunization. However, since 1965 (data for 1968-1971 are shown), measles has been subject to periodic fadeouts in Baltimore.

Thus, the U.S. may be closer to measles eradication than is generally appreciated. Inspection of the measles immunization profiles (FIGURE 14) suggests that a maximal immunization effort might at most reduce susceptibles to 50 per cent of their current level. It seems unlikely that this alone could result in eradication. Another strategy is worth consideration, namely a containment program mounted during the seasonal trough of measles, since only about 10 per cent of annual cases are now occurring during the 4 month interval,

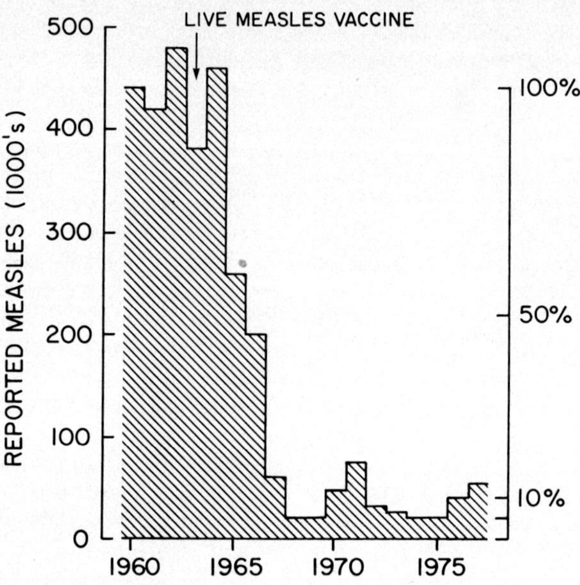

FIGURE 13. Reported measles, U. S., 1960-1976. After CDC (37).

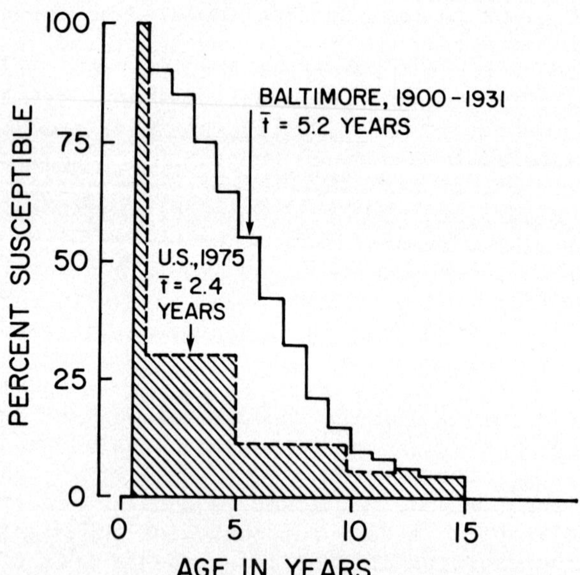

FIGURE 14. Age-specific susceptibility to measles. Pre-vaccine era: based on measles age distribution for Baltimore, Maryland, 1900-1931 (9). Post-vaccine era: based on measles immunization survey and measles age distribution, U.S., 1975 (37).

TABLE 4
MEASLES: HYPOTHETICAL NUMBER OF SUSCEPTIBLES BEFORE AND AFTER IMMUNIZATION PROGRAM

Population	Susceptibles before Immunization	Susceptibles after Immunization
9,000,000	840,000	390,000
900,000	84,000	39,000[a]
500,000	47,000	22,000[a]
300,000	28,000	13,000[a]

[a]Predicted fadeouts.

TABLE 5
PERPETUATION OF MEASLES BEFORE AND AFTER MEASLES IMMUNICATION PROGRAMS. After Yorke (8).[a]

Year	No. of Reports in Low Month	
	New York	Baltimore
1958	47	14
1959	97	22
1960	43	11
1961	123	19
1968	11	0
1969	39	0
1970	31	0
1971	39	0

[a]Measles vaccine has been widely used in U.S. since 1963.

TABLE 6
HUMAN VIRAL INFECTIONS WHICH ARE CANDIDATES FOR ERADICATION IN THE USA: A SPECULATIVE LIST

Infection	Predicted Eradicability	Eradication Advisable
Poliomyelitis	+	+
Measles	±	+
Hepatitis A	+	+
Hepatitis B	?	+
Rubella	±	No
Mumps	±	No

August through November.

<u>Future Prospects for Eradication</u>. The principles discussed above can be used to conjecture about the possible eradication of other human viral infections (TABLE 6).

Rubella and mumps are two infections for which vaccines exist. However, it is highly questionable whether it would be advisable or worthwhile to attempt eradication of either of these agents, at the present time.

Of the infections for which vaccines will probably become available in the future, the most logical candidate for eradication (in the U.S.) is hepatitis A. Hepatitis B also deserves consideration, since its ability to persist is offset by its very limited communicability. However, further elucidation of the pathogenesis, immunology, and mode of transmission of hepatitis B is a prerequisite for serious evaluation of the feasibility of eradication.

ACKNOWLEDGMENTS

We would like to express our warm appreciation to Dr. Lila Elveback who suggested the collaboration which resulted in this paper.

REFERENCES

1. Yorke, J.A., Hethcote, H.W., and Mold, A. (1978). *J. Sex. Trans. Dis.*, in press.
2. Henderson, D.A. (1976). *Scientific American* 235, 25.
3. Matumoto, M. (1969). *Bact. Rev.* 33, 404.
4. Hope Simpson, R.E. (1954). *Lancet* 2, 1299.
5. Bartlett, M.S. (1957). *J. Roy. Stat. Soc. Ser. A.* 120, 48.
6. Bartlett, M.S. (1960). *J. Roy. Stat. Soc. Ser. A.* 123, 37.
7. Black, F.L. (1966). *J. Theor. Biol.* 11, 207.
8. Yorke, J.A., and London, W.P. (1973). *Am. J. Epidemiol.* 98, 469.
9. Hedrick, A.W. (1933). *Am. J. Hyg.* 17, 613.
10. Black, F.L., Hierholzer, W.J., Pinheiro, F.D., et al. (1974). *Am. J. Epidemiol.* 100, 230.
11. Black, F.L. (1975). *Science* 187, 515.
12. Skinhoj,, P. (1977). *Am. J. Epidemiol.* 105, 99.
13. Panum, P.L. (1940). "Observations Made during the Epidemic of Measles on the Faroe Islands in the Year 1846.". American Publishing Association, New York.
14. Paul, J.H., and Freese, H.L. (1933). *Am. J. Hyg.* 17, 517.
15. Paffenbarger, R.S., and Bodian, D. (1961). *Am J. Hyg.* 74, 311.
16. Holdenried, R., ed. (1966). "Viruses of Laboratory Rodents." NCI Monograph No. 20, U.S. Govt. Print. Office, Washington, D.C.
17. Fenner, F. (1949). *J. Immunology* 63, 341.
18. Fenner, F. (1948). *J. Hygiene* 46, 383.
19. Briody, B. (1959). *Bact. Rev.* 23, 61.
20. Fenner, F. (1949). *Aust. J. Exp. Biol. Med. Sci.* 27, 45.
21. Greenwood, M., Bradford Hill, A., Topley, W.W.C., and Wilson, J. (1936). "Experimental Epidomiology." Special Report Series No. 209, Medical Research Council, H.M.S.O., London.
22. Fenner, F. (1948). *Brit. J. Exp. Path.* 29, 69.
23. Fenner, F. (1949). *Aust. J. Exp. Biol. Med. Sci.* 27, 19.
24. Robinson, G.W., Nathanson, N., and Hodous, J. (1971). *Am. J. Epidemiol.* 94, 91.
25. Meyer, H.M., Brooks, B.E., Doublas, R.D., and Rogers, N.G. (1962). *Am. J. Dis. Child.* 103, 307.
26. Bailey, N.T.F. (1975). "The Mathematical Theory of Infectious Diseases and its Applications." Hafner Press, New York.

27. Serfling, R.E. (1952). *Human Biol.* 24, 245.
28. Abbey, H. (1952). *Human Biol.* 24, 201
29. Dietz, K. (1975). in "Epidemiology" D. Ludwig and K. Cooke, eds., SIAM, Philadelphia, 104.
30. London, W.P., and Yorke, J.A. (1973). *Am. J. Epidemiol.* 98, 453.
31. Soper, H.E. (1929). *J. R. Stat. Soc.* 92, 34.
32. Center for Disease Control. (1977). Poliomyelitis Surveillance Summary 1974-1976.
33. Center for Disease Control. (1977). Morbidity and Mortality Weekly Report 26, 424.
34. Hall, W.J., Nathanson, N., and Langmuir, A.D. (1957). *Am. J. Hyg.* 66, 214.
35. Center for Disease Control (1971). Poliomyelitis Surveillance Report No. 56.
36. Center for Disease Control (1971). Poliomyelitis Surveillance Summary 1970.
37. Center for Disease Control (1977). Measles Surveillance Report No. 10, 1973-1976.

Subacute Sclerosing Panencephalitis (SSPE) An Epidemiologic Review [1]

Neal A. Halsey, M.D.,[2] John F. Modlin, M.D.,[2]
J. T. Jabbour, M.D.[3]

Center for Disease Control, Atlanta, Georgia 30333
and Department of Neurology, University of Tennessee
Center for Health Sciences, Memphis, Tennessee 38103

Abstract

Measles infections clearly contribute to the development of subacute sclerosing panencephalitis (SSPE). However, epidemiologic data from several countries suggest that other factors play an important role in the pathogenesis of SSPE. Geographic clusterings of higher incidence rates have been documented, and rates are higher in rural areas.

The incidence in males is 2-4 times higher than in females. Racial differences in incidence have been noted but without a consistent pattern worldwide. Lower socioeconomic status, large family size, and crowded home living conditions have been correlated with SSPE. Close exposure to birds (particularly sick birds) has been shown be more common in children with SSPE than in matched controls in 2 different case-control studies. No other animal exposures have been consistently correlated with SSPE.

Several lines of investigation have indicated that the rate of SSPE following measles vaccine is significantly lower than the rate following natural measles.

A hypothesis is proposed which would explain the various factors known to correlate with SSPE. Available evidence suggests that an as yet unknown factor, possibly a second infectious agent, may be important in the pathogenesis of SSPE.

1. This work was supported in part by Food and Drug Administration Contract No. 223-75-1105
2. Immunization Division, Bureau of State Services, Center for Disease Control, Atlanta, Georgia 30333
3. Department of Neurology, University of Tennessee Center for Health Sciences, Memphis, Tennessee 38103

ISBN 0-12-668350-6

Introduction

The association between measles and subacute sclerosing panencephalitis (SSPE) was first made by Connolly et al in 1967 (1). Subsequently, several other investigators have confirmed that measles virus plays a critical role in the development of this slow virus infection. Elevated measles antibody titers are found in the serum of patients with SSPE, and measles-specific antibodies are found in the cerebrospinal fluid (CSF) (1-5). Measles-like viral particles have been noted in electron-microscopic preparations of brain tissue of SSPE patients; fluorescent-antibody stains have indicated that these antigens are similar or identical to measles virus (1,6-11). Additionally, several investigators, using co-cultivation techniques, have isolated a measles-like virus from brain biopsies or postmortem specimens of patients with SSPE (12-18).

However, questions linger about the possibility that other infectious agents contribute to the pathogenesis of SSPE (19-22). In addition, other phenomena remain to be explained: (i) What predisposes some children to develop a latent infection with the measles virus? (ii) How and where does the virus persist in the body for several years before causing a progressive panencephalitis? (iii) What triggers the virus to cause a progressive panencephalitis? (iv) Is live measles vaccine capable of causing SSPE? (v) Does live measles vaccine administered to a child who has had measles increase the risk of developing SSPE?

Epidemiologic studies carried out in recent years provide answers to some questions and possible clues to the answers of others. These studies will not be reviewed in detail. Instead, the unique epidemiology of SSPE will be illustrated with appropriate examples.

Measles and Measles Vaccine

In a review of 350 SSPE cases in the United States, Modlin et al found that 83.4% of the patients were known to have had measles (23). The average interval between measles and onset of SSPE was 7 years. Although the majority of children in the early 1960's contracted measles, differences at the time of measles infection have been noted between those who did and did not contract SSPE. Forty-six percent of children with SSPE had measles infection before they were 2 years of age. Detels et al, in a case-control study, found that children with SSPE had measles at younger ages than their playmate controls (24). We have recently completed another case-control study involving 52 children with SSPE and 96 controls matched for age, sex, and race

(25,26). Children with SSPE were significantly more likely to have had measles than either playmate or hospital controls. When measles infection had occurred, it had been at significantly younger ages in children with SSPE.

Based on an estimated 90% underreporting of measles cases in the United States each year (27), the risk of SSPE following measles can be estimated. In a previous publication we reported that between 5.2 and 9.7 cases of SSPE ultimately developed for every million children who had probably had measles in a given year (23). Additional reporting of cases in the past 3 years has increased this rate to between 6 and 22 cases per million children who had had measles in a given year (unpublished data).

Measles vaccines were introduced in 1963 and by 1966 were used extensively. Following the widespread use of these vaccines, the incidence of measles decreased by over 90% in the United States. Likewise, the incidence of SSPE, especially measles-related cases of SSPE, has also decreased in recent years (Figure 1), starting approximately 7 years after the decrease in measles incidence. A cohort analysis shows the incidence of SSPE by age for children born in different years (Figure 2). Each line represents the mid-year of a 3-year birth cohort. For children born prior to 1965, the highest incidence of SSPE occurred between 4 and 14 years of age. Children born since the time of widespread measles vaccine usage (1966) have had a lower incidence of SSPE at all ages.

Figure 1

469 CASES OF SUBACUTE SCLEROSING PANENCEPHALITIS, BY YEAR OF ONSET, UNITED STATES, 1960-1977

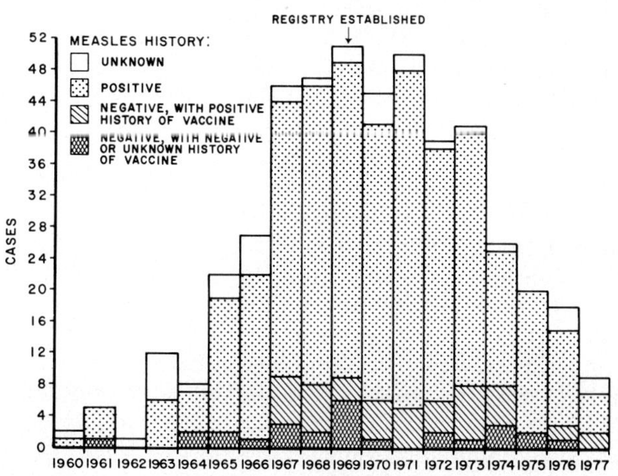

Figure 2

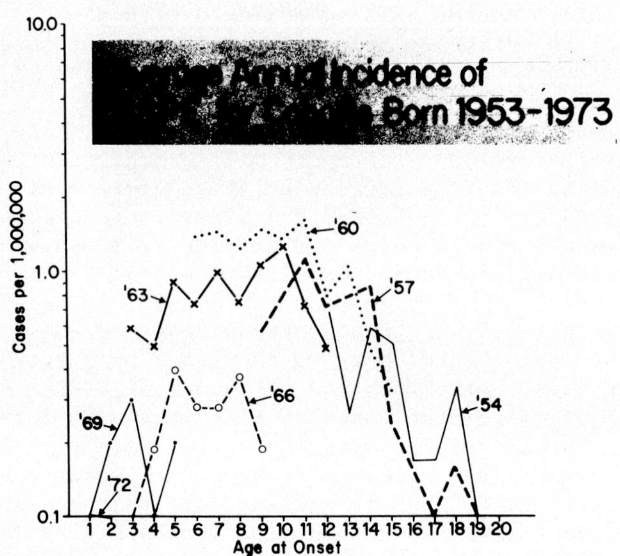

Of persons vaccinated against measles in 1964-1970, between 0.48 and 1.13 recipients (who had no history of measles) per million doses distributed subsequently contracted SSPE. The rate following natural measles was 5-50 times higher. The average interval between vaccination and onset of SSPE was 3.3 years, but there was no association with any particular age at vaccination.

In our case-control study, children with SSPE were significantly less likely to have received measles vaccine than their controls. The SSPE patients who had no history of measles, but had received measles vaccine, had been vaccinated at about the same age as the control groups (mean of 4.2 years vs. 4.3 years). Having received measles vaccine after having had measles was noted more in the control groups than in children with SSPE.

Therefore, administering measles vaccine at an early age or after natural measles is not associated with an increased risk of SSPE. Also, there is reason to believe that at least some of the children with vaccine-associated cases actually had had natural measles infections which were responsible for causing SSPE. Of the 6 children with SSPE in the case-control study who had received measles vaccine but had no history of measles, at least 4 of them had had more than 1 rash illness in the first 2 years of life, or had been exposed to a sibling with measles several years before receiving measles vaccine. We are aware of 17 cases of SSPE in which the parents reported that their child had neither had measles nor measles vaccine. All of these children who have been tested have had measles antibodies in the serum and CSF. Presumably, they had measles when less than 1 year of age. They evidently either had subclinical infections, or the infection was modified by passively acquired maternal antibody.

Worldwide Incidence

A national registry for SSPE cases occurring in the United States was established in 1969 in Memphis, Tennessee, by 1 of the authors (JTJ) (28). Based on the 475 cases reported with onset in 1960-1977, the average annual incidence of SSPE is 0.13 cases per million population (29). This rate is somewhat higher than that observed in the United Kingdom and Japan, but lower than the rates observed in most other countries (Table 1). The rates shown in Table 1 were derived from reports in the literature, and all were adjusted to show the average annual rate per total population. The authors did not necessarily intend their report to be an accurate estimate of the incidence in these countries. Therefore, these rates must be accepted as only crude approximations of the true incidences of SSPE in these countries. Many variables affect case ascertainment, and several of these were not controlled in these studies. For instance, the high rate observed in Israel may be related to the almost complete case ascertainment obtained through computer records of all medical care. The highest rate (7.7 per million population aged 5-20) was reported from the northern half of North Island, New Zealand, during the years 1956-1969. However, the total population for this area was not given and therefore the rate cannot be compared with those for other countries. Also, since this was a report of a clustering of cases, the average incidence may be somewhat lower than noted here.

There is no apparent association of SSPE incidence with latitude or any other geographic factor in any particular area of the world.

Geographic Clustering Within Countries

In the United States a significantly higher incidence of SSPE has been noted in the Southeastern states and the Ohio River Valley (28,29). This clustering is not secondary to reporting artifact, and no satisfactory explanation has been offered to explain this phenomenon. Similar geographic clusterings have been noted in South Africa (33), New Zealand (34), Belgium (35), and the United Kingdom (31).

Table 1. Average Annual Incidence of SSPE in 13 Countries

Country	Ref	Years Studied	No. Cases	Pop. in Millions	Ave. Cases Per Million Population Per Year
North Island, New Zealand	34	1956-59	27	(see text)	
Israel	37	1966-73	52	3.33 (1970)	1.95
Algeria	*	1967-74	21	1.9 (1970)	1.6
Quebec, Canaca			12	5.7 (1965)	1.0
N.E. Region, North Scotland	32	1952-61	4	?	<1.00
Iran	**	1976	15	33.0	0.45
Finland	39	1960-69	20	4.65 (1965)	0.43
Belgium	35	1939-61	79	9.6 (1965)	0.28
South Africa	33	1955-74	79	17.8 (1965)	0.22
(Cape Province only)	33	1970-74	30	10.6 (1965)	(0.57)
Brazil (State of Sao Paulo)	38	1954-66	25	13.0 (1960)	0.15
U.S.A.		1960-77	475	203.2 (1970)	0.13
United Kingdom	30, 31	1961-75	49	55.5 (1970)	0.06
Japan	47	1966-76	45	103.5 (1970)	0.04

*Baugermough A: Personal communication
**Mirchamsy H: Personal communication

Rural-Urban Residence

Higher rates of SSPE in children from rural areas have been noted in Belgium (35), Brazil (38), Latin America (36), and the United States (29). Progressively lower rates of SSPE occur with increasing population density in the United States (Figure 3). The incidence of SSPE in children residing on rural farms is almost 6 times that of children living in inner city areas.

Figure 3

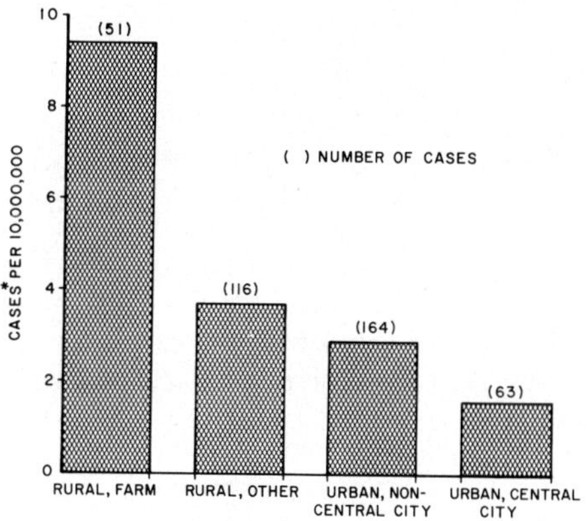

Age at Onset

SSPE has been noted in infants as young as 2 months of age (40) and in persons as old as 32 years (41). The median age at onset in the United States is 9 years, and 85% of patients were between 5 and 14 years of age (29).

Sex

A male predominance of cases has been noted in every country where SSPE has been studied (29,30,31,33,35,37,38). The incidence in males in the United States is 2.3 times higher than in females. The male-female ratio is higher (4.5:1) for children with onset at less than 5 years of age and lower (1.5:1) for persons with onset after 14 years of age (29).

No satisfactory explanation has been offered for the higher incidence in males or the varying sex ratio with age at onset.

Race

The overall incidence in whites under 20 years of age in the United States is 4 times higher than in blacks (29). When standardized for rural-urban residence the difference between races is less noticeable, but the rates for whites still exceed those for blacks. In South Africa the highest rates occur in children of mixed racial marriages (33). In Israel higher rates of SSPE have been observed in Sephardic Jews and Arabs than in Ashkenazic Jews (37). Since no consistent racial patterns of increased incidence have been noted, other factors (such as rural-urban residence, socioeconomic status) must also be important in explaining the observed racial differences.

Animal Exposures

The geographic clustering and higher incidence in rural areas imply that environmental factors probably play a role in the pathogenesis of SSPE. Additionally, since other viral infections have been postulated in this disease, it is possible that exposure to certain animals could be related to the acquisition of these viruses. Detels *et al* found that children with SSPE had been exposed to dogs with canine distemper and fowl significantly more often than their matched controls (24). Similarly, other investigators have observed exposure to illnesses in birds or other animals shortly before the onset of SSPE (2,35,38). We did not find any differences in exposure to dogs, cats, horses, cows, sheep, or a variety of other household pets in our case-control study. Although children with SSPE were more likely to have had contact with pigs ($p<.05$), only one-third of all cases had a history of such exposure. However, a highly significant difference ($p<.001$) in rates of close exposure to birds was noted. Sixty-five percent of patients had been exposed to 1 or more birds prior to the onset of SSPE. In addition, patients were significantly more likely to have been exposed to sick birds. The most common types of birds involved in the exposures were household caged birds (canaries, parakeets, parrots, etc.). However, exposure to farmyard birds (chickens, ducks, geese, etc.) was almost as frequent. The meaning of these exposures is not apparent. They might represent the transmission of an infectious agent to children that in some way contributes to the pathogenesis of SSPE.

Other Factors

No seasonal trend has been noted for the recognized onset of SSPE in the United States or Latin America (Figure 4) (38). This is not surprising, since the onset of disease is often insidious and hard to pinpoint accurately.

Figure 4

416 CASES OF SSPE, BY MONTH OF ONSET OF SYMPTOMS, UNITED STATES, 1960-1976

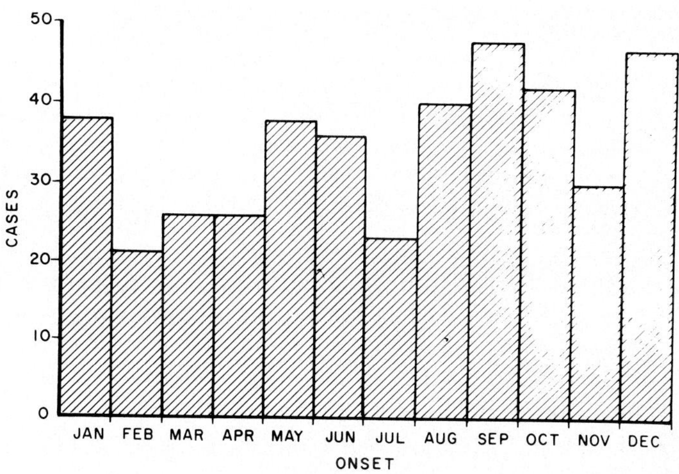

Several persons with SSPE have been noted to have had either significant head injuries (32,42,43,44) or central nervous system (CNS) infections (37,45,46) prior to the onset of SSPE. Thirty-eight percent of the 52 children with SSPE in our case-control study had suffered a significant head injury, 20% had had at least 1 seizure in infancy, and 8% had had either meningitis or encephalitis prior to the onset of SSPE. The inflammation or possible bleeding associated with these disorders may allow entry into the brain of cells carrying a latent virus, or they may release a latent infection in the CNS.

Children with SSPE tend to come from large families who live in crowded homes and who are of lower socioeconomic status (26,37). These factors are known to be associated with the development of measles infections at an early age.

Previous investigations have suggested that chickenpox occurring at about the time of administration of measles or inactivated poliomyelitis vaccines contaminated with SV40 virus might predispose to SSPE (24,34). No significant differences in the frequency of childhood infections or receipt of vaccines other than measles was observed in our case-control study.

Summary

A variety of factors has been shown to be correlated with an increased frequency of SSPE infections: sex, race, certain geographic areas, rural areas, measles infections at an early age, close exposure to birds, CNS insults, large family size, and lower socioeconomic status. A uniform hypothesis to incorporate all of these factors would have to assume that several coincidental events must occur in order for a child to develop SSPE. First, measles infection must occur at an early age (large families, lower socioeconomic status, crowded home conditions) in a host predisposed to develop a latent infection (sex and race). The geographic clustering, increased incidence in rural areas, and exposure to birds may be related to a factor(s) which either predisposes the host either to develop latent infection with measles or to a factor(s) responsible for releasing the latent infection. The CNS injuries or insults could allow the introduction of leukocytes bearing a viral (or other infectious) agent into the CNS.

Further laboratory studies are in order to look for infectious agent(s) other than measles that may be necessary for the development of SSPE.

References

1. Connolly, J.H., Allen I.V., Hurwitz, L.J., Millar, J.H.D.: Measles-virus antibody and antigen in subacute sclerosing panencephalitis. Lancet 1: 542-544, 1967.
2. Legg, N.J.: Virus antibodies in subacute sclerosing panencephalitis: A study of 22 patients. Br Med J: 350-352, 1967.
3. Brody, J.A., Detels R., Sever J.L.: Measles antibody titers in sibships of patients with subacute sclerosing panencephalitis and controls. Lancet 1:177-178, 1972.
4. Link H., Panelius M., Salmi, A.A.: Immunoglobulins and measles antibodies in subacute sclerosing panencephalitis. Demonstration of synthesis of oligoclonal IgG with measles antibody activity within the central nervous system. Arch Neurol 28:23-30, 1973.
5. Sever J.L., Krebs H., Ley A., Barbosa L.H., Rubinstein D: Diagnosis of subacute panencephalitis. The value and availability of measles antibody determinations. JAMA 228:604-606, 1974.

6. Tellez-Nagel I., Harter, D.H.: Subacute sclerosing leukoencephalitis: Ultrastructure of Intranuclear and introcytoplasmic inclusions. Science 154:899,901, 1966
7. Dayan A.D., Gostling J.V.T., Greaves J.L., Stevens, D.W., Woodhouse, M.A.: Evidence of a pseudomyxovirus in the brain in subacute sclerosing leucoencephalitis. Lancet 1:980-981, 1967.
8. Freeman, J.M., Magoffin, R.L., Lennette, E.H., Herndon, R.M.: Additional evidence of the relation between subacute inclusion-body encephalitis and measles virus. Lancet 2:129-131, 1967.
9. ter Meulen V., Enders-Ruckle G., Muller D., Joppich G. Immunhistological microscopical and neurochemical studies on encephalitis. III. Subacute progressive panencephalitis. Virological and immunhistological studies. Acta Neuropath 12:244-259, 1969.
10. Oyanagi S., ter Meulen V., Katz M., Koprowski H: Comparison of subacute sclerosing panencephalitis and measles viruses: An electron microscope study. J Virol 7:176-187, 1971.
11. Oyanagi S., Koprowski H: Subacute sclerosing panencephalitis: Structures resembling myxovirus necleocapsids in cells cultured from brains. Nature 222:888-890, 1969.
12. Horta-Barbosa L., Fuccillo D.A., London, W.T., Jabbour, J.T., Zeman W., Sever, J.L.: Isolation of measles virus from brain cell cultures of two patients with subacute sclerosing panencephalitis. Proc Soc Exp Biol Med 132:272-277, 1969.
13. Chen T.T., Watanabe I., Zeman W., Mealey J.: Subacute sclerosing panencephalitis propagation of measles virus from brain biopsy in tissue culture. Science 163:1193-1194, 1969.
14. Payne, F.E., Baublis, J.V., Itabashi, H.H.: Isolation of measles virus from cell cultures of brain from a patient with subacute sclerosing panencephalitis. N Engl J Med 281:585-589, No. 11, 1969.
15. ter Meulen V, Muller D, Katz M, Kackell MY, Joppich G: Immunohistological microscopical and neurochemical studies on encephalitis. IV. Subacute sclerosing (progressive) panencephalitis. Histochemical and immunohistological findings in tissue cultures derived from SSPE brain biopsies. Acta Neuropath 15:1-10, 1970.
16. Kettyls G.D., Dunn, H.G., Dombsky N., Turnbull, I.M.: Subacute sclerosing panencephalitis: Isolation of a measles-like virus in tissue culture of brain biopsy. Canad Med Assoc J 103:1183-1184, 1970.

17. Ueda S., Okuno Y., Okuno Y., Hamamota Y., Ohya H.: Subacute sclerosing panencephalitis (SSPE); Isolation of a defective variant of measles virus from brain obtained at autopsy. Biken J 18:113-122, 1975.
18. Raine, C.S., Feldman L.A., Sheppard R.D., Bornstein, M.B.: Subacute sclerosing panencephalitis virus in cultures of organized central nervous tissue. Laboratory Investigation 28:627-640, 1973.
19. Oyanagi S., ter meulen V., Muller D., Katz M, Koprowski H: Electron microscopic observations in subacute sclerosing panencephalitis brain cell cultures: Their correlation with cytochemical and immunocytological findings. J Virol 6:370-379, 1970.
20. Koprowski H., Barbanti-Brodano G., Katz M: Interaction between papova-like virus and paramyxovirus in human brain cells: A hypothesis. Nature 225:1045-1047, 1970.
21. Barbanti-Brodano, Oyanagi S, Katz M., Koprowski H.: Presence of two different viral agents in brain cells of patients with subacute sclerosing panencephalitis. Proc Soc Exp Biol Med 134:230-1970.
22. ter Meulen V: Subacute sclerosing panencephalitis and measles virus. Ninth International Congress on Tropical Medicine and Malaria, October 13-21, 1973, Athens, Greece, Volume 1.
23. Modlin, J.F., Jabbour, J.T., Witte, J.J., Halsey, N.A.: Measles, measles vaccine, and subacute sclerosing panencephalitis: Pediatrics 59:505-512, 1977.
24. Detels R., Brody, J.A., McNew J., Edgar A.H.: Further epidemiological studies of subacute sclerosing panencephalitis. Lancet 23:11-14, 1977.
25. Center for Disease Control: Subacute sclerosing panencephalitis. Morbidity and Mortality Weekly Rep 26:309, 1977.
26. Halsey N.A., Modlin, J.F., Eddins, D.L., Dubey L., Jabbour, J.T.: Risk factors in subacute sclerosing Panencephalitis. (In preparation)
27. Center for Disease Control: Measles Surveillance Report No. 10, 1973-1976. Issued July, 1977.
28. Jabbour, J.T., Duenas, D.A., Sever, J.L., Krebs, H.M., Horta-Barbosa L: Epidemiology of subacute sclerosing panencephalitis (SSPE). A report of the SSPE registry. JAMA 220:959-1972.
29. Modlin, J.F., Halsey, N.A., Eddins, D.L., Conrad, J.L., Jabbour, J.T., Chien, L., Robinson H.: Epidemiology of subacute sclerosing panencephalitis. (in press)
30. Dick G. Epidemiology. Register of cases of subacute sclerosing panencephalitis. Br Med J 3:359-360, 1973.

31. Dick G.: Register of cases of subacute sclerosing panencephalitis (SSPE). Br Med J 3:238, 1975.
32. Clark N.S., Best, P.V.: Subacute sclerosing (inclusion body) encephalitis. Arch Dis Child 39:356-362, 1964.
33. Mackenzie D.J.M., Kipps A., McDonald R: Subacute sclerosing panencephalitis in Southern Africa. S.A. Med J 49:2083-2086, 1975.
34. Baguley D.M., Glasgow G.L.: Subacute sclerosing panencephalitis and Salk vaccine. Lancet 2:763-765, 1973.
35. Canal N., Torck P: An epidemiological study of subacute sclerosing leucoencephalitis in Belgium. J Neurol Sc: 1:380-389, 1964.
36. Cuba J.M., Ducharne, P.P., Geurra R.R., Uribe J.C.S.: Subacute sclerosing panencephalitis in Latin America. Cooperative study. Presented at the General Program, 4th Pan American Congress of Neurology, Mexico, October 12-17, 1975.
37. Soffer D., Rannon L., Alter M., Kahana E., Feldman S.: Subacute sclerosing panencephalitis: An epidemiologic study in Israel. Amer J Epidemiol 103:67-74, 1976.
38. Canelas H.M., Juliano O.F., Lefevre A.B., De Assis, J.L., Tognola, W.A., DeJorge, F.B., Fonseca, L.C., Xavier-Lima A: Subacute sclerosing leucoencephalitis: An epidemiological clinical and biochemical study of 31 cases. Arq Neuro-Psiquiat 25:255-268, 1967.
39. Donner M., ja Matti Haltia H.H.: Subakuutti Sklerosoiva Panenkefaliitti. Duodecim 85:541-553, 1969.
40. Dayan, A.D., Cumings, J.N.: An infantile case of subacute sclerosing panencephalitis with an abnormal ganglioside pattern in the brain. Arch Dis Child 44:187-196, 1969.
41. Cape C.A., Martinez, A.J., Robertson, J.T., Hamilton, R., Jabbour, J.T.: Adult onset of subacute sclerosing panencephalitis. Arch Neurol 28:124-126, 1973.
42. Gordon, J.V.: Subacute inclusion encephalitis in South Australia. Med J Australia 1:118-124, 1963.
43. Metz H., Gregoriou M, Sandifer, P.: Subacute sclerosing panencephalitis. A review of 17 cases with special reference to clinical diagnostic criteria. Arch Dis Child 39:554-557, 1964.
44. Castleman B., Scully, R.E., McNeely, B.U.: Case records of the Massachusetts General Hospital. New Engl J Med 290:273-279, 1974.

45. Greenfield J.G.: Encephalitis and encephalomyelitis in England and Wales during the last decade. Brain 73:141-166, 1950.

46. Dubois-Dalq M., Coblentz J.M., Pleet, A.B.: Subacute sclerosing panencephalitis. Unusual nuclear inclusions and lengthy clinical course. Arch Neurol 31:355-363, 1974.

47. Okuno Y., Nakao R, Ishida N., Konno T., Mizutani H., Sato T., Isomura S., Ueda S., Kitamura I., Kaji M.: An epidemiologic study of subacute sclerosing panencephalitis in Japan, 1976. In press

PERSISTENT MuLV INFECTION IN WILD MICE[1]

Murray B. Gardner, Vaclav Klement,[2] Suraiya Rasheed, John D. Estes, Robert W. Rongey, Mary F. Dougherty, and Martin L. Bryant[3]

Department of Pathology
University of Southern California School of Medicine
Los Angeles, California 90033

ABSTRACT Although most wild mouse colonies are free of infectious MuLV, and have only a very low incidence of spontaneous lymphoma, persistent infection with these endogenous viruses and an increased incidence of lymphoma has been observed in several different feral mouse populations in southern California. In one such population (LC), about 90% of the wild mice have a congenital maternal infection and lifelong MuLV viremia with specific humoral immune tolerance. The infectious MuLVs consist of distinct amphotropic and ecotropic classes with different natural distributions and pathogenicities. The amphotropic MuLV are most prevalent and high titered; they cause lymphoma while the ecotropic MuLV cause both lymphoma and a neurogenic paralysis. The lymphomas generally arise in the spleen, are of null or B cell origin and are accompanied by leukemia. The paralysis is caused by a direct viral nonimmunogenic injury to motor neurons, mainly in the lower spinal cord. These viruses and their diseases can be readily transmitted as a congenital or venereal infection to susceptible wild and laboratory mice and, by experimental inoculation, to laboratory rats. Viruses and associated diseases are eliminated by the selective breeding of nonviremic females.

[1]Supported by contract number NO1 CP-5-3500 within the Virus Cancer Program of the National Cancer Institute.

[2]Department of Pediatrics and Microbiology, University of Southern California School of Medicine, Los Angeles, California 90033.

[3]Present address: Meloy Laboratories, Springfield, Virginia 22121.

ISBN 0-12-668350-6

INTRODUCTION

Most wild mouse colonies (e.g. Bouquet Canyon = BQ) are largely free of infectious murine leukemia virus (MuLV) and have only a very low incidence of spontaneous lymphoma late in life (1-3). Persistent infection with MuLV is, however, a feature of several different populations of wild mice (*Mus musculus domesticus*) located in southern California. Most thoroughly studied have been the feral mice habitating a squab farm near Lake Casitas (LC) in south-western Ventura County. The natural history of MuLV in these LC mice has been reviewed elsewhere (4-6) and is only summarized here. This paper will outline new information on the tissue distribution and the natural and experimental transmission of the LC MuLV and will show that infectious MuLV and associated diseases are eliminated by selective breeding of nonviremic LC mice.

Summary of LC Wild Mouse Natural History. The major features of LC wild mice are shown in Table 1. The outstanding characteristic of these mice is their high level of infectious endogenous MuLV. The presence of complete amphotropic MuLV proviral sequences in all wild and inbred mice tested (7) indicates the endogenous nature of this virus. These viruses are transmitted primarily as a maternal congenital infection transplacentally and via milk, resulting in lifelong persistent viremia, systemic infection with virus replication concentrated in the spleen and specific immune tolerance (8, 9). The evidence for immune tolerance is summarized in Table 2 and the evidence of preservation of general immune competence is given in Table 3. Subtle defects in cellular immune function may, of course, yet be found in viremic LC mice.

The viremic LC mice are unusually prone to develop lymphoma (10) and a lower limb paralytic disease (11, 12). Lymphoma occurs with an overall incidence of about 18% between 8 and 36 months of age and paralysis occurs in about 12% between 8 and 18 months of age. About 2% of LC mice develop both lymphoma and paralysis. Other tumors, e.g. breast carcinomas, hepatomas, lung adenomas, sarcomas, arise in about 4% late in life (13). The lymphomas arise in the spleen, are of null or B cell origin and early undergo leukemic distribution. The paralysis is caused by direct nonimmunogenic and noninflammatory viral injury to the central nervous system with damage focused upon the motor neurons in the anterior lateral horns of the lower spinal cord (12, 14, 15). The similarities between this neurologic disease in wild mice and the subacute spongioform encephalopathies in man, sheep, and mink are shown in Table 4.

TABLE 1

MAJOR FEATURES OF LC WILD MICE[a]

---

1. Congenital maternal MuLV infection, lifelong viremia, specific immune tolerance; general immunity intact.

2. High infectious expression of endogenous MuLV.

3. Prone to lymphoma and paralysis.

4. Several classes of endogenous MuLV; amphotropic (most prevalent), ecotropic, and xenotropic.

5. Complete amphotropic proviral sequences are detected in all wild and laboratory mice[b].

6. MuLVs are essential determinants of lymphoma and paralysis; amphotropic → lymphoma; ecotropic virus → lymphoma and paralysis; xenotropic virus → strongly repressed (? function).

7. Pathology: lymphomas - splenic null or B cell origin; paralysis - direct viral nonimmunogenic injury to lower motor neurons.

8. Stable features over past eight years.

---

[a]See references 4-6.

[b]See reference 7.

TABLE 2

EVIDENCE FOR TOLERANCE TO ENDOGENOUS MuLV IN LC MICE[a]

1. Lifelong high titered systemic MuLV infection with viremia.
2. Lack of natural or vaccine inducible free antibody by neutralization, RIA, IIF, CF, cytotoxicity tests, or of disease prevention via active viral immunization.
3. Lack of circulating or deposited immune complexes or related immunopathology.

[a]See reference 8 and 9.

TABLE 3

EVIDENCE FOR INTACT GENERAL IMMUNE COMPETENCE IN LC MICE

1. Normal longevity, vigor, and general health.
2. Normal inflammatory response to other pathogens.
3. Normal humoral response to sheep red blood cells[a].
4. Normal humoral immune response to other MuLV, (MSV[M]).
5. Normal mitogenic and mixed lymphocyte responses.
6. Normal tumorigenic response to 3-methylcholanthrene.

[a]See reference 9.

TABLE 4

SUBACUTE SPONGIOFORM ENCEPHALOPATHIES

	Wild mice[a]	Man, sheep, mink[b]
Slow progressive fatal course	Yes	Yes
Neuronal loss, gliosis, spongiosis	Yes	Yes
Absent inflammatory changes	Yes	Yes
Site of major CNS damage	Anterior lateral horns of lower spinal cord	Upper CNS, cerebellum
Transmissible agent	Ecotropic MuLV	Unconventional viruses
Immune reponse	Absent	Absent
Genetic predisposition	Yes	Yes
Natural transmission	Congenital infection	? vertical chromosomal; horizontal
Experimental disease	Latent period shorter	Latent period shorter
Virus replication outside of CNS	Yes	Yes
Associated disease	Lymphoma	None

[a]See references 4-6.

[b]See reference 30. This group of diseases, also called transmissible virus dementias, includes Jakob-Creutzfeldt, Kuru, Scrapie and mink encephalopathy.

There are two major classes of infectious MuLV, amphotropic and ecotropic, present in these wild mice (16-19); the amphotropic virus is the more prevalent and high titered and ecotropic virus is readily isolated only from the naturally paralyzed LC mice. Xenotropic virus, although present, is apparently seldom expressed in LC or BC mice (6).

Experimental transmission to newborn laboratory mice of purified virus clones shows that the amphotropic viruses induce lymphoma while the ecotropic viruses induce both lymphoma and paralysis (20). The other tumor types are not directly related to these viruses.

Although the great majority (85-90%) of LC mice are congenitally infected and show persistent lifelong viremia, a small but constant minority (10-15%) of these mice remain nonviremic for most or all of their natural lifespan. Most of these mice are the progeny of nonviremic mothers but a few arise from viremic mothers and remained uninfected despite the continued exposure to virus in their mothers' milk and virus shed from their infected littermates. The mechanisms for this resistance are unknown. It does not appear to be due to a humoral immune response to MuLV since this was not detected. Nor is it the result of $Fv\text{-}1^b$ gene expression since these nonviremic mice, as well as all other wild mice tested, have been $Fv\text{-}1^{nn}$ genotype and the LC amphotropic and ecotropic viruses have all been N-tropic (17-19). Other MuLV restrictive genes, as yet unidentified, are probably segregating in this population. One such gene, the $Fv\text{-}4^r$ allele, detected in some Japanese wild mice (21) and in the G strain of Japanese laboratory mice (22) is currently under study. The nonviremic LC mice are fully susceptible to infection and induction of lymphoma and paralysis when inoculated as newborns with amphotropic and ecotropic MuLV. As the nonviremic LC mice age beyond 12 months a small number ($\sim$10%) become viremic but none of these have yet been observed to develop lymphoma or paralysis. These late converters probably represent activation and spread of endogenous virus as a result of lessening with age of the effectiveness of the MuLV restrictive genes.

These features have been remarkably constant over the past 8 years in LC mice trapped in the wild and in LC mice bred in the laboratory. Natural selection against this genetic trait of high MuLV expression has apparently not occurred because the associated diseases develop only after maturity has been reached and reproduction largely completed and the overall vigor, longevity and reproductive capacity of the younger infected mice is generally unimpaired.

Major differences obviously exist between the high leukemia strains of wild and inbred mice, as summarized in Table 5.

TABLE 5

COMPARISON OF LC AND AKR MOUSE MODELS

	AKR	LC
Epidemiology	Lymphoma - 90% by 1 year; no sex predilection	Lymphoma - 20% by 3 years; female predilection
Pathology	Thymic origin	Splenic origin
Thymus antigen	$Thy\text{-}1^a$	$Thy\text{-}1^a$; $thy\text{-}1^b$
H-2	$H\text{-}2^k$	No $H\text{-}2^k$
Virus transmission	Vertical, chromosomal	Maternal congenital infection
Virus properties	Ecotropic, xenotropic and recombinants	Ampho-, eco- and xenotropic; No known recombinants
	Ecotropic - not neurotropic	Ecotropic - neurotropic and lymphomagenic
Autogenous MuLV humoral immunity	Detectable	Not detectable
Genetic resistance	$Fv\text{-}1^b$; $H\text{-}2^b$	$Fv\text{-}1^b$; others undefined
Control of virus and disease	Passive immunization	Passive immunization;
		Foster nursing;
		Inbreeding of $Fv\text{-}1^b$ allele
		Selective breeding of nonviremic females

RESULTS AND DISCUSSION

Maternal Transmission of MuLV. The importance of maternal congenital infection upon spread of MuLV in LC mice is shown in Table 6. When the LC mothers were viremic before breeding, regardless of the serum virus status of the LC fathers, a total of 160 of 204 (78%) of the F1 weanlings were viremic. Most (27 of 38 = 70%) of the initially negative progeny in this group became viremic within several months, i.e. usually around 6 months of age. The high prevalence of viremia was constant in successive litters. When the nonviremic LC mothers were bred, again regardless of the serum virus status of the LC fathers, only 2 of 215 (∿1%) of the progeny were viremic. None (0 of 125) of these offspring in three successive litters became viremic during the first year but between 12 and 20 months of age 5 of 44 (11%) progeny tested became viremic. Almost all (39 of 41) of the nonviremic LC breeders also remained nonviremic for 9-18 months of observation.

MuLV Expression in LC Crosses with Uninfected Wild or Laboratory Mice. MuLV expression in F1 progeny of crosses between LC wild mice and uninfected strains of wild and laboratory mice was also strongly influenced by the viremic status of the LC mothers (Table 7). In crosses of viremic LC females with uninfected BQ, NIH Swiss or C57L males, 82 of 135 (61%) of the hybrid progeny were viremic at 2-4 months of age and many (19 of 29) of the initially negative mice were positive on later testing. No infectious virus was transmitted when the LC female was nonviremic and the BQ, NIH Swiss and C57L controls were always free of infectious virus. When progeny of the LC x NIH Swiss cross were foster nursed on NIH Swiss mothers, viremia was almost eliminated (2 of 41) in the progeny. This illustrates the importance of milk spread MuLV. Matings of viremic LC males with BQ or NIH Swiss females, by contrast, gave rise to almost totally nonviremic hybrid progeny. None of 127 F1 progeny from these crosses were infected at weanling age and only 2 of 104 became viremic on subsequent testing. However, when viremic LC males were mated with C57L females, 48 of 88 (55%) of the weanling progeny were viremic and half of those initially negative became positive by 6-12 months of age.

TABLE 6

MATERNAL INFLUENCE UPON MuLV EXPRESSION IN LC WILD MICE

Viremia in parental mice		Number of mice in breeding pairs		Number of litters	Viremia in F1 progeny		
Female	Male[a,b]	Female	Male		2-4 months	Percent	6-12 months[c]
+	+	9	4	23	97/127[d]	76	19/26
+	-	5	4	14	63/77	82	8/12
-	+	8	6	13	2/68	3	0/35
-	-	21	12	28	0/147	0	0/90

[a]Viremia determined by fluorescent focus assay on SC-1 cells (24). The average serum infectious virus titer in adult LC mice is $10^{3.5}$ infectious units/ml (9). The lower limit of sensitivity for this assay is $10^{1.2}$ infectious units/ml. + = viremic; - = nonviremic.

[b]These breeding pairs were wild trapped and mated only after determining their serum virus status. Only those LC mice that were free of detectable serum virus at trapping and again 3 months later were considered nonviremic. This was necessary because about one-third of the initially negative mice became detectably viremic during this interval. These were mostly smaller and presumably younger mice who may not have lived long enough for the serum virus titer to have reached a detectable level.

[c]These mice were nonviremic at 2-4 months of age.

[d]Number positive/number tested.

TABLE 7

MuLV EXPRESSION IN RECIPROCAL CROSSES BETWEEN LC WILD MICE AND NONVIREMIC STRAINS OF WILD AND LABORATORY MICE

Female x male	MuLV viremia in LC parent[a]	Number of mice in breeding pairs		Number of litters	MuLV viremia in F1 progeny		
		Female	Male		2-4 mo.	%	6-12 mo.
LC x BQ[b]	+	5	5	5	44/65[c]	68	8/14
LC x NIH Swiss[d]	+	12	9	17	16/32	50	11/15
LC x NIH Swiss (foster nursed on NIH Swiss)	+	4	2	9	2/41	5	0/26
LC x C57L[d]	+	7	3	7	23/38	58	NT[e]
LC x C57L	-	1	1	1	0/10	0	0/9
BQ[d] x LC	+	4	4	11	0/68	0	1/45
NIH Swiss x LC	+	16	14	19	0/59	0	0/59
C57L x LC	+	11	9	15	48/88	55	7/14
C57L x LC	-	2	1	3	0/26	0	0/26

[a]Same as footnote 'a' in Table 4.

[b]BQ = Bouquet Canyon wild mice, low MuLV expressors infected with indigenous polyoma virus.

[c]Number positive/number tested.

[d]Control BQ (25 mice), C57L (14 mice) and NIH Swiss (25 mice) were negative for viremia.

[e]NT = not tested.

Indirect Extrachromosomal Transmission in Utero of C57L Females from LC Males. Venereal infection of C57L females with MuLV from LC male partners was proven by isolation of virus from the sera, vaginal swabs and spleen extracts of C57L females after these matings. About 50% of the first-born hybrid litters were viremic, increasing to 90% in the second and third-born litters and it could be shown that virus was isolated from the progeny only after the C57L mother became venereally infected. After breeding with viremic LC females, several C57L males also became infected. Neither of 2 C57L females mated with nonviremic LC males became infected and their progeny were also uninfected. By contrast, this horizontal venereal infection apparently did not occur in matings between viremic LC mice and uninfected LC or BQ wild mice or NIH Swiss laboratory mice. Sera and vaginal plugs remained MuLV negative on these mice even after repeated matings with the same viremic LC males that had infected the C57L females.

Lymphoma-Paralysis Occurrence in Hybrid Progeny of LC Wild Mice x NIH Swiss and C57L Laboratory Mice. Although only a limited number of these F1 hybrid progeny have been observed beyond the age at which lymphoma and paralysis begin to occur in LC mice, i.e. about 10 months, it is evident that both diseases occurred only in the viremic offspring. Among 54 viremic offspring observed beyond 10 months of age, lymphoma developed in 8 and paralysis in another 2, whereas among 57 nonviremic hybrids followed for a similar period, neither disease was observed. The lymphomas occurred between 10 and 15 months of age in 3 of the (C57L x LC) progeny and in one of the (LC x C57L) progeny, and in 4 of the LC x NIH Swiss progeny. In each instance the lymphoma arose in the spleen and was pathologically identical to the naturally occurring disease in LC mice. Amphotropic virus was recovered from the spleen extracts of all 8 lymphomatous mice and, in addition, ecotropic virus was isolated from 3 of these same extracts. Lower motor neuron paralysis with pathology identical to the naturally occurring disease in LC mice was observed at 12 months of age in 2 of the (C57L x LC) viremic progeny but was not observed in the nonviremic offspring. Amphotropic and ecotropic virus were both recovered from the spleen extracts of these 2 hybrids. In addition, splenic lymphomas yielding amphotropic MuLV occurred between 11 and 15 months of age in 3 C57L breeders, 2 females and 1 male, each of which had become viremic after mating with infected LC partners. Thus, it is apparent that both amphotropic and ecotropic classes of MuLV are potentially transmissible from viremic LC mice to susceptible laboratory mice by congenital or venereal epigenetic means and that these viruses can induce the same diseases, lymphoma and paralysis, that occur naturally in LC wild mice.

Experimental Transmission of Wild Mouse MuLV to Rats. Earlier studies with various field virus preparations showed that newborn laboratory mice of Fv^{nn} genotype were susceptible to serial infection and induction of lymphoma and paralysis (6, 11, 12). In addition, one field virus isolate (strain 1504) (23) was readily transmitted to newborn laboratory rats and induced lymphoma in about 60% after 10 months. Further serial passage of these rat lymphoma extracts to more newborn rats and NIH Swiss mice led to the induction of lymphoma and paralysis in both rodents. The latent period for these diseases was longer in rats than in mice and only the paralytic disease in mice occurred with a frequency of greater than 30% (about 50%). On the third *in vivo* passage, amphotropic virus was recovered *in vitro* from the sera and tumors of lymphomatous mice and rats and ecotropic and amphotropic virus were both recovered from the sera, spleen, and CNS of the paralyzed mice and rats. Pathologically and virologically the induced diseases in mice and rats were remarkably similar to the naturally occurring diseases. These results confirm that both amphotropic and ecotropic MuLV of wild mouse origin are infectious and pathogenic for laboratory rats under experimental conditions.

Natural Prevalence of Ampho- and Ecotropic MuLV in LC Mice. Based upon the XC plaque test for detection of ecotropic virus in SC-1 cells and the fluorescent focus assay for detection of amphotropic virus in SC-1 and mink cells (24), a survey of the natural prevalence of these viruses in fresh tissues and sera of LC mice is shown in Table 8. The amphotropic virus is more prevalent than ecotropic virus in sera and spleen of normal and tumor bearing LC mice. Only in paralyzed LC mice is ecotropic virus, together with amphotropic virus, isolated readily from sera, spleen, and central nervous system tissue. These findings are consistent with the earlier observation that LC mice destined for eventual paralysis had high level serum p30 antigenemia at time of trapping (8) and with the experimental transmission studies which show that only the ecotropic viruses are paralytogenic (20). It should be noted that if these tissue extracts had been blind passaged in SC-1 cells, the prevalence of ecotropic virus detection would likely have been higher. This is because a small amount of ecotropic virus initially present would have been amplified in mouse cells to detectable levels, as was shown previously (17).

REFERENCES

1. Gardner, M. B., Officer, J. E., Rongey, R. W., Estes, J. D., Turner, H. C., and Huebner, R. J. (1971). Nature (Lond.) 232, 617.
2. Gardner, M. B., Henderson, B. E., Rongey, R. W., Estes, J. D., and Huebner, R. J. (1973). J. Natl. Cancer Inst. 50, 719.
3. Gardner, M. B., Henderson, B. E., Menck, H., Parker, J., Estes, J. D., and Huebner, R. J. (1974). J. Natl. Cancer Inst. 52, 979.
4. Gardner, M. B., Klement, V., Rasheed, S., Rongey, R. W., Brown, J. C., Pike, M., Henderson, B. E., and Huebner, R. J. (1976). Bibl. Haemat. 43, 204.
5. Gardner, M. B. (1977). In "Microbiology 1977" (D. Schlessinger, ed.), pp. 498-501. Amer. Soc. Microbiol., Washington, D.C.
6. Gardner, M. B. (1978). In "Current Topics in Microbiology" (in press).
7. Chattopadhyay, S. K., Hartley, J. W., Lander, M. R., Kramer, B. S., and Rowe, W. P. (1978). J. Virol. (in press).
8. Gardner, M. B., Klement, V., Rongey, R. R. (*sic*), McConahey, P., Estes, J. D., and Huebner, R. J. (1976). J. Natl. Cancer Inst. 57, 585.
9. Klement, V., Gardner, M. B., Henderson, B. E., Ihle, J. N., Estes, J. D., Stanley, A. G., and Gilden, R. V. (1976). J. Natl. Cancer Inst. 57, 1169.
10. Gardner, M. B., Henderson, B. E., Estes, J. D., Menck, H., Parker, J. C., and Huebner, R. J. (1973). J. Natl. Cancer Inst. 50, 1571.
11. Officer, J. E., Tecson, N., Estes, J. D., Fontanilla, E., Rongey, R. W., and Gardner, M. B. (1973). Science 181, 945.
12. Gardner, M. B., Henderson, B. E., Officer, J. E., Rongey, R. W., Parker, J. C., Oliver, C., Estes, J. D., and Huebner, R. J. (1973). J. Natl. Cancer Inst. 51, 1243.
13. Gardner, M. B., Henderson, B. E., Estes, J. D., Rongey, R. W., Casagrande, J., Pike, M., and Huebner, R. J. (1976). Cancer Res. 36, 574.
14. Andrews, J. M., and Gardner, M. B. (1974). J. Neuropathol. Exp. Neurol. 33, 285.
15. Oldstone, M. B. A., Lampert, P. W., Lee, S., and Dixon, F. J. (1977). Amer. J. Pathol. 88, 193.
16. Rasheed, S., Gardner, M. B., and Chan, E. (1976). J. Virol. 19, 13.
17. Bryant, M. L., and Klement, V. (1976). Virology 73, 532.

18. Hartley, J. W., and Rowe, W. P. (1976). J. Virol. 19, 19.
19. Rasheed, S., Toth, E., and Gardner, M. B. (1977). Intervirology 8, 323.
20. Gardner, M. B., Klement, V., Henderson, B. E., Casagrande, J., Bryant, M. L., Dougherty, M. F., and Estes, J. D. (1978). In "Proc. VI Perugia Quad. Internatl. Conf. Cancer. Tumors of Early Life in Man and Animals" (in press).
21. Odaka, T., Ikeda, H., Moriwaki, K., Mutsuzawa, A., Mizuno, M., and Kondo, K. (1978). (in press).
22. Suzuki, S. (1975). Jap. J. Exp. Med. 45, 473.
23. Gardner, M. B., Officer, J. E., Rongey, R. W., Charman, H. P., Hartley, J. W., Estes, J. D., and Huebner, R. J. (1973). Bibl. Haemat. 39, 335.
24. Klement, V., and Nicolson, M. O. (1977). In "Methods in Virology" (K. Maramorosch and H. Koprowski, eds.), pp. 59-108. Academic Press, New York.
25. Hartley, J. W., Wolford, N. K., Old, L. J., and Rowe, W. P. (1977). Proc. Natl. Acad. Sci. USA 74, 789.
26. Troxler, D. H., Boyar, J. K., Parks, W. P., and Scolnick, E. M. (1977). J. Virol. 22, 361.
27. Gardner, M. B., Klement, V., Henderson, B. E., Meier, H., Estes, J. D., and Huebner, R. J. (1976). Nature (Lond.) 259, 143.
28. Gardner, M. B., Klement, V., Estes, J. D., Gilden, R. V., Toni, R., and Huebner, R. J. (1977). J. Natl. Cancer Inst. 58, 1855.
29. Gardner, M. B., Dougherty, M. F., and Estes, J. D. (1978). In "Proc. Internatl. Symp. Comp. Leuk. Res." (in press).
30. Gajdusek, D. C. (1977). Science 197, 943.
31. Rubin, H., Cornelius, A., and Fanshier, L. (1961). Proc. Natl. Acad. Sci. USA 47, 1058.
32. Hughes, W. F., Watanabe, D. H., and Rubin, H. (1963). Avian Dis. 7, 154.

GENETICALLY TRANSMITTED VIRAL GENES OF RODENTS AND PRIMATES

George J. Todaro, Robert Callahan,
Charles J. Sherr, Raoul E. Benveniste,
and Joseph E. De Larco

Laboratory of Viral Carcinogenesis
National Cancer Institute
National Institutes of Health
Bethesda, Maryland 20014

INTRODUCTION

In recent years a number of different endogenous retroviruses have been found in rodents and primates. There are four distinct classes of endogenous reverse transcriptase containing viral genes in rodents; each is present in multiple copies in the cellular DNA (1). In primates, five different reverse transcriptase containing viruses have been found, each of which also has been shown to be endogenous; i.e., present in normal primate cellular DNA (2). Related species have related virogene sequences. Three of these primate viruses have been discovered in our laboratory in the past year. They are a type D virus from langur monkeys (3), a type C virus from a stumptail monkey, *Macaca arctoides* (2) and a type C virus from the New World species, *Aotus trivirgatus* (owl monkey) (4). Table 1 summarizes the origin and transmission of primate retroviruses. The first candidate primate viruses were isolated from leukemic and normal gibbons and from a woolly monkey fibrosarcoma (5,6,7). It was subsequently shown that they are not endogenous primate viruses, but rather are derived from the species *Mus* and have been transmitted horizontally into the primates (8,9).

Some of the viruses are easily activated from some primate species but not from others. Species exist which show a high probability of activation and others which show a low probability of activation. The former are the Old World baboon type C virus group (10) and the New World squirrel monkey type D virus group (11). As indicated in Table 2, it is seen from the frequency of activation that baboon and squirrel monkey tissues readily release virus, while in other cases the primate viruses represent single isolates from one cell line or one cocultivation experiment. For example, the virus from the stumptail monkey, MAC-1, represents a single

ISBN 0-12-668350-6

TABLE 1
Origin and Transmission of Primate Retroviruses[a]

Virus	Isolation	Endogenous in Primates	Origin
Gibbon	Gibbons (GALV, G-Br, SEATO, SSAV)	No	Mice (*Mus* ssp.)
Baboon	Baboon tissues and cell lines (M28, M7, PP-1, etc.)	Yes	Baboons
MPMV and Langur	Rhesus tissues	Yes	Langurs
	Langur cell line (PO-1-Lu)	Yes	Langurs
Squirrel monkey	Squirrel monkey tissues and cells (SMRV, M534, M706, etc.)	Yes	Squirrel monkeys
Owl monkey	Owl monkey kidney (OMC-1)	Yes	Owl monkeys
Macaque	Stumptail monkey spleen (MAC-1)	Yes	Macaques

isolate from one cocultivation experiment. Repeated experiments using the same cells have not succeeded in activating virus. Seven months were required for infectious virus to be isolated. Clearly then, there is a long period in which the virus is replication defective. The continued presence of the second cell in the cocultivation experiment, in this case a human tumor cell line, A549, appears to be required to change the replication defective virus into a replication competent virus. Table 3 shows the time course of appearance of virus from the macaque cell cocultivation experiment. Note that it only came up on the human carcinoma cells and was only evident after 210 days of cocultivation.

TABLE 2
Retroviruses from Primate Tissues[a]

Primate	# Tested	# Positive	% Positive
Old World			
Baboon	22	17[b]	77
Gelada	7	4[b]	57
Mangabey	10	0	0
Patas	7	0	0
Vervet	16	0	0
Macaque	65	1	2
Colobus	12	0	0
Langur	9	1	11
New World			
Squirrel	17	14	82
Woolly	8	0	0
Marmoset	16	0	0
Spider	12	0	0
Owl monkey	9	1	11

[a]Tissue from spleen, kidney, lung, liver, and, in some cases, thymus, was cocultivated with indicator cells. With the exception of two osteosarcomas, a lymphosarcoma, and a breast carcinoma from rhesus monkeys, all tissues from monkeys and higher apes were from nontumored animals. Finely minced tissue was added to flasks containing 1×10^6 indicator cells. In approximately 50% of the experiments, cells were treated with 100 μg/ml of 5-iododeoxyuridine and 2 μg/ml of Polybrene for the first 24 hr of cocultivation. The cells were subcultured once every 7 to 14 days and the supernatant fluids were tested for viral-specific polymerase every 2 weeks beginning at the fourth week of cocultivation (10). The experiments were considered negative if no viral enzyme was seen throughout the 3-4 months that the experiments were maintained (12). All positive cultures were tested for type C viral polymerase which could be inhibited by antisera to reverse transcriptase

TABLE 7
Mouse Sarcoma Virus Transformed Mouse Cells: Effect on EGF and MSA Receptors[a]

Cell Line	^{125}I cpm/10^6 cells EGF	MSA
3T3-M	2,830	1,230
Mo-MSV transf. (S^+L^-)	140	1,671
NIH/3T3	1,480	1,550
Mo-MSV transf. (S^+L^-)	<50	1,310
BHK/21 cl 13	2,220	860
H-MSV transf.	130	1,240
BH-RSV transf.	1,505	1,190

[a]The binding assays were performed on subconfluent cell culture monolayers in 35 mm cluster plate wells. Approximately 5 x 10^5 cells were seeded 24 hr prior to binding [^{125}IEGF]. The cells were washed twice with binding buffer (25). Bindings were initiated by adding 1.0 ml of binding buffer containing 2 ng of [^{125}I]EGF (86,400 cpm) or 0.5 ng of [^{125}I]MSA (24,500 cpm). After incubation for 40 min at 37° C, unbound radioactive ligand was removed and the cells washed four times with cold binding buffer. The amount of ligand bound to the cells was measured by lysing the washed monolayers and transferring the lysate to counting vials. The radioactivity present was determined using a Beckman 300 series gamma counter at 67% efficiency. All data given are the averages of duplicate assays and have been corrected for nonspecific binding. The value of <50 cpm could not be distinguished from the values obtained for nonspecific binding.

TABLE 8
Cell Transformation and EGF Receptor Availability[a]

	Mouse cells (3T3, BALB/3T3, C3H 10T1/2)				
	Untransformed parental clones	SV40	Polyoma	MSV	Chemical
Clones with decreased EGF binding	0/35	0/8	0/4	12/12	5/47
Percent of clones with decreased EGF binding	0	0	0	100	11

[a]Summary of data on EGF receptors present on clonal lines of Swiss 3T3, BALB/3T3, C3H 10T1/2 and various transformed subclones derived from these cells. Cell lines with levels of EGF binding of less than 20% of those found in the untransformed control cell cultures were considered to have shown decreased receptor availability. Logarithmically growing cells, 24 hours after seeding at 2-5 x 10^5 cells/plate on 20 cm^2 plastic petri dishes, were tested with ^{125}I-EGF at room temperature for two hours, as described (25).

activity coelute, appearing to be different activities of the same proteins. In addition, the SGFs have the ability to induce normal anchorage dependent cells to grow in agar as progressively forming colonies and to show the morphologic properties of transformed cells in monolayers (27). The SGF-induced phenotypic transformation is reversible; when the factor is no longer supplied the cells resume their normal morphology and are no longer able to grow in agar. In spite of the similarities to EGF, SGF is not immunologically related to EGF. In radioimmunoassays or in indirect percipitation experiments, anti-mouse EGF antibody has no effect on the SGF growth stimulating activity or its EGF competing activity.

We conclude that the murine sarcoma viruses induce cells, either directly or indirectly, to produce new peptide growth hormones that phenotypically change normal cells to transformed cells in a reversible fashion. The continued presence of such factors would provide a continued internal stimulation of cell division. Since normal cells respond to SGF by displaying the transformed phenotype, transformed cells producing the factors would not be able to respond to the growth stimulatory substances they may themselves be producing. The sarcoma genes are needed to produce permanent phenotypic changes in fibroblasts, but are not needed for viral replication. Like the viral genes they are derived from genetically transmitted elements already contained in the host cell DNA. The possibility that the transforming proteins are polypeptide growth hormones is suggested by the data using mouse sarcoma virus transformed cells. However, the model of endogenous growth factor production as a mechanism of cell transformation may also be applicable to cells transformed by DNA viruses and possibly to a subclass of those transformed by chemical carcinogens.

REFERENCES

1. Callahan, R., and Todaro, G. J. (In press). In "Workshop on the Origins of Inbred Mice" (H. Morse, ed.), Academic Press, New York.
2. Todaro, G. J., Benveniste, R. E., Sherwin, S. A., and Sherr, C. J. (1978). Cell 13, 775.
3. Todaro, G. J., Benveniste, R. E., Sherr, C. J., Schlom, J., Schidlovsky, G. and Stephenson, J. (1978). Virology 84, 189.
4. Todaro, G. J., Sherr, C. J., Sen, A., King, N., Daniel, M. D., and Fleckenstein, B. (1978). Proc. Natl. Acad. Sci. USA 75, 1004.
5. Theilen, G. H., Gould, D., Fowler, M., and Dungworth, D. L. (1971). J. Natl. Cancer Inst. 47, 881.
6. Kawakami, T. G., Huff, S. D., Buckley, P. M., Dungworth, D. L., Snyder, S. P., and Gilden, R. V. (1972). Nature New Biol. 235, 170.
7. Todaro, G. J., Lieber, M. M., Benveniste, R. E., Sherr, C. J., Gibbs, C. J. Jr., and Gajdusek, D. C. (1975). Virology 67, 335.
8. Lieber, M. M., Sherr, C. J., Todaro, G. J., Benveniste, R. E., Callahan, R., and Coon, H. G. (1975). Proc. Natl. Acad. Sci. USA 72, 2315.
9. Benveniste, R. E., Callahan, R., Sherr, C. J., Chapman, V., and Todaro, G. J. (1977). J. Virol. 21, 849.
10. Benveniste, R. E., Lieber, M. M., Livingston, D. M., Sherr, C. J., Todaro, G. J., and Kalter, S. S. (1974). Nature 248, 17.
11. Heberling, R. L., Barker, S. T., Kalter, S. S., Smith, G. C., and Helmke, R. J. (1977). Science 195, 289.

12. Lieber, M. M., Benveniste, R. E., Livingston, D. M., and Todaro, G. J. (1973). Science 182, 56.
13. Todaro, G. J., Sherr, C. J., Benveniste, R. E., Lieber, M. M., and Melnick, J. L. (1974). Cell 2, 55.
14. Sherr, C. J., and Todaro, G. J. (1974). Virology 61, 168.
15. Benveniste, R. E., and Todaro, G. J. (1977). Proc. Natl. Acad. Sci. USA 74, 4557.
16. Chopra, H. C., and Mason, M. M., (1970). Cancer Res. 30, 2081.
17. Ahmed, M., Korol, W., Schidlovsky, G. Vidrine, J., and Mayyasi, S. (1973). Proc. Amer. Assoc. Cancer Res. 14, 34.
18. Callahan, R., Benveniste, R. E., Sherr, C. J., Schidlovsky, G., and Todaro, G. J. (1976). Proc. Natl. Acad. Sci. USA 73, 3579.
19. Wilson, S. H., Bohn, E. W., Matsukage, A., Lueders, K. K., and Kuff, E. L. (1974). Biochem. 13, 1087.
20. Benveniste, R. E., Heinemann, R., Wilson, G. L., Callahan, R., and Todaro, G. J. (1974). J. Virol. 14, 56.
21. Benveniste, R. E., and Todaro, G. J. (1976). Nature 261, 101.
22. Schlom, J., Hand, P., Teramoto, Y. A., Callahan, R., Todaro, G. J., and Schidlovsky, G. (In press). J. Natl. Cancer Inst.
23. Stehelin, D., Varmus, H. E., Bishop, J. M., and Vogt, P. K. (1976). Nature 260, 170.
24. Frankel, A. E., and Fischinger, P. J. (1976). Proc. Natl. Acad. Sci. USA 73, 3705.
25. Todaro, G. J., De Larco, J. E., and Cohen, S. (1976). Nature 264, 26.
26. Todaro, G. J., De Larco, J. E., Nissley, S. P., and Rechler, M. M. (1977). Nature 267, 526.
27. De Larco, J. E., and Todaro, G. J. (In press). Proc. Natl. Acad. Sci. USA.

REGULATION OF EXPRESSION OF AN ENDOGENOUS AVIAN LEUKOSIS VIRUS[1]

Paul E. Neiman,[2] Mark Groudine,[3] Maxine Linial, and Robert Eisenman

Fred Hutchinson Cancer Research Center, Seattle, Washington 98104

ABSTRACT This report concerns viral and host cell functions which are important for the regulation of endogenous oncoviruses. The genetically transmitted endogenous virus of chickens, RAV-0, is represented in the DNA of normal chicken cells at a level of about two copies per haploid genome. Most chicken embryo fibroblasts (CEF) do not release virus particles. In these cells the accumulation of viral RNA in the cell appears to be under the control of autosomal dominant loci including one called *GS* specifying the synthesis of a viral polypeptide termed P120. In chick embryos in which *GS* controlled the level of viral RNA in fibroblasts, we found that *GS* did not exclusively control viral RNA levels in red blood cells (RBCs). New presumptive viral polypeptides were found in RBCs independent of *GS* controlled P120 production. Like P120, these RBC proteins contained antigenic determinants shared with viral internal structural proteins encoded in the viral gene *gag*.

We found that endogenous viral DNA sequences in nuclei from both chick RBCs and CEF were resistant to digestion with staphylococcal nuclease. By this criterion endogenous viral sequences appear to exist in nucleosomes. A little more than one-half of the total viral information was sensitive to digestion by DNase I independent of the level of viral RNA in these cells. Therefore, at least one viral genome equivalent is in the DNase I sensitive (transcriptionally active) conformation. However, by this criterion, *GS* does not appear to regulate viral gene expression at the level of nucleosome structure.

[1]This work was supported by grants CA15704, CA18282, CA20068, and CA20525 from the National Cancer Institute and IN-26(R) from the American Cancer Society.

[2]Scholar, Leukemia Society of America.

[3]Special Fellow, Leukemia Society of America.

ISBN 0-12-668350-6

We also found that nucleotide sequences in viral genes themselves are important for the regulation of endogenous viruses. The intracellular restriction of growth of RAV-0 on CEF could be transmitted to recombinant viruses carrying unique genetic markers of both RAV-0 and exogenous nonrestricted sarcoma viruses. The RNA of both RAV-0 and the recombinants contain clustered nucleotide sequences unique to the genome of the endogenous virus.

INTRODUCTION

The prototype of the genetically transmitted endogenous virus of chickens is called Rous associated virus type-0 (RAV-0) (1). Molecular hybridization studies using labeled RNA sequences from the single stranded 35S viral genome as a probe have detected complete or nearly complete sets of viral sequences in the DNA of normal chicken embryos at the level of one to two copies per haploid genome regardless of the level of viral gene expression in these cells (2). RAV-0 shares over 80% of its genomic nucleotide sequences with horizontally transmitted exogenous chicken leukosis viruses (3) which are more oncogenic than RAV-0 and which do not appear to be regulated by chicken cells in the same manner as the endogenous virus. In this communication we will summarize our current data concerning viral and cellular functions which determine the expression of these endogenous viral genes. We have investigated four different aspects of control of endogenous viral gene expression. These include observations first at the level of viral RNA and, second, protein gene products in cells derived from different tissues of chick embryos. Third, we examined the relationship between viral RNA content in these cells and the conformation of viral chromasomal subunits (nucleosomes). Finally, we have asked whether sequences unique to endogenous viral genomes may play a role in the susceptibility of these agents to regulation by the cell.

RESULTS

Expression of Endogenous Viral Genes in Normal Chicken Cells. Current knowledge of endogenous avian viral gene expression is based upon studies in cultured chick embryo fibroblasts (CEF). Such cells may lack evidence of viral gene expression. Alternatively they may synthesize polypeptides antigenically related to viral internal structural proteins encoded in the viral gene *gag* called gs proteins

TABLE 1
GENES REGULATING ENDOGENOUS VIRUS EXPRESSION[a]

Trait	Locus/Alleles	Phenotype
Spontaneous release of endogenous virus[b]	*V*, *v*	V^+, V^-
Coordinate expression of gs antigen and envelope	*GS*, *gs*	chf^+gs^+, chf^-gs^-

[a]These are just the traits we have analyzed in this study.

[b]There may be unlinked *V* loci in different inbred lines of chickens (7).

(4) and/or viral envelope glycoproteins called chick helper factor (chf) (5), or may release infectious RAV-0 (6, 7). Autosomal dominant genes which regulate the phenotypes analyzed in this study are described in Table 1 according to the nomenclature suggested by Harriet Robinson (8).

In CEF of the V^- and gs^-chf^- phenotypes, viral RNA levels are very low (less than one copy per cell) in comparison with those in cells with positive phenotypes (9, 10). Little is known regarding viral expression in other cell types. We have therefore begun to examine viral RNA and protein synthesis in different cells. Table 2 lists the viral RNA levels in CEF as well as in chick red blood cells (RBC) and a transformed chicken lymphoblast line called MSB-1 which is chf^-gs^- and free of exogenous leukosis virus (11). In each case viral RNA was detected by hybridization of a single strand cDNA probe [prepared by copying viral 35S subunit genomic RNA with reverse transcriptase using limit digest calf thymus DNA primers (12)] to an excess of total cellular RNA according to previously described methods (13, 14).

The table demonstrates two important points. First, although the activity of the *GS* locus controls the level of viral RNA accumulation in CEF [as has been previously observed (9, 10)] there is no evidence that *GS* exclusively controls viral RNA in RBCs from embryos of the same genotype. Viral RNA could also be detected in chf^-gs^- MSB-1

TABLE 2
VIRAL RNA LEVELS IN UNINFECTED V^- CHICKEN CELLS

Cell[a] type	Genotype	$CR_0t_{\frac{1}{2}}$[b]	Percent viral RNA[c]	Copies/cell[c]
CEF	*gs/gs*	$>5 \times 10^4$	$<10^{-5}$	<1
CEF	*GS/GS*	2.5×10^2	0.008	100-150
CEF	*GS/gs*	1.0×10^3	0.004	25-50
RBC	*gs/gs*	7.5×10^2	0.008	30-60
RBC	*GS/GS*	5.0×10^2	0.008	50-100
MSB-1		2.0×10^3	0.0004	5-10

[a]Chicken cells of the chf^+gs^+ phenotype were from inbred line 6 embryos of the *GS/GS* genotype from the USDA Regional Poultry Research station at East Lansing, Michigan. The gs^-chf^- cells of presumptive *gs/gs* genotype were obtained from embryos of the H and N flock, Redmond, Washington. RBC are nucleated red blood cells from the same embryos from which CEF were cultured. MSB-1 cells are phenotypically chf^- and gs^- and their genotype is unknown.

[b]$CR_0t_{\frac{1}{2}}$ is the concentration of total cell RNA (Moles nucleotide/liter) x time in seconds for one half of the maximum hybridization of the cDNA probe to occur.

[c]The percent of viral RNA in total cellular RNA and number of viral RNA copies was calculated by comparison of the $CR_0t_{\frac{1}{2}}$ of purified viral RNA (1.6×10^{-2}) according to a previously described method (10).

cells, although one cannot tell whether this relates to the transformed phenotype or the fact that these are lymphoid cells. The second main point is that CEF from gs^+chf^+ heterozygote embryos (obtained from progeny of crosses with "true breeding" homozygotes from the H and N flock) showed an intermediate level of viral RNA. These observations suggest that *GS* is a "cis" acting co-dominant allele which acts to either promote or permit accumulation of viral RNA in fibroblasts, while the presence of the *gs* allele at the same locus signals a failure of transcription in fibroblasts, but not in other cell types such as RBCs.

Tissue specific differences in regulation can also be discerned from an examination of the protein products of endogenous viral genes in the two different cell types. Little or no detectable viral protein occurs in chf^-gs^- fibroblasts, while the principal viral polypeptide which accumulates chf^+gs^+ CEF under the control of the *GS* allele (in addition to the envelope glycoprotein) is a 120,000 dalton protein (P120) carrying antigenic specificities of both the small internal structural (*gs*) virion proteins (p27, p19, p12) and reverse transcriptase (15). Tryptic peptide analysis indicates that this composite molecule is an authentic viral polypeptide missing portions of both the precursor polyprotein for the *gs* proteins (lacks p15) and reverse transcriptase. It is not cleaved or processed further to any recognized viral protein. We therefore compared the viral proteins synthesized in RBCs from chf^+gs^+ and chf^-gs^- embryos with those of CEF from the same embryos. Cell lysates labeled with ^{35}S-methionine were incubated with antisera to either the *gag* gene *gs* proteins or to envelope glycoproteins, and precipitated proteins were analyzed by SDS-polyacrylamide gel electrophoresis (PAGE) as previously described (15). Viral proteins were not detected in chf^-gs^- CEF. In Figure 1 the migration of immunoprecipitated proteins from chf^+gs^+ CEF were compared with that of RBCs.

The figure shows that in addition to P120 (the cellular *gag*-like gene product observed in CEF of the same phenotype) chf^+gs^+ RBCs contain at least two additional putative viral proteins precipitated by the anti-*gs* antiserum designated X and Y. These same molecules appear in about equal quantities in the chf^-gs^- RBCs, but P120 does not. No detectable envelope proteins were found in RBCs using this technique. Although definitive identification and characterization of the RBC proteins by tryptic mapping has yet to be carried out, the experiment suggests that, while the *GS* allele controls its characteristic gene product (P120) in these hematopoietic cells, some additional viral proteins (and presumably viral RNAs) are also produced. These products are tissue specific (in that they are not seen in fibroblasts), and involve at least part of the cellular equivalent of the same viral structural gene (*gag*), but do not appear to be under the control of the *GS* locus.

Relationship of Endogenous Viral Gene Expression to Viral Chromatin Structure. Recent studies have demonstrated

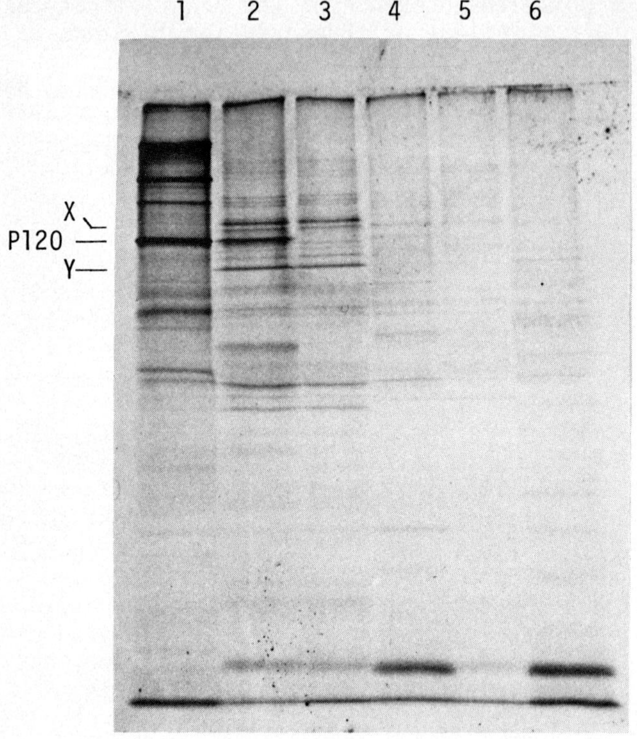

FIGURE 1. SDS-Polyacrylamide gel electrophoresis of *gag* proteins in chick embryo fibroblasts and red blood cells. Gels were prepared and run, and proteins located by radioautography as previously described (15). Lane 1: proteins from chf$^+$gs$^+$ CEF immunoprecipitated with antisera to gs proteins. Lane 2: the same immunoprecipitation with RBC proteins from the same chf$^+$gs$^+$ embryos. Lane 3: the same immunoprecipitation experiment with proteins from RBCs from chf$^-$gs$^-$ embryos. Lanes 4 and 5: RBC proteins from chf$^+$gs$^+$ and chf$^-$gs$^-$ embryos respectively immunoprecipitated with antiserum to viral envelope glycoproteins. Lane 6: RBC proteins from chf$^+$gs$^+$ embryos incubated with pre-immune serum.

that while both transcriptionally active and inactive cellular genes are organized in chromatin subunits (nucleosomes) the nucleosomes of some if not all active genes are differentially sensitive to digestion by DNase I (16). It seemed reasonable to ask whether the sort of regulation described above was reflected in nucleosome conformation (as detected by selective nuclease digestion). The detailed description of the characterization of viral chromatin subunit structure is beyond the scope of this paper, and will be described in detail elsewhere. The observations derived from these studies germane to this discussion are listed in Table 3. In this analysis DNA was extracted from RBC and CEF of chf^+gs^+ or chf^-gs^- phenotypes and compared with DNA from nuclei of the same cells which had been digested to 20% acid solubility with either staphylococcal nuclease or DNase I. Viral sequences in each DNA preparation were quantitated by saturation hybridization with a 10 to 20 fold excess of

TABLE 3
NUCLEASE SENSITIVITY OF ENDOGENOUS VIRAL SEQUENCES IN CHICKEN CELL CHROMATIN[a]

Cell	Genotype	Copies/ haploid genome[b]	After staph nuclease	After DNase I
CEF	*gs/gs*	2	2	.9
CEF	*GS/GS*	2	2	.9
RBC	*gs/gs*	2	2	.9
RBC	*GS/GS*	2	2	.9
RSV infected[c] CEF	*gs/gs*	3	3	.9

[a]The method of isolating nuclei, enzyme treatment, and extraction of residual DNA has been described (16).

[b]Copy number was estimated from the radioactivity hybridized at saturation (above background observed with uninfected quail cell DNA) based upon specific activity of the viral RNA probe and the mass ratio of cellular to viral genomes (about 2×10^6). Preparation of labeled viral RNA probes and the conditions of hybridization which allow RNA-DNA hybridization but not DNA-DNA reassociation have been described (22).

[c]RSV transformed CEF prepared as previously described (17).

^{125}I labeled viral genomic RNA over proviral DNA in the reaction mixture.

The copy number in the table represents an estimate of the total mass of viral sequences assuming a single strand viral genome molecular weight of about 3×10^6 daltons. We detected about two copies or viral equivalents per haploid genome in both RBCs and CEF and no observed evidence of loss of copies after digestion of the DNA in isolated nuclei to 20% acid solubility with staphylococcal nuclease. This observation suggests that endogenous viral genes are organized in nucleosomes in the same fashion as other cellular genes. About 55% of the viral sequences are lost from both cell types after digestion of nuclei to the same extent with DNase I, suggesting that about half of the viral information is found in nucleosomes in the DNase I sensitive, or presumptively active, conformation. Relevant to this discussion is the fact that this ratio of sensitive to resistant viral nucleosomes is not altered by the activity of the *GS* locus in CEF or RBCs. Thus we failed by this technique to detect any evidence that the alterations in gene expression controlled by *GS* or observed between RBCs and CEF are reflected in nucleosome structure. All cell types which are expressing viral gene functions have, however, at least one genome mass equivalent in the sensitive or "active" conformation. For example, as shown in Table 3 the additional viral sequences detected in exogenously infected cells appeared to be sensitive to DNase I.

Unique Functions in Endogenous Viral Genomes Which Are Involved in Regulation by the Cell. As mentioned above, the endogenous viruses released from uninfected CEF under the influence of the *V* gene, RAV-0, are highly homologous to the genomes of exogenous leukosis viruses. Nevertheless this virus is regulated by CEF far more efficiently than are the exogenous agents even when RAV-0 is introduced in the cells as an exogenous agent. For example, on most CEF the growth of RAV-0 is strikingly restricted at an intracellular level in comparison with that of exogenous viruses (17, 18). We wondered whether despite the sequence homology there was some unique genetic element which rendered RAV-0 subject to such restriction. If so, this trait might appear in genetic recombinants. Most recombinants between RAV-0 and exogenous viruses are not restricted in their growth on CEF (17). Recently, however, we have isolated a series of cloned recombinants resulting from mixed infection of quail cells with RAV-0 and Prague strain of Rous sarcoma virus subgroup C (PR-RSV-C) which were selected for the transforming properties of the sarcoma virus and the unique host range markers of the endogenous virus (subgroup E) by previously described methods (17). Table 4 shows the growth of the endogenous virus and representative recombinant sarcoma viruses, called PR-RSV-E, on chf^-gs^- CEF. Because the assays used for sarcoma viruses are different than those of leukosis viruses like RAV-0, we also show the growth of a nonrestricted recombinant subgroup E leukosis virus, RAV-60.

TABLE 4
GROWTH OF ENDOGENOUS VIRUS
AND RECOMBINANTS ON CEF

Virus	Virus titer[a] on CEF cell type	
	Permissive[b]	Restricting[c]
RAV-0	$2x10^5$	10^2-10^3
RAV-60	$>10^6$	$>10^6$
PR-RSV-C	$\sim10^6$	$\sim10^6$
PR-RSV-E 95b	$1x10^5$	$4x10^2$
PR-RSV-E 97c	$8.8x10^5$	$3.3x10^5$

[a]RAV-0 (spontaneously released from line 7 cells) and RAV-60 (obtained from H. Hanafusa) titers are given in interference units (IU) per ml. CEF were infected at an m.o.i. of 0.01 and virus produced after two cell passages was titered by interference assay on permissive pheasant cells as described previously (Linial and Neiman, 1976). Sarcoma virus were titered directly by focus formation on indicated cell types. Restriction is a measure of decreased transformation.

[b]Permissive cells (for RAV-0 replication) are either L15 C/C or K28xL15 C/0 cells obtained from H. Robinson which yielded equivalent titers for each virus.

[c]Restricted cells are C/0 CEF from the H&N flock.

The data indicate that in CEF from certain embryos both RAV-0 and PR-RSV-E clone 95B are restricted in their growth while a second recombinant sarcoma virus is not. Thus restriction depends upon a function in the RAV-0 genome which can be transferred to the recombinant progeny of a cross with an exogenous virus lacking this property. Further, this property is independent of the subgroup E host range marker of RAV-0 (a function of the viral envelope encoded in the viral gene *env*). Recently experiments by Harriet Robinson (19) clearly show that this viral trait (restriction) depends upon a recessive host cell function(s) which seems widespread among chickens. This is seen in the differences shown in the table between restricting and nonrestricting cells. At present we have no clear indication as to the nature and location of the required viral function in the viral genome. We have, however, determined that the majority of the small fraction of nucleotide sequences present in the genome of RAV-0 and missing in exogenous viruses is concentrated within the 3' terminal 1000 nucleotides of the endogenous viral RNA genome (20). Most of these sequences appear in the genome of the restricted recombinant PR-RSV-E clone 95B. The largest segment was found just to the 5' side of the region containing the transforming gene *src* corresponding to the 3' end of the region thought to code for viral envelope glycoprotein. A second small segment was observed within the 3' terminal 500 nucleotides of the recombinant genome (20). We do not know as yet whether there is any systematic difference in either the content or arrangement of RAV-0 specific sequences in unrestricted recombinants.

DISCUSSION

Integrated animal virus genomes appear to provide a special opportunity to elucidate regulatory mechanisms affecting animal cell genes. The genetically transmitted endogenous viral genes of the chicken seem an attractive model, because they do not appear to exist in multiple copies, and molecular probes needed for analysis can be obtained with relative convenience. Thus this avian system may be simpler to dissect than many of the mammalian models where multiple related viral genomes are present. Even in this chicken model, however, the situation seems increasingly complex. In this paper we have reviewed our current data with respect to the mechanism of regulation of these genes. It seems clear from an examination of just two cell types that the genetic loci which control endogenous viral RNA levels and protein products in embryo fibroblasts are either operating in a tissue specific way or are not unique and additional controls are operative in hemopoetic cells of the same embryos. Although these observations are consistent with the notion that endogenous viral genes may be playing a role in differentiation, there is in fact no evidence at present to distinguish whether this differential viral gene expression in CEF and RBCs is either a cause or an effect of the differentiated state. It is also important to note that we presently lack unambiguous proof that the immunologically detected proteins in chick RBCs are authentic viral gene products translated from the viral RNA detected in these cells. If for purposes of discussion we can assume that the proteins specifically immunoprecipitated by the anti-*gs* antiserum in chick RBCs are coded for by endogenous viral genes, it seems reasonable to conclude that the same viral cistron(s) can be expressed in two different protein products in the same cell type (although not necessarily in the same cell). That is, in RBCs from embryos of the *GS/GS* genotype we observed both P120 and additional proteins (yet to be fully characterized) carrying shared antigenic determinants specified by the viral gene *gag*. Further, as shown in Figure 1 these two sets of *gag* gene products appear independently in other cells ($gs^{-}chf^{-}$ RBCs and $chf^{+}gs^{+}$ fibroblasts). Whether these represent products of two different viral *gag* cistrons (corresponding perhaps to the two copies) or whether they are products of the same structural gene remains to be determined. If the latter were true (as supported by Table 3 showing a single DNase I sensitive copy/haploid genome) one would have to suspect that the cellular genes controlling endogenous viral *gag*-like gene expression operate at the level of RNA processing to create alternative messengers rather than as simple promoters of transcription.

We examined the DNase I sensitivity of nucleosomes in these different cell types because the structural alterations detected by this enzyme seem to reflect transcriptional controls for such cellular chicken genes as globin and ovalbumin (16). However, it should be pointed out that recent experiments indicate that the nucleosomes of the globin genes remain DNase I sensitive in mature RBCs despite the fact that these genes are no longer transcribed in these cells (16). This observation is also true for ovalbumin genes in oviduct cells from which hormone stimulation has been withdrawn (21). Thus, while DNase I sensitive nucleosome conformation may be a requirement for transcription, it does not necessarily mean the gene is being actively transcribed at the time of assay. In this study we have always found at least one genome equivalent of viral information in

THE ENDOGENOUS AND ACQUIRED PROVIRUSES OF MOUSE MAMMARY TUMOR VIRUS[1]

Harold E. Varmus, J. Craig Cohen, Peter R. Shank, Gordon M. Ringold[2], Keith R. Yamamoto[2], Robert Cardiff[3] and Vincent L. Morris[4]

Department of Microbiology, University of California, San Francisco, California 94143

ABSTRACT The mouse mammary tumor virus (MMTV) can be transmitted through the milk of females from inbred mouse lines with a high incidence of tumors or inherited as proviruses found endogenous to all tested mice. The endogenous proviruses are present in 3-4 copies per haploid genome in most strains of laboratory mice, and they are distributed over at least two or three chromosomes, including chromosome number 4. Most virus-induced mammary tumors contain a increment of copies of MMTV DNA sufficient for detection by measurement of hybridization kinetics with radiolabeled viral cDNA. However, this technique is not sufficiently sensitive to detect the small increment in copies of MMTV DNA present in virus-infected mammary glands.

We have used restriction endonucleases which cleave the DNA of MMTV at one site (Eco RI) and at several sites (Pst I) to study infection and tumorigenesis in BALB/c mice foster nursed by virus-producing C3H females and to describe the organization of MMTV DNA endogenous to various strains of mice. Proviruses newly-acquired during infection can be distinguished from endogenous proviruses by the nature of fragments generated with these enzymes; thus digestion of DNA from

[1]This work was supported by grants from the American Cancer Society and the U.S. and Canadian National Institutes of Health.

[2]Department of Biochemistry and Biophysics, UCSF, San Francisco, California, 94143.

[3]Department of Pathology, UCD, Davis, California, 95616.

[4]Department of Bacteriology, University of Western Ontario, London, Ontario, Canada.

ISBN 0-12-668350-6

virus-induced mammary tumors and virus-infected lactating mammary glands with *Pst* I yields fragments (2.5 and 0.6 x 10^6 Mr) unique to the acquired proviruses. DNA from BALB/c mice not exposed to virus and from livers of tumor-bearing animals does not yield these fragments, documenting the specificity of the assay and the tissue tropism of the virus. *Pst* I fragments specific for the proviruses of MMTV strains normally transmitted in milk are found, however, upon digestion of liver DNA from GR mice and from a single colony of C3H mice, suggesting that proviruses of horizontally transmitted MMTV have occasionally entered the germ line.

The new MMTV proviruses in mammary tumors and infected lactating mammary glands are linked to cellular DNA within a limited region of the viral DNA; the location of this region, based upon mapping studies with linear and closed circular forms of unintegrated MMTV DNA, suggests that the orientation of the proviruses is probably co-linear with the viral RNA genome. Comparisons of many mammary tumors and studies of lactating mammary glands with a large proportion of infected cells indicate that a large number of sites in the cellular genome can accommodate a new provirus; the acquired proviruses are rarely, if ever, found in tandem with each other or with endogenous proviruses. Similar conclusions about sites of integration have been reached in parallel studies of rat hepatoma cells, heterologous host cells which lack endogenous MMTV-specific sequences and have been infected in culture with MMTV.

Since cleavage of DNA from each mammary tumor with *Eco* RI generates a unique set of virus-specific fragments, despite the large number of available sites for integration, we propose that the tumors are derived from one or few cells among the many infected cells in a mammary gland. This proposal implies that neoplastic conversion of MMTV-infected mammary cells is a relatively infrequent event.

INTRODUCTION

Although the murine mammary tumor virus (MMTV) resembles other, more easily studied RNA tumor viruses in many respects, a number of atypical features of this virus command particular attention: (i) it causes carcinomas, rather than the leukemias and sarcomas produced by most other tumor viruses; (ii) the rate of synthesis of viral RNA is profoundly affected by glucocorticoid hormones; and (iii) various strains of the virus are transmitted in certain inbred mice either as infectious agents in milk or as DNA proviruses endogenous to the germ line (see refs. 1-3 for reviews). Despite the attractions of MMTV as a model system for studying carcinogenesis, hormonal regulation, and endogenous viruses, considerable logistical difficulties have impeded progress: (i) there is no simple assay *in vitro* for the biological action of the virus, preventing a traditional genetic approach to its functions; (ii) the virus appears reluctant to infect cells in culture; and (iii) until recently, the most accessible source of virus was the milk of lactating female mice, a less than optimal source for biochemical studies.

In the past few years, some of these problems have been obviated by the development of mammary tumor cell lines which produce adequate amounts of virus in tissue culture fluids, by successful infection of a variety of cultured cell lines by MMTV, and by the application of biochemical and immunological techniques with improved sensitivity and specificity to the definition of virus-specific macromolecules (2).

Our laboratories have been working for several years towards a description in molecular terms of the phenomena which make MMTV uniquely interesting. Thus, we have sought to know the genetic composition of the genome of MMTV, the physical organization of virus-specific DNA endogenous to normal mice, the mechanisms for replication and expression of viral information in an infected cell, and the elements required for steroidal regulation of viral gene expression.

In several of our previous studies, we have depended upon the techniques of molecular hybridization to enumerate virus-specific DNA and RNA present in uninfected and infected cells. Recently our ability to describe viral nucleic acids has been augmented by the use of additional tools---restriction endonucleases, gel electrophoresis, and the DNA transfer method (4), among others. In this brief review of our recent progress, we outline our efforts to define MMTV proviral DNA endogenous to normal murine cells; to determine the structure and organization of unintegrated viral DNA in infected cells; to distinguish endogenous viral DNA from proviruses acquired by infection; and to assess the specificity of the process which inserts viral DNA covalently into the genome of infected cells. In addition, we consider the implications of our findings for an understanding of genetic transmission of MMTV, tumor formation, and regulation of viral gene expression. Full accounts of the recent information summarized here will be published elsewhere.

A SURVEY OF MOUSE STRAINS AND MMTV

Multiple copies of MMTV-specific DNA are present in the germ line of all tested members of the genus Mus, including inbred strains of laboratory mice, wild mice, and the Asian mice, M. caroli and M. cervicolor (5). In contrast, we have observed little if any annealing between MMTV cDNA and the DNA from a large number of other animals, including rodents closely related to mice (6). Since all mice harbor MMTV DNA, we have focused our efforts upon a small number of inbred strains which offer certain inducements to further examination. These strains and some of their properties are listed in Table I.

(i) The BALB/c strain has a low tumor incidence, particularly in American colonies; although it contains approximately 7 copies of MMTV DNA per diploid genome, there is very little spontaneous expression of endogenous viral DNA in this strain. Virus has not been observed in milk from BALB/c animals, the amount of virusspecific RNA in lactating mammary glands is extremely low, and there is no evidence that the rare mammary tumor in American BALB/c mice is virus-associated. Nevertheless, BALB/c mice are extremely susceptible to infection by MMTV, either by parenteral injection or by foster-nursing with virus-producing females from other strains, such as C3H. Sublines of BALB/c f C3H mice offer an excellent opportunity to examine the natural history of infection by MMTV, since BALB/c mice provide a genetically-matched control unexposed to milk-borne virus.

(ii) The C3H strain is a prototypic strain with a high tumor incidence dependent upon infection with virus carried in milk. However, if the C3H animals are foster-nursed by virus-negative animals (e.g., BALB/c), they do have a significant rate of mammary tumors late in life; a strain of MMTV believed to be genetically transmitted is associated with those tumors.

(iii) The A strain is very similar to the C3H strain with respect to milk-borne and genetic transmission of MMTV, although tumor incidence in this strain is more highly dependent upon hormonal factors (e.g., the incidence is low in virginal females). We have used this strain principally because of the availability of somatic cell hybrids formed between macrophages of A strain mice and Chinese hamster lung cells (see below).

(iv) The GR strain commands particular attention because its high tumor incidence is not dependent upon transmission of milk-borne virus. When GR newborns are nursed by virus-negative animals, they nevertheless produce virus and develop mammary tumors early in life; on the basis of

TABLE I

Mouse	Tumor incidence[1]	Exposed to Virus in milk	Viral DNA/cell[2] Liver	Lactating mammary glands	Mammary tumors
BALB/c	Low	No	6–7	7	7
BALB/cfC3H	High	Yes	6–7	7	10–30
C3H	High	Yes	7–8	N.T.	N.T.
C3Hf BALB/c	Low	No	N.T.	N.T.	N.T.
A	High	Yes	8	N.T.	N.T.
Af	Low	No	N.T.	N.T.	N.T.
GR	High	Yes	9–14	N.T.	12–16
GRf	High	No	N.T.	N.T.	N.T.

[1]The information is derived largely from data cited in refs 1 and 3.

[2]The numbers indicate the genome equivalents of MMTV DNA measured in the listed tissues by the rate of annealing of labeled viral cDNA to cellular DNA in the presence of differentially-labeled unique sequence DNA as an internal rate standard (ref 5 and unpublished data of authors).

N.T. = Not tested.

genetic studies, it has been proposed that tumorigenesis is determined by one or more dominant genes which segregate in Mendelian fashion.

Studies of the kinetics of annealing of labeled, virus-specific cDNA to a vast excess of unlabeled cellular DNA indicate that, with the exception of GR mice, these inbred strains and others contain about 6-8 copies of virus-specific DNA per diploid cell in uninfected tissues (e.g., liver) (5). The GR strain carries additional virus-specific DNA (5); in a later section, we describe experiments which indicate that at least some of this extra DNA includes proviruses for the virus strain also transmitted in the milk.

CHARACTERIZATION OF ENDOGENOUS MMTV DNA

We are, of course, curious about the function of viral DNA endogenous to these various mouse strains. Although our curiosity is far from satisfied, it has been possible to obtain some preliminary information about the genetic content of the endogenous DNA, its distribution among chromosomes, and its organization within chromosomes.

(i) Genetic content of endogenous MMTV DNA. We have been unable to demonstrate major differences using molecular hybridization between viral sequences transmitted genetically in mice with low tumor incidence and those transmitted by viruses present in the milk of animals with high tumor incidence. For example, we have compared the RNA of an endogenous MMTV strain, isolated originally from C3Hf mice, with the RNA from virus transmitted in the milk of GR and other mice, and we found no detectable differences in competition hybridization tests (7). In addition, we have been unable to identify a portion of viral cDNA which can anneal solely to the DNA from tissues infected with the milk-borne strains of virus (5), even when the cDNA was prepared in a fashion which insures that it will uniformly represent the genome of milk-borne virus. Using labeled RNA as a hybridization reagent, Schlom and his colleagues have made the conflicting claim that as much as 20% of the sequences present in milk-borne virus genomes is absent from the DNA of uninfected tissues (8). However, their results might reflect subtle differences distributed throughout the genomes of endogenous and acquired viruses rather than a difference in numbers or kinds of genetic elements. At present, this issue remains unresolved. As discussed below, we are able to distinguish resident viral DNA from that acquired during infection by the use of restriction endonucleases;

however, the generation of fragments of different size from resident and acquired proviruses does not necessarily imply a major difference in genetic composition, since minor differences in base sequence can produce major changes in the sizes of digestion products.

(ii) Chromosomal distribution of endogenous MMTV DNA. Kozak and Ruddle have recently developed a series of somatic cell hybrids by fusion of mouse (A strain) macrophages with Chinese hamster lung cells (9). Since these hybrids segregate murine chromosomes, they are useful for making linkage assignments for murine markers. In a collaborative study, we have recently shown that endogenous MMTV DNA is located on more than a single chromosome, including chromosome 4 (10). Several hybrid clones were assessed for murine chromosomes (by karyotype and presence of isozymes) and for MMTV-specific DNA (by annealing with viral cDNA). Fortunately, clones containing only one mouse chromosome, number 4, were found to have viral DNA, permitting an assignment of at least one copy of endogenous viral DNA to that chromosomal pair. Other clones which lacked chromosome 4 also contained viral DNA, indicating that viral DNA was on more than one chromosome. (This finding contrasts with the observation that the multiple copies of endogenous proviruses of cats are on a single chromosome (11).)

(iii) Organization of endogenous MMTV DNA. Until recently, it was difficult to predict whether the several copies of endogenous viral DNA detected by hybridization kinetics were likely to be composed of intact proviruses resembling those acquired during infection; of tandem repeats of viral DNA; or of scattered, subgenomic fragments of virus-specific DNA. Although our information remains limited, preliminary analyses with restriction endonucleases suggest that the endogenous MMTV DNA is probably in the form of a small number of proviruses located at separate positions in the host genome. There seems, moreover, to be some striking similarities among the physical maps of the several endogenous proviruses and of the proviruses acquired by infection. For example, Eco RI cleaves the DNA of the milk-borne viruses at a single internal site; when used to cleave cellular DNA from various mouse strains, it produces 6 to 8 fragments of sizes which suggest that each of 3-4 proviruses per haploid cell yields two virus-specific fragments. This analysis can be simplified by the use of DNA from the somatic cell hybrids discussed earlier; thus, DNA from hybrids containing only chromosome 4 yield two of the eight Eco RI fragments observed with liver DNA from A strain mice, and DNA from hybrids lacking chromosome 4 yield the other six fragments.

Further experiments will be required to develop maps of the viral DNA endogenous to various mouse strains. However, even at this early stage, some conclusions are apparent: (i) the virus-specific products of endonuclease digestion of DNA from adult tissues are relatively simple, suggesting that the organization and position of the endogenous DNA is stable during development of an individual mouse; (ii) the digestion products are the same in individuals from a single mouse colony or strain, but they vary significantly among different inbred strains of mice, suggesting considerable genetic heterogeneity within or around the relevant loci; and (iii) analysis with restriction enzymes, as described more fully below, can be used to show that milk-borne MMTV may also be transmitted genetically in certain mouse strains.

MAPPING THE RESTRICTION ENDONUCLEASE SITES ON UNINTEGRATED MMTV DNA

As a necessary prelude to the use of restriction enzymes to examine the integration of proviral DNA, we have generated physical maps of the sites cleaved by various restriction enzymes within viral DNA (12). Such mapping poses major problems in the study of any retroviral DNA, since the number of copies of viral DNA per cell is small (13) and the synthesis of full-sized viral DNA *in vitro* is difficult (14). In the case of MMTV, further difficulties are encountered, since the supply of infectious virus is limited and cultured cells are notoriously refractory to infection. We have observed, however, that MMTV-infected rat and mink cells contain unintegrated viral DNA, even many months after the initial infection, when grown in the presence of glucocorticoids (12,15,16). We have therefore obtained our substrate for mapping of restriction endonuclease sites from such chronically-infected heterologous cells. The viral DNA is present, however, in relatively small quantities (e.g., 1 to 20 copies per cell, in addition to the few copies of integrated proviral DNA); to eliminate a requirement for purity, we have detected viral DNA and fragments thereof by the procedure developed by Southern (4). After electrophoresis in agarose gels, the DNA was denatured *in situ* and transferred by elution directly to a sheet of nitrocellulose. The DNA fixed to the filter was then annealed with ^{32}P-labeled virus-specific DNA (synthesized *in vitro*, using viral 35S RNA as template and oligomers of calf thymus DNA as primers); the virus-specific fragments were then visualized as bands on autoradiograms, after exposure of the filter to sensitive x-ray film in the presence of intensifying screens (17). We estimate that this procedure permits the detection of as little as 10 pg (ca. 10^6 complete molecules).

Initial analysis of the unintegrated DNA from infected cells by these methods confirmed our earlier demonstration of both linear and closed circular forms of viral DNA (15). The linear DNA is approximately the same length as a subunit of the viral RNA (9 kilobases (kb)), and it has unique ends (i.e., it is not randomly permuted). Surprisingly, the closed circular DNA is present in two size classes; the predominant form (7.8 kb) is smaller than the linear DNA, and the second form (9.0 kb) is identical in size to the linear DNA. We have shown that linear DNA is the precursor to closed circular viral DNA in cells infected with avian sarcoma virus (18); hence we presume that the smaller circles are formed by deletion during circularization. As indicated in Figure 1, our mapping studies show that the deleted sequences are derived from one or both ends of the linear DNA. Similar, though smaller, deletions have been observed in circular DNA from avian sarcoma virus (19) and murine leukemia virus infected cells (29), but the significance of this phenomenon in the life cycle of the virus is not apparent.

We have generated physical maps of the sites in viral DNA cleaved by eight restriction endonucleases (12; Fig. 1). To order these sites, we made use of three principal strategies: comparison of digests of linear and large and small circular DNA; sequential digestions with two or more enzymes; and hybridization with reagents specific for defined regions of the viral genome. In addition, we identified three enzymes (*Hpa* I, *Sal* I, and *Sma* I) which do not cleave viral DNA.

The mapping studies were performed with virus strains derived from mammary tumors from C3H and GR mice. We have previously found that the genomes of these viruses are indistinguishable by competition hybridization tests, which are relatively insensitive to minor differences in sequence (7). However, the viruses differ with respect to biological activity and immunological characteristics (1,3), and we have demonstrated their non-identity by oligonucleotide "fingerprinting" (21). The mapping studies confirm the high degree of similarity of these strains, since all but one of the eleven enzymes examined fail to distinguish between the DNAs (12,22). The difference between the strains is, however, confirmed by the presence of an addition *Xho* I site in the DNA of MMTV from C3H mice.

correspond to the 3' terminus of the RNA (12), it seems likely that proviral DNA is, in fact, colinear with viral RNA; however, until more detailed mapping is performed, we can only conclude that integration occurs at some point encompassed by this fragment.

HOW MANY SITES IN THE HOST GENOME CAN ACCOMMODATE NEW MMTV PROVIRAL DNA?

If a limited region of viral DNA is used for recombination with cellular DNA, the size of digestion fragments containing both viral and cellular DNA will depend upon the distance from the end of the provirus to a recognition site in cellular DNA for the enzyme employed. In this way the cellular sequences flanking proviruses can be compared to obtain a provisional evaluation of the number of positions in the host genome capable of receiving a provirus. We have performed this type of analysis with the endonuclease *Eco* RI, which recognizes a single site in viral DNA. Thus, as shown in Figure 4, each provirus should yield two fragments containing appreciable portions of viral DNA. When this experiment is performed with DNA isolated from separate mammary tumors, each bearing multiple proviruses acquired by infection, several results would be theoretically possible (Figure 4):

(i) If the requirements for integration were highly specific, a limited number of sites might be available in the host genome; in this case, a similar series of *Eco* RI fragments containing viral sequences would be generated from each sample of tumor DNA.

(ii) If a very large number of sites in the host genome were suitable substrates for integrative recombination, DNA from tumors composed of many independently infected cells would yield too many *Eco* RI fragments to permit resolution of individual fragments as bands in our tests.

(iii) If integration of viral DNA occurred at many sites in host DNA, discrete bands representing resolved *Eco* RI fragments could still be detected if the tumors were each composed mainly of the descendants of a small number of infected cells. In this case, of course, the pattern of bands would differ significantly from tumor to tumor.

(iv) Although not shown in the diagram, multiple copies of viral DNA could be integrated in tandem at one or more sites in cellular DNA; in this case, cleavage with *Eco* RI would then produce viral fragments the size of a complete provirus (ca. 6×10^6 Mr).

We have recently examined DNA from a large number of mammary tumors from BALB/cfC3H mice in this fashion (22). In all cases, the Eco RI digest contains the 5-6 fragments also found in uninfected BALB/c DNA; these presumably result from cleavage of endogenous proviruses, and their conservation suggests that new proviruses are not commonly inserted adjacent to endogenous viral DNA. The most striking finding is the appearance of a unique pattern of additional fragments in the digest of DNA from each tumor (cf. patterns on the right in Figure 4). Few, if any, of these additional bands are precisely the size of a complete provirus, indicating that acquired viral DNA is rarely arranged in tandem.

We conclude from these complex patterns that many sites in the mouse genome can accept an acquired provirus. The same conclusion must be drawn from Eco RI digestions of DNA from lactating mammary glands known to contain an average of 1-2 new proviruses per cell. In the latter case, no distinct virus-specific fragments other than those ascribed to endogenous viral DNA can be identified.

Similar conclusions about the sites of integration have been reached in parallel studies of rat hepatoma cells infected in culture with MMTV (28,29). Again, in all cases, the proviral DNA appears to be colinear with viral RNA, even in clones of infected cells with only 1-2 copies of viral DNA and no detectable synthesis of viral RNA. Analysis of DNA from clones which vary markedly with respect to synthesis of viral RNA and its regulation by glucocorticoid hormones demonstrates a multitude of integration sites in the host genome; therefore it is not yet possible to correlate transcriptional activity or hormonal responsiveness with cellular sequences flanking the proviruses.

Although a large number of sites is available for integration of proviral DNA into the DNA of either the natural host or a heterologous host for MMTV, it is nevertheless premature to conclude that viral DNA is inserted randomly in the host genome. It is possible, for example, that the integrative mechanism recognizes a set of short, highly reiterated sequences in cellular DNA. Direct sequencing of the cellular and viral DNA at several integration sites will probably be required to elucidate this problem.

ONCOGENIC TRANSFORMATION MAY BE A RARE CONSEQUENCE OF INFECTION BY MMTV

Since the sites occupied in the DNA of individual mammary tumors can be mapped with our tests, we presume that each tumor must be composed largely of the descendants of one or a few infected mammary gland cells (cf. patterns on the right in Figure 4). This presumption has been challenged in two further experiments. (i) When Eco RI digests of DNA from several tumors were mixed, we obtained a heterogeneous pattern (cf. patterns in the middle in Figure 4), indicating that the number of available sites was too large to permit resolution of most of the Eco RI fragments from an uncloned population of cells. (ii) When a single mammary tumor was subdivided and transplanted in new animals, the patterns of Eco RI fragments from the DNA in all transplanted tumors were identical to each other and to the pattern obtained with DNA from the primary tumor.

In a previous section we described evidence that a significant proportion of cells in a lactating mammary gland is infected with MMTV. The observation that tumors are composed of relatively homogeneous populations of cells then suggests that only a few of the large number of infected mammary cells are destined to be progenitors of tumor cells. This finding may correlate with the failure to observe altered behavior of MMTV-infected cells in culture. The long latency and infrequency of neoplastic conversion suggests that the mechanism of oncogenesis by MMTV is more akin to that of certain leukosis viruses than to the rapidly-acting sarcoma viruses of birds and rodents. Obviously a potentially complex interplay of viral and host genetic features, hormonal stimulation, and immunological defenses might influence the probability of successful neoplastic conversion in this system. Recent progress in the definition of such modulating factors encourages our conviction that the mechanism of mammary tumorigenesis in the mouse may soon be described in molecular terms.

REFERENCES

1. Nandi, S., and McGrath, C.M. (1973). Adv. Cancer Res. 17, 353
2. Varmus, H.E. Ringold, G., and Yamamoto, K.R. (1978). In "Glucocorticoid Hormone Action" (J. Baxter and G. Rousseau, eds.), Springer-Verlag, New York, in press.
3. Hilgers, I., and Bentvelzen, P. (1978). Adv. Cancer Res. 26, 195.
4. Southern, E.M. (1975). J. Mol. Biol. 98, 503.
5. Morris, V., Medeiros, E., Ringold, G.M., Bishop, J.M., and Varmus, H.E. (1977). J. Mol. Biol. 114, 73.
6. Varmus, H.E., Stavnezer, J., Medeiros, E., and Bishop, J.M. (1975). In "Comparative Leukemia Research" Y. Ito and R.M. Dutcher, eds.), pp. 451-461. University of Tokyo Press, Basel.

7. Ringold, G.M., Blair, P., Bishop, J.M., and Varmus, H.E. (1976). Virology 70, 530.
8. Drohan, W., Kettmann, R., Colcher, D., and Schlom, J. (1977). J. Virol. 21, 986.
9. Kozak, C., Nichols, E., and Ruddle, F.H. (1975). Somatic Cell Genetics 1, 371.
10. Morris, V.L., Kozak, C., Jolicoeur, P., Ruddle, F., and Varmus, H.E. (1978), submitted for publication.
11. Benveniste, R.E., and Todaro, G.J. (1975). Nature 257, 506.
12. Shank, P.R., Cohen, J.C., Varmus, H.E., Yamamoto, K.R., and Ringold, G.M. (1978). Proc. Nat. Acad. Sci. USA, in press.
13. Weinberg, R.A. (1977). Biochem. Biophys. Acta 473, 39.
14. Verma, I.M. (1977). Biochim. Biophys. Acta 473, 1.
15. Ringold, G.M., Yamamoto, K.R., Shank, P.R., Varmus, H.E. (1977). Cell 10, 19.
16. Ringold, G.M., Shank, P.R., and Yamamoto, K.R. (1978). J. Virol., in press.
17. Swanstrom, R., and Shank, P.R. (1978). Anal. Biochem., in press.
18. Shank, P.R., and Varmus, H.E. (1978). J. Virology 25, 104.
19. Shank, P.R. _et al_., manuscirpt in preparation.
20. Yoshimura F., and Weinberg, R.A., personal communication.
21. Friedrich, R., Morris, V., Goodman, H.M., Bishop, J.M., and Varmus, H.E. (1976). Virology 72, 330.
22. Cohen, J.C. _et al_., manuscript in preparation.
23. Nandi, S., and Helmich (1974). J. Nat. Cancer Inst. 52, 1277.
24. Varmus, H.E., Quintrell, N., Nowinski, R., Medeiros, E., Sarkar, N., and Bishop, J.M. (1973). J. Mol. Biol. 79, 663.
25. Stallcup, M., personal communication.
26. Botchan, M., Topp, W., and Sambrook, J. (1976). Cell 9, 269.
27. Bishop, J.M., and Varmus, H.E. (1975). In "Cancer: A Comprehensive Treatise" (F.F. Becker, ed.), Vol. 2, pp. 3-48, Plenum Press, New York.
28. Ringold, G.M., Cardiff, R.D., Varmus, H.E., and Yamamoto, K.R. (1977). Cell 10, 11.
29. Ringold, G.M. _et al_., manuscript in preparation.

DETECTION OF VIRAL NUCLEIC ACID SEQUENCES USING *IN SITU* HYBRIDIZATION

James K. McDougall and Denise A. Galloway

Cold Spring Harbor Laboratory
Cold Spring Harbor, New York 11724

ABSTRACT Nick translated herpesvirus type 2 DNA (^{3}H) has been hybridized *in situ* to frozen sectioned tissue from human sensory ganglia and neoplastic tissue from human cervix. The results show that this method has detected viral RNA in both cases and that in the abnormal cervical tissue the viral RNA is frequently found in association with the malignant or pre-malignant cells.

INTRODUCTION

The *in situ* molecular hybridization method (1,2) results in the autoradiographic detection of specific hybrids between nucleic acid sequences in cytological preparations. Using this technique satellite DNA sequences have been localized on the chromosomes of many species, including man (3,4). Globin messenger RNA has been localized by *in situ* hybridization with *in vitro* transcribed complementary DNA (5). One class of unique sequences found in tumor cells is that representing viral genes which may be covalently integrated into the host genome or present in non-integrated plasmid form. From the extensive data available from experimentally induced transformations and tumors, it is known that where viruses are the inducing agent persistence of viral DNA sequences can almost always be demonstrated (6) by re-association kinetics methods or nitrocellulose "blotting" techniques (7). Examples of the successful use of *in situ* hybridization to detect viral DNA sequences in virus-transformed or virus-induced tumor cells have been reported by a number of workers (8-13). One of the most significant studies (9) localized Epstein-Barr virus DNA in the epithelial cells of nasopharyngeal carcinoma. EBV is normally found in lymphoid cells in Burkitt's lymphoma and in infectious mononucleosis as well as in the lymphocytes of normal individuals.

In situ hybridization therefore is of value in terms of the localization of viral DNA to particular cell types or

ISBN 0-12-668350-6

areas of tissue within a tumor. The probability of there being sufficient DNA integrated to be routinely detectable by *in situ* cytological methods is slight as in most cases there are only limited regions of the virus genome present and these are in few copies (14).

The detection of virus RNA using a radioactive DNA probe offers greater possibilities of detection as the target sequences are an amplification of the integrated viral DNA. For example, an adenovirus type 2 transformed cell line which contains an estimated 6 copies of 14% of the adenovirus genome (14), approximately half of which is transcribed to give an average of 300-400 mRNA molecules of viral origin/cell (15), has been used as a model for the detection of viral RNA by the *in situ* method (16). Virus mRNA was detected after a four-week exposure using an *in vivo* labelled (^{3}H) virus DNA probe. Detection of the virus RNA is enhanced when nick-translated virus DNA (^{3}H) with a specific activity of 10^7 cpm/ug is used.

We are investigating this method as a means of detecting viral RNA sequences in human tumors and in non-neoplastic tissues. One of the key features of the method lies in the fact that with cytological hybridization as opposed to more conventional biochemical methods it is possible to carry out the hybridization so that individual cells or sub-populations of cells within, for example, a tumor section, can be studied for their content of viral specific RNA transcripts. A major difficulty encountered in screening tumor tissue for viral nucleic acid sequences is that the proportion of tumor cells in a biopsy is unlikely to be of the order of 100% and may be much lower. If DNA and/or RNA is extracted from the tissue mass there is inevitably a dilution of the sequences which are complementary to the probe being used. The method described here is not subject to these constraints as the probe is being applied directly to tumor cells and normal cells in the same preparation. Herpesvirus transformed cells have been examined in our laboratory using the *in situ* cytological hybridization method to detect virus mRNA (17) and we are continuing these studies in experimental herpesvirus type 2-induced tumors.

Sero-epidemiological surveys have produced evidence associating herpesvirus type 2 (HSV-2) infection with premalignant changes in human cervix and carcinoma of the cervix (18-23) and there has been one published report of the detection of an HSV-2 DNA fragment in a carcinoma of the cervix (24), although other attempts have given negative results (25). Our experiments are aimed at defining the optimum techniques for hybridization of nick translated HSV-2 DNA to frozen sections in order to detect any evidence of HSV-2 RNA in neoplastic tissue from the cervix.

METHODS

The methods used are based on those previously described by Moar and Jones (16), Copple and McDougall (17) and Jones et al. (29).

Virus DNA was nick translated by the method of Maniatis et al. (30).

RESULTS

As an initial demonstration of the use of this method we have hybridized (^{3}H) HSV-2 DNA to frozen sections of ganglia removed from a human cadaver undergoing autopsy. It has been shown that HSV remains latent in the sensory ganglia (26,27) and that in situ hybridization with a complementary RNA probe to HSV DNA can detect virus genomes (8). In this case we have found autoradiographic label in association with the nucleus of thoracic ganglion cells (Figs. 1 and 2) with no hybridization to other cells in the section. We have not determined whether the RNA detected is encoded by latent HSV-1 or HSV-2 genomes (27).

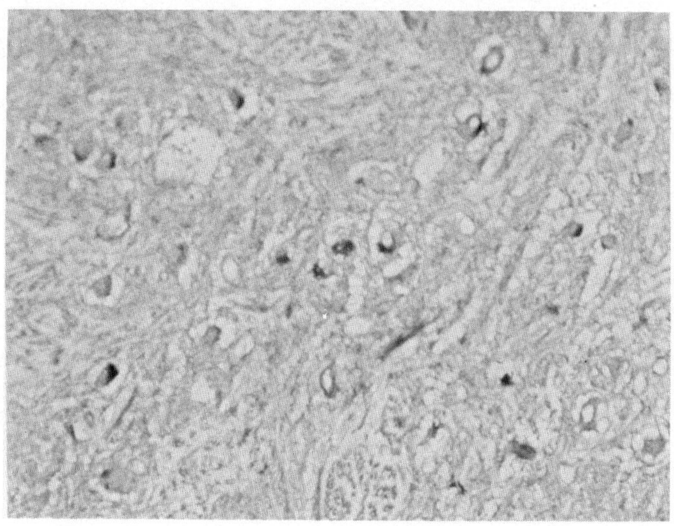

FIGURE 1. Hybridization of ^{3}H-HSV-2 DNA to human thoracic ganglia. (a) Low power magn. showing autoradiographic grains associated with ganglion cells.

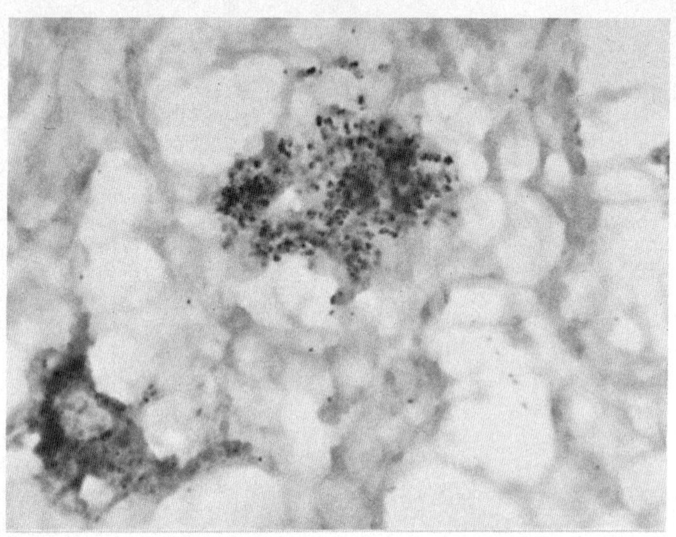

FIGURE 1. (b) High power magn. of ganglion cell. (Autoradiographs exposed 4 weeks.)

A series of 42 cervical biopsies taken for pathological diagnosis have been examined by in situ hybridization with the same virus DNA probe. The hybridizations and pathological diagnoses were carried out independently in different laboratories. The results from this series indicate that HSV-2 RNA can be detected more frequently in the frozen sections which have areas of cells undergoing pre-malignant changes than in those representing normal tissue. The autoradiographic grains are most frequently associated with neoplastic cells but can also be found on macrophages and lymphoid cells in the areas of abnormality. The probe did not hybridize to normal cells and did not hybridize to human DNA bound to nitrocellulose filters under conditions which would detect $5x10^{-5}$ug of human DNA in the probe. Figure 3 shows representative areas from autoradiographs of hybridized sections.

The results from this series of cervical tissues are summarized in Table 1. The negative tissues are from sections which do not contain any area of abnormality which can be related by pathological examination to pre-neoplastic changes. Hybridization of other viral DNA probes to cervical sections has given negative results.

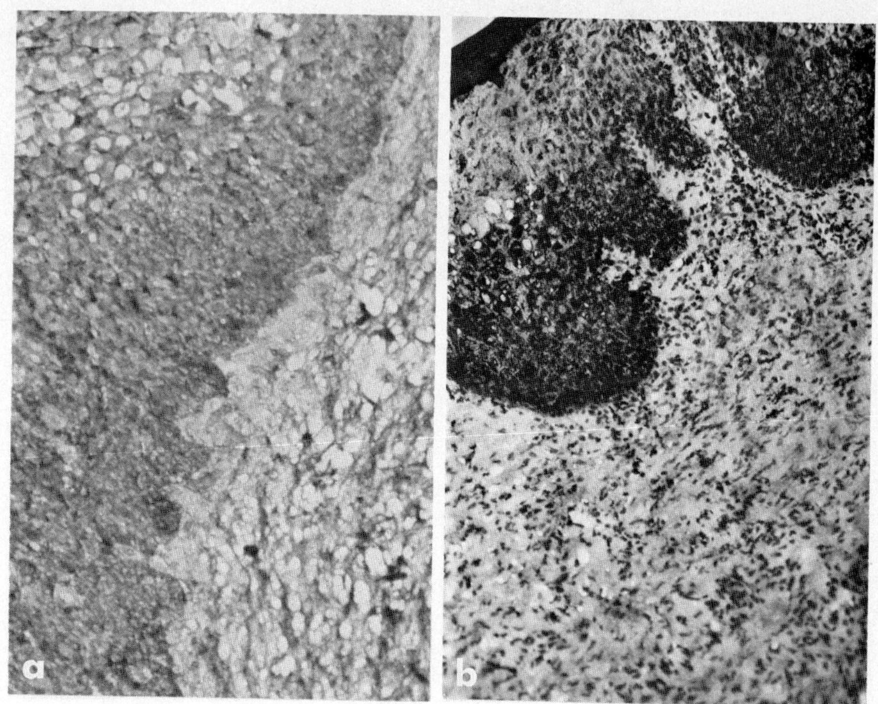

FIGURE 2. Hybridization of ^{3}H-HSV-2 DNA to frozen sections of human cervical tissue. (a) Area of epithelial tissue showing no autoradiographic grains. (b) An area of intraepithelial neoplasia showing autoradiographic grains localized in the lower half of the epithelium. (Autoradiographs exposed 4 weeks.)

TABLE 1
CERVICAL BIOPSIES (42 CASES)

Diagnosis	*In situ* hybridization with ^{3}H-HSV2 DNA	
	HSV2-RNA positive	HSV2-RNA negative
CIN I-III	10	4
Squamous Metaplasia	2	2
Negative	1	23
Total	13	29

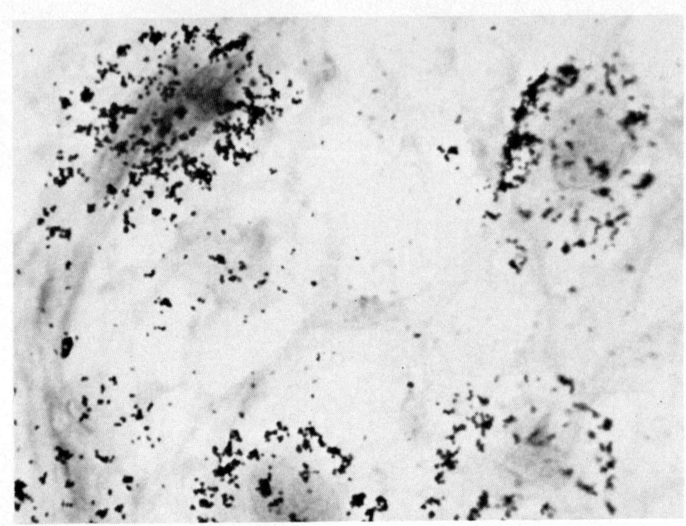

FIGURE 3. Hybridization of ^{3}H-HSV2 DNA to frozen sections of human cervical tissue. High power magn. showing autoradiographic grains mainly associated with cytoplasm. (Autoradiograph exposed 4 weeks.)

DISCUSSION

Cytological hybridization has many applications in the study of virus persistence in the form of latent or integrated genomes. Although the ability to detect viral DNA is limited to those situations where multiple copies of the virus chromosome persist or are replicated, the extension of the technique to a detection of amplified targets (RNA) is of practical value. It is premature to assume that the results from *in situ* hybridizations are more than indicative of a significant presence of HSV-2 RNA in cervical tumor cells. Infections of the genital area with HSV-2 are common and have become a significant venereal disease factor (28); antibody responses to such infections nevertheless appear to have their corollary in the severity of progression from dysplasia to invasive carcinoma of the cervix (23). While it is possible that we have detected RNA synthesized during a permissive virus replication cycle, there is no evidence from clinical records that these patients were undergoing acute clinical infection at the time of biopsy. If there is HSV-2 replication it would be surprising to find it preferentially in neoplastic cells when all cells in the sample are permissive. Further *in situ* hybridizations using restriction endonuclease derived fragments of HSV-2 DNA should identify whether the observed RNA is transcribed from limited regions of the viral genome, a phenomenon common to DNA virus transformed cells. In a prospective study virus isolation attempts and serological data will complete the picture.

ACKNOWLEDGMENTS

The collaboration of Drs. C.M. Fenoglio, M. Shevchuk and K.W. Jones is an essential part of this study which is supported by the National Cancer Institute, CA13106.

REFERENCES

1. John, H.A., Birnsteil, M.L., and Jones, K.W. (1969). Nature 233, 528.
2. Gall, J.G., and Pardue, M.L. (1969). Proc. nat. Acad. Sci. U.S.A. 63, 378.
3. Corneo, G., Ginelli, E., and Polli, E. (1970). J. mol. Biol. 48, 319.
4. Jones, K.W., Prosser, J., Corneo, G., Ginelli, E., and Bobrow, M. (1972). In "Symp. Medica Hoechst. Modern aspects of cytogenetics", F.K. Schattauer, Verlag, Stuttgart.
5. Harrison, P.R., Conkie, D., Paul, J., and Jones, K.W. (1973). FEBS Letters 32, 109.
6. Benyesh-Melnick, M., and Butel, J. (1974). In "The Molecular Biology of Cancer" (H. Busch, ed.), Academic Press.
7. Southern, E.M. (1975). J. mol. Biol. 98, 503.
8. Zur Hausen, H., and Schulte-Holthausen, H. (1972) In "Oncogenesis and Herpesviruses", 1st. Int. Symp. I.A.R.C. Lyon, p. 321.
9. Wolf, H., Zur Hausen, H., and Becker, V. (1973). Nature N.B. 244, 245.
10. Loni, M.C., and Green, M. (1973). J. Virol. 12, 1288.
11. Dunn, A.R., Gallimore, P.H., Jones, K.W., and McDougall, J.K. (1973). Int. J. Cancer 11, 628.
12. Watkins, J.F. (1973). J. gen. Virol. 21, 69.
13. Orth, G., Jeanteur, P., and Croissant, O. (1970). Proc. nat. Acad. Sci. (Wash.) 68, 1876.
14. Sambrook, J., Botchan, M., Gallimore, P.H., Ozanne, B., Pettersson, U., Williams, J., and Sharp, P.A. (1974) Cold Spring Harbor Symp. Quant. Biol. 39, 615.
15. Flint, S.J., and Sharp, P.A. (1976). J. mol. Biol. 106, 749.
16. Moar, M.H., and Jones, K.W. (1975). Int. J. Cancer 16, 998.

17. Copple, C.D., and McDougall, J.K. (1976). Int. J. Cancer 17, 501.
18. Rawls, W.E., Tompkins, W.A.F., Figueroa, M., and Melnick, J.L. (1968). Science 161, 1255.
19. Sprecher-Goldberger, S., Thiry, L., Gould, I., Fassin, Y., and Gampel, C. (1972). Amer. J. Epidemiol. 97, 103.
20. Adam, E., Kaufman, R.H., Melnick, J.L., Levy, A.H., and Rawls, W.E. (1974). Amer. J. Epidemiol. 98, 77.
21. Rawls, W.E., Tompkins, W.A.F., and Melnick, J.L. (1969). Amer. J. Epidemiol. 80, 547.
22. Nahmias, A.J., Josey, W.E., Naib, Z.M., Luce, C.F., and Duffey, A. (1970). Amer. J. Epidemiol. 91, 539.
23. Skinner, G.R.B., Whitney, J.E., and Hartley, C. (1977). Arch. Virol. 54, 211.
24. Frenkel, N., Roizman, B., Cassai, E., and Nahmias, A. (1972). Proc. nat. Acad. Sci. U.S.A. 69, 3784.
25. Zur Hausen, H., Schulte-Holthausen, H., Wolf, H., Dorries, K., and Egger, H. (1974). Int. J. Cancer 13, 657.
26. Stevens, J.G. (1975). In "Oncogenesis and Herpesviruses II", 2nd Int. Symp. I.A.R.C., Lyon, p. 67.
27. Baringer, J.R. (1975). In "Oncogenesis and Herpesviruses II", 2nd Int. Symp. I.A.R.C., Lyon, p. 73.
28. Kessler, I. (1977). Cancer 39, 1912.
29. Jones, K.W., Fenoglio, C.M., Shevchuk-Chaban, M., Maitland, N.J., and McDougall, J.K. (1978). In "Oncogenesis and Herpesviruses III", 3rd Int. Symp. I.A.R.C., Lyon (in press).
30. Maniatis, T., Kee, S.G., Efstratiadis, A., and Kafatos, F.C. (1976). Cell 8, 163.

PHYSICAL MAPPING OF VIRAL EPISOMES IN *HERPESVIRUS SAIMIRI* TRANSFORMED LYMPHOID CELLS[1]

Fred-Jochen Werner[2], Ronald C. Desrosiers, Carel Mulder, Georg W. Bornkamm[3], and Bernhard Fleckenstein

Institut fur Klinische Virologie, Universitat Erlangen-Nurnberg, W. Germany; New England Regional Primate Research Center, Harvard Medical School, Southborough, Mass. 01772; Departments of Microbiology and Pharmacology, University of Massachusetts Medical School, Worcester, Massachusetts 01605.

ABSTRACT *Herpesvirus saimiri* (*H. saimiri*) which is a highly oncogenic agent of New World primates has two unusual structural features in the virion DNA, an extreme intramolecular heterogeneity in base composition (L-DNA, 36% G+C versus H-DNA, 71% G+C) and the presence of terminal repetitive (H) DNA in a peculiar arrangement. Some virus strains are differing in their oncogenic potential; they are completely homologous in DNA-hybridizations, but show differences in the cleavage specificity for numerous restriction endonucleases. Lymphoid T cell lines have been derived from *H. saimiri* induced tumors in marmoset monkeys. Some of the lines do not produce infectious virus, and viral antigens could not be detected so far. Non-producer cell lines carry relatively large amounts of both types of *H. saimiri* DNA, L- and H- sequences. The cell lines were found to contain covalently closed circular viral DNA with a significantly higher molecular weight than virion DNA. Partial denaturation maps of episomes from one line (#1670) showed an uniform arrangement of L- and H- sequences, all partially denatured circles containing two L-DNA regions and two segments of H-DNA. Episomal L-DNA segments were correlated with the known physical cleavage maps of linear *H. saimiri* DNA by computer

[1]This work was supported by Deutsche Forschungsgemeinschaft, SFB 118, by Contract NO1 CP 81005 within the Virus Cancer Program of the National Cancer Institute, and by American Cancer Society Grant VC 197.

[2]Present address: Institut fur Molekularbiologie (Tumorforschung), Universitat Essen, West Germany.

[3]Present address: Institut fur Virologie, Universitat Freiburg, West Germany.

ISBN 0-12-668350-6

alignment of their partial denaturation histograms. This showed that the episomes are a form of defective genomes, since a 14 megadalton segment (*Sal* I C fragment) of linear L-DNA is missing.

INTRODUCTION

Herpesvirus saimiri (*H. saimiri*), an indigenous agent of squirrel monkeys, is not pathogenic in the natural host species (1, 2); however the virus is highly oncogenic in many other New World primate species (3, 4, 5, 6, 7, 8). Tamarin marmoset monkeys (*Saguinus sp.*) are the most susceptible animals for tumor induction; infection of these monkeys with *H. saimiri* invariably results in a neoplastic disease of the lymphatic system within a few weeks, irrespective of the age of the animals or the dose of inoculum. The tumors are mostly characterized as malignant lymphomas of the reticulum cell or poorly differentiated type or as acute lymphocytic leukemia.

There are obvious similarities between the oncogenesis by *H. saimiri* in New World primates and the pathogenesis of human herpesvirus (Epstein-Barr-virus) associated neoplastic diseases. The molecular biological approach of *H. saimiri* induced transformation and oncogenesis became possible, while the biochemistry of EBV transformation still poses many unresolved problems. *H. saimiri* grows lytically in permissive epithelial monolayer cells, yielding comparatively large amounts of viral structural components. *H. saimiri* is different in the genomic organization from any human or nonprimate herpesvirus. The particular structure of *H. saimiri* DNA renders it an ideal object for analysis by restriction endonucleases and electron microscopy.

RESULTS

Structure of Linear Herpesvirus Saimiri Virion DNA. *H. saimiri* particles contain two different forms of linear viral DNA molecules. They have been designated as *M-genomes* and *H-genomes*.

M-genomes (M-DNA) are the predominant type of virion DNA (9). M-DNA appears to contain the complete viral genetic information as it is infectious in cell culture. M-genomes are linear duplex DNA molecules of about 100 megadaltons (md). The molecules show two unusual structural features, (i) the presence of manyfold repetitive sequences and (ii) an extreme intramolecular heterogeneity in base composition (10). Each M-DNA molecule contains a long region of non-repetitive L-DNA (36% G+C). The length of the L-DNA region in each molecule is constant, corresponding to 71.6 md (11). The L-DNA region

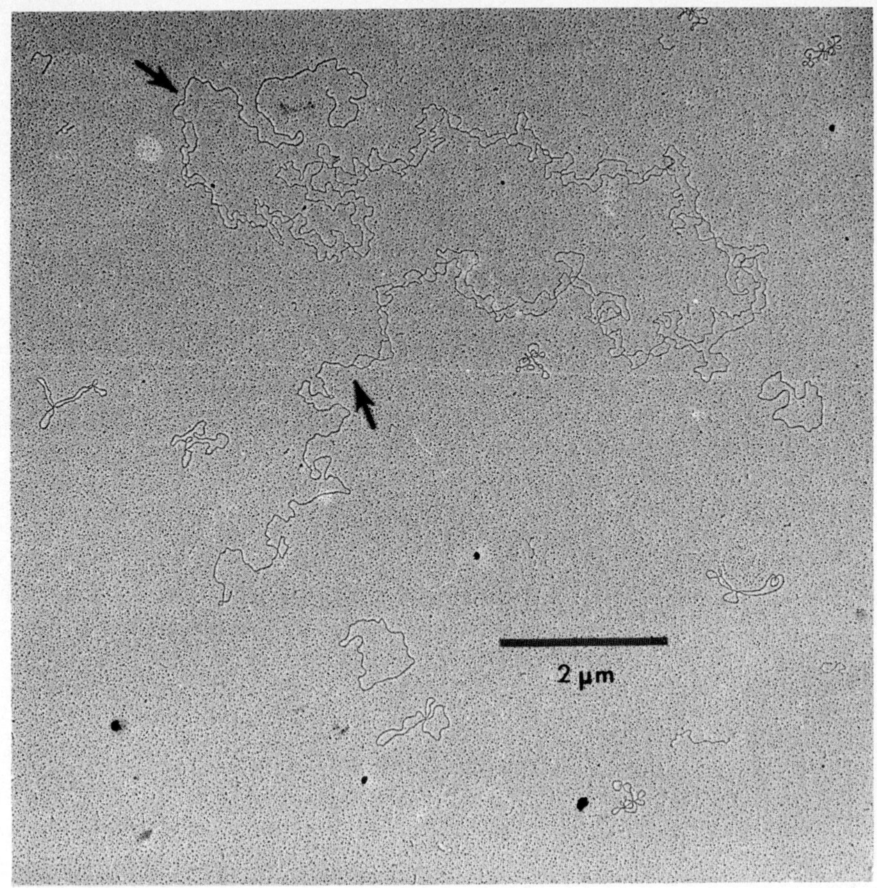

FIGURE 1. Electron micrograph of a partially denatured M-DNA molecule from *H. saimiri*. The arrows indicate the transition between terminal H-DNA segments and internal L-DNA region.

is inserted between two stretches of H-DNA with high G+C content (Figure 1, 2). Both terminal H-segments are variable in length and repetitive DNA. H-DNA consists of identical repeat units (molecular weight 830,000) in tandem arrangement, and the H-sequences from both sides of the molecule are oriented in the same direction. The length of individual terminal H-DNA segments varies between 21 um and less than 1 um. In general, molecules with a long H-DNA end at one side have a short H-DNA end at the other side, thus giving rise to a limited size heterogeneity. Orientation of M-DNA molecules by the denaturation map of the internal L-DNA region shows that the longer H-DNA end may be located at either side of the M-genome (11), and there is indication that the H-DNA segments of M-DNA molecules differ in length by integer numbers of repeat units.

A number of different *H. saimiri* strains have been isolated; some strains differing significantly in their oncogenic potential from the prime strain. Virus strains have been attenuated by continuous passage in cell culture (12, 13, 14). Corresponding DNA sequences of different *H. saimiri* strains have a high degree of base sequence homology, as they showed complete cross-reaction in DNA-DNA hybridizations. On the other hand, we found specific differences in the fragmentation pattern of different virus strains after cleavage of *H. saimiri* M-DNA with restriction endonucleases. Strain #11 of *H. saimiri* was the main object for our studies of *H. saimiri* DNA structure, since there exist several lymphoid cell lines that are derived from tumors induced by this virus (15). Restriction endonuclease *Sam* I (endo R. *Sam* I) was found to cleave four times within each repeat unit of H-DNA from *H. saimiri* #11, but this enzyme did not cleave within the entire length of the L-DNA region. Five other endonucleases cleaved L-DNA of *H. saimiri* #11 into a limited number of fragments without cleavage of H-DNA. Endo R. *Sal* I, *Bam* HI, *Kpn* I, *Xho* I, and *Eco* RI cleaved the L-region into respectively three, six, seven, eight and ten major fragments. The localization of variable length H-DNA segments at the termini of *H. saimiri* M-genomes facilitated mapping of the terminal fragments of L-DNA. For instance, cleavage by endonuclease R. *Sal* I yielded only one distinct band (fragment B) and high molecular weight DNA that did not migrate as a unique band in agarose gels; but cleavage with both endonucleases *Sal* I and *Sma* I yielded three fragments A, B, C, (47.9, 11.6, and 10.3 md, respectively) plus the 4 *Sma* I bands from H-DNA, and the high molecular weight and heterogeneous band disappeared. Thus, the order of the three *Sal* I fragments must be A, B, C. Cleavage maps for the enzymes *Bam* HI, *Kpn* I and *Eco* RI could be constructed by preparative isolation of restriction fragments from agarose gels and recleaving with one or two more enzymes.

H-genomes. A minor proportion of *H. saimiri* DNA which represents about 10-15% of the total virion DNA appears as long DNA molecules that consist of repetitive H-sequences exclusively (Figure 2). Those molecules were designated as H-genomes. H-DNA sequences of H-genomes are identical with the H- sequences of H-genomes; this could be shown by cross hybridizations (9) and by comparative cleavage with restriction endonucleases (11).

H-genomes are defective, as is expected from the low genetic complexity; correspondingly, no infectious DNA was found in the high density fractions of CsCl gradients where H-DNA is banding (16). In contrast to the defective genomes of herpes simplex virus and many viruses of other groups, we have no indication that H-genomes accumulate in *H. saimiri* stocks depending on the multiplicity of infection in subsequent cell culture passages. H-genomes appear immediately after transfection with purified M-DNA. This indicates that H-genomes are created efficiently from the repetitive H-DNA ends of M-genomes in each cycle of replication.

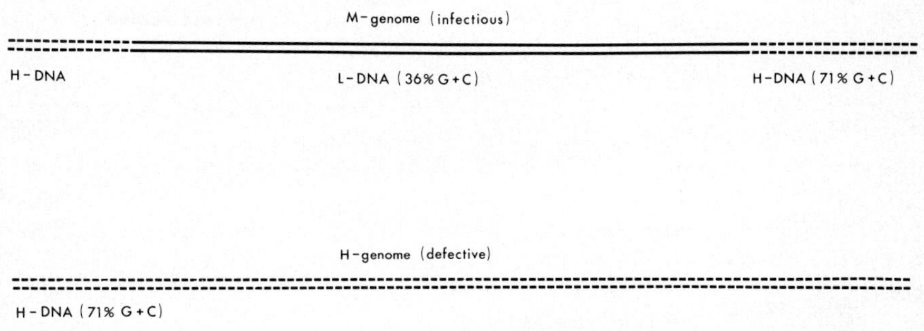

FIGURE 2. Schematic representation of the two types of *H. saimiri* virion DNA.

Presence of Episomal DNA in Herpesvirus-transformed Cell Lines. A number of lymphoid T-cell lines have been established from neoplastic cells in infiltrated organs and blood of tumor bearing marmoset monkeys (15, 17, 18). *H. saimiri* could readily be isolated from these cell lines by cocultivation with permissive monolayer cells. After prolonged passage, some of the cell lines (e.g. cell lines 70N2 and #1670) lost the ability to synthesize infectious virus. The "nonproducer" cells contain a high number of viral genome copies, though viral antigens have not been detected (18). We could show that these *H. saimiri* transformed lymphoid cells contained DNA sequences that persist in an episomal form (19).

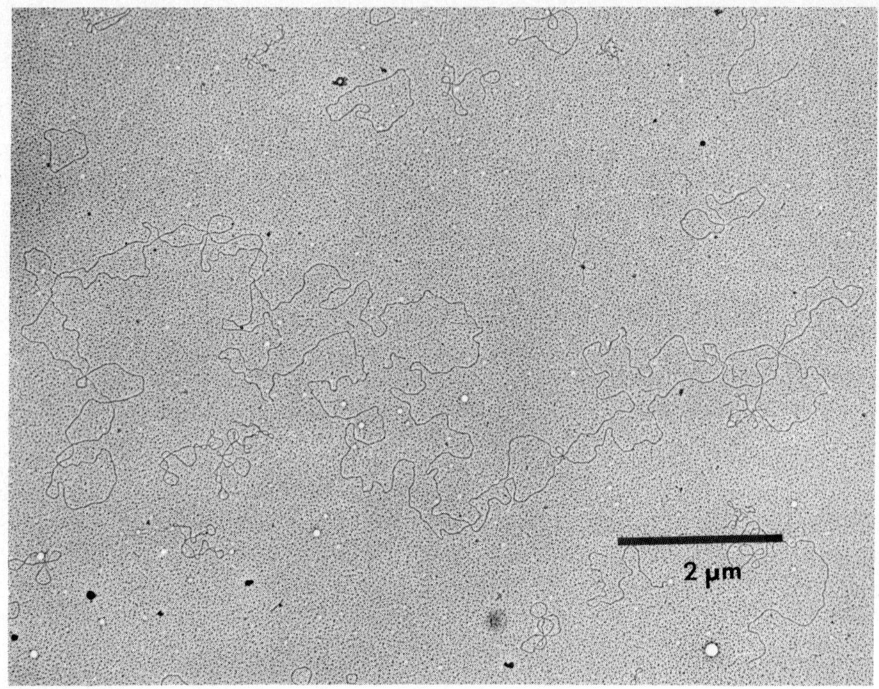

FIGURE 3. Electron micrograph of a relaxed circular DNA molecule from a _H. saimiri_ transformed lymphoid tumor cell line (#1670). The small circular molecules are bacteriophage PM2-DNA added as size marker.

Covalently closed circular DNA was isolated from extracts of _H. saimiri_ transformed cell lines by three subsequent centrifugation steps, (i) isopycnic banding in CsCl gradients, (ii) sedimentation in glycerol gradients, and (iii) density centrifugation in CsCl ethidium bromide. Alternatively, subsequent equilibrium centrifugations in CsCl ethidium bromide were employed as the sole method of fractionation. In the electron microscope, part of the circular molecules appeared as superhelices, part as relaxed circles (Figure 3). The size of episomes from _H. saimiri_ transformed tumor cell lines was found to be significantly larger than that of linear DNA molecules in virus particles. While the contour length of M-genomes is equivalent to approximately 100 md, the episomal DNA of cell line 70N2 and line #1670 had a length corresponding to 119.7 ± 2.5 md and 131.5 ± 3.6 md molecular weight, respectively. The standard deviations of these contour length determinations are compatible with the assumption that the episomal DNA molecules from #1670 and

70N2 tumor cell lines are homogeneous classes of molecules. This is in contrast to the size heterogeneity of the linear M-genomes of *H. saimiri* virus particles.

Arrangement of L-DNA and H-DNA Sequences in Episomal DNA. The following experiments were aimed at finding a correlation between the episomal DNA from tumor cell lines and the defined DNA sequences of linear *H. saimiri* M-genomes. Since the episomes are larger than virion DNA, the question arose whether and how viral L- and H- sequences could be arranged to form this type of circular molecule.

As the first approach, we chose the method of partial denaturation mapping in the electron microscope, since it allows to distinguish easily between stretches of L-DNA and H-DNA. The large circular DNA from tumor cell line #1670 was partially denatured with 7 M sodium perchlorate and 1.04 M formaldehyde at neutral pH (7.5) and low temperature (30°C). Under these conditions, L-DNA sequences of linear molecules would be denatured to 90% (11), whereas H-DNA would remain fully duplex.

The denaturation patterns of all 19 episomes in which single-stranded regions were found showed an uniform arrangement of partially denatured (L) and duplex (H) stretches. Apparently, all these circles contained two regions of viral L-DNA (57.8 ± 1.9 md and 33.7 ± 1.4 md) and two H-DNA regions 22.8 ± 1.9 md and 17.2 ± 0.8 md, respectively) (Figure 4).

As a next step, the long and short L-DNA regions of these episomes were correlated with each other by comparison of their denaturation maps. The L-DNA segment were aligned successively into optimal fitting using CD 3300 and Cyber 122 computers (Figure 5). The histograms showed that the shorter L-regions were a subset of the longer L-regions, and both L-regions in each circle were oriented in the same polarity. This indicated that the episomes of tumor cell line #1670 are a form of defective genomes containing L-DNA at a complexity of 58 md only. This is equivalent to 81% of the L-region of the complete M-genomes.

Correlation Between the L-DNA Sequences in Episomes and Physical Gene Maps of Linear Virion DNA. The methods of partial denaturation mapping have been used further to find the correspondence between the L-segments in viral episomes and distinct regions in the cleavage map of the linear M-DNA. As described above, cleavage of a M-DNA molecule with endo R. *Sal* I yields one very large L-DNA piece (fragment A, 49.7 md), contiguous with a duplex H-DNA stretch. An endo R *Sal* I digest of M-DNA was partially denatured, and the denaturation histograms of *Sal* I fragment A were fitted into optimal correspondence with the denaturation histograms of L-DNA from

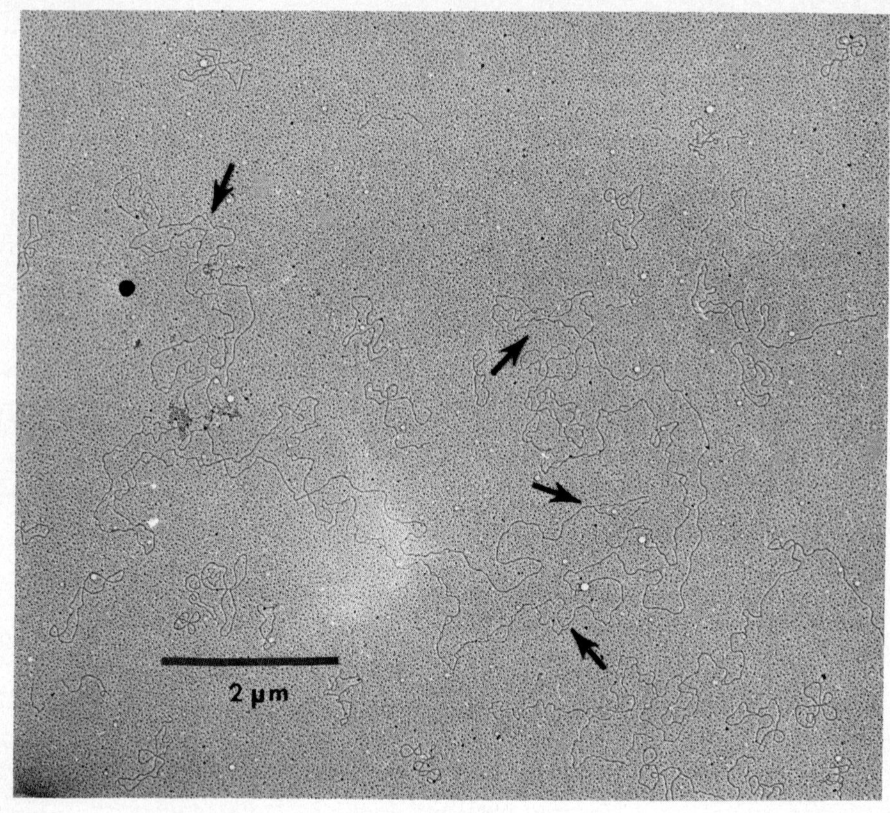

FIGURE 4. Electron micrograph of partially denatured episome from cell line #1670. The arrows indicate the transitions between H- and L- sequences. The small circles are PM2-DNA.

the linear M-DNA molecules (Figure 5). In the following calculation process, the episomal L-segments were aligned with the histograms from the L-DNA of linear genomes and its Sal I A fragment. The short L-region of #1670 cells fitted best with the left half of the L-region of M-genomes (0.00 - 0.47 fractional length) and was completely comprised within Sal I fragment A. The long L-region of #1670 episomes represented also the left side of the physical map, since it was extending from 0.00 to 0.81 fractional length of virion L-DNA after optimal alignment (Figure 5, 6). This model implies that the right terminal 19% of the virion L-DNA region (0.81 - 1.00 fractional length) are missing in #1670

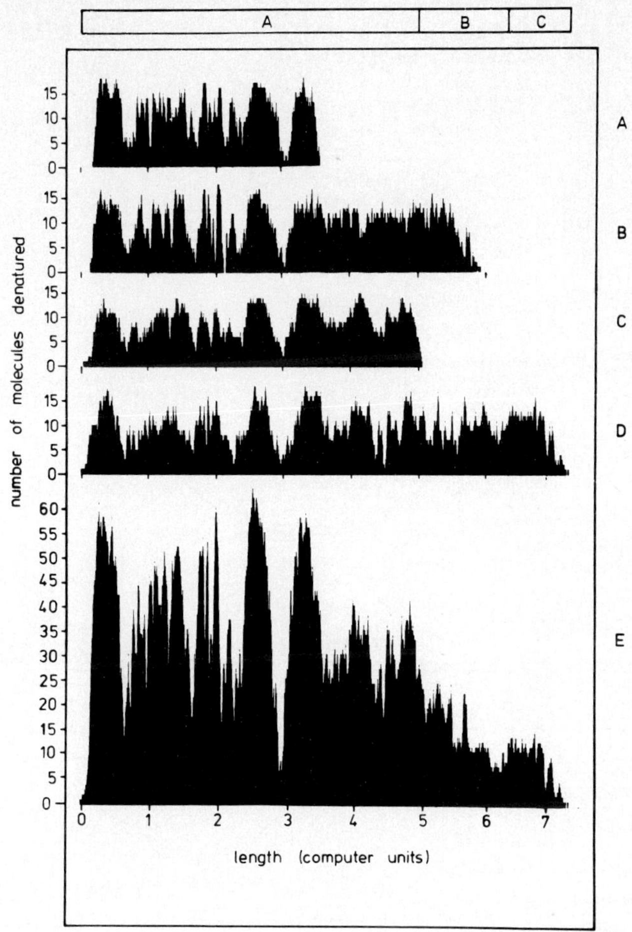

FIGURE 5. Optimal arrangement of the denaturation maps of L-DNA from linear virion M-genomes and L-DNA segments from episomes from H. saimiri transformed cells #1670. Partial denaturation histograms represent (A) short L-regions from episomes of #1670 tumor cell line; (B) long L-regions of these episomes; (C) endo R. Sal I fragment A of the M-genome, and (D) the complete L-region of a M-genome.

episomes. The missing segment corresponds to the Sal I C fragment of H. saimiri strain #11 and to the adjacent two thirds of the fragment B. The model further indicates that the virion L-segment from fractional length 0.47 to 0.81 is present once in each episome, and the L-sequences between 0.0 and 0.47 fractional length are represented in duplicate.

Further, it implicates that an #1670 episome contains only one cleavage site for restriction endonuclease Sal I which is located in the longer L-region.

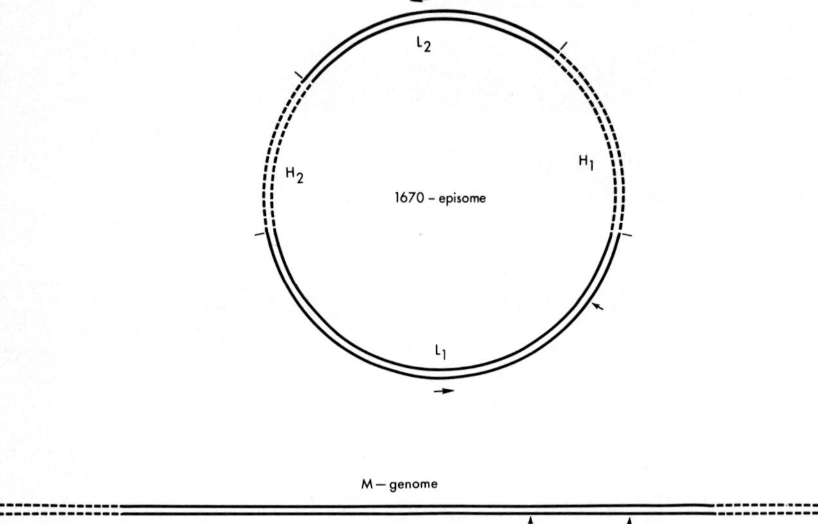

FIGURE 6. Schematic illustration of the relative representation and orientation of H. saimiri DNA sequences in episomes from lymphoid tumor cell line #1670. Part of the L-DNA sequences (fractional length 0.0 - 0.47) is present in duplicate, part (fractional length 0.47 - 0.81) is present once, and part of the L-DNA (fractional length 0.81 - 1.00) is missing in each episome. One Sal I cleavage site is preserved in the episome.

CONCLUSIONS

Herpesvirus saimiri is different in its genome structure from all other herpesviruses described before. The complete viral genomes (M-genomes) revealed two structural features that have not been observed in the DNA of any other animal virus. About 30% of the DNA are highly repetitive sequences and the molecules show an extreme intramolecular heterogeneity in base composition. We do not know what the function of the terminal repetitive DNA in *H. saimiri* genomes is and whether it is related to the high oncogenic potential of this virus. Certainly, terminal tandem repeats facilitate circularization of H. saimiri DNA, and this may be an essential mechanism for DNA replication and persistence of the virus in normal lymphoid cells and tumor cells.

First evidence that the DNA of herpesviruses persists in transformed lymphoid cells in a non-integrated form came from EBV transformed cells (20, 21). Episomal viral DNA was found in EBV transformed tumor cell lines from Burkitt's lymphomas (22) and in nasopharyngeal carcinoma cells (23). Episomal EBV DNA in transformed human cells has a molecular weight which is identical or only slightly smaller than linear viral genomes from virus particles. In contrast, the viral episomes of H. saimiri transformed tumor cell lines are significantly larger than linear viral genomes from virus particles. In contrast, the viral episomes of *H. saimiri* transformed tumor cell lines are significantly larger than virion DNA. The peculiar structure of these episomes indicates that they must have originated by covalent linkage of two M-DNA molecules. The episomal status may be the primary mechanism of herpesvirus persistence. If there are no viral genomes of different structure in integrated or free form, we can assume that part of the left side of the *H. saimiri* DNA region codes for proteins that are not required to maintain the state of transformation in lymphoid cells. In this case, part of a herpesvirus genome would be enough to transform, and this may apply to other herpesvirus transformed cells or tumor cells as well. Identification of L-DNA segments in episomes of other transformed non-producer cell lines from *H. saimiri* induced tumors may provide more information on specific L-DNA segments involved in the transformation process.

ACKNOWLEDGEMENT

The excellent technical assistance of Ingrid Muller is greatly appreciated.

REFERENCES

1. Melendez, L. V., Daniel, M. D., Hunt, R. D., and Garcia, F. G. (1968). Lab. Anim. Care 18, 374.
2. Falk, L. A., Wolfe, L. G., and Deinhardt, F. (1972). J. Natl. Cancer Inst. 48, 1499.
3. Melendez, L. V., Hunt, R. D., Daniel, M. D., Garcia, G., and Fraser, C. E. O. (1969). Lab. Anim. Care 19, 378.
4. Hunt, R. D., Melendez, L. V., King, N. W., Gilmore, C. E., Daniel, M. D., Williams, M. E., and Jones, T. C. (1970).

J. Natl. Cancer Inst. 44, 447.

5. Melendez, L. V., Hunt, R. D., Daniel, M. D., Fraser, C. E. O., Garcia, G. G., and Williamson, M. E. (1970) Int. J. Cancer 6, 431.
6. Hunt, R. D., Melendez, L. V., King, N. W., and Garcia, F. G. (1972). J. med. Prim. I, 114.
7. Rangan, R. S., Martin, L. N., Enright, F., and Abee, C. (1977). J. Natl. Cancer Inst. 59, 165.
8. Deinhardt, F. W., Falk, L. A., and Wolf, L. G. (1974). In "Advances in Cancer Research" Vol. 19 (G. Klein and S. Weinhouse, eds.), pp. 167-205. Academic Press, New York.
9. Fleckenstein, B., Bornkamm, G. W., and Ludwig, H. (1975). J. Virol. 15, 398.
10. Fleckenstein, B., and Wolf, H. (1974). Virology 58, 55.
11. Bornkamm, G. W., Delius, H., Fleckenstein, B., Werner, F.-J., and Mulder, C. (1976). J. Virol. 19, 154.
12. Schaffer, P. A., Falk, L. A., and Deinhardt, F. (1975). J. Natl. Cancer Inst. 55, 1242.
13. Falk, L. A., Wright, J., Deinhardt, F., Wolfe, L., Schaffer, P., and Benyesh-Melnick, M. (1976). Cancer Res. 36, 707.
14. Daniel, M. D., Fleckenstein, B., Silva, D., and Hunt, R. D. Manuscript in preparation (1978).
15. Falk, L. A., Wolfe, L. G., Marczyska, B., and Deinhardt, F. (1972). Bacteriol. Proc. 191.
16. Fleckenstein, B., and Bornkamm, G. W. (1976). In "Oncogenesis and Herpesviruses II" (G. de The, H. zur Hausen, and E. M. Epstein, eds.), pp. 145-149. IARC Press, Lyon.
17. Rabson, A. S., O'Conor, G. T., Lorenz, R. L., Kirschstein, F. Y., Legallais, F. Y., and Tralka, T. S. (1971). J. Natl. Cancer Inst. 46, 1011.
18. Fleckenstein, B., Muller, I., and Werner, J. (1977). Int. J. Cancer 19, 546.
19. Werner, F.-J., Bornkamm, G. W., and Fleckenstein, B. (1977). J. Virol. 22, 794.
20. Nonoyama, M., and Pagano, J. S. (1972). Nature (London) New Biol. 238, 169.
21. Tanaka, A., and Nonoyama, M. (1974). Proc. Natl. Acad. Sci. 71, 4658.
22. Lindahl, T., Adams, A., Bjursell, G., Bornkamm, G. W., Kaschka-Dierich, C., and Jehn, U. (1976). J. Mol. Biol. 102, 511.
23. Kaschka-Dierich, C., Adams, A., Lindahl, T., Bornkamm, G. W., Bjursell, G., Klein, G., Giovanella, B. C., and S. Singh. (1976). Nature (London) 260, 302.

INTEGRATION OF MOLONEY LEUKEMIA VIRUS INTO THE GENOME OF BALB/Mo MICE: GENETIC MAPPING OF THE VIRUS LOCUS (Mov-1) AND VIRUS GENE AMPLIFICATION DURING VIRUS EXPRESSION AND TRANSFORMATION

Rudolf Jaenisch, Michael Breindl, Johannes Doehmer*,
and Klaus Willecke*

Heinrich-Pette-Institut für experimentelle Virologie und Immunologie an der Universität Hamburg, Martinistrasse 52, 2000 Hamburg 20, Germany
and
*Institut für Zellbiologie, Universitäts-Klinikum Essen, Hufelandstrasse 55, 4300 Essen 1, Germany

ABSTRACT Mice genetically transmitting the exogenous Moloney leukemia virus (BALB/Mo) have been previously derived. These animals carried one copy of Moloney virus DNA (M-MuLV) in their germ line and were heterozygous for the M-MuLV locus.

Experiments were performed to investigate whether homozygosity at the M-MuLV locus would be compatible with normal development. Animals heterozygous for the M-MuLV locus were mated (♀ (+-) x ♂ (+-)) and the genotype of the offspring was analyzed. Molecular hybridization and genetic experiments identified three classes of offspring carrying two copies (++), one copy (+-) or no (--) M-MuLV-specific DNA sequence in their DNA. When mated with normal females, males of the first class (++) transmitted the virus to 100% of their offspring, males of the second class (+-) to 50% and offspring fathered by males of the third class (--) were virus negative. These results demonstrate that homozygosity at the M-MuLV locus has no detectable effect on normal development of the animals.

Homozygous BALB/Mo mice were used to genetically map the M-MuLV locus. Embryo fibroblasts were fused to established Chinese hamster cells and somatic cell hybrids were selected. Segregation of mouse chromosomal markers in the hybrids was correlated to the loss of M-MuLV-specific sequences as detected by molecular hybridization. Of 15 isozymes located on different mouse chromosomes only triosephosphate isomerase segregated syntenic with the M-MuLV gene, suggesting that the virus was integrated on chromosome No. 6. This was confirmed by sexual genetic experiments analyzing segregation of

ISBN 0-12-668350-6

specific copies was found in enlarged spleens, thymus or tumors of leukemic animals. The concentration of virus-specific RNA sequences rapidly increased with the age of the animals and reached almost maximal levels at 3-4 weeks of age.

These results suggest that somatic amplification of M-MuLV-specific DNA sequences occurs in two steps,one coincident with virus expression in preleukemic mice and a second step coincident with leukemic transformation.

DISCUSSION

Integration of exogenous RNA tumor virus sequences into the chromosomal DNA of infected cells has been demonstrated in many systems (1,30,31). The exogenous Moloney leukemia virus has been established as an endogenous virus in mice by infection at the 4-8 cell preimplantation stage and integration of viral specific information into single blastomeres (5). The subline of mice derived from such an infection transmitted the M-MuLV gene in only one chromosomal locus, and this virus gene was maintained in the colony by paternal transmission from heterozygous males mated with normal females. The integration of M-MuLV into the paternal chromosome complement had no detectable effect on normal embryonal and postnatal development of the animals. Further experiments were performed to determine whether homozygosity at the M-MuLV locus was compatible with normal development. Molecular hybridization and genetic experiments were used to identify homozygous offspring derived from matings of heterozygous parents (6; Table 1). The results indicated that homozygosity at the M-MuLV locus has no effect on the normal development of the mice. Furthermore, it was observed that the development of leukemia is not influenced by the genotype of the mice because mice homozygous or heterozygous for the M-MuLV locus or normal animals infected at birth with virus developed disease at similar rates. The latter animals carry M-MuLV-specific sequences only in their target tissue and never transmit the virus genetically (5,6).

We have previously shown that BALB/Mo mice express virus-specific RNA sequences or virus-specific proteins in target tissues only, whereas the virus gene remains repressed in non-target tissues (4). Similarly, a somatic virus gene amplification during the process of leukemogenesis is observed in target tissues only (5,21). It has been suggested that the observed tissue specificity of expression and transformation of different RNA tumor viruses may be due to different but virus-specific integration sites into the host chromosomes (22). As a first attempt to approach this question, we performed experiments to identify the mouse chromosome carrying the M-MuLV locus. It was of special interest to investigate whether M-MuLV was integrated into chromosome 15, since a trisomy of this chromosome has been shown to occur frequently in M-MuLV or AKR virus-induced lymphomas (23,24).

Somatic cell hybrids between BALB/Mo fibroblasts and hamster cells were analyzed for segregation of specific mouse chromosomes and M-MuLV-specific DNA sequences. Our preliminary experiments analyzing isozyme markers on 15 mouse chromosomes showed that only triosephosphate isomerase segregates syntenic with M-MuLV-specific sequences. The gene coding for this enzyme has been assigned recently to mouse chromosome No. 6 (20), and thus our experiments suggested that M-MuLV in BALB/Mo mice is located on this chromosome. This preliminary assignment was confirmed by sexual genetic experiments. We used two independently segregating markers, wa-1 on chromosome 6 and bt on chromosome 15, to study possible linkage of virus induction in BC-1 animals. The results in Table 4 show that M-MuLV was linked to the wa-1 marker and unlinked to the bt marker. The frequency of recombinants suggested that the M-MuLV locus is approximately 28 map units away from the wa-1 locus on chromosome No. 6. On the basis of these data we propose to call this gene <u>Mov-1</u> denoting Moloney virus in BALB/Mo mice.

The first structural gene comprised of the proviral DNA sequences of an endogenous virus, the AKR leukemia virus, was assigned to the <u>AKv-1</u> locus on chromosome No. 7 (12,13). Another, unlinked proviral genome in AKR mice, AKv-2, has not been localized yet. The similarities between the AKR virus-induced and the M-MuLV-induced disease should be emphasized here. Both AKR and Moloney virus induce a thymus-derived leukemia with a frequent chromosome 15 trisomy. In both cases a somatic amplification of virus-specific sequences is observed during leukemogenesis (5,21) and in both cases virus gene expression seems to be restricted to the target organs (4,25). The only two virus structural genes mapped so far which induce thymus-dependent leukemia, <u>Mov-1</u> and <u>AKv-1</u>, are clearly non-allelic. We consider two alternatives to explain the molecular events during virus-induced thymus-dependent leukemogenesis. In the first hypothesis, leukemia viruses can integrate at a few virus-specific chromosomal sites to induce tissue-specific virus expression and virus gene amplification, which is followed by transformation of specific target cells (22). Alternatively, the virus could integrate at one or a few out of a large number of possible integration sites. These original integration sites would not be specific for the virus, but the observed specificity of expression and transformation may depend on specific secondary integrations which occur during leukemogenesis. Only those lymphatic cells which carry additional virus sequences at transformation-specific sites might be selected for during the process of leukemogenesis.

In order to decide between these hypotheses, it will be necessary to develop new mouse lines which carry M-MuLV in their germ line and to map the integration sites. It also would be interesting to study if AKv-2 is integrated at a site homologous to Mov-1.

As another approach to answer these questions we have started to analyze the somatic amplification of M-MuLV-specific sequences during the early and late stages of virus expression. The results in Table 5 suggest two steps of amplification. The first step is observed in spleen and thymus of preleukemic animals up to 8 weeks of age and the second amplification step is observed in tumor tissues. The first amplification appears to be related to M-MuLV expression as molecular hybridization experiments have shown that spleens of 3-4-week-old animals express almost maximal levels of virus-specific RNA. The second transformation-specific amplification may or may not involve the integration of a recombinant virus isolated from lymphomas of AKR and BALB/Mo mice (26, 27). Preliminary experiments using Eco R-1 to digest DNA isolated from livers and from tumor tissues of BALB/Mo animals suggest that amplification is not in tandem to the Mov-1 sequences but involves new integration sites (28). Using these methods, it should be possible to decide whether the new integration of M-MuLV during virus expression and transformation occurs at specific or multiple sites in the mouse genome. Thus we may be able to define the specific step in the chain of molecular events which finally leads to virus-induced thymus-dependent leukemia.

ACKNOWLEDGMENTS

We thank Dr. John Bilello for critical comments and J. Dausman, D. Grotkopp, E. Otto, R. Lange, D. Schütz and R. Fesel for expert technical assistance. The studies of somatic cell hybridizations were initiated when J.D. and K.W. were at the Institute of Genetics, Köln. The work received the following financial support to R.J.: A research grant from the Deutsche Forschungsgemeinschaft and a research contract from the National Cancer Institute (NOI-CP-71008). The Heinrich-Pette-Institut is financially supported by Freie und Hansestadt Hamburg and by Bundesministerium für Jugend, Familie und Gesundheit, Bonn. Support to K.W. came from a research grant from the Ministerium für Forschung, Nordrhein-Westfalen.

REFERENCES

1. Todaro, G., Benveniste, R., Callahan, R., Lieber, M., and Sherr, C. (1974). Cold Spring Harbor Symp. Quant. Biol. 39, 1159.
2. Rowe, W. (1972). J. Exp. Med. 136, 1272.
3. Aaronson, S., and Stephenson, J. (1976). Biochim. biophys. Acta 458, 323.
4. Jaenisch, R., Fan, H., and Croker, B. (1975). Proc. Nat. Acad. Sci. 72, 4008.
5. Jaenisch, R. (1976). Proc. Nat. Acad. Sci. 73, 1260.
6. Jaenisch, R. (1977). Cell 12, 691.
7. Rowe, W., and Sato, H. (1973). Science 180, 640.
8. Gazdar, A., Oie, H., Lalley, P., Moss, W., Minna, J., and Francke, U. (1977). Cell 11, 949.
9. Lilly, F. (1970). J. Nat. Cancer Inst. 45, 163.
10. Lilly, F., and Pincus, Th. (1973). Adv. Cancer Res. 17, 231.
11. Lemons, R., O'Brien, S., and Sherr, C. (1977). Cell 12, 251.
12. Rowe, W., Hartley, J., and Bremner, T. (1972). Science 178, 860.
13. Chattopadhyay, S., Rowe, W., Teich, N., and Lowy, D. (1975). Proc. Nat. Acad. Sci. 72, 906.
14. Nichols, E., and Ruddle, F. (1973). J. Histochem. Cytochem. 21, 1066.
15. Davidson, R., O'Malley, K., and Wheeler, T. (1976). Somatic Cell Gen. 2, 271.
16. Breindl, M., Jaenisch, R., Doehmer, J., and Willecke, K. In preparation.
17. Davies, P., and Willecke, K. (1977). Molec. gen. Genet. 154, 191.
18. Kozak, C., Nichols, E., and Ruddle, F. (1975). Somatic Cell Gen. 1, 371.
19. Strand, M., August, T., and Croce, C. (1976). Virology 70, 545.
20. Leinwand, C., Kozak, C., and Ruddle, F. (1978). In press.
21. Berns, A., and Jaenisch, R. (1976). Proc. Nat. Acad. Sci. 73, 2448.
22. Jaenisch, R., and Berns, A. (1977). In "Concepts in Mammalian Embryogenesis" (M. Sherman, ed.), pp. 267-314. MIT Press, Cambridge and London.
23. Dofuku, R., Biedler, J., Spengler, B., and Old, C. (1975). Proc. Nat. Acad. Sci. 72, 1515.
24. Klein, G. Personal communication.
25. Strand, M., August, T., and Jaenisch, R. (1977). Virology 76, 886.

26. Hartley, J., Wolford, N., Old, L., and Rowe, W. (1977). Proc. Nat. Acad. Sci. 74, 789.
27. Vogt, M., and v. Griensven (1978). Submitted.
28. Berns, A. Personal communication.
29. Frankel, A., Gilbert, J., and Fischinger, P. (1977). J. Virol. 23, 492.
30. Varmus, H., Guntaka, R., Deng, C., and Bishop, M. (1974). Cold Spring Harbor Symp. Quant, Biol. 39, 987.
31. Padgett, T., Stubblefield, E., and Varmus, H. (1977). Cell 10, 649.

QUALITATIVE STUDIES OF THE ENDOGENOUS PROVIRUS IN THE CHICKEN GENOME

L. M. Souza and M. A. Baluda

UCLA School of Medicine and Molecular Biology Institute
Los Angeles, California 90024

ABSTRACT We have studied the organization of endogenous proviral DNA in the cellular genome of a few chicken lines. Chickens expressing two basic phenotypes for endogenous viral genes, nonvirus-producers (gs^-chf^- SPAFAS) and RAV-0 producers (lines 100 and 7 X 15) were used. The experimental approach consisted of: 1) digestion of cellular DNA with site specific endonucleases, 2) separation of fragments according to size by agarose gel electrophoresis, 3) DNA transfer to nitrocellulose paper by the method of Southern (1), and 4) autoradiographic detection of proviral sequences after hybridization with ^{125}I-labeled AMV or RAV-0 RNA. Analysis of the autoradiographs revealed that the endogenous proviral sequences are integrated in at least two different sites in the nuclear DNA of either virus-producing or non-producing cells. At least one of these sites is not the same in the two phenotypes. It was also found that some AMV sequences not related to RAV-0 are present in all the uninfected chicken lines studied.

INTRODUCTION

Endogenous proviral DNA has been detected in the nuclear DNA of all chicken cells and consists mostly of RAV-0 sequences (2,3,4). RAV-0 is spontaneously produced by cultured cells from some lines of chickens (5,6). There is at least 85% homology between RAV-0 and endogenous proviral DNA in the SPAFAS gs^-chf^- embryonic cells used in this study (7,8). Each haploid chicken cell genome also contains approximately 13.5 X 10^6 daltons of RAV-0 related sequences (9), but their composition and organization are not known.

Embryos from RAV-0-producing or nonproducing chickens contain approximately the same cellular concentration of proviral sequences in their nuclear DNA (10). The expression of these viral sequences varies greatly from one line of chickens to another (11) with the two most extreme cases seen in gs^-chf^- cells (virtually no viral gene expression) (11,12), and in RAV-0-producing cells (line 100 and line 7 X 15). Whether

ISBN 0-12-668350-6

structural gene defects or cellular inhibition are responsible for minimal viral gene expression in gs^-chf^- cells is not known. As a first approach to this problem, we have analyzed the topology of endogenous proviral DNA sequences in cellular DNA from virus-producing and nonproducing strains of chicken.

MATERIALS AND METHODS

Chicken Embryos. Two lines of SPAFAS gs^-chf^- (C/E and C/O) were kindly provided by Dr. W. Farrow (Life Sciences Research Laboratories, St. Petersburg, Florida). Line 100 and line 7 X 15 were kindly provided by Dr. L. Crittenden (U. S. Department of Agriculture, East Lansing, Michigan).

DNA Preparation. High molecular weight DNA (>3.0 X 10^7 daltons) was isolated either from cultured cells or from groups of six 16-day old chicken embryos by a modification of the technique as described (13).

RNAs. AMV was purified from the plasma of leukemic chickens kindly provided by Dr. J. Beard (Life Sciences Research Laboratories, St. Petersburg, Florida). RAV-0 produced by cultured cells from line 7 X 15 was concentrated and partially purified by continuous flow centrifugation in a Beckman CF-32 rotor. Both 70S and 35S RNAs were prepared as described previously (8). Ribosomal RNAs were a gift from E. Choi (UCLA).

Iodination. Viral and ribosomal RNAs were iodinated by a modified Commerford method (14) using 5 μg of RNA and one-to-three mCi of ^{125}I (Amersham Searle) in 20 μl of 0.1M Na Acetate, pH 4.3 and 0.8 mM $TlCl_3$. A minimum specific activity of 1-4 X 10^8 cpm per μg of iodinated RNA has been estimated assuming full recovery of the input RNA (5 μg).

Sucrose Velocity Sedimentation of ^{125}I RNA. ^{125}I-labeled RNA, either 70S or 35S was sedimented in a 50-20% sucrose velocity gradient (Figure 1) to select iodinated RNA with a sedimentation coefficient of 8S or greater. The RNA is fragmented during iodination and shows a peak at approximately 13S. This selection greatly reduces the background radioactivity in the autoradiographs as compared to unfractionated ^{125}I RNA. This procedure also can eliminate tRNA when 70S viral RNA was the starting material. ^{125}I RNA of 6-9S was also selected from such a gradient and the poly(A)-containing molecules were purified twice by oligo dT chromatography as described (15).

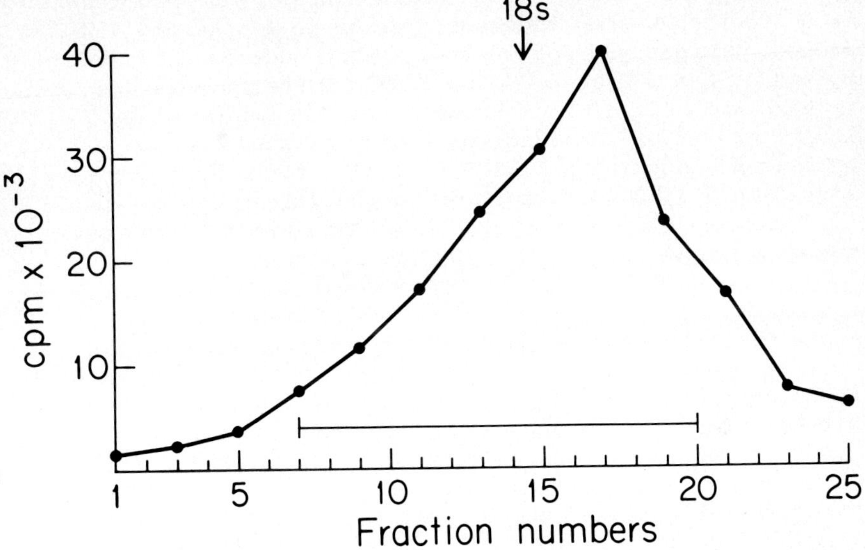

FIGURE 1. Sucrose Velocity Sedimentation of Iodinated 35S RNA. Bracket indicates fractions selected for hybridization.

Restriction Enzymes. Restriction endonuclease Hind III was purchased from Boehringer Mannheim and Eco Rl was a gift from M. Komaromy (UCLA). High molecular weight chicken DNA (20 μg per experiment) was digested with 30-60 units of enzyme for 3 hours at 37°C. Hind III and Eco Rl digests were done as previously described (16,17). Phage λ DNA (1 μg) was treated with approximately 1/20 as much enzyme under the same conditions as the chicken DNA.

Gels and Transfer. Either 0.7 or 1.2% horizontal agarose gels (Sigma Chemical) 1 cm thick with sample wells having 1 cm^2 surface area were run for 20-24 at a constant current of 40 ma. Each sample well contained either enzyme treated chicken DNA (20 μg) or λ DNA (1 μg). After gel electrophoresis, the DNA fragments were transferred by the Southern technique to nitrocellulose paper (1).

Hybridization and Autoradiography. Hybridization with ^{125}I-RNA was carried out in hybridization troughs as described by Southern (1) under the following conditions: Two filters in each trough were hybridized with approximately

0.1 μg of ^{125}I-RNA and 1.5 mg of mouse RNA in 3 ml of 4 X SSC containing 0.1% SDS at 68°C for 14 hours. Eight-to-ten filters are then washed together: twice in 1 liter of 4 X SSC at room temperature for 10 minutes and once in the same volume of 2 X SSC at 68°C for 1 hour. The filters are then treated with 20 μg/ml of RNase A and 20U /ml of RNase T_1 in 100 ml of 2 X SSC at 37°C for 1 hour, washed once in 1 liter of 2 X SSC containing 0.1% SDS at 37°C for 1 hour and finally rinsed in 2 X SSC before allowing the filters to air dry.

For autoradiography, Kodak X-Omat XR-5 film was presensitized and exposed at -70°C as reported (18).

Molecular Weight Determinations. Hind III treated λ DNA was electrophoresed in parallel with the enzyme treated chicken DNAs. The λ DNA track was cut out, stained with Ethidium bromide, illuminated with a UV light and photographed with a ruler next to the gel. The λ DNA bands served as standards to calculate the molecular weight of the bands which appeared in autoradiographs of the chicken DNA blots. Using this technique, the estimated molecular weight of bands ranging from 1.5 to 5.0 X 10^6 could vary by 5-10%, but the estimate for bands above or below those limits in a 0.7% agarose gel could be off by 25% or more. Thus, molecular weights should be interpreted accordingly.

To check whether an excess of heterologous bands might affect the mobility of fragments of specified size, the follow -ing control was carried out: Hind III digested λ DNA (1 μg) was either mixed with 20 μg of similarly treated chicken DNA or electrophoresed alone. After ethidium bromide staining, there was no apparent difference in the mobility of the λ DNA bands whether run alone or mixed with chicken DNA (data not shown).

RESULTS

Hybridization with AMV RNA Probes. AMV RNA was used initially because it is more readily available than RAV-0 RNA. Hybridization of ^{125}I-70S AMV RNA to Southern blots prepared from gs^-chf^- or line 100 DNA digested with either Eco R1 or Hind III revealed the band patterns seen in Figure 2A. Eco R1 treated gs^-chf^- and line 100 DNAs showed 10 and 8 bands, respectively. Six of the bands appear to have the same mobility in both DNAs (approximately 14.0, 5.5, 3.1, 2.4, 1.4 and 1.2 X 10^6 daltons. Four bands appear in the gs^-chf^- track but not in the line 100 track (20, 6.8, 4.2 and 1.0 X 10^6 daltons). Two bands which appear in the line 100 track are not present in the gs^-chf^- track (2.0 and 1.9 X 10^6 daltons).

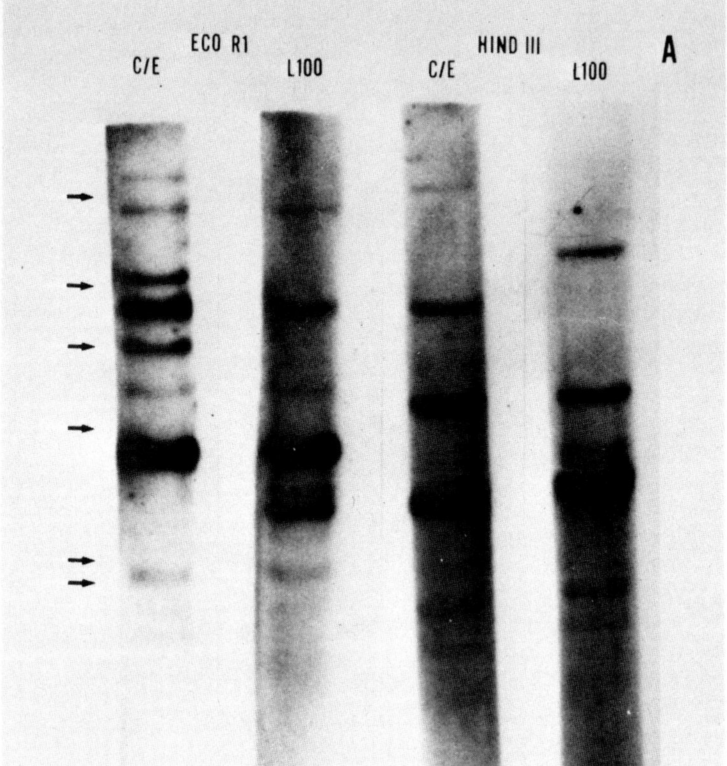

FIGURE 2A. C/E and C/O DNAs Extracted from Cultured Cells. After digestion with either Eco Rl or Hind III, the DNAs were electrophoresed in a 0.7% agarose gel and hybridized with ^{125}I-labeled 70S AMV-RNA.

After digestion with Hind III, both gs^-chf^- and line 100 DNAs showed six bands. A subset of four bands appear to be the same in both DNAs (3.2, 2.2, 1.2 and 1.0 X 10^6 daltons). Two bands in each DNA are different (18.0 and 5.7 X 10^6 daltons) for gs^-chf^- DNA and (8.0 and 2.5 X 10^6 daltons) for line 100 DNA.

Viral 70S RNA is known to be contaminated with other cellular RNA species (19,20,21,22). To determine whether or not some of the bands detected by autoradiography were the results of hybridization by contaminating RNA, twice purified 35S AMV RNA was used as the probe (Figure 2B). Some of the bands are less visible than in Figure 2A, but otherwise the patterns are identical. Thus, under our hybridization conditions, 70S and 35S viral RNA give the same autoradiographic

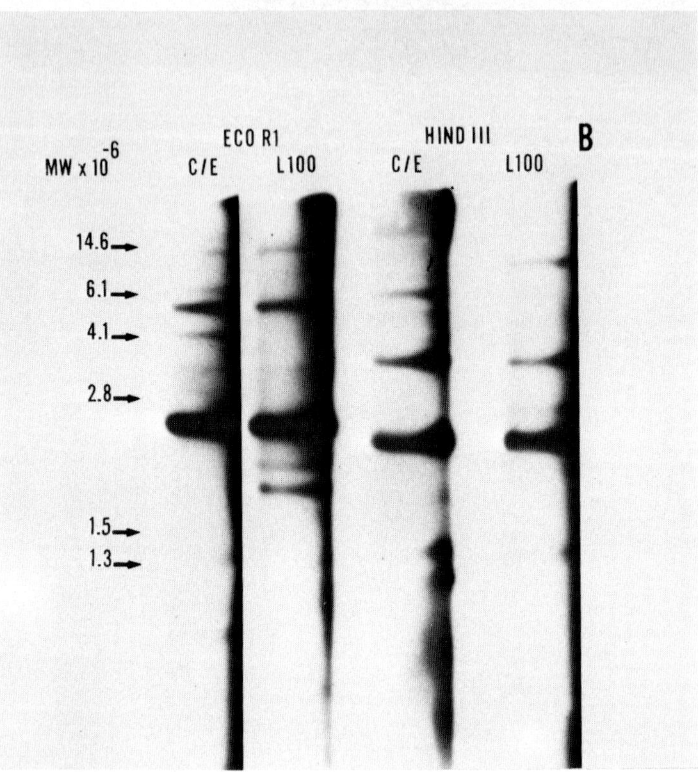

FIGURE 2B. After enzymatic digestion, the DNAs were run in a 1.2% agarose gel. The hybridization probe was ^{125}I 35S AMV RNA.

banding pattern.

An Eco Rl restriction endonuclease map of Prague-Rous sarcoma virus subgroup C (PR-RSV-C) made by Shank et al. (personal communication) shows that there are two internal cleavage sites which yield a proviral DNA fragment of approximately 2.5 X 10^6 daltons. If we assume these sites also exist in RAV-0 which has 80% homology to PR-RSV-C, then the 2.4 X 10^6 daltons band seen in Figures 2A and 2B probably consists entirely of viral information. It doesn't appear that the other Eco Rl restriction sites present in PR-RSV-C exist in RAV-0. Therefore, the other bands in the Eco Rl digest could represent fragments which contain 3' or 5' viral sequences attached to adjacent cellular sequences, viral genes not part of a complete provirus, or cellular sequences homologous to viral sequences. Thus, the endogenous provirus appears to be integrated in at least two different sites in the chicken genome.

To investigate which bands might contain DNA sequences homologous to the 3' end of the viral genome, 6-9S poly(A) selected ^{125}I AMV RNA served as the hybridization probe (Figure 3). A surprising number of bands appeared to contain viral 3' end sequences including the intense band at 2.2 X 10^6 daltons. One of the bands present in the line 100 Hind III digest, but absent in the gs$^-$chf$^-$ hybridized to the 3' end probe. This finding suggests that one provirus might be integrated at a different site in a virus-producing line.

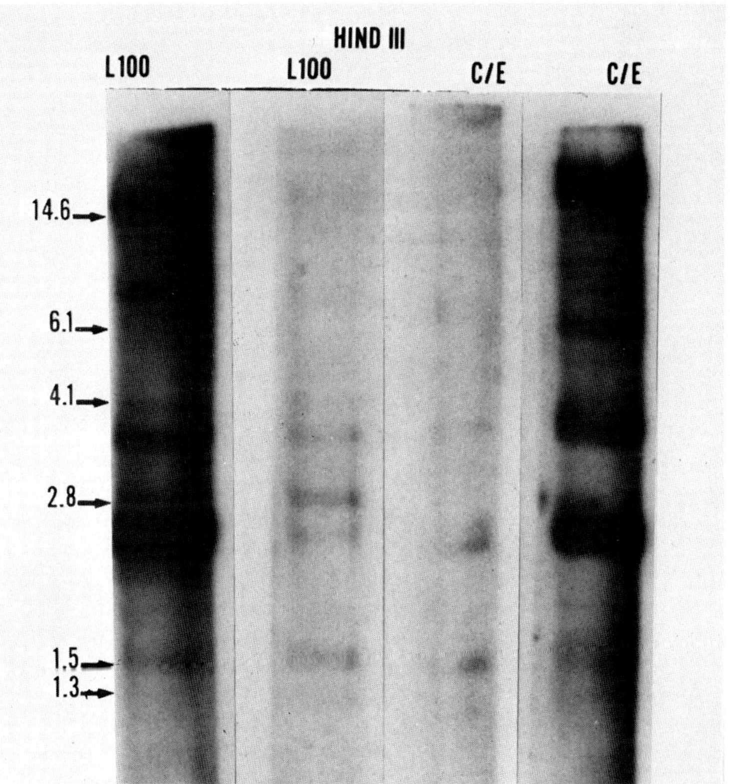

FIGURE 3. Either poly(A) selected 6-9S fragments, or total ^{125}I 70S AMV RNA was hybridized to gs$^-$chf$^-$ (C/E) or L100 Hind III digested DNA fragments. The second and third tracts show hybridization to 3' end RNA.

<u>Hybridization with a RAV-0 Probe</u>. A subset of bands seen in Eco Rl and Hind III digests with an AMV probe are also detected by hybridization with ^{125}I 70S RAV-0 RNA (Figure 4). A band of 1.3 X 10^6 daltons in Eco Rl treated gs^-chf^- (C/E and C/0 lines) DNA not apparent in the autoradiograph became visible after longer exposures. There is a readily distinguishable difference in the pattern of two bands between gs^-chf^- chickens and the two RAV-0-producing lines (100 and 7 X 15). Also, line 7 X 15 contains one more band than the other three lines.

In Hind III digests of the various chicken DNAs, at least one new band (7.0 X 10^6 daltons) is visible with the RAV-0 probe that was not apparent with an AMV probe. Thus, each viral probe seems capable of detecting some proviral sequences not detected by the other.

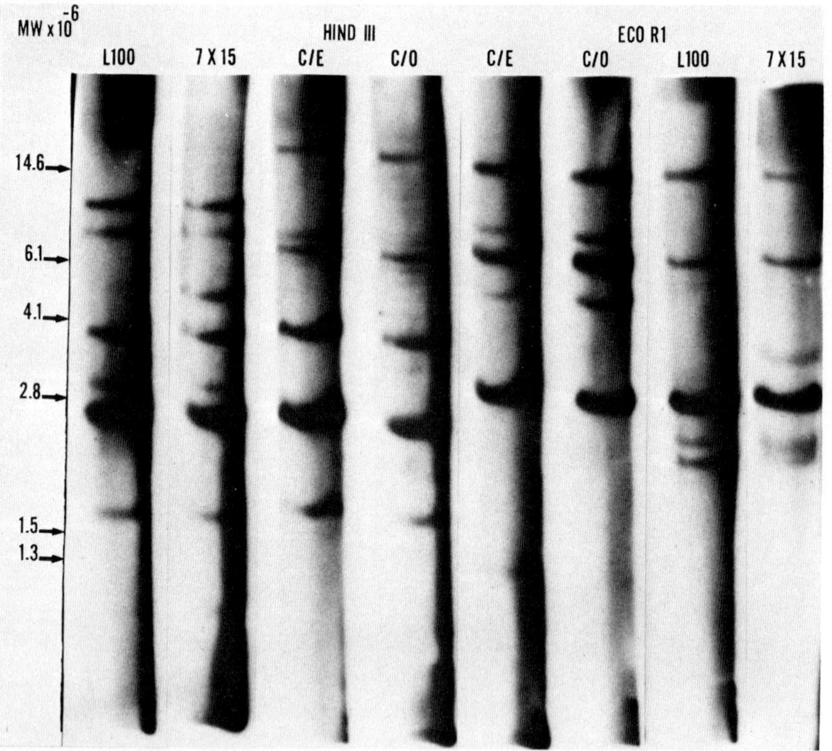

FIGURE 4. DNAs were extracted from a pool of six embryos each of C/E, C/0, and 7 X 15 chicken lines. Line 100 DNA was extracted from cultured cells. The DNAs were treated with Eco Rl or Hind III and analyzed using a ^{125}I 70S RAV-0-RNA probe.

Band Pattern of Chicken Ribosomal Genes. Autoradiographs obtained by hybridization of ^{125}I ribosomal RNA probes were analyzed for two reasons. First, there have been many reports of ribosomal RNA contamination in 70S viral RNA (20, 21,22) and we wanted to determine whether any of the bands generated by the viral RNA probes were due to contamination by ribosomal RNA. Secondly, the band pattern of chicken, murine and human ribosomal genes after digestion with Eco Rl has already been published (16,23), and this type of information can be used to determine whether the experimental DNA has been maximally digested by the restriction endonucleases. The ribosomal banding patterns generated by Eco Rl and Hind III treatment of chicken DNA are shown in Figure 5. It appears that slight contamination of the 28S RNA probe with 18S RNA causes the appearance of faint bands at approximately 14 X 10^6 daltons. Work already done indicates that the high molecular weight band in Eco Rl digested mouse DNA doesn't contain any 28S rRNA sequences (16). The 18S ribosomal probe hybridizes mostly to bands at approximately 14.0 and 5.5 X 10^6 daltons and to a few other intermediate bands in an Eco Rl digest. The 28S rRNA probe hybridizes mostly

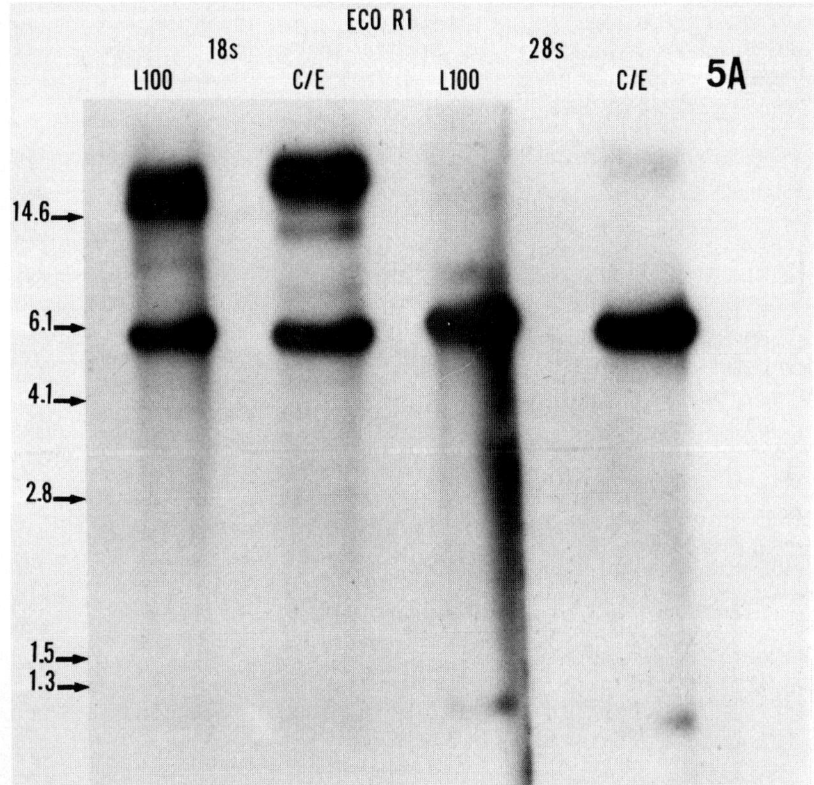

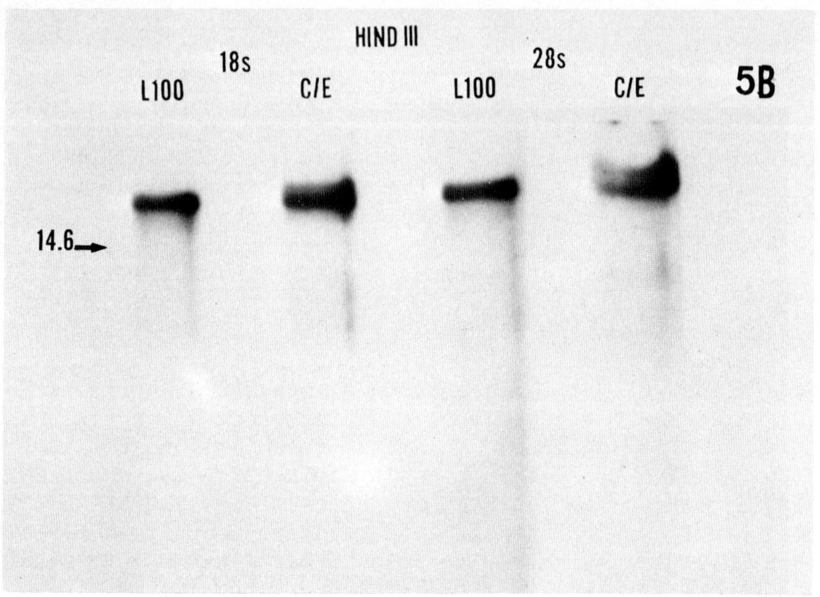

FIGURE 5A and B. ^{125}I-labeled 18S or 28S ribosomal RNAs were used as hybridization probes to detect the ribosomal banding pattern generated by either (A) Eco Rl or (B) Hind III digestion of C/E or line 100 DNA.

to the 5.5 X 10^6 daltons fragment and to a fragment at 1.3 X 10^6 daltons. Comparison of the 18S rRNA band patterns obtained with DNAs from gs$^-$chf$^-$ and line 100 chickens shows some similarities as well as some differences. The differences are probably due to heterogeneity in the nontranscribed spacers between ribosomal repeats. This phenomenon has been discussed (16), comparing ribosomal genes patterns in various mouse strains.

Three bands detected by AMV or RAV-0 ^{125}I RNA in Eco Rl treated gs$^-$chf$^-$ DNA (12.0, 5.0, and 1.3 X 10^6 daltons) are located at the same positions as bands detectable with ribosomal RNA probes. As shown in the following experiments, however, ribosomal RNA contamination can be ruled out in our band patterns due to the addition of an excess of mouse RNA in the hybridization mixture. In Hind III digested line 100 DNA, all the ribosomal genetic information is present in a fragment(s) of 20 X 10^6 daltons or larger, visible if mouse RNA is omitted from the hybridization mixture (Figure 5B). But, if mouse RNA is added with the iodinated viral RNA as in Figure 2A, this band disappears. Therefore, it appears that the bands seen in the Eco Rl digest of chicken DNA using viral RNA probes represent proviral DNA sequences homologous to the viral genomes.

DISCUSSION

The endogenous proviral sequences as detected by a RAV-0 probe, appear to be integrated in at least two separate locations. However, without knowledge of the restriction sites of Eco Rl or Hind III in the RAV-0 genome, we cannot determine the order of the fragments in the complete provirus, which sequences might represent partial genomes, nor which sequences might be cellular but homologous to viral genes. Both gs^-chf^- and RAV-0-producing chicken DNAs show a set of similar bands and a set of different bands. Since virus-producing cells should contain at least one complete provirus, the alteration in band pattern might be related to virus production.

The AMV probe selected for 3' end sequences hybridized to a relatively large number of the proviral DNA bands present in the Hind III digest. This includes the intense band at 2.2×10^6 daltons which we believe represents an internal fragment in the provirus. A similar finding also occurred with Eco Rl digested DNA (data not shown). There are four possible explanations for the observed results: i) the probe may still contain contaminating sequences from other parts of the viral genome, ii) the 3' end of AMV RNA, or of the integrated provirus, may be permuted, iii) a Hind III site may exist in the 3' end region of the provirus generating two fragments which contain homology to the probe, or iv) there may be sequences homologous to the 3' end of the RNA elsewhere in the viral genome. At present, we cannot rule out any of these alternatives.

Hind III treated line 100 DNA does not show any possible overlapping ribosomal and proviral DNA bands. Hind III digested gs^-chf^- DNA may have one band at the same position for ribosomal and viral sequences, but because of imprecise separation of fragments larger than 10×10^6 daltons this is not certain. After Eco Rl digestion, three bands in DNA from gs^-chf^- cells and two bands in DNA from line 100 cells appear at similar locations using either viral or ribosomal RNA probes. There may be two possible explanations: i) fragments of the same length could be fortuitously generated in both proviral and ribosomal genes with Eco Rl, or ii) viral and ribosomal genes may be closely linked and some enzyme generated fragments may contain both viral and ribosomal sequences.

There is at least one concentrated area of nonhomology between RAV-0 and AMV as demonstrated by the presence of a RAV-0 band not detected by AMV in a Hind III digest. Work done comparing the homology between RAV-0 and exogenous avian retroviruses show that the 3' proximal sequences contain most of the nonhomology between viruses (24). Therefore, it is reasonable to suppose that most of the nonhomology in AMV and RAV-0 may lie in their envelope genes. There are also AMV specific sequences in all the uninfected chicken DNA studied which were not detected with the RAV-0 probe. It is possible that one or more of these cellular sequences could contain information that represents a potential viral "luk gene". More specific probes are needed to locate and characterize the area(s) of nonhomology between RAV-0 and AMV RNA, as well as to understand the nature of the AMV specific sequences detectable in normal chicken cells.

ACKNOWLEDGMENTS

We would like to thank Dr. S. Hughes for technical advice. This investigation was supported by USPHS research grant CA-10197 and by National Research Service Award CA-09056 from the National Cancer Institute.

REFERENCES

1. Southern, E. M. (1975). J. Mol. Biol. 98: 503.
2. Baluda, M. A. (1972). Proc. Nat. Acad. Sci. USA 69: 576.
3. Wright, S. E. and Neiman, P. E. (1974). Biochemistry 13: 1549.
4. Shoyab, M. and Baluda, M. A. (1976). J. Virology 17: 106.
5. Vogt, P. K. and Friis, R. R. (1971). Virology 43: 223.
6. Crittenden, L. B., Smith, E. J., Weiss, R. A. and Sarma, P. S. (1974). Virology 57: 128.
7. Neiman, P. E. (1973). Virology 53: 196.
8. Shoyab, M. and Baluda, M. A. (1975). J. Virology 16: 1492.
9. Chen, C. Y. and Baluda, M. A. (this publication).
10. Hefti, E. and Baluda, M. A. (1978) Virology, in press.
11. Wang, S. Y., Hayward, W. S. and Hanafusa, H. (1977). J. Virology 24: 64.
12. Hayward, W. S. and Hanafusa, H. (1976). Proc. Nat. Acad. Sci. USA 73: 2259.
13. Gross-Bellard, M., Oudet, R. and Chambon, P. (1973). Eur. J. Biochem. 36: 32.

14. Commerford, S. L. (1971). Biochemistry 10: 1993.
15. Bantle, J. A., Maxwell, I. A. and Hahn, W. E. (1976). Ana. Biochem. 72: 413.
16. Arnheim, N. and Southern, E. M. (1977). Cell 11: 363.
17. Polisky, B., Green, P., Garfin, D. E., McCarthy, B. J., Goodman, H. M. and Helling, R. B. (1974). Proc. Nat. Acad. Sci. USA 72: 3310.
18. Laskey, R. A. and Mills, A. D. (1975). Eur. J. Biochem. 56: 335.
19. Ikawa, Y. J., Ross, J. and Leder, P. (1974). Proc. Nat. Acad. Sci. USA 71: 1154.
20. Bonar, R. A., Sverak, L., Bolognesi, D. P., Langlois, A. J., Beard, D. and Beard, J. W. (1967). Cancer Res. 27: 1138.
21. Obara, T., Bolognesi, D. P. and Bauer, H. (1971). Int. J. Cancer 7: 535.
22. Fidanian, H. M., Drohan, W. N. and Baluda, M. A. (1975). J. Vir. 15: 449.
23. McClements, W. and Skalka, A. M. (1977). Science 196: 195.
24. Neiman, P. E., Das, S., McDonnell, D. and McMillin-Helsel, L. (1977). Cell 11: 321.

QUANTITATIVE STUDIES ON INTEGRATED PROVIRAL DNA IN NORMAL AND SCHMIDT RUPPIN-ROUS SARCOMA VIRUS TRANSFORMED CHICKEN CELLS

C. Y. Chen and M. A. Baluda

UCLA School of Medicine & Molecular Biology Institute
Los Angeles, California 90024

ABSTRACT We have quantitated endogenous proviral DNA in uninfected gs$^-$chf$^-$ chicken embryonic fibroblasts (CEF) and the additional exogenous proviral DNA in transformed CEF infected with Schmidt-Ruppin Rous sarcoma virus subgroup A (SR-RSV-A). The concentration of proviral DNA integrated per haploid cell genome was calculated from the maximal concentration of H^3-labeled 35S viral RNA that hybridized to DNA from normal or SR-RSV-A transformed chicken cells. Noninfected cells were shown to contain 13.5 X 10^6 daltons of double-stranded proviral DNA (approximately two copy equivalents of the RAV-0 genome) per haploid genome. In SR-RSV-A transformed cells only one exogenous provirus was integrated in addition to the endogenous sequences. We have also analyzed the reassociation kinetics of proviral DNA in randomly sheared cellular DNA fragments of specified size, 2.6 X 10^6 or 4.2 X 10^6 daltons (double-stranded) to determine the reiteration frequency of the cellular DNA sequences adjacent to the integrated SR-RSV-A provirus. The reassociation of viral sequences was followed by monitoring the proviral DNA that remained single-stranded after increasing time intervals of DNA renaturation. The SR-RSV-A specific proviral DNA sequences, i.e., sequences not shared with RAV-0, were titrated by hybridization with H^3 -labeled SR-RSV-A RNA after saturation with unlabeled RAV-0 RNA. The results indicate that in transformed cells, the single copy of SR-RSV-A provirus is integrated between cellular DNA sequences reiterated approximately 1000 times and there is at least one set of reiterated cellular sequences between the exogenous and endogenous proviruses.

INTRODUCTION

Infection of chicken cells with avian retroviruses results in the integration of a limited number of exogenous proviruses in the cell genome (1-5). The small number of exogenous proviruses which can integrate is independent of the

ISBN 0-12-668350-6

analyzed the renaturation kinetics of 250 base pairs fragments of nonrepetitive chicken DNA and E. coli DNA under identical conditions. The normal gs^-chf^- chicken cells used in this study contain 13.5 X 10^6 daltons of double-stranded proviral DNA per haploid cellular genome. The same value was obtained using either one of the four viral RNA probes after correcting for the homology that exists between RAV-0 and the other three viruses, i.e., 70% for AMV and 78% for the non- src portion of SR-RSV-A, determined independently from hybridization competition between the viral RNAs in DNA excess (17; also Chen & Baluda, unpublished data). It appears therefore that the uninfected gs^-chf^- C/E chicken cells used in this study contain approximately two copy equivalents of the RAV-0 genome per haploid cell genome.

Similar hybridization experiments were carried out with DNA from SR-RSV-A transformed CEF and the same H^3-labeled 35S viral RNAs, except AMV RNA, were used. The data shown in Figure 2 and Table I demonstrate that SR-RSV-A transformed cells contain only one integrated exogenous provirus in addition to the endogenous proviral DNA.

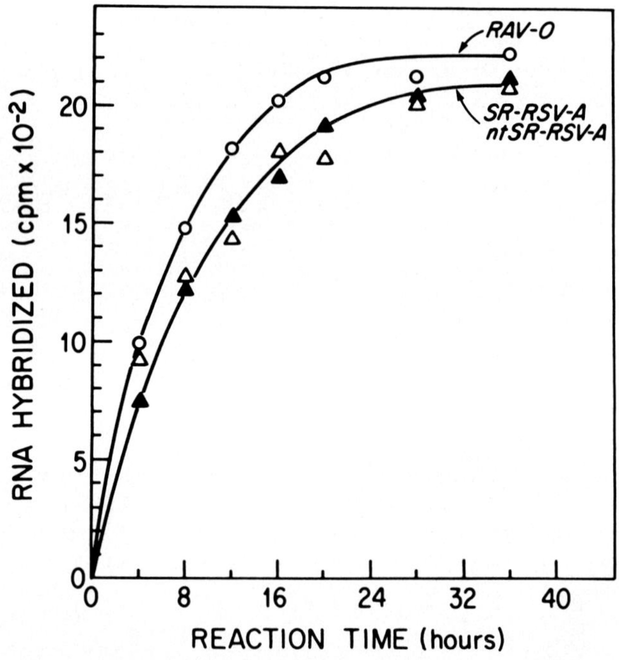

FIGURE 2. Time-dependent saturation Curves of Proviral DNA in SR-RSV-A Transformed Cells. DNA was extracted from the Hirt pellet of transformed CEF 3 weeks or longer after infection with freshly cloned SR-RSV-A. Hybridization was carried out as described in Figure 1.

Size and Homogeneity of DNA Fragments to be Used in Reassociation Kinetics Analysis of Proviral DNA. DNA fragments with a sedimentation coefficient of 15S or 18S were selected from the preparative DNA velocity gradients described in Materials and Methods. The size uniformity of the fragments was analyzed by alkaline sucrose gradient velocity sedimentation using H^3-labeled M13 circular single-stranded DNA as 18S marker. The larger fragments showed a relatively narrow sedimentation profile (Figure 3A) indicating a fairly uniform size and had a median sedimentation coefficient of 18S. This corresponds to single-stranded linear DNA molecules of approximately 2.1×10^6 daltons. The smaller fragments (Figure 3B) showed a sedimentation profile that had the same distribution as the M13 marker indicating a uniform size and had a median sedimentation coefficient of 15S corresponding to single-stranded linear molecules of 1.3×10^6 daltons.

Proviral DNA Reassociation Kinetics. Only approximately 10% of the genome of our White Leghorn chickens consists of repetitive DNA sequences interspersed with nonrepetitive sequences and these sequences are much longer than in other higher eukaryotes tested (Baluda, unpublished results; 22). On the average, the length of the repeated sequences is 4000 bases and that of the unique sequences 36,000 bases. Because of this arrangement, it is possible to have relatively long DNA fragments without repeated sequences. Also, the viral genome appears to contain only two terminal redundant sequences (23-25). Therefore, if cellular DNA fragments of appropriate size and smaller than the viral genome are randomly generated, viral sequences will be present with adjacent cellular sequences on some fragments or alone on other fragments. The reassociation rate of a given fragment depends upon the repetitiveness of some of its sequences in the chicken DNA. Since the viral sequences are of low reiteration frequency, only viral sequences on fragments with reiterated cellular sequences will reassociate at an accelerated rate. The reassociation of viral sequences was followed by monitoring the viral DNA that remained single-stranded after increasing time intervals of DNA renaturation. The single-stranded DNA fragments were separated from renatured or partially renatured fragments by hydroxylapatite fractionation. Viral DNA sequences in the single-stranded fractions were quantitated by filter hybridization in an excess of H^3-labeled 35S RNA from SR-RSV-A after prehybridization in an excess of either yeast RNA or 70S RAV-0 RNA (18). Prehybridization with yeast RNA permits the detection of endogenous plus SR-RSV-A proviral DNA whereas prehybridization with RAV-0 RNA allows the detection of SR-RSV-A specific sequences only.

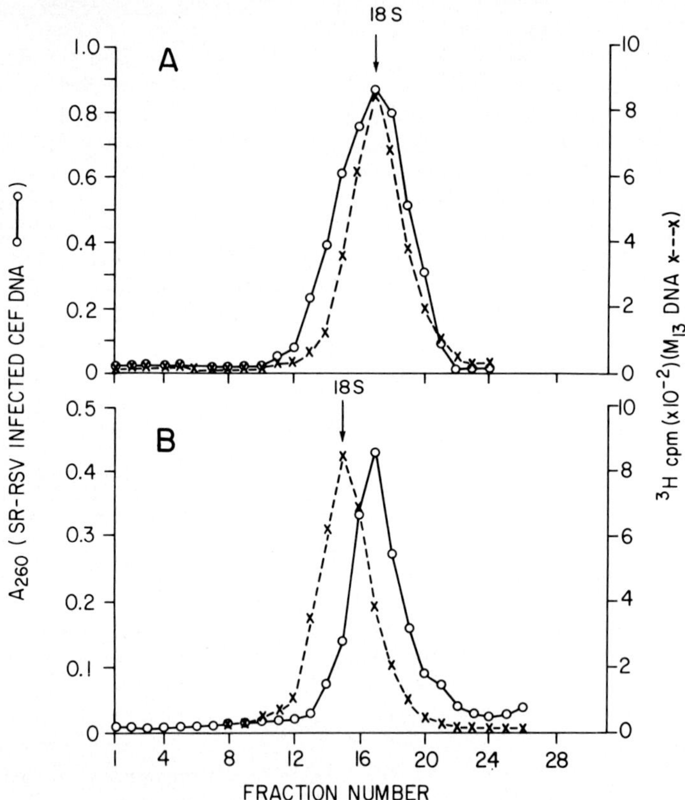

FIGURE 3. Alkaline Sucrose Velocity Sedimentation Profile of Size Selected DNA Fragments from SR-RSV-A Transformed CEF. DNA fragments with a sedimentation coefficient of 15S or 18S were obtained from a preparative DNA gradient and their size distribution was checked by alkaline sucrose velocity sedimentation. H^3-labeled M13 DNA, kindly given by Dr. Dan Ray, UCLA, served as sedimentation marker (18S). The DNA was dissolved in 0.2 ml of 0.3N NaOH, 0.5M-NaCl, and 0.01M EDTA, and layered on top of 4.8 ml of a linear 5-20% sucrose gradient containing 0.3N NaOH, 0.5M NaCl, and 0.01M EDTA. A) The 18S fragments were sedimented in a SW-65 rotor at $4^{o}C$ at 60,000 rpm for 2 hours. B) The 15S fragments were centrifuged in the same manner for 3 hours. Fractions were collected from the bottom of the tube, diluted to 0.7 ml with distilled water, and absorbance at 260 nm was measured. After neutralization, H^3-radioactivity was measured. o, Absorbance profile of transformed CEF DNA; x, Radioactivity profile of H^3-labeled M13 DNA.

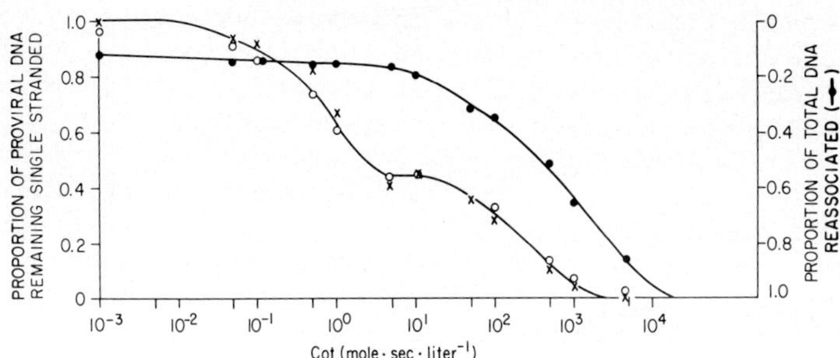

FIGURE 4. Reassociation Kinetics of Denatured Cellular and Proviral DNA Fragments of 1.3 X 10^6 Daltons. After denaturation at 100°, the 15S fragments were allowed to reassociate at 60° to different Cot (mol. sec. liter^{-1}) values as previously described (14). For Cot values below 10^0, the DNA concentration was 10-100 μg/ml in 0.05M phosphate buffer, pH 6.8 and for Cot values above 10^0 the DNA concentration was 200 μg/ml in 0.4M phosphate buffer. Data are expressed in equivalent Cots standardized to a monovalent cation concentration of 0.18M (21). Single-stranded DNA was separated from double-stranded and partially double-stranded DNA by hydroxylapatite column chromatography (14). The DNA fragments remaining single-stranded at different Cot values were dialyzed against 0.1 X SSC, denatured and immobilized on nitrocellulose filters (12). Four filters were made with single-stranded DNA at each Cot value. Two filters were prehybridized with 15 μg/ml of yeast RNA and the other two filters were prehybridized with 15 μg/ml of 70S RAV-0 RNA. This was followed by a second hybridization with H^3-labeled 35S SR-RSV-A RNA. Unfractionated DNA from SR-RSV-A transformed CEF was hybridized in the same manner to determine the concentration of proviral DNA in total cellular DNA. The hybridization conditions were described in Materials and Methods. Mouse embryonic DNA served as control for non-specific binding of H^3-labeled SR-RSV-A RNA. The renaturation of cellular DNA (●——●) was also monitored and used to make appropriate corrections for proviral DNA enrichment; (o——o), Reassociation of endogenous and SR-RSV proviral DNA; (x——x), Reassociation of SR-RSV-A specific proviral sequences.

Figure 4 shows the renaturation profile of single-stranded DNA fragments of 1.3 X 10^6 daltons from SR-RSV-A transformed cells. Of the cellular DNA, 15% renatures as repetitive and the rest as unique with a $Cot_{1/2}$ of approximately 650. The renaturation profiles of total and SR-RSV-A specific proviral DNAs are similar with accelerated renaturation starting at Cot 10^{-1} and reaching a transient plateau at 50-60% renaturation between Cot 2 and Cot 50. Beyond Cot 50, there is another reassociation phase which approaches completion at Cot 5 X 10^3. Since there is only one SR-RSV-A provirus integrated per haploid genome, the observed rapid renaturation of approximately 55% of the proviral DNA must be caused by its covalent linkage to reiterated cell sequences adjacent to both ends.

Figure 5 shows the renaturation profile of single-stranded DNA fragments of 2.1 X 10^6 daltons. Approximately 20% of the cellular DNA renatures as repetitive. The unique DNA has an apparent $Cot_{1/2}$ of about 120. The fraction of total and SR-RSV-A specific proviral DNAs which renatures rapidly is larger than that observed with the smaller fragments. A transient plateau at 70-80% renaturation is reached between Cot 1 and Cot 50. A later renaturation phase approaches completion at Cot 5000. These data indicate that the endogenous as well as SR-RSV-A proviral DNAs are integrated as double-stranded pieces equal to or longer than 4.2 X 10^6 daltons and that the adjacent cellular sequences appear to be reiterated 1000-2000 times.

As an additional check of the reassociation kinetics tech-nique, the concentration of proviral DNA was measured in the double-stranded and partially double-stranded molecules present in the plateau region, i.e., between Cots 1 and 50, of Figures 4 and 5. The data shown in Table 2 indicate that all the proviral DNA can be accounted for in the single-stranded and double-stranded hydroxylapatite fractions with a recovery of 92-104%.

DISCUSSION

The gs^-chf^- White Leghorn chickens studied here appear to contain 13.5 X 10^6 daltons of double-stranded endogenous proviral DNA integrated per haploid cell genome. Although there is a certain degree of uncertainty caused by potential errors in measurements of DNA content per filter and H^3-RNA specific activity, this value has been reproducible in several experiments and with different viral RNA probes. Other lines of White Leghorn chickens including spontaneous RAV-0 producers such as lines 100 or 7 X 15 also appear to contain

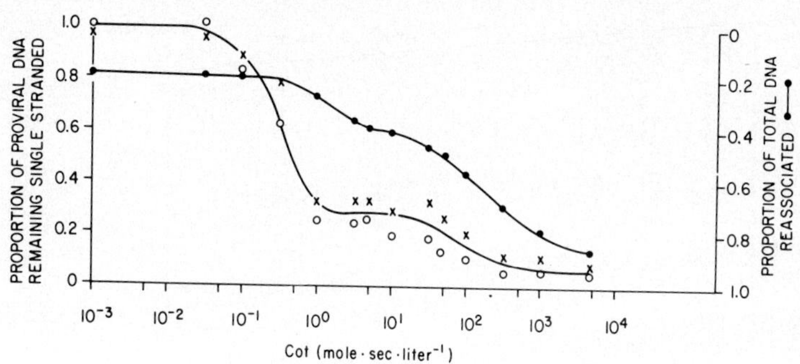

FIGURE 5. Reassociation Kinetics of Denatured Cellular and Proviral DNA Fragments of 2.1 X 10^6 Daltons. The experimental conditions were similar to those described in Figure 4. (●——●), Renaturation of cellular DNA; (o——o), Reassociation of endogenous and SR-RSV proviral DNA; (x x), Reassociation of SR-RSV-A specific proviral sequences.

a similar concentration of endogenous sequences (10; Hefti & Baluda, unpublished). This corresponds to at least two copy equivalents of the RAV-0 genome per haploid cell genome assuming the molecular weight of the RAV-0 single-stranded RNA genome to be 3 X 10^6. At lesst 85% of the RAV-0 genome is present in uninfected chicken cells (15,16) and nearly all the endogenous proviral DNA appears to be homologous to RAV-0 (16,17,26). Therefore, at least one complete, or nearly complete, copy of the RAV-0 genome must be present but several possibilities exist, e.g., one entire copy plus several partial copies or two full copies plus homologous cellular sequences. A choice among these alternatives must await qualitative analysis of proviral DNA fragments generated by restriction endonucleases. Some light will be shed on this subject in the accompanying paper (Souza & Baluda).

The quantitation of exogenous proviral DNA in SR-RSV-A transformed cells is more straightforward because i) the hybridization to endogenous sequences and putative cellular src sequences in uninfected cell DNA can be subtracted through the use of nt-SR-RSV-A and SR-RSV-A; ii) only an average of 1.1 copy equivalent of SR-RSV-A was integrated in transformed cells, and iii) all cells from which DNA was extracted were transformed. It appears therefore unlikely

TABLE II

TITRATION OF PROVIRAL DNA IN DOUBLE-STRANDED HYDROXYLAPATITE FRACTION

DNA Size	Cot	% DNA Renatured	Total Proviral DNA[a]			SR-RSV Specific Proviral DNA[b]		
			cpm hybridized per 100 μg DNA	% in DS fraction	% in SS fraction[c]	cpm hybridized per 100 μg DNA	% in DS fraction	% in SS fraction[c]
15S	1	15.8	1492 ± 117[d]	32	66	657 ± 89	42	61
	5	17.7	2217 ± 59	53	41	720 ± 91	52	44
	10	18.9	1902 ± 26	48	45	611 ± 49	47	45
	50	31.4	1490 ± 126	63	37	464 ± 44	59	37
18S	1	26.4	1869 ± 52	67	32	637 ± 49	69	26
	3	36.1	1433 ± 54	70	33	496 ± 35	73	25
	5	39.0	1250 ± 8	66	33	449 ± 11	71	26
	10	40.7	1272 ± 36	70	28	450 ± 74	75	19
	30	46.3	1046 ± 32	65	33	409 ± 34	77	19
	50	50.0	1147 ± 70	77	26	402 ± 35	82	14
Unfractionated			742 ± 31			245 ± 23		

a Filters prehybridized with 15 μg/ml yeast RNA. Total proviral DNA = endogenous plus SR-RSV-A. DS = double-stranded; SS = single-stranded.

b Filters prehybridized with 15 μg/ml 70S RAV-0 RNA.

c Figures obtained from Figure 4 and Figure 5.

d Mean of 2 or 3 filters ± standard deviation.

that some cells did not contain exogenous proviral DNA and that others contained more than one copy. Nevertheless, to eliminate this possibility individual clones of transformed cells have been prepared but, unfortunately, their proviral DNA content has not yet been determined. The presence of only one integrated exogenous provirus does not necessarily indicate that the integration site is, or is not, specific. Again, the answer must await analysis of proviral DNA fragments generated by restriction endonucleases from different clones of transformed cells. However, if the integration is not specific, it would appear that there is some mechanism for provirus induced inhibition of integration. Another possibility might be that transformed cells with only one integrated exogenous provirus have a selective growth advantage.

The integration of endogenous proviral DNA between repetitive cellular DNA sequences (18) allows for a rough estimate of the size of the integrated pieces, and for testing whether exogenous proviruses are similarly integrated. Unfortunately, the fact that two, or slightly more, copy equivalents of RAV-0 DNA are present per haploid cell genome and our lack of knowledge about its chemical complexity and arrangement in the cellular genome make models tenuous. Nevertheless, making the reasonable assumption that the endogenous proviral DNA is integrated in 2-4 large pieces, the reassociation kinetics of proviral DNA fit best a model in which the SR-RSV-A provirus is singly integrated between cellular sequences repeated 1000-2000 times. The endogenous proviral DNA could be in close proximity or distant. Thus, SR-RSV-A appears to integrate at a location different from that of AMV which seems to integrate very close to endogenous proviral DNA and without interjacent repetitive cellular sequences. This is not unexpected since AMV and SR-RSV-A have in their RNA genome different redundant terminal sequences which may be involved in the recognition of specific attachment sites in cellular DNA (23,27,28).

ACKNOWLEDGEMENTS

This work was supported by USPHS grant CA-10197 from the National·Cancer Institute.

REFERENCES

1. Baluda, M. A. and Nayak, D. P. (1970). Proc. Nat. Acad. Sci. USA 66: 329.
2. Rosenthal, P. N., Robinson, H. L., Robinson, W. S.,

Hanafusa, T. and Hanafusa, H. (1971). Proc. Nat. Acad. Sci. USA 68: 2336.

3. Neiman, P. E. (1972). Science 178: 750.
4. Shoyab, M., Evans, R. M. and Baluda, M. A. (1974). J. Virol. 14: 47.
5. Varmus, H. E., Heasley, S. and Bishop, J. M. (1974). J. Virol. 14: 895.
6. Ali, M. and Baluda, M. A. (1974). J. Virol. 13: 1005.
7. Baluda, M. A., Shoyab, M., Ali, M., Markham, P. D. and Drohan, W. N. (1976), p. 311 in R. Neth (ed.) Modern Trends in Human Leukemia, J. F. Lehmans, Munich.
8. Khoury, A. T. and Hanafusa, H. (1976). J. Virol. 18:383.
9. Tsuruo, T. and Baluda, M. A. (1977). J. Virol. 23: 533.
10. Baluda, M. A. (1972). Proc. Nat. Acad. Sci. USA 69: 576.
11. Markham, P. D. and Baluda, M. A. (1973). J. Virol. 12: 721.
12. Neiman, P. E. (1973). Virology 53: 196.
13. Vogt, P. K. and Friis, R. R. (1971). Virology 43: 223.
14. Weiss, R. A., Friis, R. R., Katz, E. and Vogt, P. K. (1971). Virology 46: 920.
15. Shoyab, M. and Baluda, M. A. (1975). J. Virol. 16: 1492.
16. Wright, S. E. and Neiman, P. E. (1974). Biochemistry 13: 1549.
17. Shoyab, M. and Baluda, M. A. (1976). J. Virol. 17: 106.
18. Evans, R. M., Baluda, M. A. and Shoyab, M. (1974). Proc. Nat. Acad. Sci. USA 71: 3152.
19. Shoyab, M., Dastoor, M. N. and Baluda, M. A. (1976). Proc. Nat. Acad. Sci. USA 73: 1749.
20. Vogt, P. K. (1969). In Fundamental Techniques in Virology, Acad. Press, New York, p. 198.
21. Hirt, B. (1967). J. Mol. Biol. 26: 365.
22. Colbert, D. A., Edwards, K. and Coleman, J. R. (1976). Differentiation 5: 91.
23. Haseltine, W. A., Maxam, A. M. and Gilbert, W. (1977). Proc. Nat. Acad. Sci. USA 74: 989.
24. Schwartz, D. E., Zamecznik, P. C. and Weith, H. L. (1977). Proc. Nat. Acad. Sci. USA 74: 994.
25. Evans, R. M., Shoyab, M., Drohan, W. N. and Baluda, M. A. (1977). J. Virol. 21: 942.
26. Neiman, P. E., Purchase, H. G. and Okazaki, W. (1975). Cell 4: 311.
27. Stoll, E., Billeter, M. A., Palmenberg, A. and Weissmann, C. (1977). Cell 12: 57.
28. Wang, L. H., Duesberg, P. H., Robins, T., Yokota, H. and Vogt, P. K. (1977). Virology 82: 472.

DEFINING TRANSFORMING (onc) GENES AND GENE PRODUCTS OF AVIAN ACUTE LEUKEMIA AND CARCINOMA VIRUSES

Peter Duesberg, Pamela Mellon, Tony Pawson,[*]
G. S. Martin[*], Klaus Bister[†], and Peter Vogt[†]

Department of Molecular Biology, Virus Laboratory,
and Department of Zoology,
University of California, Berkeley, California 94720,[*]
and Department of Microbiology,[†]
University of Southern California, School of Medicine,
Los Angeles, California 90033

ABSTRACT The RNA species of the defective avian acute leukemia virus MC29 and of the defective avian carcinoma virus MH2 were compared to the RNAs of other avian tumor viruses by gel electrophoresis, fingerprinting of RNase T_1-resistant oligonucleotides, RNA-cDNA hybridization and in vitro translation. Electrophoresis and fingerprint analyses indicated that each virus (MC29 and MH2) contained a specific 28S RNA species of 5700 nucleotides. Hybridization of viral RNA with cDNA transcribed from other avian tumor viruses showed that each 28S RNA contained 1500-2000 nucleotides of specific sequences and 3500-4000 nucleotides of sequences shared with other avian tumor viruses. Src gene-related sequences of Rous sarcoma virus were not found in MC29 or MH2 RNA. The specific sequences of MC29 RNA were mapped between 3000 and 4500 nucleotides from the 3' poly(A) end of the RNA. This was accomplished by ordering the oligonucleotides which represent specific sequences relative to the poly(A) of the RNA end using fingerprint analyses of poly(A)-tagged RNA fragments. In vitro translation of MC29 RNA generated a major 110,000 dalton protein and minor 49,000 and 31,000 proteins. The 110,000 dalton protein shared protein sequences with the gene products of the avian tumor viral gag gene, which maps at the 5' end of independently replicating viruses. Since a gag gene-related oligonucleotide is also found near the 5' end of MC29 RNA, which does contain a defective gag gene, we propose that the 110,000 MC29 protein is translated from the 5' end of MC29 RNA and includes sequences of the defective gag gene as well as MC29 specific sequences. A possible interpretation of our data suggests that the onc gene of MC29 (and MH2) includes specific RNA sequences and is translated into a polyprotein which is initiated at the defective gag gene.

ISBN 0-12-668350-6

INTRODUCTION

Different Transforming (*onc*) genes in RNA Tumor Viruses. Different members of the avian RNA tumor virus family may cause specific forms of cancer in the same animal. For example Rous sarcoma virus causes mostly sarcomas; avian myelocytoma virus, MC29, causes mostly acute leukemias and avian carcinoma virus, MH2, causes mostly carcinomas (1,2,3,4,5). These viruses transform cells with a unique gene, termed *onc* (short for oncogenicity). The *onc* gene is not necessary for the virus to replicate. Consequently some RNA tumor viruses lack an *onc* gene altogether. In order to replicate RNA tumor viruses need three other genes: *gag*, coding for internal structural proteins, *env*, coding for envelope glycoproteins and *pol*, coding for viral DNA polymerase (6).

We ask here whether viruses causing specific viral cancers carry specific *onc* genes which might account for their disease specificity. Indirect evidence suggests that the *onc* genes of different transforming viruses of the same taxonomic group may differ from each other because (i) the *src* genes of avian sarcoma viruses are not found in avian lymphatic leukemia viruses and transformation-defective (td) viruses (8,9,10,12) or in the defective, acute leukemia virus MC29 (13) and because (ii) different defective sarcoma and acute leukemia viruses of the murine tumor virus family contain specific sequences not shared with those of related helper viruses (14,15,16). A direct answer depends on the definition of viral *onc* genes as well as viral replicative genes. These genes, in particular *onc*, have not been sufficiently defined either genetically or biochemically, in most virus strains except in the case of nondefective (nd) Rous sarcoma virus.

Definition of *src*, the *onc* gene of nd Rous sarcoma virus. Nd RSV segregates spontaneously transformation-defective (td) deletion mutants whose RNA is 12-15% smaller than the RNA of nd RSV. Hence *onc* gene-specific sequences totaling 1500 nucleotides were defined by subtracting the 8500 nucleotides of the RNA of the td virus from nd RSV RNA, which measures 10,000 nucleotides (7,8,9). In addition the same *onc* gene-specific sequences were defined as segregating with sarcomagenicity in recombinants formed between nd RSV and various lymphatic leukemia viruses (10). The *onc* gene of RSV has been termed *src* (for sarcomagenicity), to distinguish it from possible different *onc* genes present in other transforming viruses of the avian RNA tumor virus family. The *src* gene has been mapped near the 3' end of nd RSV RNA (10,12).

Problems in Defining the *onc* Genes of Defective Viruses, in Particular MC29 and MH2. Defining the *onc* gene of these viruses is complicated because MC29 and MH2 are defective, as are most other cancer causing RNA tumor viruses (1). They lack most or all viral genes necessary for replication but they must contain _onc_ genes, because they are highly oncogenic in animals and in cell culture (4,17). Because of their defectiveness it has not been possible to isolate _onc_-deletion mutants from MC29 and MH2 as yet. Such deletions would be hard to detect because they would neither be able to replicate nor to transform cells having lost their only diagnostic marker, i.e., their _onc_ genes. In addition recombination between the _onc_ genes of MC29 and MH2 and the _onc_ gene of other avian tumor viruses, in particular with _src_ has not been described. Thus neither deletion nor recombination analysis can be used to define the _onc_ genes of MC29 and MH2. Therefore more indirect approaches had to be used to determine whether they contain specific _onc_ genes. The first approach was to determine biochemically whether their RNAs contained the well defined _src_ gene of the Rous family. The second approach was to investigate whether the RNAs of MC29 and MH2 contain specific RNA segments unrelated to the conserved replicative genes of the avian tumor virus family (10,18) which would qualify as candidates for specific _onc_ genes. The third approach was to translate MC29 RNA _in vitro_ and to explore whether the translation products obtained are related to the _src_ gene product of RSV, which is thought to be a polypeptide of 60,000 daltons (19).

It was found that neither MC29 nor MH2 RNA were related to the _src_ gene of RSV. Each RNA contained specific sequences, defined as being unrelated to all avian tumor virus RNAs tested, which measured approximately 1500-2000 nucleotides. In the case of MC29 the specific sequences mapped about 3000 to 4500 nucleotides from the 3' end of MC29 RNA which measures about 5700 nucleotides (13). The major _in vitro_ translation product of MC29 RNA was a protein of 110,000 daltons, which shared some peptides with the conserved _gag_ gene products of avian tumor viruses. Since the only _gag_-related oligonucleotide of MC29 mapped near the 5' end of MC29 RNA, it is concluded that the MC29-specific 110,000 protein is translated from the 5' end of the RNA. Assuming colinearity between RNA and protein, it would appear that this protein must also include peptides translated from MC29-specific RNA sequences.

We conclude that MC29 and MH2 contain specific _onc_ genes. One possible interpretation of our data is that the _onc_ gene of MC29 includes a specific sequence mapping in the 5' half of the RNA which is translated into a polyprotein together with _gag_ gene-related sequences.

RESULTS

MC29 and MH2 Contain Specific 28S RNA Species. A 28S RNA species that is physically and chemically distinct from the

FIGURE 2. *Autoradiographs of RNase T_1-digested viral [^{32}P]RNA components after 2-dimensional electrophoresis-homochromatography (fingerprinting). Preparation of viral RNA components and conditions of fingerprinting have been described (7,8,12,13). Numbers identify distinct large RNase T_1-resistant oligonucleotides, or spots consisting of more than one oligonucleotide. Cap designates the 5' terminal capped oligonucleotide. The following RNAs were fingerprinted: (A) The 28S MC29 RNA prepared electrophoretically MC29(RPV) RNA (see Fig. 1 and (13)), (B) the 28S RNA of MC29(MCAV-B), electrophoretically prepared from MC29(MCAV-B) RNA as for Fig. 1, (C) the 34S MCAV-B RNA electrophoretically prepared from MC29 (MCAV-B) as in (B), (D) the 60-70S RNA of MCAV-A prepared from virus propagated in chicken fibroblasts, (E) the 28S RNA of MH2(RAV-7), prepared as for (A,B), (F) the 28S RNA of MH2 (MH2AV-Z and C) also prepared as for (A,B), (G) the 34S RAV-7 RNA prepared from MH2(RAV-7) RNA as described for (C) and (H) the 50-70S RNA of MH2AV-A and C. A mixture of these two helper viruses was propagated in chicken fibroblasts.*

(MCAV-B) share many, in particular the most characteristic large oligonucleotides. Homologous oligonucleotides were given the same numbers in Fig. 2A and B. In addition there are oligonucleotides not shared by the two 28S RNAs. These may include oligonucleotides of 28S MC29 RNAs differing from each other due to spontaneous mutations or to recombination with helper virus and may also include oligonucleotides from contaminating helper virus RNAs.

Similar experiments exploring the relationships of 28S RNAs in different pseudotypes of MH2, e.g., MH2(MH2AV-A+C) and MH2(RAV-7) indicated that the 28S RNAs of different virus complexes are also closely related sharing most of their large T_1-oligonucleotides (Fig. 2, Table II). These experiments confirm previous proposals that the RNA genomes of MC29 and MH2 are specific 28S RNA species (5,13). In accord with previous analyses we note that neither 28S MC29 nor 28S MH2 RNA contain the highly conserved oligonucleotides of the src gene of avian sarcoma viruses (5,13).

Fingerprint analyses of the 34S helper virus RNAs isolated from the original stocks of MC29 i.e., MCAV-A and that of MH2AV-A and C, isolated from MH2 are also shown in Fig. 2. Partial sequence analysis of large RNase T_1-resistant oligonucleotides is reported in Tables III and IV. Comparison suggests that the 28S MC29 RNA shares very few (i.e., two to four; see Tables I and III) large oligonucleotides with the 34S RNAs of the respective original helper viruses. No large oligonucleotides have been found in 28S MH2 RNA as yet which are shared with the helper viruses MH2AV-A and C (Tables II and IV).

TABLE I
COMPOSITION OF T_1-OLIGONUCLEOTIDES OF 28S MC29 RNA[a]

28S MC29 Spot no.[b]	RNase A digestion products	PR-B[c] Spot nos.	RPV[c] Spot nos.
1	2U,3C,G,2(AC),(AU),2(AAC)		
2	7U,7C,G,(AC),3(AU),(AAC),(AAU)		
3	8U,2C,G,(AAC),<1(AAAN)		
4	5U,8C,2(AC),(AU),(AAC),(AAG),(AAAN)		1
5	3U,5C,1.5(AC),(AU),(AAU),(AAAC),A_4G)		
6	2U,8C,3(AU),(AG),(AAAU)		
7a	2U,5C,G,2(AC),(AU),(AAC)		
7b	3U,9C,(AC),(AAG)		
8a	U,2C,G,2(AC),(AAC),(AAAC)		
8b	2U,4C,(AC),(AG),2(AAU),(AAAC)		
9	5U,6C,(AAG),(A_5C)	8	5
10	3U,2C,(AC),2(AU),(AG),(AAU)		
11	2U,4C,G,2(AC),(AU),(AAC)	14	
12	U,4C,2(AC),(AU),(AG),(AAC)		
13	U,3C,G,2(AC),(A_4N)		
15	4U,4C,G,1.5(AC),2(AU),(AAU)		
18	5U,3C,G,(AU),(AAU)		
C	G,(AC),(AU),(AAU),(AAAN)		

[a]*Elution of oligonucleotides from DEAE-cellulose thin-layers used for fingerprinting and analysis of RNase A-resistant fragments have been described (13).*

[b]*Numbers refer to spots shown in Fig. 2A and B. The composition of most of the oligonucleotides of 28S MC29 RNA from MC29 (RPV) has been described previously (5,13).*

[c]*See footnotes of Tables 3 and 4 and reference 13.*

TABLE II
COMPOSITION OF T_1-OLIGONUCLEOTIDES OF 28S MH2 RNA[a]

Spot no.[b]	RNase A digestion products
3	3U,4C,G,3(AU),2(AAU),(AAU)
4 (2x)	2U,8C,G,3(AC),2(AU),3(AAC),(AAG)
5 (2x)	U,7C,G,(AC),2(AAC),(AAG),(AAAN)
7	5U,7C,G,(AC),2(AU)
10	4U,2C,G,3(AU),(AAU)
11	8U,15C,G,(AC),3(AU),(AAG),4(AAAN)
12	10U,5C,G,(AU)

[a]*Determined in Table I. Numbers are similar to those reported previously (5) and due to scarcity of material available not entirely quantitative.*

[b]*Numbers refer to spots shown in Fig. 2E and F. The oligonucleotides of 28S RNA from MH2(RAV-7) and MH2(MH2AV-A and C) were the same as far as analyzed.*

TABLE III

COMPOSITION OF T_1-OLIGONUCLEOTIDES OF MCAV-A RNA[a]

MCAV-A		MC29[c]	PR-B[c]
Spot no.[b]	RNase A digestion products	Spot no.	
1	7U,5C,(AC),(AU),(AG),(AAC),(AAU)		
2	6U,10C,2(AC),(AU),(AAC),(AAG)		4
3a	4U,7C,G,(AC),(AU),(AAAC)		6
3b	3U,5C,(AC),(AAU),(AAAC),(A_4G)	5	
4	6U,6C,G,3(AC),3(AU)		
5	8U,9C,G,2(AC),(AU),(AAG)	9	8
6	3U,6C,G,3(AC),2(AU)		
7	2U,5C,G,2(AC),(AU),(AAC)	11	14
8	4C,2(AC),(AU),(A_4G)		
10	U,5C,3(AC),(AAG)		18
11	3C,(AU),(AG),(A_5N)		20b
13	6U,2C,G,3(AU),(AAU)		
15	8U,5C,G,(AU)		
18	5C,G,4(AC)		
21	5U,G,(AC),2(AU)		
22	5U,3C,G,(AU),(AAU)	18	18
23	3U,2C,G,2(AU),(AAU)		
24	5C,<2(AC),(AG)		

[a]*As in Table I.*

[b]*Numbers refer to oligonucleotide spots shown in Fig. 2D.*

[c]*Oligonucleotides of the same composition were found previously in PR-RSV-B (10,12) and MC29 (see Table I and ref. 13).*

However further analyses of MH2 RNA may reveal homologous oligonucleotides. Analysis of MH2 RNA has been hampered by difficulties in propagating in macrophages sufficient radioactive virus for biochemical analysis of RNA.

The Relationship of 28S MC29 and 28S MH2 RNAs to the RNAs of other Avian Tumor Viruses Measured by Hybridization. To determine whether 28S MC29 and 28S MH2 RNA contain src-specific nucleotide sequences of avian sarcoma viruses and to investigate their relationship to the RNAs of helper-independent avian RNA tumor viruses, the RNAs were hybridized to various

TABLE IV

COMPOSITION OF T_1-OLIGONUCLEOTIDES OF MH2AV-A AND C RNA[a]

MH2AV-A and C		PR-B[c]
Spot no.[b]	RNase A digestion products	Spot no.
1	7U,10C,G,3(AC),(AAC)	
2	5U,10C,2(AC),(AU),(AAC),(AAG)	4
3	6U,10C,G,(AC),(AU),(AAAC)	6
4	8U,6C,2(AC),2(AU),(AG),(AAC),(AAU)	
5	3U,6C,G,2(AC),(AU)	
6	6U,6C,G,2(AC),4(AU)	
7 (2x)	4U,3C,2(AC),2(AU),(AG),(AAU),(A_7NN)	
8	6U,6C,(AAG),(A_5N)	
9	6U,9C,G,(AAC)	
10	8U,3C,4(AU),(AG),(AAU)	
11	6U,2C,G,4(AU),(AAC)	
12	U,2C,G,(AC),(A_5NN)	
13	4U,3C,G,2(AU),(AAAN)	
14	5U,4C,C,(AC),(AAAC),(AAAG)	
15	4U,4C,(AC),(AAAC),(A_4G)	
16	2U,5C,G,3(AC),(AU),(AAC)	

[a]*Determined as in Table I.*

[b]*Numbers refer to oligonucleotide spots shown in Fig. 2H.*

[c]*Oligonucleotides of the same composition were found in PR-RSV-B (10,12).*

cDNAs. All hybridizations were carried out with an excess cDNA and at different cDNA to RNA ratios to reach plateau values of maximal hybridization. Under our conditions, maximal hybridization of 28S MC29 RNA with homologous MC29(RPV) cDNA and of 30-40S Prague RSV-B(PR-B) RNA with homologous cDNA was about 93% (Table V). To determine whether 28S MC29 and MH2 RNAs contain src-specific sequences each RNA was first hybridized to cDNA from PR-B which contains src and then to cDNA from td PR-B which lacks src. It is seen in Table V that approximately the same percentage (62 to 66%) of each RNA was hybridized by each cDNA. The MC29 and MH2 RNA sequences hybridized by PR-B and td PR-B cDNA were not additive, because a mixture of these two cDNAs did not hybridize more than each by itself (Table V). The PR-B cDNA used was shown to include src-specific sequences, because it was able to hybridize 13% more PR-B RNA than td PR-B cDNA (Table V). This was the expected difference, because the src gene corresponds to about 13% of the viral RNA (8). It follows that MC29 RNA and MH2 RNA are about 62 to 66% related to the RNAs of PR-B and td

TABLE V
HYBRIDIZATIONS[a] OF VIRAL RNAs WITH VIRAL cDNAs[b]

RNA	cDNA	% Hybridization at various cDNA/RNA ratios mean ± S.D.					
		1:1	5:1	20:1	50:1	100:1	150:1
28S MC29[c]	MC29(RPV)		65±6	80±4	88±1	93±7	96
	RPV		46±3	61±0	66±3	69±4	
	PR-B		40±2	54±7	61±8	61±8	
	tdPR-B		50±2	55±6	61±6	62±7	
	PR-B+tdPR-B		50	62±6	60	61	
	MCAV-A		55	60±3	61		
	PR-B+MCAV-A		61	60±5			
34S RPV	MCAV-A		64	65			
	MC29(RPV)		48	81	80	80	
	tdPR-B		35	58			
34S MCAV-A	MCAV-A	79	91±6	94±5			
	MC29(RPV)			49	80	80	
	RPV				75		
	PR-B			66	70		
28S MH2[c]	PR-B			60.5±1	64	66.5±5	
	tdPR-B			58±5	59.5	65.5±2	
	PR-B+tdPR-B			62±3			
	MC29(RPV)			61.5±5	61.5	64±1	
	MC29(RPV)+PR-B[d]			67±3		70.5±6	
34S MH2AV	PR-B			79.5±8			
	tdPR-B			79±8	78		
	PR-B+tdPR-B			86.5±2			
	MC29(RPV)			70±5	75		
	MC29(RPV)+PR-B[d]			84.5±4			
PR-B[d]	PR-B			93±0	93±1		
	tdPR-B			81	81		
	PR-B			94±0	92±1		
	tdPR-B			79	81		

[a]*Each reaction mixture contained about 1 ng of ^{32}P RNA (2000 cpm/ng) and 5-100 ng of ^{3}H cDNA (50 cpm/ng). Hybridizations were in 4 μl of 70% deionized formamide/0.3 M NaCl/0.03 M Na citrate/1.5 mM Na phosphate, pH 7.0/0.05% $NaDodSO_4$ at 40° for 12 hr. Percentage nuclease-resistance is expressed as the radioactivity recovered in aliquots digested with nuclease relative to that found in undigested aliquots. Each value is the mean of two or three experiments using, in some cases, independent preparations of RNA and cDNA. Digestion was with*

RNases A (5 μg/ml), T_1 (10 units/ml) and T_2 (10 units/ml for 30 min at 40° in 0.3 M NaCl/0.03 M Na_3 citrate, pH 7.0. The background of nuclease-resistance of an aliquot heated at 100° in 0.01 M Na^+ was <0.5%. Many of these data are from refs. 5 and 13.

[b]*cDNA was prepared as described previously (13).*

[c]*Prepared by gel electrophoresis from MC29(RPV) (13) as shown for Fig. 1.*

[d]*Shown electrophoretically to be free of tdPR-B RNA (7).*

PR-B and lack src-specific sequences. At least 31% (i.e., 93 minus 62) or 1800 nucleotides of each RNA would appear to be specific for MC29 or MH2, respectively. This is considered a minimal estimate because each electrophoretically prepared 28S RNA was contaminated with 10 to 20% of degraded helper virus RNA (c.f., Fig. 1). Helper virus RNAs including 34S RPV RNA, prepared as a 60-70S RNA complex from RPV (13), 34S MH2AV-A and C RNA, prepared electrophoretically from MH2(MH2AV-A and C) as in Figs. 1 and 2 and 34S MCAV-A RNA prepared as a 60-70S RNA complex from MCAV-A as in Fig. 2, were 60 to 80% homologous to the RNAs of PR-B, td PR-B and of other avian tumor viruses (Table V).

To test whether the specific sequences of MC29 and MH2 RNA (defined as those which did not hybridize with PR-B or td PR-B cDNA) are related to each other, 28S MH2 RNA was hybridized to MC29(RPV) cDNA. About 64% of the RNA was hybridized (Table V). If annealed with MC29(RPV) and PR-B cDNAs, about 70% of 28S was hybridized (Table V). It follows that at least 30% of the 28S MH2 RNA is unrelated to MC29 and PR-B RNA.

It has been argued that (defective) transforming viruses are generated by recombination of a helper virus with an unknown (defective) virus preexisting in the cell or with cellular genetic elements (16,20,21). The helper virus isolated in the original stock of a defective transforming virus such as MC29 and MH2 would appear to be a likely candidate for one parent of such a recombinational event. It would then be expected that the defective recombinant virus shares more sequences with its progenitor than with other possible helper viruses. Thus subtraction from MC29 or MH2 RNAs of nucleotide sequences shared with the RNAs of their original helpers would give a minimal estimate of specific sequences in MC29 or MH2 RNA. It is seen in Table V, that 28S MC29 RNA is hybridized by MCAV-A cDNA to approximately the same extent, i.e., 61%, as by cDNA prepared from other avian tumor viruses. It is concluded that MC29 is not more closely related to its original helper than to other avian tumor viruses tested. This conclusion is consistent with the above result that MC29 and MCAV-A RNAs have different fingerprint patterns. Since MH2 and MH2AV also have different fingerprint patterns, the same may be true for the relationships of MH2 RNA, MH2AV RNA and other avian tumor virus RNAs.

Mapping Sequences of MC29 RNA which Hybridize with DNAs Complementary to RNAs of Other Avian Tumor Viruses and of Sequences which are MC29-Specific. If compared by RNA-cDNA hybridization, the 28S MC29 and MH2 RNAs contain 30 to 40% specific sequences and 60 to 70% sequences in common with other avian tumor virus RNAs. By contrast, fingerprinting indicates that most of the large T_1-oligonucleotides of 28S MC29 and MH2 RNAs are not shared with other avian tumor virus RNAs. The apparent discrepancy reflects the different sensitivities of the two methods: Fingerprinting detects specific oligonucleotides not only in RNA sequences which are totally unrelated but also in sequences which differ by only a few percent of their nucleotides and which would be closely related if compared by RNA-cDNA hybridization.

In order to map the specific sequences of MC29 RNA, defined as those which do not hybridize with cDNAs of other avian tumor viruses, the following strategy was used: First all large T_1-oligonucleotides of 28S MC29 RNA (shown in Fig. 2 and 4 and Table I) were mapped relative to the 3' poly(A)-end of the RNA on the basis of fingerprint analyses of poly(A)-tagged RNA fragments of discrete size classes (12). The resulting oligonucleotide map is shown in Fig. 3.

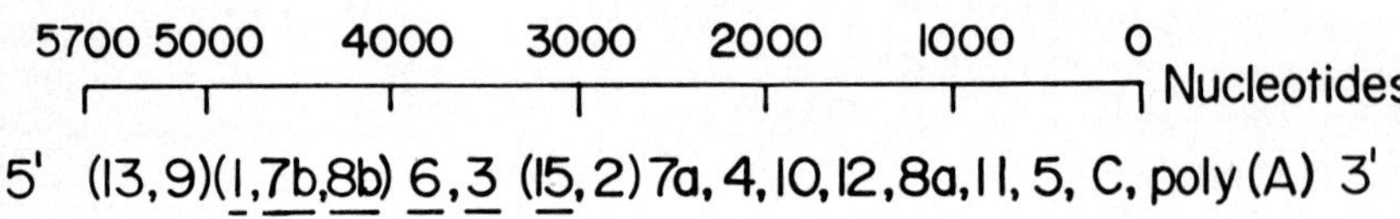

FIGURE 3. *Oligonucleotide map of 28S MC29 RNA. The oligonucleotides of 28S RNA prepared from MC29(RPV) were analyzed as described (12). The numbers represent the large T_1-oligonucleotides depicted in Fig. 2A. Their composition is recorded in Table I. The upper scale denotes the approximate length of the RNA in nucleotides. The bottom line depicts the 110,000 protein translated in vitro from MC29 RNA, in a position reflecting colinearity with the 5' 60% of the viral RNA as deduced from data described in text.*

Oligonucleotide C, which maps near the 3' end of MC29 RNA is not seen in the fingerprints derived from 28S MC29 RNA shown in Fig. 2, because it can only be detected as a distinct spot in short poly(A)-tagged RNA fragments of low complexity, which are not shown here. This oligonucleotide is found as a highly conserved sequence near the 3' end of all avian tumor virus RNAs tested by us (10,11,12,18).

Next, the oligonucleotides of MC29-specific RNA segments and of RNA segments shared between MC29 and other avian tumor virus RNAs were identified. For this purpose MC29(RPV) [^{32}P] RNA was hybridized to MC29(RPV) cDNA, that had been prehybridized to an excess of unlabeled RPV and PR-B RNA. Therefore, all but the MC29-specific sequences of the above cDNA, were already heteroduplexes making it a MC29-specific cDNA when it was mixed with MC29(RPV) [^{32}P]RNA. MC29(RPV) [^{32}P]RNA-cDNA hybrid formation was restricted to MC29-specific cDNA. In the first step MC29-specific cDNA was prepared by incubating 2 μg of MC29(RPV) cDNA with 15 μg RPV RNA and 12 μg PR-B RNA for 12 hr at 40° in 10 μl 50% formamide containing 0.45 M NaCl, 0.045 M Na citrate and 0.01 M Na-PO_4-buffer pH 7.0. This prehybridization of cDNA with RNA was under moderately stringent conditions to allow hybridization to occur between sequences that are not completely homologous (27). In the second step 1.5 μg of MC29(RPV) [^{32}P]RNA (specific activity 5 × 10^6 cpm/μg) was added and hybridization was for 1 hr in 30 μl of the above formamide buffer. After digestion for 30 min at 40° in 200 μl of 0.3 M NaCl, 0.03 M Na citrate containing 5 μg/ml RNase A and 50 units/ml RNase T_1, the resistant hybrid was isolated from the void volume of a Biogel P100 column (12 × 0.6 cm) equilibrated in 0.1 M NaCl, 0.01 M Tris pH 7.4, 1 mM EDTA and 0.2% SDS. After 3 phenol extractions in the presence of 30 μg carrier yeast tRNA, the hybrid was ethanol-precipitated, heat-dissociated in buffer of low ionic strength, digested with RNase T_1 and subjected to fingerprint analysis (Fig. 4) as described previously (12,18). The oligonucleotides from MC29-specific RNA segments so identified are underlined in the oligonucleotide map shown in Fig. 3. Similar methods have been used by others to identify specific sequences in tumor virus RNAs (23,24,27).

Oligonucleotides of RNA sequences that MC29 has in common with PR-B and RPV RNAs were obtained by hybridizing 0.25 μg of electrophoretically purified (13) 28S MC29 [^{32}P]RNA (specific activity 2 × 10^6 cpm/μg) with 1 μg of PR-B and 1 μg of RPV cDNA for 12 hr under the conditions described previously (13). The resulting hybrid was treated with RNase T_1 only, rather than with RNases A and T_1, in order to preserve small MC29-specific oligonucleotide segments which are part of larger MC29 polynucleotide segments hybridized with the cDNAs of the other

RNA tumor viruses. Thus mismatches involving oligonucleotide segments with less than two Gs would register as complete hybrids in our conditions. The T_1-oligonucleotides of the resulting hybrid are shown in Fig. 4. They represent RNA sequences that MC29 shares with PR-B and RPV and they

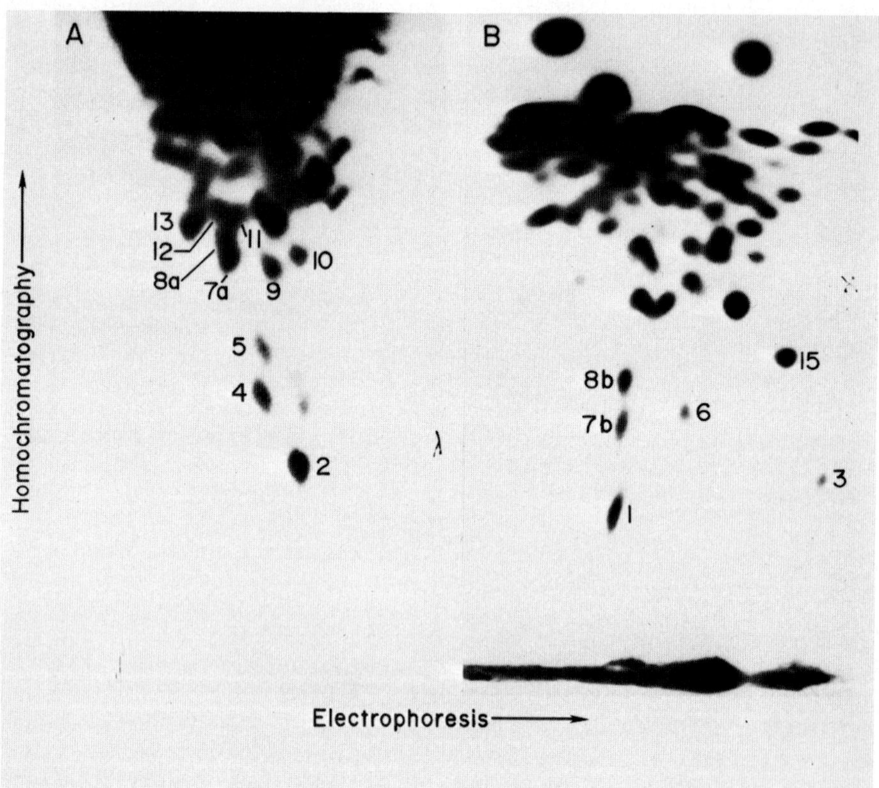

FIGURE 4. *T_1-oligonucleotides of MC29 RNA segments which hybridize with DNAs complementary to other avian tumor virus RNAs (A) and of MC29-specific RNA segments (B). (A) 28S MC29 [^{32}P]RNA was prepared as described for Fig. 1 and previously (13). After fragmentation and removal of poly(A)-tagged fragments for oligonucleotide mapping (see Fig. 3 and ref. 12) the RNA was hybridized with PR-B and RPV cDNAs. The RNA of the RNase T_1-resistant hybrid was prepared as described in the text and fingerprinted as described in Fig. 2. (B) 50-70S MC29(RPV) [^{32}P]RNA (13) was hybridized with MC29(RPV) cDNA which had been prehybridized with an excess of non-radioactive PR-B and RPV RNA. Therefore MC29(RPV) [^{32}P]RNA-cDNA hybrid formation was restricted to MC29-specific cDNA. After digestion with RNases A and T_1 the RNA of the resistant hybrid was*

fingerprinted. Further details are described in the text. Large oligonucleotides in (A) and (B) were analyzed as described for Table I and numbered identically with their compositional counterparts of MC29 RNA shown in Fig. 2A.

are not underlined in Fig. 3. Two previously unresolved fingerprint spots of MC29 RNA which contain two oligonucleotides i.e., nos. 7 and 8 were each resolved into one oligonucleotide derived from a specific (7b and 8b) and another (7a and 8a) derived from a common sequence of MC29 RNA (see Figs. 3, 4 and Table I). This resolution of spots 7 and 8 was independently confirmed by analyses of poly(A)-tagged RNA fragments carried out to derive the oligonucleotide map shown in Fig. 3. In these experiments spots 7 and 8 consisted of oligonucleotides 7a and 8a, when short poly(A)-fragments were analyzed and of both oligonucleotides when long fragments were analyzed.

It can be seen in Fig. 3 that the oligonucleotides of MC29-specific RNA (underlined) cluster approximately between 3000 and 4500 nucleotides from the poly(A)-end in a contiguous map segment and oligonucleotides derived from RNA that MC29 has in common with PR-B and RPV RNAs (not underlined) map at the 5' end and on the 3' half of the viral RNA.

In Vitro Translation of MC29 RNA. To understand the nature of the polypeptides encoded by MC29 RNA we have compared them with those encoded by two other classes of independently replicating avian tumor viruses (i) nd RSV, which contains src as well as gag, pol and env and (ii) td RSV and RPV which contain gag, pol and env but may lack an onc gene (5). Viral RNA monomers, prepared by heat-dissociation of 50-70S RNA complexes and subsequent selection for poly(A) containing species on oligo(dT) cellulose (12), were translated in the mRNA-dependent rabbit reticulocyte lysate (28). Translation products programmed by any of the above RNAs were analyzed directly and after immunoprecipitation with antibody against all gag gene-proteins or with antibody against the major gag gene-protein, p27 (25). 50-70S MC29 RNA was prepared from a mixture of MC29(RPV) and its helper RPV (13). To identify the polypeptides encoded by MC29 RNA the products of the mixed MC29 and RPV RNAs were compared with those of RPV RNA, td RSV and nd RSV by electrophoresis on discontinuous SDS-polyacrylamide slab gels (25). As can be seen in Fig. 5, MC29 RNA is translated predominantly into a 110,000 dalton polypeptide. In addition two smaller proteins of approximately 49,000 and 31,000 daltons (not seen in Fig. 5, but clearly discernible in other gels not shown here) also appear to be MC29-specific translation products. These polypeptides were synthesized with considerably lower efficiency than the 110,000 protein. The

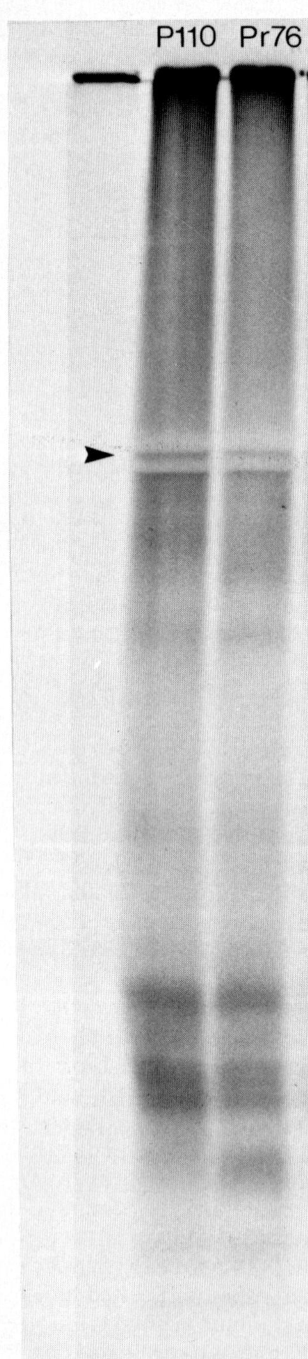

FIGURE 6. *250 μl of rabbit reticulocyte lysate primed with a mixture of MC29(RPV) RNAs or td RSV PR-B RNA (at 40 μg/ml) was electrophoresed on a preparative 10% polyacrylamide gel. The gel was briefly fixed, dried and autoradiographed. The P110 MC29-specified protein, and td PR-B RSV-specified Pr76 were excised from the gels and the protein eluted from the gel slices (25). 250 μg of β-galactsidase was added to each eluate as carrier, and the proteins were precipitated with TCA and digested with Staphylococcus Aureus V8 protease as described by Cleveland et al. (26) at a concentration of 75 μg/ml. The products of this digestion were analyzed on a 15% polyacrylamide gel. The tracks containing the digestion products of the P110 and Pr76 proteins are indicated. The position of the protease, which causes an artefactual band, is indicated by an arrow.*

that these portions must have a more acidic isoelectric point than Pr76, because the 110,000 MC29 protein is more acidic than Pr76 (Pawson and Radke, unpublished).

In an effort to determine whether the 110,000 MC29 protein is translated from full-sized 28S MC29 RNA or from subgenomic RNA fragments, poly(A)-tagged RNA fragments of different size classes were translated (25). It was found that the 110,000 protein was only efficiently translated from full-sized RNA (data not shown).

DISCUSSION

Synthesis of the Data on the RNA and Proteins of MC29 and MH2. The findings that different pseudotypes of MC29 and MH2 contained physically and chemically very similar or identical 28S RNAs, but 34S RNAs that varied with the respective helper virus, proved the notion that the 28S RNAs are MC29- or MH2-specific. Each 28S RNA contained 30-40% specific nucleotide sequences, which failed to hybridize with the cDNAs of all other avian tumor viruses tested, and 60-70% of common sequences, which hybridized with these cDNAs.

The sequences that MC29, MH2 and other avian tumor viruses have in common are nearly indistinguishable, if compared by hybridization, but are distinct in each viral RNA if analyzed by the more sensitive method of fingerprinting, which detects single base changes. Finally, *src* gene-related sequences were not detected in 28S MC29 or MH2 RNA. We conclude that MC29 and MH2 contain specific *onc* genes.

In the case of 28S MC29 RNA, specific sequences, identified by the large T_1-oligonucleotides they contain, mapped about 3000 to 4500 nucleotides from the 3' poly(A)-end of the RNA which measures about 5700 nucleotides. The common sequences at the 5' end of the RNA included one oligonucleotide, no. 9, previously identified as a conserved element of the *gag* gene of other avian tumor viruses (18). The common sequences of the 3' half of MC29 RNA included one oligonucleotide, no. 11, found previously at the *env-src* border in other avian tumor viruses and the C oligonucleotide which is a highly conserved oligonucleotide that maps near the 3' end of avian tumor viruses (12,18). *In vitro* translation of MC29 RNA generated one major 110,000 dalton protein product, and two minor proteins of 49,000 and 31,000 daltons. The 110,000 MC29 protein included protein sequences serologically related to the *gag* gene of other avian tumor viruses. Since one *gag* gene-related oligonucleotide was found near the 5' end of MC29 RNA and since the *gag* gene of avian tumor viruses maps near the 5' end of the viral RNA (18), it appears plausible that the *gag* gene-related portion of the 110,000 MC29 protein was translated from the 5' end of the RNA. Assuming colinearity of viral RNA and the 110,000 protein we deduce that most of the remaining portion of the protein is translated from MC29-specific sequences of MC29 RNA, as depicted in Fig. 3.

What Is the *onc* Gene of MC29 and MH2? A definitive answer to this question can only be given if genetic variants of the *onc* genes of these viruses become available. Nevertheless it appears plausible to postulate that the specific sequences of MC29 and MH2 are part of or identical with their *onc* genes, since neither MC29 nor MH2 contain *src* gene-related sequences and since each virus has a specific oncogenic spectrum. Two classes of specific sequences may be considered as candidates for viral *onc* genes; those which do not hybridize with cDNAs of other avian tumor viruses and which in MC29 RNA cluster between 3000 and 4500 nucleotides from the 3' end and those which are interspersed as small specific oligonucleotides among common RNA segments that hybridize with cDNAs of other avian tumor viruses.

The finding that the large, specific segment of MC29 RNA is probably translated together with sequences of a defective *gag* gene, present in MC29 RNA, implies that the 110,000 MC29 protein is involved in transformation. Therefore it would appear from our data that both large specific segments of MC29 RNA as well as sequences which MC29 shares with other avian tumor viruses but which also include small specific oligonucleotides may be involved in transformation. Postulating an intracellular role for the 110,000 MC29 protein, such as transformation, is also consistent with the failure to find this protein in mature virus (unpublished). Further analyses are necessary to determine the possible function of the 49,000 and 31,000 MC29 proteins. These proteins may be related to the 110,000 MC29 protein or maybe specific considering that 28S MC29 RNA may code for approximately 200,000 daltons of protein.

Besides being possibly involved in transformation nucleotide sequences of MC29 and MH2 RNAs which are shared with independently replicating avian tumor viruses must play roles in virus replication. These would include terminal sequences of the RNA essential for transcription of RNA to DNA, nucleotide sequences that allow specific association with viral proteins essential for packaging of the viral RNA into the coat of the helper virus (e.g., murine helper viruses fail to package MC29 RNA (K. Bister and P. K. Vogt in preparation)), and sequences which allow two 28S RNA monomers to form a 50-70S dimer structure which may be essential for virus replication.

Our data, that viruses with specific oncogenicity carry specific *onc* genes does not exclude a role of other viral genes, including those of the helper virus, in determining the oncogenic spectrum of a defective transforming virus.

For example, the oncogenic spectrum of a defective virus should be greatly influenced by the env gene of its helper virus which provides the envelope glycoproteins for the defective virus. Since the cellular receptors for viral envelope glycoproteins differ greatly among different animals (1) and even among different target cells of the same animal (29), different helper viruses may deliver the same onc genes into specific target cells and thus cause a different form of cancer.

ACKNOWLEDGMENTS

We thank Carlo Moscovici for cultures infected with MC29 and Lorrine Chao for assistance with these experiments. This work was supported by NIH Research Grants CA 11426, CA 13213 and CA 17542 from the National Cancer Institute. K. B. was the recipient of a fellowship from the Deutsche Forschungsgemeinschaft, Bonn-Bad Godesberg, Germany.

REFERENCES

1. Tooze, J. (Ed.) (1973). "The Molecular Biology of Tumor Viruses". Cold Spring Harbor Laboratory, Cold Spring Harbor, New York.
2. Ivanov, X., Mladenov, Z., Nedyalkov, S., and Bozlakov, S. (1965). Dakl. Bulog. Akad. Nauk. 18, 543.
3. Begg, A. M. (1927). The Lancet Vol. 1, 912.
4. Vogt, P. K., Bister, K., Hu, S.S.F., and Hayman, M. J. (1978). In Oncogenic Viruses and Host Cell Genes" (Oji Int. Seminar on Friend Virus and Friend Cells) (Y. Ikawa, ed.), Academic Press, New York. In press.
5. Duesberg, P. H., Vogt, P. K., Bister, K., Troxler, D., and Scolnick, E. M. (1978). In "Oncogenic Viruses and Host Cell Genes" (Oji. Int. Seminar on Friend Virus and Friend Cells) (Y. Ikawa, ed.), Academic Press, New York. In Press.
6. Baltimore, D. (1975). Cold Spring Harbor Symp. Quant. Biol. 39, 1187.
7. Duesberg, P. H., and Vogt, P. K. (1973). Virology 54, 207.
8. Lai, M. M-C., Duesberg, P. H., Horst, J., and Vogt, P. K. (1973). Proc. Natl. Acad. Sci. USA 70, 2266.
9. Stehelin, D., Guntaka, R. V., Varmus, H. E., and Bishop, J. M. (1976). J. Mol. Biol. 101, 349.
10. Wang, L. H., Duesberg, P. H., Mellon, P., and Vogt, P. K. (1976). Proc. Natl. Acad. Sci. USA 73, 1073.
11. Wang, L. H., Duesberg, P. H., Kawai, S., and Hanafusa, H. (1976). Proc. Natl. Acad. Sci. USA 73, 447.
12. Wang, L. H., Duesberg, P. H., Beemon, K., and Vogt, P. K. (1975). J. Virol. 16, 1051.

frequency. (2) The env gene can be divided into two portions: the 3'-half sequences are relatively constant while the 5'-half sequences are diverged among different viral strains. We did not find any evidence for the presence of spacer gene between the env and src genes.

RESULTS

Size and Location of the src and env Genes. To define the location and size of the src gene, we studied the heteroduplex molecules formed between genome-length complementary DNA (cDNA) made from nondefective Schmidt-Ruppin strain of Rous sarcoma virus of subgroup D (SR-D) and 35S RNA of the corresponding tdSR-D viruses. The rationale of this approach has been described previously (5,6). As can be seen in Fig.1A a deletion loop (1.98 ± 0.24 kb) could be observed at 0.6 kb from the 3'-end of the RNA genome. We conclude that the src gene is located at 0.6 kb from the 3'-end and spans 1.98 kb in length. This estimate is very close to the figures obtained by oligonucleotide and heteroduplex mapping using Prague strain of Rous sarcoma virus (3,5).

To define the size and location of the env gene, we made use of the finding by Tal et al (7) that the env gene of the ring-necked pheasant virus (RPV) has very little homology to that of chicken leukosis and sarcoma viruses. The remainder of the RPV genome is, however, largely homologous to the

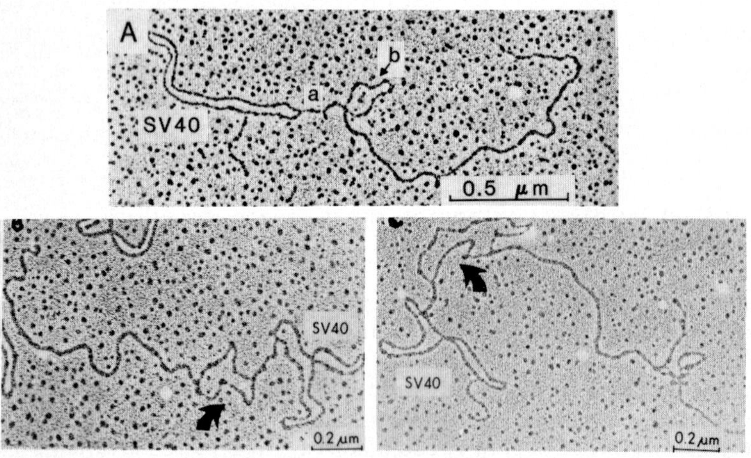

FIGURE 1. Electron micrographs of heteroduplex molecules formed between (a) SR-D cDNA and tdSR-D RNA. (B) RPV cDNA and tdSR-D RNA and (C) RPV cDNA and SR-D RNA. The arrows depict the deletion loop (A) and the substitution loops (B and C).

chicken virus genome (7). We therefore studied heteroduplex molecules formed between genome-length cDNA made from RPV and the 35S RNA of tdSR-D. Since neither RPV nor tdSR-D has src gene, the nonhomologous region in these heteroduplex molecules would be expected to reside only in the env gene. As can be seen in Fig. 1B, a substitution loop with two arms of 1.55 kb and 1.70 kb respectively was observed at 0.6 kb from the 3'-end of the RNA genome. We therefore conclude that the env gene starts at 0.6 kb from the 3'-end in leukosis or td viruses and spans 1.55 - 1.7 kb in length. This study was extended to the heteroduplex molecules formed between RPV cDNA and the 35S RNA of SR-D which contains the src gene. As can be seen in Fig. 1C, a substitution loop with two arms of 3.5 kb and 1.5 kb is located at the same site (0.6 kb from the 3'-end) as that observed in the heteroduplex between RPV cDNA and tdSR-D RNA. The long arm appears to be the sum of the src and env genes of SR-D. These data suggest that the env gene is located immediately next to the 5'-end of the src gene. There is no evidence for the presence of long "spacer" sequences between the src and the env genes, although a short stretch of such sequences can not be ruled out.

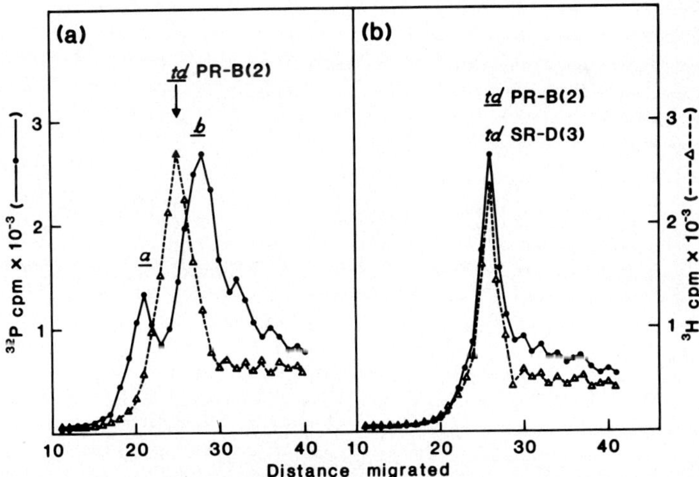

FIGURE 2. Polyacrylamide gel electrophoreses of the RNAs from ^{3}H-uridine-labeled tdPR-B (Isolate 2) and (a) ^{32}P-labeled PR-C and tdB77 and (b) ^{32}P tdSR-D (Isolate 3).

Many td Viruses Retain 25% of the src Gene. Previous studies indicated that most of the td viruses have deleted 13-15% of the avian sarcoma virus genome, and these deleted sequences appear to span most of the temperature-sensitive lesions affecting the src functions (8,9,10). These td viruses have a smaller 35S RNA of the same size termed class b (11). This suggests that the src deletion of this size occurs at high frequency or that it is selected for during virus growth. Recently the isolation of td viruses which retained part of the src sequences has been described (6,12,13,14). To determine whether these partial deletions of the src gene are of general occurrence and whether deletion can take place randomly within the src gene, we examined the size of the 35S RNA of various td viruses isolated independently by picking plaques or by end-point dilution. Among the 30 independently isolated tdB77 examined, only one has a slightly bigger RNA. On the other hand, two of the 5 tdSR-Ds and 5 of 10 tdPR-Bs examined have an RNA bigger than class b. All of these td viruses with larger RNA have a deletion which is 25-30% smaller than that in regular td viruses (Fig. 2a). There is no difference in the size of the RNA among these partially deleted viruses. One example is shown in Fig. 2b. To determine whether the extra sequences in these td viruses came from the src sequences, we studied the heteroduplex molecules formed between cDNA from sarcoma virus and the 35S RNA of td viruses with larger RNA. As shown in Fig. 3, the src deletion loop in these heteroduplex molecules is smaller than that with regular td viruses (Fig. 1A), but is located at the same distance (0.6 kb) from the 3'-end. It is therefore concluded that the residual src sequences have come from the 5'-end of the src gene. Since all of the td viruses with extra sequences have retained identical amounts of RNA sequences, it suggests that there is a favorite internal deletion site within the src gene.

The env Gene is Composed of a 5'-Variable Region and a 3'-Conserved Region. To determine the internal genetic

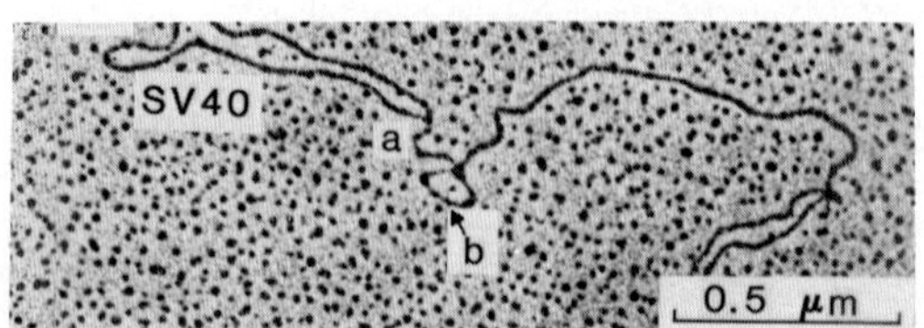

FIGURE 3. An electron micrograph of heteroduplex between SR-D cDNA and tdSR-D (Isolate 3) RNA. The arrow locates the deletion loop.

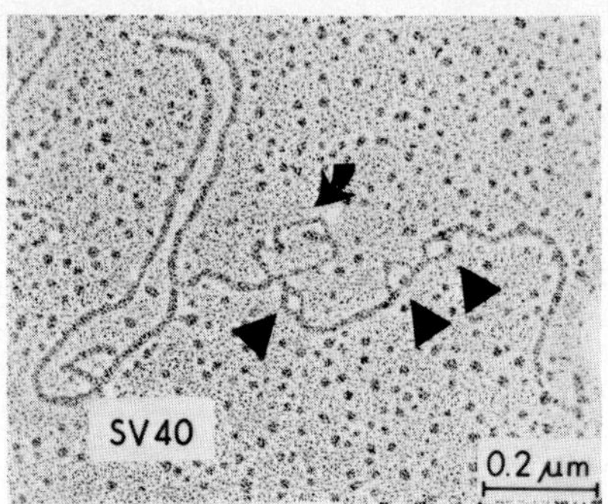

FIGURE 4. Heteroduplex between RPV cDNA and tdSR-D RNA spread with 65% urea + formamide (6). The arrow locates the src deletion loop. The triangles locate the substitution loops.

organization of the env gene, we studied the heteroduplex molecules between RPV cDNA and tdSR-D RNA under slightly less stringent conditions, i.e., 65% urea-formamide (6). This allowed related but not identical sequences to form homologous hybrids. As shown in Fig. 4, the env region as determined in Fig. 1 is roughly separated into two halves. The 5'-half of the env region consists of two small substitution loops. These two loops were fused into a loop of 0.6 - 0.8 kb in most of the molecules. The 3'-half was on the contrary homologous except at the extreme 3'-end where a small nonhomologous loop of 0.3 kb was occasionally observed. Thus the 5'-half of the env gene is more diverged between RPV and chicken viruses while the 3'-half is more homologous.

To determine whether this genetic organization of the env gene is of general occurrence, we have studied the heteroduplex molecules formed between different strains of avian oncoviruses. A summary is presented in Fig. 5. We conclude that the 5'-half of the env gene is quite variable while the 3'-half is relatively conserved among avian oncoviruses. A small region at the extreme 3'-end of the env gene is also slightly variable.

DISCUSSION

We have determined the relative map positions of the src and env genes on the RNA genome by heteroduplex mapping. The results obtained by this method are consistent with those obtained by oligonucleotide mapping (3,4). However, both of these approaches were based on the assumptions that the deletion mutants used have deleted the entire src or env genes.

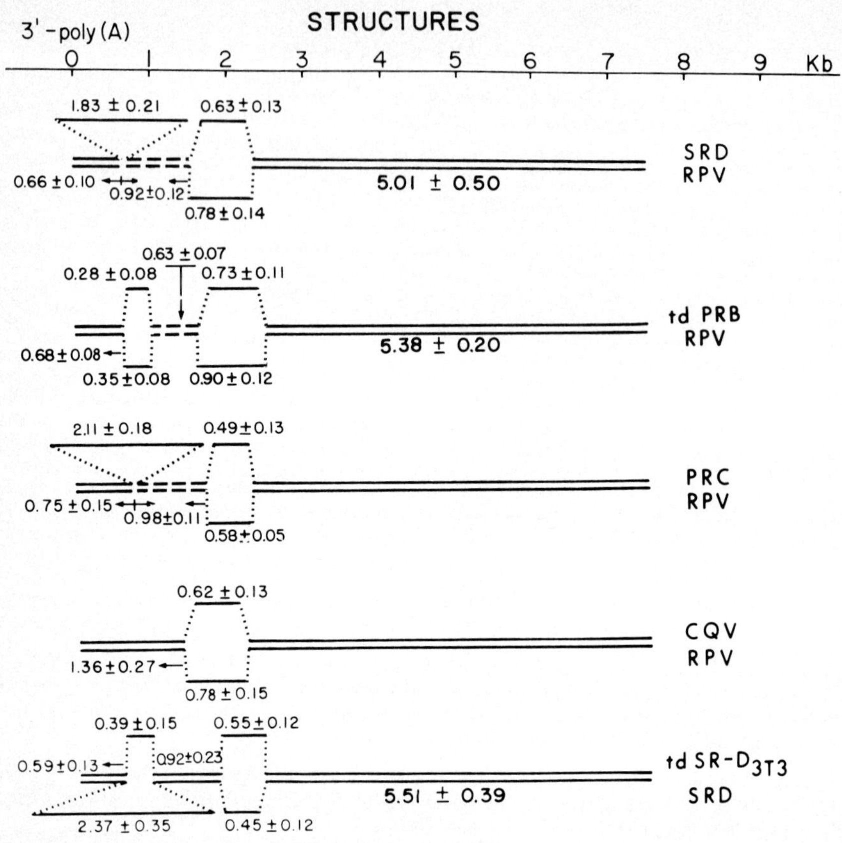

FIGURE 5. Schematic drawing of heteroduplex spread with 65% urea-formamide. The env gene covers the region between 0.6 kb and approximately 2.4 kb from the 3'-end. CQV: Chinese quail virus, tdSR-D_{3T3}: td viruses derived from SR-D passaged through 3T3 cells.

For mapping the env gene we further assumed that the region of nonhomology between RPV and chicken oncoviruses does not exceed the limits of the env gene (7). Lacking detailed genetic data, these two assumptions can not be substantiated with certainty. Therefore, our data could only be considered as rough genetic maps. It is interesting to note that although the previous studies (3) have suggested the presence of spacer sequences between the env and src deletions, our data did not support this claim. The lack of long spacer sequences between these two genes was further suggested by recent studies of heteroduplex between tdSR-D and Bryan high titer strain Rous sarcoma virus [RSV(-)] (unpublished observation). The reason for this discrepancy is unclear.

The isolation of several td viruses which retained extra RNA sequences equivalent to 25% of the src gene suggests that this type of deletion occurs much more frequently than previously realized. Since these viruses were independently isolated and included viruses of different strains (SR-D, PR-B and B77), it suggests that there may be an internal site within the src gene at which deletion takes place at higher frequency. Among these viruses two have been studied in detail. The residual src sequences were all derived from the 5'-end of the src gene.

The finding that the 3'-end of the env gene is relatively conserved while the 5'-end is quite variable could be of general occurrence and could have wider significance. This result suggests that in the viral envelope glycoproteins, the N-termini, which are coded for by the 5'-half of the env gene, are variable and could be the site of type-specific antigenic markers. On the other hand, the C-terminal portions, derived from the 3'-half of the env gene, of the viral glycoproteins are less variable and could be embedded in the lipid bilayer. The nonhomology at the extreme 3'-end of the env region may also reflect a type-specific function of the glycoprotein which could be an interaction with internal virion proteins, important in the assembly of the virus.

ACKNOWLEDGMENT

We wish to thank Mrs. Cynthia Yung and Mrs. Jean Wang for excellent technical assistance.

REFERENCES

1. Baltimore, D. (1975). Cold Spring Harbor Symp. Quant. Biol. 39, 1187.
2. Vogt, P.H., and Hu, S.S.F. (1977). Ann.Rev. Genet. 11, 203.
3. Wang, L-H., Duesberg, P.H., Kawai, S., and Hanafusa, H. (1976). Proc. Natl. Acad. Sci. USA 73, 447.
4. Duesberg, P.H., Wang, L.H., Mellon, P., Mason, W.S., and Vogt, P.K. (1976). In "Animal Virology" (C.F. Fox, ed.), pp 107-125. Academic Press, New York.
5. Junghans, R.P., Hu, S., and Davidson, N. (1977). Proc. Natl. Acad. Sci. USA 74, 477.
6. Lai, M.M.C., Hu, S.S.F., and Vogt, P.K. (1977). Proc. Natl. Acad. Sci. USA 74, 4781.
7. Tal, J., Fujita, D.J., Kawai, S., Varmus, H.E. and Bishop, M.J. (1977). J. Virol. 21, 497.

8. Duesberg, P.H., and Vogt, P.K. (1973). J. Virol 12, 594.
9. Neiman, P.E., Wright, S.E., McMillin, C., and MacDonnel, D. (1974). J. Virol. 13, 837.
10. Bernstein, A., MacCormick, R., and Martin, G.S. (1976). Virology 70, 206.
11. Duesberg, P.H., and Vogt, P.K. (1970). Proc. Natl. Acad. Sci. USA. 67, 1673.
12. Stone, M.P., Smith, R.E., and Joklik, W.K. (1975). Cold Spring Harbor Symp. Quant. Biol. 39, 859.
13. Kawai, S., Duesberg, P.H., and Hanafusa, H. (1977). J. Virol. 24, 910.
14. Yoshida, M., and Ikawa, Y. (1977). Virology 83, 444.

THE PROVIRAL DNA OF VISNA VIRUS: SYNTHESIS AND PHYSICAL MAPS OF PARENTAL AND ANTIGENIC MUTANT DNA[1]

Janice E. Clements, Opendra Narayan and Diane E. Griffin[2]

Departments of Neurology, Johns Hopkins University
Baltimore, Maryland 21205

ABSTRACT The synthesis of proviral DNA of visna virus was measured at various intervals after inoculation of sheep cell cultures at a multiplicity of 1 pfu/cell. The DNA from the infected cells was fractionated by the Hirt procedure (1) into low and high molecular weight DNA and each fraction was quantitated for infectivity by plaque assay using the calcium phosphate transfection technique (2). The appearance of infectious DNA in the low m.w. fraction was biphasic apparently reflecting two rounds of low m.w. DNA synthesis. The amount of infectious DNA in the high m.w. fraction increased with time, reaching its maximum level at the time of peak virus production. Using the DNA from the Hirt supernatant and the blotting technique of Southern (3), a restriction endonuclease maps of visna viral DNA (strain 1514) have been constructed. Since visna virus changes antigenically in the presence of neutralizing antibody, both in sheep and in cell culture, a comparison was made of the restriction maps of the DNA from visna virus strain 1514, its naturally antigenic variant LV1-1 and an antigenically distinct strain D1-2.

INTRODUCTION

Visna virus is an exogenous retrovirus of sheep. The virus resembles RNA tumor viruses in morphology (4), RNA content (5, 6), protein composition (7, 8) and reverse transcriptase activity (9). In contrast to the tumor viruses, visna virus causes a productive lytic infection in cultures of sheep cells (10). Studies have shown that proviral DNA is synthesized in infected cell cultures (11) and this DNA is infectious for cell cultures (12).

Visna virus differs from the tumor viruses in that it causes a persisent infection and chronic neurological disease

[1]The work was supported by Training Grant 5T32-NS7000-03.
[2]Investigator, Howard Hughes Medical Institute.

ISBN 0-12-668350-6

of sheep (13, 14). Recent studies (15) have shown that in animals persistently infected with visna virus, antigenically altered viruses can be isolated from peripheral blood lymphocytes. This provides a mechanism for viral persistence in the presence of neutralizing antibody.

The amount of viral DNA synthesized in either tissues of persistently infected animals or in lytically infected cell cultures is unknown. This study was undertaken to examine the synthesis of infectious proviral DNA in infected sheep cell cultures to understand the viral replication cycle in culture. The proviral DNA in the Hirt supernatant thus supplied us with the DNA to begin the physical mapping of the viral genome and the comparison of parental and antigenic mutant viruses.

MATERIALS AND METHODS

Cell Cultures and Virus Strain: Primary cultures of sheep choroid plexus (SCP) cells were prepared and grown as described previously (16). Visna virus, strain 1514, was used and the virus was assayed in SCP cells either by plaque assay or by limiting dilution in microtiter plates (16).

Inoculation of Cell Cultures: Cells in $150cm^2$ plastic flasks were inoculated with visna virus as previously described (16) at a multiplicity of 1 pfu/cell. The multiplicity was confirmed by infectious center assays of inoculated cells after cultures had been incubated for 4 h at 37^oC. Cultures were sampled at various intervals after inoculation (Fig. 1). At each time interval, 8 flasks were used. Virus titers were determined on the supernatant fluid from one flask at each time period.

Isolation, Fractionation and Extraction of DNA: The medium was discarded from the flasks, the cells were washed once with 25-50ml of Hanks salts without phenol red and the cells in each flask were lysed with 5ml of 0.6% sodium dodecyl sulphate (SDS), 0.02M Tris-HCl (pH 7.4) and 0.005M EDTA. The DNA was then fractionated according to the method of Hirt (1). The supernatant and precipitate fractions were separated by centrifugation for 1 h at 15,000 rpm in a Sorvall RC-5 centrifuge equipped with an SS-34 rotor. The supernatant was treated for 2 h at 37^oC with 50μg/ml RNase A (heated at 80^oC for 10 min to inactivate DNase), made 1% in SDS and digested with 1 mg/ml pronase for 2 more hours at 37^oC. The solution was deproteinized by 2 phenol extractions and the phenol was removed by extraction with ethyl ether. The DNA was concentrated by ethanol precipitation. The pellet was redissolved in 3/4 the original sample volume of 0.05M NaCl, 0.02M Tris-HCl (pH 8.0) and 0.005M EDTA (TES). The sample was made 0.5% SDS and treated with 1 mg/ml pronase at 37°C for 16-18 h. Following phenol and ether extraction, the samples was dialyzed against TES and the sample was treated with RNase (as above). The RNase was removed by treatment with 1 mg/ml pronase in 1% SDS for 2 h at 37°C, followed by phenol and ether extraction and ethanol precipitation.

Assay of Infectious DNA: Assay for infectious DNA was performed as described by Graham and Van Der Eb (2). The extracted DNA was diluted in Hepes buffered salone (8.0 g/l Na Cl, 0.37 g/l KCL, 0.125 g/l Na_2HPO_4O (BDH Chemicals Ltd., Poole, England), 1 g/l dextrose, 5g/l N-2-hydroxyethylpiperazine-N-2-ethane sulfuric acid, pH 7.05) containing 40 μg/ml salmon sperm DNA (Calbiochem, San Diego, Calif.). Calcium chloride was added to a final concentration of 0.125M, and the mixture was allowed to stand at room temperature for 15-20 min. Aliquots (0.5ml) of the DNA and calcium phosphate were added to confluent monolayers of SCP cells in $60mm^2$ tissue culture dishes which contained 4.5ml of Leibowitz medium supplemented with 1% lamb serum (heated at 56°C for 60 min) (Grand Island Biological Company, Grand Island, N.Y.). The dishes were incubated for 5 h at 37°C. The medium was then removed and the plates washed with Hanks solution without $CaCl_2$ and overlayed with 0.5% agarose in MEM supplemented with non-essential amino acids, vitamins and 1% heated lamb serum. The plates were incubated for 3 weeks at 37°C, and the plaques were visualized by removing the agarose, fixing the cells with methanol and staining with 0.1% crystal violet in 20% ethanol. To insure that the infectivity was measured within the linear range, each DNA sample was assayed at 4 concentrations (initially at 2 fold dilutions and repeated at different concentrations, if the dose response curve was not linear).

Digestion of DNA with Restriction Endonucleases: Restriction endonucleases were purchased from Bethesda Research Laboratories, Inc. Each reaction mixture contain 5-10 μg of Hirt supernatant DNA extracted as above, and for all the digestions except those with Eco R1, 20mM Tris (pH 7.5), 7mM Mg Cl_2, 4mM 2-mercaptoethanol, Eco R1 digestions contained 100 mM Tris (pH 7.5), 50mM NaCl, 5mM $MgCl_2$ and 2mM 2-mercaptoethanol. All reactions were carried out at 37°C except for Taq I which were done at 50°C for 5 hours. ^{32}P labelled adenovirus 2 DNA, (from G. Ketner) was used as a molecular weight standard.

Agarose Gel Electrophoresis: Electrophoresis of DNA was carried out in 1% and 1.4% agarose slab gels in a buffer containing 40mM Tris-HCl, pH 7.8, 5mM sodium acetate and 1mM EDTA at 1 v/cm for 15 hr. The agarose gels were 29 cm wide and 12 cm long and were supported by a 3 cm thick 6% acrylamide plug.

Transfer of DNA to Nitrocellulose Filter: After electrophoresis DNA was denatured and transfered to a nitrocellulose filter by the method of Southern (3) using the modifications as described by Ketner and Kelly (17).

Hybridization: Visna viral RNA was iodinated by the method of Commerford (18). The specific activity of the RNA was 5×10^7 - 2×10^8 cpm/μg. The nitrocellulose filters were hybridized with ^{125}I labelled visna RNA according to the method of Gillespie and Speigelman (19). The solution contained 0.04μg/ml, ^{125}I labelled visna RNA and 200μg/ml of yeast tRNA in 2 x SSC. The filters were incubated at 67°C for 36 h. They were then washed with hot (65°C) 0.6% SDS in 2 x SSC for 4-5 h treated with pancreatic RNase (25μg/ml in 2 x SS at 37°C, for 1 h) washed with 2 x SSC and dryed. The filters were then placed in contact with Kodak x-ray film (XR-1), an Ilford fast tungstate intensitying screen and left at -70°C for a suitable exposure time (3-10 days).

RESULTS AND DISCUSSION

Comparison of the Synthesis of Infectious Low and High Molecular Weight DNA with Virus Replication: Cultures of SCP cells were infected at a moi of 1; viral infectivity of supernatant fluids and DNA content of the cells were determined at various intervals after inoculation (p.i.) Peak viral titers occurred at 96 h p.i. (Fig. 1) and the cultures had degenerated by 120 h p.i.

The Hirt fractionation was used to separate low and high molecular weight DNA from the infected cultures of SCP cells. The biological activity of this DNA was measured by the transfection technique of Graham and Van Der Eb (4). DNA from uninfected cultures of SCP cells had no infectivity at any concentration tested. Infectivity in the low and high molecular weight DNA fractions (Fig. 1) was detected at the same time after inoculation, although at this time (8 h p.i.) there was less in the low molecular weight DNA fraction. The appearance of infectious low molecular weight DNA appeared biphasic, the first phase probably representing synthesis of proviral DNA by virus in the inoculum. The second phase, occurring at 24 h, may have resulted from the synthesis of proviral DNA after reinfection of the cells with progeny virus, since the growth cycle of visna virus in culture is 20-24 h. This second phase may also have resulted from DNA synthesis by newly synthesized viral RNA-dependent DNA polymerase (RTase).

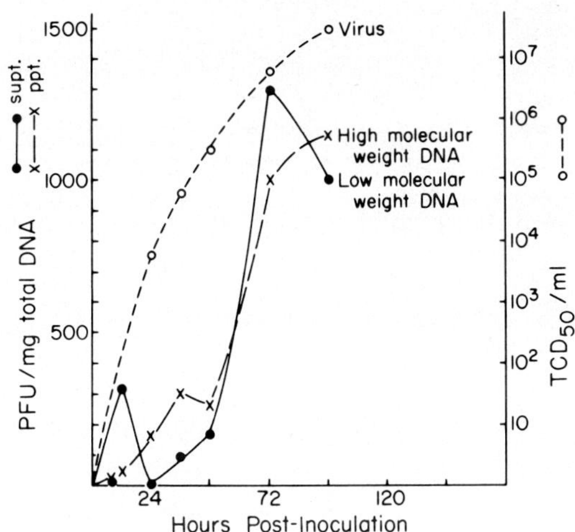

Figure 1. Temporal appearance of infectious virus and infectious DNA in sheep cells.

Infectious DNA was detected in the high molecular weight fraction at 8 h p.i. (Fig. 1). The appearance of this DNA paralleled the growth cycle of visna virus, and maximum levels of infectious high molecular weight DNA coincided with peak production of virus. During the early stage of infection, the level of high molecular weight DNA remained low until the first phase of low molecular weight infections DNA appeared. It then increased rapidly. This suggests that the low molecular weight DNA is a precursor to the high molecular weight. The appearance of the two DNA's at the same time after inoculation, presumably reflects rapid conversion from the low to high molecular weight form.

Structure and Preliminary Physical Map of Low Molecular Weight DNA: From studies in our laboratory, the low molecular weight viral DNA sedimented in sucrose gradients with a major peak at 18-19s and a minor peak at 30s. These sedimentation coefficients are consistent with double stranded linear molecules or nicked double stranded circular molecules of 6 x 10^6. Thus, this DNA represented the entire genome of visna virus whose RNA has a complexity of 3-3.5 x 10^6 and was used for physical mapping studies (20). The low molecular weight infectious DNA from 72 h p.i., was used for the endonuclease mapping sutdies. After electrophoresis of this DNA in agarose gels, transfer, and hybridization, a single band was detected which migrated with the same mobility as linear double stranded bacteriophage PM2 DNA which has a molecular weight of 6.2 x 10^6 (21). When the visna DNA was denatured by boiling, subjected to electrophoresis, and transferred, a single band was detected by hybridization at a molecular weight of 3.1 x 10^6. These data along with subsequent partial restriction endonuclease digests strongly suggest that the major form of proviral DNA at this time after infection is a double stranded linear molecule of molecular weight 6.1 x 10^6.

When this visna DNA was digested to completion by endonuclease Eco R1, three fragments were obtained (Fig. 2).

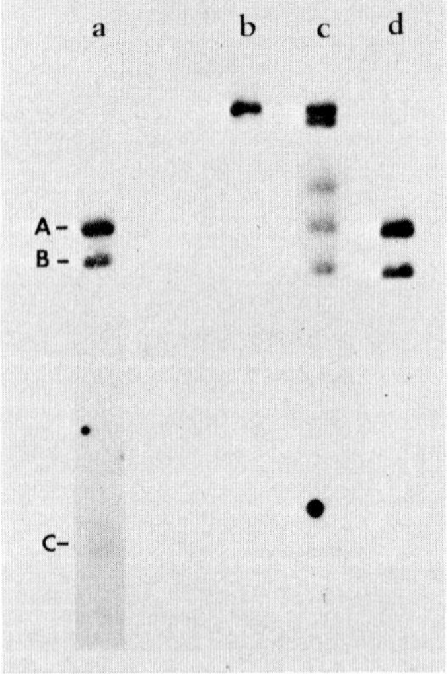

Figure 2. Restriction endonuclease cleavage fragments obtained with Eco R1; (a) complete digest, (b) undigested DNA, (c) partial digest, (d) complete digest.

Partial digestion of the DNA with Eco R1 resulted in 5 bands. From the partial and complete endonuclease R1 digests only a linear molecule is consistent, thus the fragments were ordered on a linear molecule (Fig. 3).

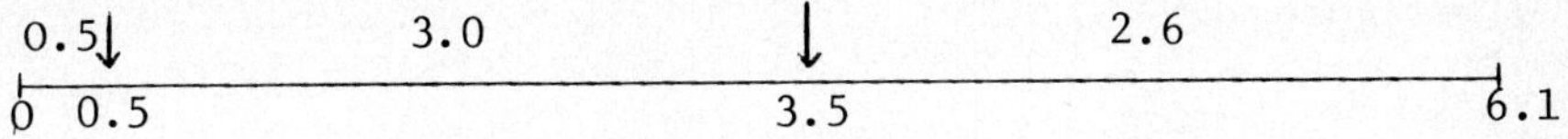

Figure 3. Eco R1 cleavage map of Visna Virus DNA.

Visna virus changes antigenically in the presence of neutralizing antibody, both in vivo and in vitro (15). To determine whether these antigenic changes could be detected by the restriction endonuclease mapping technique, three strains of visna virus DNA were compared: Visna virus 1514, our prototype, LV1-1 an antigenic mutant derived in vivo from 1514, and D1-2 a strain of visna virus which is antigenically distinct from 1514. The DNAs from these three strains were digested with endonucleases Eco R1, BAM I and a combination of Eco R1 and BAM I (Fig. 4). The Eco R1 digestion patterns were identical. However, D1-2 differs from 1514 and LV1-1 when digested with BAM I. From the double digestions, the region of 1514 and D1-2 which differs was located near the left end of the Eco R1 map.

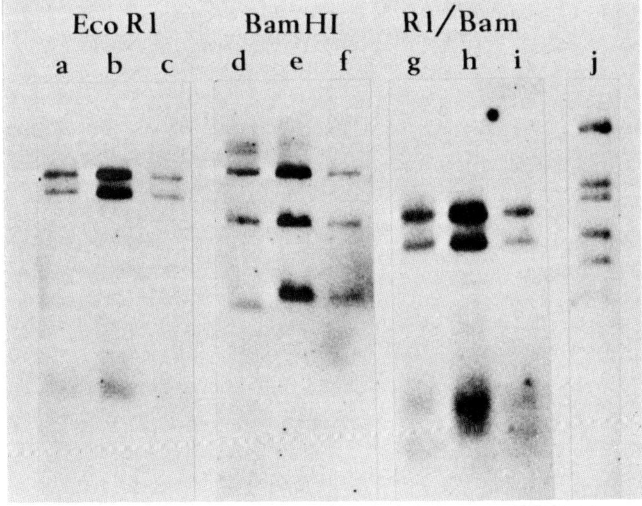

Figure 4. Restriction endonuclease digest of the DNAs of strains 1514, LV1-1 and D1-2. Lanes a-c, were digested with Eco R1, d-f with BAM I and g-1 with BAM I and Eco R1. In each group of digests, the viral DNA's (left to right) are D1-2, LV1-1 and 1514. Lane j contains an Eco R1 digest of ^{32}P adenovirus 2 DNA.

The visna strains 1514 and D1-2 differ most markedly in their antigenic characteristics. The size of their DNAs are indistinguishable but they yield different fragments upon digestion with BAM I. Thus, these viruses probably differ in only one region of the genome, perhaps the region which codes for the major glycoproteins (gp 135). Viruses with altered antigenic properties could arise by point mutation, recombination or rearrangement of the viral nuclei acid. While all of these changes could be detected by the methods used here, it is unlikely that a single point mutation would alter the viral restriction pattern. Thus, we feel the change is probably a consequence of recombination or rearrangement. These events need not occur at a high frequency, since neutralizing antibody directed against the parental virus provides a strong selective pressure for virus with altered antigenicity. If this change in viral antigenicity is shown to be a consequence of alteration in the gp 135 protein, then the region of the visna DNA containing the variable BAM I site probably codes for this protein.

When DNAs from the three strains were digested with restriction enzyme, Taq I which cleaves the DNA at several sites differences between LV1-1 and 1514 were found. However, because of the difficulty in ordering these maps by partial digestion products, we have not yet localized these changes on the endonuclease map of visna virus. Thus, antigenic changes in visna virus can be detected by restriction endonuclease mapping and it should be possible to map the region which undergoes alteration during antigenic change.

ACKNOWLEDGMENTS

We thank Nancy D'Antonio and Linda Kelly for excellent technical assistance and Richard Johnson, Thomas Kelly and Gary Ketner for editorial assistance.

REFERENCES

1. Hirt, B. (1967). J. Mol. Biol. 26, 365.
2. Graham, F.L. and Van Der Eb, A. J. (1973). Virology 52, 456.
3. Southern, E. M. (1975). J. Mol. Biol. 98, 503.
4. Thormar, H. (1961). Virology 14, 463.
5. Haase, A., Garapin, A. C., Faras, A. J., Taylor, J. M. and Bishop, J. M. (1974). Virology 57, 259.
6. Harter, D. M., Schlom, J. and Spiegelman, H. (1971). Biochem. Biophys. Acta 250, 435.

7. Mountcastle, W. E. and Harter, D. H. (1972). Virology 47, 542.
8. Haase, A.T. and Baringer, J. R. (1974). Virology 57, 238.
9. Lin, F. H. and Thormar, H. (1970). J. Virol. 6, 702.
10. Sigurdson, B., Thormar, H. and Palsson, P. A. (1960. Arch. ges Virusforsch. 10, 368.
11. Haase, A. T. and Varmus, H. E. (1970). Nature NB 245, 435.
12. Haase, A. T., Traynor, B. L. and Ventura, P. E. (1976). Virology 70, 65.
13. Sigurdsson, B., Palsson, P. A. and Grimsson, H. (1957). J. Neuropath. Exp. Neurol. 16, 389.
14. Sigurdsson, B. and Palsson, P. A. (1958). Brit. J. Exp. Path. 39, 519.
15. Narayan, O., Griffin, D. E., and Chase, J. (1977). Science 197, 376.
16. Narayan, O., Griffin, D. E. and Silverstein, A.M. (1977). J. Infect. Dis. 135, 800.
17. Ketner, G. K. and Kelly, T. J. (1976). P.N.A.S. 73, 1102.
18. Commerford, S. L. (1971). Biochemistry 10, 1993.
19. Gillespie, D. and Spiegelman, S. (1965). J. Mol. Biol. 12, 829.
20. Beeman, K. L., Faras, A. J., Haase, A. T., Duesberg, P. H. and Maisel, J. E. (1976). J. Virol. 17, 525.
21. Espejo, R. T., Canelo, E. S. and Sinsheimer, L. (1971). J. Mol. Biol. 56, 597.

A COMPARATIVE STUDY OF NUCLEOTIDE SEQUENCES AT THE 3' TERMINI OF RIBONUCLEIC ACIDS FROM VESICULAR STOMATITIS VIRUS AND ITS DEFECTIVE INTERFERING PARTICLES

Jack D. Keene, Manfred Schubert, Martin Rosenberg, and Robert A. Lazzarini

Laboratories of Molecular Biology, National Institute of Neurological Communicative Disorders and Stroke
and
National Cancer Institute, National Institutes of Health, Bethesda, Maryland 20014

ABSTRACT VSV and DI particle RNAs were labeled at their 3' ends using RNA ligase and ^{32}P cytidine 3',5' bis diphosphate. The RNAs were subjected to partial digestion with alkali and analyzed by oligonucleotide fingerprinting in two dimensions. VSV and DI particle RNAs were found to have complete sequence homology for the first 8 bases from the end with 3 mismatched nucleotides in the following 4 positions. There is again complete homology for the next 5 bases. The locations of purine residues within the sequence were confirmed by partial digestion with RNase T_1 and RNase U_2 and separation by size on 20% acrylamide gels. The latter method also indicates that sequences of VSV and DI particle RNAs diverge beyond 17 nucleotides from the 3' termini.

INTRODUCTION

The replication of VSV RNA is frequently curtailed by defective interfering particles which arise in cells following high multiplicity infections or undiluted passage of virus particles (1). The mechanisms of generation of defective particles are not established, however, it has been suggested that defective deletion mutants which have 3' and 5' terminal complementary sequences arise by copy-back synthesis on a nascent minus

ISBN 0-12-668350-6

sense RNA during transcription from the plus strand intermediate (2,3). It has also been suggested that interference of viral replication might result from a competition between the VSV and DI particle RNAs for limited quantities of viral replicase with the resultant loss of full replicative potential by the infectious virus (4). Both of these models postulate a role for VSV RNA polymerase in the generation and interference by DI particles.

Since available evidence suggests that the transcription and replication of VSV involves linear templates and products, the nucleotide sequences at the 3' termini of the VSV and DI particle RNAs would be expected to play a role in binding polymerase or initiating synthesis by polymerase and thus might be important in the generation and interference mechanisms. Attempts to probe the 3' ends of DI RNAs with the small leader RNA which is transcribed from the 3' end of VSV RNA have suggested that DI_{LT} (whose RNA is derived from the 3' portion of the genome) has complete 3' terminal sequence homology with the infectious viral RNA for about 68 nucleotides (5). The DI particle RNAs from the 5' portion of the genome were capable of forming only small and spurious RNA-RNA duplexes with the VSV leader, however. In addition, the putative panhandle RNAs from the DI particles of both sendai (2) and VSV (6) hybridize to the 5' ends of their respective infectious viral genomes but not to the 3' end. Thus, these investigators have concluded that there is no homology detectable by nucleic acid hybridization methods between the 3' ends of the infectious and defective viral RNAs.

We have determined the precise nucleotide sequence for 17 bases from the 3' termini of the San Juan and Mudd-Summers strains of VSV and from two DI particles which were derived from them. We conclude that infectious and defective viral RNAs have 14 homologous bases at their 3' termini with only three mismatched bases in the first 17 positions. Beyond these 17 bases there is extensive divergence of sequence. These findings provide an explanation for the inability of VSV leader to form stable hybrids with these DI particle RNAs and the inability of panhandle RNA to hybridize to infectious viral RNA. Furthermore, it is clear from this study that the 3' ends of DI particle RNAs are not derived directly from the 3' end of VSV RNA but are derived from somewhere else in the molecule or from an extragenomic source.

These findings are thus compatible with the suggestions that DI particle RNAs arise by a copy-back synthesis on the 5' end of the DI minus strand and that interference is the result of competition by VSV and DI particle RNAs for limited amounts of viral replicase.

METHODS

Virus Growth and Preparation. VSV and DI particles were grown in BHK_{21} cells as described (7). Virus particles were purified by PEG precipitation and twice banded isopycnically.

RNA Extraction and Labeling. RNA was extracted and purified as described (7). Isolated RNA was labeled at it's 3' end using modifications of the method of England et.al.(8) with RNA ligase from T_4 infected *E.coli* at 830 units/ml in the presence of 50 mM Hepes, pH 8.0, 8 mM $MgCl_2$, 3 mM dithiothreitol, 5 uM ATP, 67 nM RNA and 3 uM ^{32}P cytidine 3',5' bis diphosphate (specific activity 1000-2000 C/mM) in a final reaction volume of 60 ul. Ligated RNA was purified by chromatography over Sephadex G-150 and used for partial digestion.

Partial Digestion and Sequence Analysis. Terminally labeled RNAs were subjected to partial digestion using modifications of the methods of Donis-Keller *et.al*. (9). Fragments generated by partial digestion with RNase T_1, RNase U_2 and alkali were analyzed by 20% acrylamide gel electrophoresis and/or oligonucleotide fingerprinting according to the method of Sanger *et.al*. (10). Fragments of RNA were identified by autoradiography with Kodak X-Omat film.

RESULTS

Sequence studies were conducted on terminally labeled RNAs by partial digestion in alkali as described in Materials and Methods and separation of the RNA fragments in two dimensions using electrophoresis in the first dimension and homochromatography in the second dimension. Figure 1 shows the array of labeled fragments of RNA from VSV San Juan (SJ) and the DI_T that arose from the San Juan parental stock.

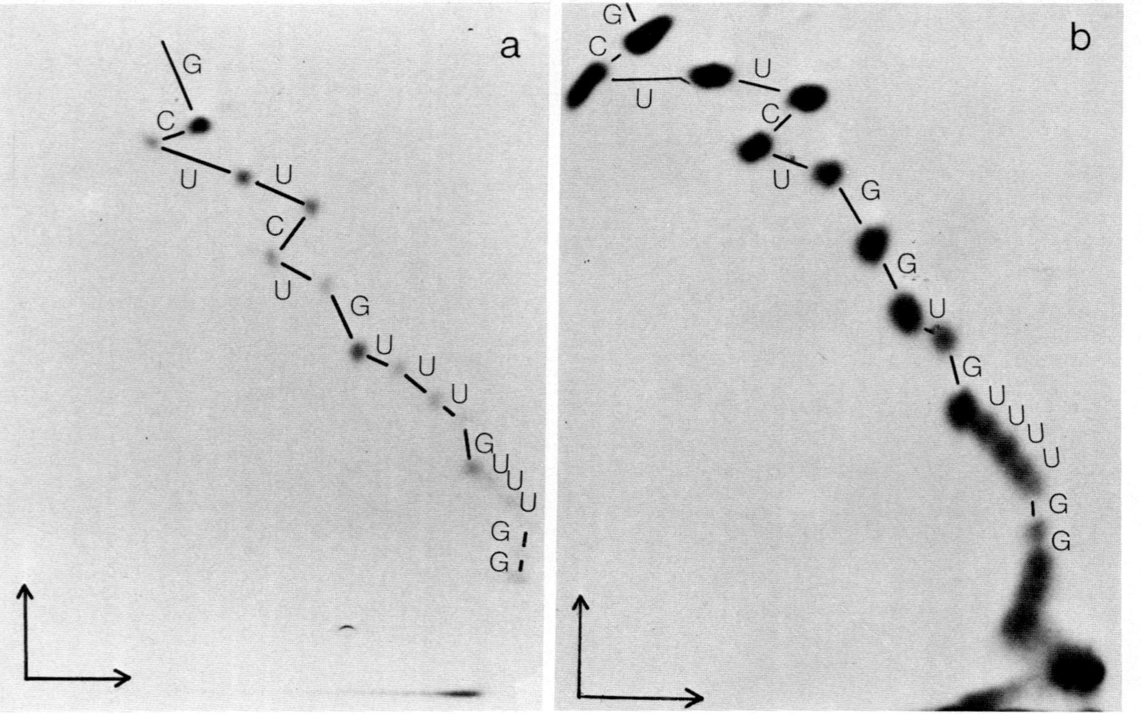

Figure 1. Oligonucleotide fingerprint analysis of 3' ^{32}P RNA following partial digestion in alkali. Samples were adsorbed directly onto cellulose acetate strips and subjected to high voltage electrophoresis in the first dimension (horizontal arrow). Following transfer to DEAE cellulose thin-layer plates, fragments were resolved by homochromatography (vertical arrow) and identified by autoradiography. a) VSV_{SJ} b) DI_T.

It is clear that the first eight nucleotides yield identical fingerprints indicating complete homology over this tract.

When the same analysis was performed using the Mudd-Summers (MS) isolate of the VSV Indiana serotype and DI_{611} derived from the MS strain, an identical fingerprint was obtained (figure 2) as for the infectious VSV_{SJ} and the DI_T, respectively. We conclude that the RNAs from all four particles are identical for the first 8 nucleotides from the 3' end. The chromatograms shown in figures 1 and 2 were developed with solutions designed to maximally mobilize the labeled fragments. As a result, very small fragments at the terminus don't appear in these figures. However, other chromatograms which were developed with weaker homochromatographic solutions clearly demonstrated that the termini were in each case ...CpGpU-OH, as was shown previously (11).

The precise sequence of nucleotides on these fingerprints was determined by analysis of the relative position of each fragment. In the vertical dimension (homochromatography), each additional nucleotide moves a fragment's position toward the bottom of the chromatographic plate, while the composition of the added base, in general, determines the relative position to the left or right hand side in the first dimension. Thus, addition of a C residue retards migration during electrophoresis, resulting in an apparent shift to the left as shown in figures 1 and 2. Addition of a U residue, on the other hand, accelerates migration in this dimension and produces an apparent shift to the right of the chromatographic plate. A and G residues are intermediate between C and U in their affect upon the rate of migration of an oligonucleotide. The addition of an A residue has little affect on the mobility of an RNA fragment in the first dimension, while G slightly accelerates the rate of migration. As shown in figures 1 and 2, the first 8 nucleotides at the 3' termini of these VSV and DI particle RNAs are 5'...pGpUpCpUpUpCpGpU-OH 3'.

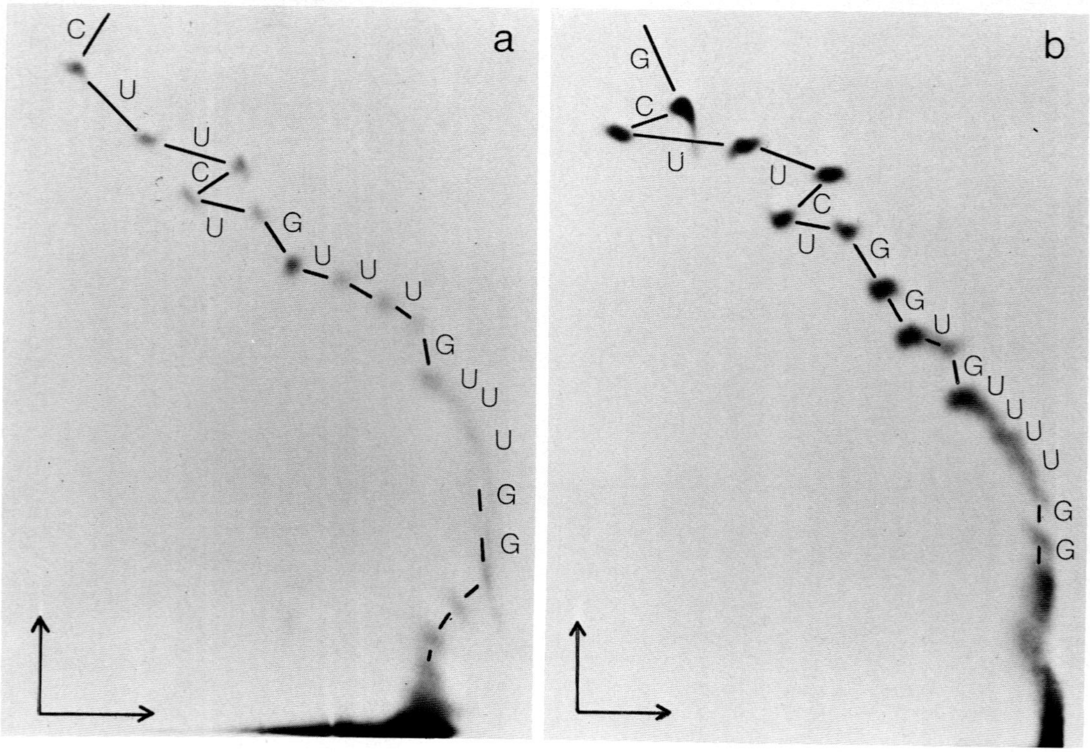

Figure 2. Oligonucleotide fingerprint analysis of 3'-^{32}P RNA following partial digestion in alkali as described in the legend to figure 1. a) VSV_{MS} b) DI_{611}.

In order to unequivocally identify the major spots on the homochromatograms, spots 1 through 6 were eluted and analyzed by high voltage electrophoresis at pH 1.7 and pH 3.5 on DEAE paper (data not shown). From this analysis, it was determined that fragment 1 was UpCp, fragment 2 was GpUpCp, etc. In each case, this sequence corresponds to the predicted position on the chromatograms.

In some fingerprints a faint shadow for each major spot appeared at the right and followed the array of spots down the plate. This shift is predicted if a single C residue was removed from each fragment. These faint spots result from the fact that RNA ligated to p*Cp is labeled in the penultimate phosphate (...Up*Cp). Upon degradation with alkali or any nuclease which cleaves the sequence Up*Cp, two sets of fragments can be generated, one set ending in ...Up* and another ending in Up*Cp. Fragments terminating in Up* are generated by two cleavages, while the major fragments (terminal p*Cp) were each derived by a single cleavage. For the study reported here, this second array of fragments was not apparent on oligonucleotide fingerprints unless the alkali digestions were extreme.

Beginning with fragment 9 in the DI particle RNAs, there is sequence divergence from the infectious viral RNAs (figure 1b and 2b). In position 9, the defective particle RNA has a G residue, while the VSV RNA has a U. This divergence continues for the next three bases since both the DI_T and DI_{611} RNAs contain the sequence 5'...UpGpUpGp...3', and the VSV parental RNAs have the sequence 5'...GpUpUpU...3'. Thus, within these four positions, there are three mismatched bases, all of which are G-->U or U-->G conversions.

Following the short stretch of three mismatched bases, there is complete homology between the defectives and their infectious parental viruses in the next 5 positions with the sequence 5'...GpGpUpUpUp...3' (figures 1 and 2).

The sequence for the 3' termini of VSV and the DI particle RNAs was confirmed by 20% acrylamide-urea gel electrophoresis of partial digests using RNase T_1 and U_2. This gel system allows the separation of these digestion products by size. RNase T_1 makes specific cleavages next to G residues, while RNase U_2 cleaves next to G and A residues under the conditions used in these

DISCUSSION

During viral replication, several kinds of events might result in the generation of deletion mutants. The selection and amplification that is required for a particle to emerge restricts the kinds of particles to those whose RNAs can be replicated. Thus, DI particle RNAs of VSV which have descrete origins of replication could emerge by a selection process that allows the propagation of molecules which possess replicase initiation sites that are compatible with the the viral replicase. The 3' terminus of the RNAs of the infectious and defective particles is one region in which to expect the origin of replication since available evidence suggests that these replicative processes occur on linear ribonucleoprotein molecules. Therefore, the DI_{LT} whose RNA is homologous to the 3' portion of the VSV genome, could have originated by a different mechanism than that which generated DI particles with RNAs from the 5' portion of the genome, but have been selected and propagated because the full replicase initiation sequence is present (presumably, without mismatching) (5).

The presence of mismatched bases in the region of homology between VSV and DI particle RNAs provides an explanation for the inability of VSV leader RNA to form stable hybrids with the 3' end of these DI particle RNAs from the 5' portion of the genome and the inability of the putative panhandle RNAs from the DI particle to form stable hybrids with the 3' ends of infectious particle RNAs (6). The small regions with mismatched bases (positions 9, 11, 12) would be susceptible to degradation by RNase and the remaining duplexed strands would probably not remain attached to one another because of their small size (8 bases). Because of this mismatching and the divergence beyond position 17, it is clear that the 3' ends of these DI particle RNAs are not derived directly from the 3' end of the VSV RNA. Instead, they must be derived from elsewhere in the molecule. The most comprehensive model to date maintains that DI particles from the 5' portion of the genome are generated by "copy-back" synthesis on a nascent replicative intermediate which results in the formation of 3' and 5' complementary termini with the 3' end of the DI minus strand. The data presented here is totally compatible with this model since a basic assumption of the model is that some sequence differences exist at the 3' ends of VSV and the DI RNA. The establishment of the precise origin of the 3' terminus of DI particle RNA will have to await additional nucleotide sequence information.

The sequence information presented here, in accordance with the above model, predicts that the 5' and 3' ends of VSV plus and minus strands should have complementarity for 17 nucleotides at the ends and mismatched bases in positions 9, 11 and 12. Terminal complementarity allows the full length plus and minus RNAs to conserve polymerase initiation sites at their 3' ends. In addition, since the 3' terminus of the DI RNA would be identical to that of the VSV plus strand, any competitive advantage or disadvantage that the DI particle might have over the minus strand during replication would also be provided for the plus strand. It is interesting to note that a mixture of VSV plus and minus RNAs would have the 3' terminal DI particle sequence present only on the plus strands while a mixture of DI particle plus and minus RNAs would have this sequence represented on each molecule.

Nucleotide sequence homology at the 3' termini of these RNAs is compatible with a linear RNA replicative scheme for VSV and the DI particles. Furthermore, since autointerference involves the replication of DI genomes at the expense of infectious viral genomes, these homologous 3' terminal sequences are candidates for a site of competition, perhaps involving the replicase enzyme. One might speculate that the three mismatched bases in the terminal seventeen nucleotides of the DI RNA are either tolerated by the replicase or perhaps confer slight advantages for replicase binding and/or initiation.

ACKNOWLEDGMENTS

We thank Drs. Olke Uhlenbeck and Tom England for communicating their ligation method prior to publication.

REFERENCES

1. Huang, A.S. (1973). Ann. Rev. Micro. 27, 101.
2. Leppert, M., Kort, L., and Kolakofsky, D. (1977). Cell 12, 539.
3. Huang, A.S. (1977). Bact. Rev. 41, 811.
4. Perrault, J., and Holland, J.J. (1972). Virology 50, 159.
5. Colonno, R.J., Lazzarini, R.A., Keene, J.D., and Banerjee, A.K. (1977). Proc. Nat. Acad. Sci. (US) 74, 1884.
6. Perrault, J., and Leavitt, R. (1978). J. Gen. Viro. 38, 35.
7. Stamminger, G.M., and Lazzarini, R.A. (1974). Cell 3, 85.
8. England, T.E., Gumport, R.I., and Uhlenbeck, O.C. (1977). Proc. Nat. Acad. Sci. (US) 74, 4839.
9. Donis-Keller, H., Maxam, A., and Gilbert, W. (1977). Nucleic Acids Res. 4, 2527.
10. Sanger, F., Brownlee, G.G., and Barrell, B.G. (1965). J. Mol. Bio. 13, 373.
11. Keene, J.D., Rosenberg, M., and Lazzarini, R.A. (1977). Proc. Nat. Acad. Sci. (US) 74, 1353.

ARE VIROIDS AUTOINDUCING REGULATORY RNAs?

T. O. Diener

Plant Virology Laboratory, Plant Protection Institute, FR, SEA
U.S. Department of Agriculture, Beltsville, MD 20705

ABSTRACT Viroids are low molecular weight RNAs that are present in certain plants afflicted with specific maladies. Viroids are not detectable in healthy individuals of the same species but, when introduced into such individuals, they replicate and cause the appearance of the characteristic disease syndrome. Viroids are the smallest known agents of infectious disease; they are responsible for a number of destructive diseases of crop plants. Electron microscopy of native viroids discloses a uniform population of seemingly double-stranded rods of about 50 nm length. After denaturation, two types of molecules are evident: a majority of linear, single-stranded molecules, about 110 nm long, and a minority of covalently-closed circular molecules, about 140 nm long. Both types of molecules are infectious. Although formation of hairpin structures and collapsed circles indicate the presence of large regions of intramolecular complementarity, thermal denaturation and other properties of viroids demonstrate that they are not regularly base-paired structures. Viroids are not translated in several cell-free protein-synthesizing systems and no new proteins are detectable in infected plants. Replication of viroids occurs in the cell nucleus and is actinomycin D-sensitive. Sequences complementary to one viroid are present in the DNAs of uninfected and infected host plants. Based on these properties, a speculative model of viroid replication and pathogenesis will be presented. Viroids are considered as abnormal regulatory RNAs that are able to interfere directly with gene regulation in their hosts.

INTRODUCTION

Viroids are low molecular weight RNAs (*ca.* 10^5d.) that can be isolated from certain species of higher plants afflicted with specific maladies, but not from healthy individuals of the same species (1). When introduced into such individuals, viroids replicate and cause the appearance of the characteristic disease symptoms. In certain other plant species,

ISBN 0-12-668350-6

however, viroids may replicate without producing obvious disease symptoms. Viroids are the smallest known agents of infectious disease (2); so far, they are known to exist only in higher plants (3).

The first viroid came to light in attempts to isolate and characterize the agent of the potato spindle tuber disease, which for many years had been assumed to be of viral etiology (4). Diener and Raymer (5) reported that the infectious agent of this disease is a free RNA and that viral nucleoprotein particles apparently are not present in infected tissue. Later, sedimentation and gel electrophoretic analyses conclusively demonstrated that the infectious RNA has a very low molecular weight (1) and that therefore the agent basically differs from conventional viruses. Five additional plant diseases, citrus exocortis (6), chrysanthemum stunt (7), cucumber pale fruit (8), chrysanthemum chlorotic mottle (9), and hop stunt (10) are now known to be caused by viroids, and recent evidence suggests that a sixth, coconut cadang-cadang, also may be (11).

In this report, recent evidence on the molecular structure of viroids, on viroid replication, and on viroid-host cell relationships will be presented, and possible mechanisms of viroid synthesis and pathogenesis will be discussed.

RESULTS

Purified potato spindle tuber viroid (PSTV) (12) migrates during electrophoresis in polyacrylamide gels as a single, homogeneous component and infectivity distribution in the gels coincides with the position of this component (13, 14). Exposure of purified PSTV to ultraviolet radiation and subsequent determination of residual infectivity levels indicated that the inactivation dose for PSTV is 90 times as large as the dose for a conventional plant virus (tobacco ringspot virus) (15). Furthermore, two-dimensional RNA fingerprinting of ribonuclease T_1 digests of PSTV revealed a pattern of complexity compatible with an RNA of 250 to 350 nucleotides (16). These results indicate that purified PSTV is composed of a single molecular species, and not of a population of RNA molecules of similar length but different nucleotide sequence.

Molecular Structure of Viroids. Electron microscopy of purified PSTV reveals a uniform population of rods (*ca.* 50 nm long) with widths similar to that of double-stranded DNA (17). Although these observations suggest double-strandedness of PSTV, other results are not compatible with this idea. For example, upon heating, PSTV melts at lower temperatures (13) than genuine double-stranded RNA (see below), and the viroid mostly elutes from hydroxyapatite at a lower phosphate buffer concentration (18, 19) than does double-stranded RNA. On the basis of these and other results, we concluded that native PSTV is a single-stranded RNA molecule with a hairpinlike configuration and with extensive regions of intramolecular base-pairing (17).

After denaturation of PSTV by heating in the presence of formaldehyde, however, the RNA reproducibly separates into two distinct bands upon electrophoresis in high concentration polyacrylamide gels under denaturing conditions (14). This observation was unexplained, but results obtained recently by electron microscopy of denatured viroids suggest a reasonable explanation.

McClements and Kaesberg (20, 21) first discovered the presence of circular as well as linear molecules in preparations of purified denatured PSTV. In two independent isolates of PSTV, the proportion of circular to linear molecules was found to be about 1:4. Under partially denaturing conditions, PSTV consisted of four types of structures: 1. Undenatured, seemingly double-stranded rods about 50 nm long; 2. partially denatured hairpin structures resembling tennis rackets; 3. completely denatured circular molecules; and 4. completely denatured linear molecules (21). The presence of circles and the complete absence of hairpins after formaldehyde treatment suggest that the circular molecules are covalently closed rather than held together by invisibly short base-paired segments (21). Measurements indicated that the circumference of the circular molecules is 140 ± 10 nm and the length of the linear molecules 110 ± 15 nm (21). The difference in length between the circular and linear forms of PSTV indicates that the linear form is not simply a nicked circle.

In confirmation of earlier work based on infectivity studies (22), electron microscopic observations after nuclease treatment showed that PSTV is sensitive to endonucleolytic digestion by pancreatic ribonuclease, but that all forms of PSTV are resistant to exonucleolytic digestion by snake venom phosphodiesterase (21). Because pretreatment of PSTV with alkaline phosphatase did not alter its resistance to exonuclease, it appears that the 3'-terminus of linear PSTV has an unusual composition.

The results of RNA fingerprinting, done with a mixture of the two components, make it unlikely that the circular and linear PSTV structures are two distinct RNA species. More likely, the two structures represent two stages of maturity of PSTV. In the presence of formamide and urea, PSTV can be separated into two fractions by gel electrophoresis (23). One fraction contains predominantly circular molecules; the second fraction contains exclusively linear molecules. Bioassay on tomato demonstrated that both circular and linear molecules are infectious (23). These results conflict with a report by Sänger *et al.* (24) in which viroids are considered to be exclusively covalently closed circular RNA molecules and in which linear molecules are claimed to be rare (0.5 - 1.0%) and representing nicked circles (24). The authors' electron micrograph depicting denatured viroid reveals, aside from some fully extended circular molecules, a majority of forms resembling "tennis rackets" or "balloons" and some linear molecules. Sänger *et al.* consider all these structures to be the result of partial renaturation of circular molecules; however, several of the linear molecules are longer than is compatible with a collapsed circle (which can be no longer than one-half the contour length of the fully extended circular molecule). Furthermore, McClements and Kaesberg (21) observed that denaturation of (at least) the linear molecules always begins at the closed or loop end of the hairpin and proceeds toward the open end, and they detected no Y-shaped molecules; they concluded that the double helical structure is less stable at the loop end than at the open end of the hairpin (21). In view of these results, an unknown number of the racketlike and balloonlike structures shown by Sänger *et al.* (24) may in fact represent partially denatured (or renatured) linear molecules and not partially collapsed circles. It can be estimated that the proportion of linear molecules in the denatured viroid preparation shown by Sänger *et al.* (24) was at least 20% but could have been as high as 60%. Thus there does not appear to be a basic contradiction between the results of McClements and Kaesberg (21) and those of Sänger *et al.* (24).

Present knowledge thus clearly indicates that viroids have a novel and unique structure. Circular single-stranded RNA molecules have not been reported previously. Also, the very high degree of intramolecular complementarity, which results in collapsed circles and hairpins with the appearance of double-stranded RNA molecules, is unusual. It is difficult to believe that this unique structure of viroids does not have important biological significance.

Physical and Chemical Properties of Viroids. Early attempts to determine the molecular weight of PSTV by sedimentation and gel electrophoretic analyses (based on infectivity distribution) led to a value of 50,000 M_r (1). Later when purified PSTV became available, gel electrophoresis of formylated PSTV gave an estimated molecular weight of 75,000 to 85,000 M_r (14). It is now clear that unambiguous molecular weight determinations were frustrated by the unknown and evidently unique structure of viroids, which rendered all RNA standards of known molecular weight inappropriate. Far more dependable molecular weight estimates have now been obtained by high- and low-speed equilibrium sedimentation of purified viroids (24) with the following results: for citrus exocortis (CEV), 119,000 ± 4,000; for PSTV, 127,000 ± 4,000; and for cucumber pale fruit viroid, 110,000 ± 5,000.

In 0.01 X SSC (SSC = 0.15 *M* NaCl - 0.015 *M* Na citrate, pH 7.0), PSTV displays a total hyperchromic shift of about 24% with a T_m of 50° (13). The thermal denaturation curve indicates that PSTV is not a regular base-paired structure, such as double-stranded RNA, but the relatively narrow temperature range for denaturation suggests the presence of extensive regions of base-pairing in the molecule.

On the basis of quantitative thermodynamic and kinetic studies on thermal denaturation, Henco *et al.* (25) concluded that viroids contain an uninterrupted double helix of 52 base-pairs, as well as several short double-helical stretches, and they proposed a tentative model for the secondary structure of viroids.

The presence of a long, uninterrupted double helix in PSTV or CEV is incompatible with the results of experiments in which treatment of unlabelled PSTV or ^{125}I-labelled CEV preparations with double-strand-specific *Escherichia coli* RNase III affected neither the electrophoretic mobility of PSTV or ^{125}I-CEV (26), nor the infectivity of PSTV (Diener, unpublished). For RNA to be cleaved by RNase III, it must contain either an extended region of perfect double-stranded RNA (25 or more base-pairs) or a highly specialized RNA sequence (27); the lack of effect implies that neither viroid contains such regions.

The base composition of PSTV, as determined by the Randerath *et al.* procedure, is: G, 28.9; C, 28.3; A, 21.7; U, 20.9 mole % (28). Thus, the A/U and G/C ratios of PSTV are close to unity, and the G+C content is 57.2%.

Sänger *et al.* (24) reported that purified cucumber pale fruit viroid does not have a 5'-end accessible to γ-^{32}-phosphorylation. This viroid is not ^{3}H-labelled upon metaperiodate oxidation and ^{3}H-borohydride reduction, which suggests that it also lacks a free 3'-end. Despite this apparent exclusion of both free 3'- and 5'-ends, including the possibility of a capped 5'-terminus containing exposed 2',3'-hydroxyls, interpretation of the experimental results is somewhat ambiguous because of efficient phosphorylation and ^{3}H-labelling of a contaminating RNA which may be degraded host RNA or "nicked viroid RNA" (24).

Attempts to detect poly(A) or poly(C) stretches in PSTV by use of the *E. coli* DNA polymerase I system in the presence of oligo(dT)$_{10}$ or oligo(dG)$_{12-18}$ primers gave negative results (29). Similarly, CEV was shown not to contain poly(A) sequences, by its inability to hybridize with ^{3}H-labelled poly(U) (30). Some plant viral RNAs are known to bind a specific amino acid in a tRNA-like manner, but no such binding of CEV has been observed (31).

Viroids are of sufficient chain length to code for a polypeptide of about 10,000 M_r. *In vitro* testing for messenger RNA function of PSTV and CEV in cell-free protein-synthesising systems from wheat germ, wheat embryo, *E. coli*, and *Pseudomonas aeruginosa* showed that neither viroid has this function (31, 32). Incubation at 60° with extracts from the thermophilic *Bacillus stearothermophilus* and dimethyl sulfoxide denaturation were done in the hope that these procedures might permit ribosome recognition of RNA regions normally inaccessible *in vitro*, but no viroid messenger activity was induced (31, 32).

Viroid-host cell interactions. When viroids are introduced into susceptible cells, they are capable of autonomous replication; i.e., replication without the assistance of a helper virus (1, 33). This basic biological fact raises a number of intriguing questions. Foremost among these are the following: 1. *By what mechanisms are viroids replicated?* Since it has been demonstrated that viroids are distinct species of low molecular weight RNA which introduce only a very limited amount of genetic information into host cells, it appears that preexisting host enzymes are largely or entirely responsible for viroid replication. 2. *By what mechanisms do viroids incite diseases in certain hosts, yet replicate in a majority of susceptible plant species without discernible damage to the host?* 3. *How did viroids originate?*

Unequivocal answers to these questions are not yet possible; currently available knowledge is summarized below.

Possible Mechanisms of Replication. Bioassay of subcellular fractions from PSTV-infected tissue disclosed that appreciable infectivity is present only in the original tissue debris and in the fraction containing nuclei (22). Chloroplasts, mitochondria, ribosomes, and the soluble fraction contain no more than traces of infectious RNA (22). Furthermore, when chromatin was isolated from infected tissue, most infectivity was associated with it and could be extracted as free RNA with high ionic strength phosphate buffer (22). These and other experiments indicate that *in situ* PSTV is associated with the nuclei, and particularly with the chromatin, of infected cells. The significant amount of infectivity regularly present in the tissue debris is most probably a consequence of incomplete extraction of nuclei, but the possibility that some PSTV is associated with cell membranes cannot be ruled out.

The fact that infectious PSTV is located primarily in the nuclei of infected cells does not prove that it is synthesized there. However, experiments with an *in vitro* RNA-synthesizing system, in which purified cell nuclei from infected tomato leaves were used as an enzyme source, demonstrated that this is the case (34). It appears, therefore, that the infecting viroid migrates to the nucleus (by an as yet unknown mechanism) and is replicated there. The absence of significant amounts of PSTV in the cytoplasmic fraction of infected cells suggests that most of the progeny viroid remains in the nucleus.

Although viroids do not act as messenger RNAs in *in vitro* protein-synthesizing systems (see above), the possibility must be entertained that they are translated *in vivo*. It is conceivable that viroids are modified after infection; for example, by addition of a poly(A) sequence to the 3'-end, and that in this form they might be translated by the host's protein-synthesizing machinery. Alternatively, preexisting host enzymes might accept the infecting viroid as a template and synthesize a complementary strand which then might act as a messenger RNA. In either case a novel, viroid-specific protein should be detectable in protein preparations from infected host tissue. Comparisons of protein species in healthy and PSTV-infected tomato (35), or healthy and CEV-infected *Gynura aurantiaca* (36) did not, however, reveal qualitative differences between healthy and infected plants. It is conceivable that a viroid-specific protein exists in quantities below the limits of detection of the methods used; nevertheless the just-mentioned results, as well as the lack of viroid association with ribosomes, suggest that viroids are not translated *in vitro*. If so, one may conclude that viroids are replicated entirely by preexisting (but possibly activated) host enzymes.

The possibility must be considered that a preexisting replicase enzyme which accepts as templates a wide variety of RNA species, including viroids, occurs in uninfected plants. RNA-directed RNA polymerases have been reported to occur in apparently healthy Chinese cabbage (37) and tobacco plants (38). A similar enzyme was detected in healthy young tomato leaves, and its activity was found to be greatly enhanced by the addition of PSTV as well as several other species of RNA, but not DNA (3). PSTV is recognized as a template by bacteriophage Qβ replicase in the presence of either Mg^{2+} or Mn^{2+}. Although the *in vitro* reaction products were heterogeneous, the largest products had molecular weights of about 100,000 M_r (39).

mally repressed host genes that are activated as a result of viroid infection. Although available evidence favors this concept (see above), it is far from conclusive. Thus, future work may invalidate the model or at least necessitate major modification.

The model of Britten and Davidson (42) for gene regulation in higher cells serves as a convenient starting point. In this model, producer genes are derepressed by activator RNAs which form a sequence-specific complex with receptor genes linked to producer genes. Activator RNAs are transcribed from integrator genes which, in turn, are regulated by sensor genes (42).

Comparison of the properties of viroids with those postulated for activator RNAs shows that viroids immediately fulfill three of the four criteria: 1) They are, in the main, confined to the nucleus; 2) they are associated with chromatin; and 3) they are not detectable in ribosomes or the cytoplasm of infected cells.

The fourth criterion for activator RNAs is that they are often the product of the redundant fraction of the genome (42). Although DNA-viroid hybridization studies appear to show that this is not true for viroids (40), closer examination reveals that viroids do fulfill this criterion. Native viroids exist in the form of hairpinlike structures with extensive regions of intramolecular base-pairing (see above). It follows, as already pointed out (43), that viroids are most likely transcribed from palindromic regions of DNA; indeed, the physical properties of viroids are to a large extent those expected of the transcriptional product of an imperfect palindrome. Palindromic structures are known to exist in the genomes of higher plants (44).

Because of the rapidity with which intramolecular complementary DNA sequences reanneal, palindromic regions anneal in DNA reassociation experiments as if they were highly reiterated (44), yet each region may contain a unique or infrequent sequence. Thus, there is no inherent conflict between the observation that DNA sequences complementary to PSTV are infrequent or unique ones (40) and the postulate that viroids are activator RNAs in Britten and Davidson's sense with the implication that they are the product of the redundant fraction of the genome.

If one accepts that viroids are activator RNAs, explanation of their pathogenic properties poses little difficulty. Introduction of a viroid into a susceptible cell may result in the activation of previously silent producer genes and, in certain genetic milieus, this activation could lead to metabolic aberrations and ultimately to discernible macroscopic symptoms. However, no novel protein species has been detected in viroid-infected plants. It is more likely, therefore, that the regulating effect of viroids rather is a subtle quantitative one, resulting in overproduction of certain proteins, as has been observed. In the version of the Britten and Davidson model that assumes redundancy in receptor genes, a producer gene might normally be derepressed to a limited extent by non-viroid activator RNA complexed to one associated receptor gene; introduction of viroid could then lead to a complex with another associated receptor gene, and further activation of the producer gene might occur.

It is not known whether the mechanism of gene derepression involves, as has often been suggested, the dissociation of nuclear protein from DNA, but, as noted (21), the hairpin-like structure of viroids provides the proper configuration for specific interaction of the RNA with a protein (45).

In the discussion of the mechanism of viroid replication below, we remain within the general framework of the Britten and Davidson model but consider also ideas expressed by others, including Gierer (46) and Frenster (47).

On theoretical grounds, Gierer has postulated "that there are many regulatory proteins, each capable of activating its own cistron, because each cistron is linked to the recognition sequence for the corresponding regulatory protein. In this way, a state of regulation would be self-stabilizing, via the cytoplasm, and could be transferred to daughter cells." The same purpose could be achieved in a more direct fashion, however, if gene regulation were mediated by RNA instead of protein; i.e., if activator RNAs could induce their own synthesis: 1) The whole process could take place in the nucleus without the necessity for migration of messenger RNAs from the nucleus to the cytoplasm and of the newly synthesized regulatory proteins back into the nucleus; 2) at least in the case of viroids, a simple mechanism for the initiation of their own synthesis suggests itself.

Frenster has postulated (47) that derepressor (activator) RNA is able "to hybridize to complementary base sequences on the anticoding strand of the DNA template, thereby freeing the coding strand of DNA for messenger RNA synthesis that is both locus-specific and strand-specific." Application of this concept to viroids leads to the recognition that, because of their hairpinlike structures, the primary sequence of the RNA transcribed from the "coding" strand of the DNA would bear a close relationship to that of the viroid hybridized to the "anticoding" strand. In particular, all palindromic sequences would be faithfully reproduced and deviations from the primary sequence of the viroid would occur only in mismatched regions of the hairpin structure. Thus, a closely related, but not identical, RNA would be synthesized. Evidently, if the necessary recognition sites were located primarily within the palindromic regions of the viroid, the progeny RNA could act in the same capacity as the original viroid.

REFERENCES

1. Diener, T. O. (1971). *Virology* 45, 411.
2. Diener, T. O. (1974). *Annu. Rev. Microbiol.* 28, 23.
3. Diener, T. O., and Hadidi, A. (1977). In "Comprehensive Virology" (H. Fraenkel-Conrat and R. R. Wagner, eds.), XI, pp. 285-337. Plenum Press, New York.
4. Diener, T. O., and Raymer, W. B. (1971). *CBI/AAB Descriptions of Plant Viruses* No. 66.
5. Diener, T. O., and Raymer, W. B. (1967). *Science* 158, 378.
6. Semancik, J. S., and Weathers, L. G. (1972). *Virology* 47, 456.
7. Diener, T. O., and Lawson, R. H. (1973). *Virology* 51, 94.
8. Van Dorst, H. J. M., and Peters, D. (1974). *Neth. J. Plant Pathol.* 80, 85.
9. Romaine, C. P., and Horst, R. K. (1975). *Virology* 64, 86.
10. Sasaki, M., and Shikata, E. (1977). *Proc. Japan Acad.* 53, Ser. B, 109.
11. Randles, J. W., Rillo, E. P., and Diener, T. O. (1976). *Virology* 74, 128.
12. Diener, T. O., Hadidi, A., and Owens, R. A. (1977). In "Methods in Virology" (K. Maramorosch and H. Koprowski, eds.), 6, pp. 185-217. Academic Press, New York.
13. Diener, T. O. (1972). *Virology* 50, 606.
14. Diener, T. O., and Smith, D. R. (1973). *Virology* 53, 359.
15. Diener, T. O., Schneider, I. R., and Smith, D. R. (1974). *Virology* 57, 577.
16. Dickson, E., Prensky, W., and Robertson, H. D. (1975). *Virology* 68, 309.
17. Sogo, J. M., Koller, T., and Diener, T. O. (1973). *Virology* 55, 70.
18. Diener, T. O. (1971). In "Comparative Virology" (K. Maramorosch and E. Kurstak, eds.), pp. 433-478. Academic Press, New York.
19. Lewandowski, L. J., Kimball, P. C., and Knight, C. A. (1971). *J. Virol.* 8, 809.
20. McClements, W. (1975). Ph.D. Thesis, University of Wisconsin, Madison.
21. McClements, W., and Kaesberg, P. (1977). *Virology* 76, 477.
22. Diener, T. O. (1971). *Virology* 43, 75.
23. Owens, R. A., Erbe, E., Hadidi, A., Steere, R. L., and Diener, T. O. (1977). *Proc. Nat. Acad. Sci. U.S.A.* 74, 3859.
24. Sänger, H. L., Klotz, G., Riesner, D., Gross, H. J., and Kleinschmidt, A. K. (1976). *Proc. Nat. Acad. Sci. U.S.A.* 73, 3852.

25. Henco, K., Riesner, D., and Sänger, H. L. (1977). *Nucleic Acids Res.* 4, 177.
26. Dickson, E. (1976). Ph.D. Thesis, Rockefeller University, New York.
27. Robertson, H. D., and Hunter, T. (1975). *J. Biol. Chem.* 250, 418.
28. Niblett, C. L., Hedgcoth, C., and Diener, T. O. (1976). In "Abstracts Beltsville Symp. on Virology in Agriculture" (J. A. Romberger, J. D. Anderson, and R. L. Powell, eds.), p. 27, U.S.D.A., Beltsville, MD.
29. Hadidi, A., Diener, T. O., and Modak, M. J. (1977). *FEBS Lett.* 75, 123.
30. Semancik, J. S. (1974). *Virology* 62, 288.
31. Hall, T. C., Wepprich, R. K., Davies, J. W., Weathers, L. G., and Semancik, J. S. (1974). *Virology* 61, 486.
32. Davies, J. W., Kaesberg, P., and Diener, T. O. (1974). *Virology* 61, 281.
33. Diener, T. O., Smith, D. R., and O'Brien, M. J. (1972). *Virology* 48, 844.
34. Takahashi, T., and Diener, T. O. (1975). *Virology* 64, 106.
35. Zaitlin, M., and Hariharasubramanian, V. (1972). *Virology* 47, 296.
36. Conejero, V., and Semancik, J. S. (1977). *Virology* 77, 221.
37. Astier-Manifacier, S., and Cornuet, P. (1971). *Biochim. Biophys. Acta* 232, 484.
38. Duda, C. T., Zaitlin, M., and Siegel, A. (1973). *Biochim. Biophys. Acta* 319, 62.
39. Owens, R. A., and Diener, T. O. (1977). *Virology* 79, 109.
40. Hadidi, A., Jones, D. M., Gillespie, D. H., Wong-Staal, E., and Diener, T. O. (1976). *Proc. Nat. Acad. Sci. U.S.A.* 73, 2453.
41. Diener, T. O., and Smith, D. R. (1975). *Virology* 63, 421.
42. Britten, R. J., and Davidson, E. H. (1969). *Science* 165, 349.
43. Reanney, D. (1976). *Bacteriol. Rev.* 40, 552.
44. Walbot, V., and Dure, L. S. III. (1976). *J. Mol. Biol.* 101, 503.
45. Gralla, J., Steitz J. A., and Crothers, D. M. (1974). *Nature (London)* 248, 204.
46. Gierer, A. (1973). *Cold Spring Harbor Symp. Quant. Biol.* 38, 951.
47. Frenster, J. H. (1965). *Nature (London)* 106, 1269.

NONCODING SEQUENCES IN ADAPTIVE GENETICS

Darryl C. Reanney

Department of Biochemistry, Lincoln College

Canterbury, New Zealand

ABSTRACT Evolution in the microbial world appears to be potentiated more by the retention of homology in duplicated *noncoding* units than by nucleotide divergence (loss of homology) in duplicated *coding* genes. The presence of identical noncoding units in different replicons facilitates the exchange and reorganisation of DNA sequences. However, the adaptive value of these recombinations is limited in the absence of effective mechanisms for cell-to-cell polynucleotide passage. Data presented in this paper suggest that the transfer efficiencies and host ranges of many conjugative plasmids have been under-estimated. Noncoding sequences probably constitute the bulk of nuclear DNA. mRNA molecules from various eukaryotic systems characteristically have noncoding leader and trailer sequences while precursor mRNAs contain noncoding insertions which appear to be excised during maturation. A model is developed which suggests that existing mRNAs represent the optimised mosaics drawn from a large number of precursor possibilities, most of which have been tested and rejected by selection.

INTRODUCTION

The best known elements of the genome are the structural genes which code for proteins. The degeneracy of the genetic code allows a diversity of nucleotide sequences to generate a common protein product; also multiple amino acids can often occupy common points along the chain (18). The result is that the functional architecture of a given protein can sometimes be sustained by policing as little as 20% of the coding gene.

These observations suggest that coding genes in microorganisms are free to accumulate substitution mutations to a quite remarkable degree. Since over 90% of the bacterial chromosome consists of coding genes it follows that tests for evolutionary relatedness based on techniques such as DNA hybridisation monitor chiefly the most "noisy" component of

ISBN 0-12-668350-6

the genome. I make this point at the beginning of this paper because our image of the evolutionary process has been largely dominated by coding genes as all metabolic mechanisms are mediated by the products of these genes.

A deeper understanding of molecular genetics projects a different image; *this is the concept that much evolutionary change is channelled through those elements of the genome which do not code for proteins.* This concept inverts a conventional evolutionary axiom. In classical genetics it is assumed that new functions arise from nucleotide *divergence* (loss of homology) between duplicated coding genes, one gene retaining its ancestral function, the other accumulating hitherto forbidden changes (49). By contrast the adaptive value of noncoding units may depend on the *retention of homology* in (certain) reiterated sequences.

Noncoding Sequences in Bacteria. Where DNA molecules recombine by either "normal" (reciprocal) or by site-specific (quantal) mechanisms strong selective constraints may be placed on sequences which represent "hot spots" for such interactions. This is because in these cases selection acts directly on the genetic material itself rather than on products once or twice removed from DNA as in the case of genes specifying tRNA/rRNA and proteins respectively.

The hypothesis that recombination targets tolerate fewer changes than do structural genes is supported by heteroduplex studies on related phages (32,58). The most compelling data known to me come from work on phages P22 and λ. Despite differences in gene content and particle morphology, it is possible to obtain viable P22/λ hybrids by exploiting the bacterial and/or phage-specified recombination systems (4). Botstein & Herskowitz (4) propose that the two phage genomes are organised into functional "quanta" (within which divergence is permitted) separated by regions which undergo ready recombination. Although not stated, it is impled that stringent restraints are placed on sequences within such recombinable targets in order to preserve the homology needed for reciprocal exchange. Since λ infects *Escherichia coli* while P22 infects *Salmonella typhimurium* it appears that these homologies have been preserved since the time when *E. coli* and *S. typhimurium* diverged from a common ancestor.

Although one cannot prove that such homology regions lie outside genes, recombination within specific noncoding sequences would appear to be less disruptive than recombination within genes (52).

Such regions of homology are *fixed* with respect to other areas of the genome. The situation is different with IS units, and phages such as Mu and λ. These units can insert promiscuously into many sites by quantal recombination (13).

If IS units have been selected chiefly as "portable regions of homology" (39) or "transmissible recombination targets" (52) it follows that extensive divergence between ancestrally identical units cannot be tolerated without destroying the very feature upon which the adaptive value of these units depends. A complication is introduced by the possibility that IS units themselves may contain a coding gene, whose product may mediate translocation (30). If selection tolerates mutations in this putative gene then homology will be most strongly retained in the noncoding *terminal* regions which seem to flank IS units and transposons (39). Some idea of the length of these terminals may be gauged from the fact that whereas Tn9 and Tn10 contain complete IS units at both ends, Tn1 contains inverted repeats only 140 base pairs long (39). These repeats presumably represent the *minimal* "substrate" requirements for those proteins that mediate transposition (29).

Direct evidence for the retention of homology in DNA domains which act as recognition signals for sequence-specific proteins has come from studies on the primary structure of the λ *att* region (42); this region shows a perfect 10 base pair match between one arm of IS1 and the OP^1 domain of the *att* site (42).

Ironically enough some of the best evidence for sequence conservation in noncoding units and divergence in coding genes comes, not from bacteria, but from viruses of higher organisms. Haywood *et al.* (28) have shown that the inverted repeats at the boundaries of the L and S subunits of the DNA of two herpesviruses, HSV-1 and HSV-2, have been conserved in evolution despite a 50% sequence divergence between the two molecules. An even more striking example comes from the study of genetic RNA in the case of Brome Mosaic Virus: the genome of this virus consists of 4 RNA molecules, each having a distinct coding sequence yet each terminating in a virtually identical noncoding sequence of 161 nucleotides at the 3' end (15).

Translocation between cells. Because they are subject to severe selective constraints, noncoding units are likely to remain (relatively) unchanged while mutation rewrites other areas of DNA. Such units then remain as favourably positioned targets for recombinational events (52). These elements, which comprise such a minute fraction of the cell's genetic content, appear to potentiate the bulk of adaptive responses in bacteria.

As I have emphasised elsewhere, (53) recombinational interactions between common sequences distributed within and among replicons in one cell can amplify and redistribute existing genes but - at least within short time intervals - further novelty is unlikely to be generated. *The wider adaptive value of IS units and related elements depends crucially on the existence of specialised mechanisms for the transfer of replicons between cells* (53). In my view the survival strategy of the bacterial cell (below) has been organised around this fact.

Conservatism and Change in Evolution. The survival strategy of the bacterial cell is schematized in Fig. 1 (53).

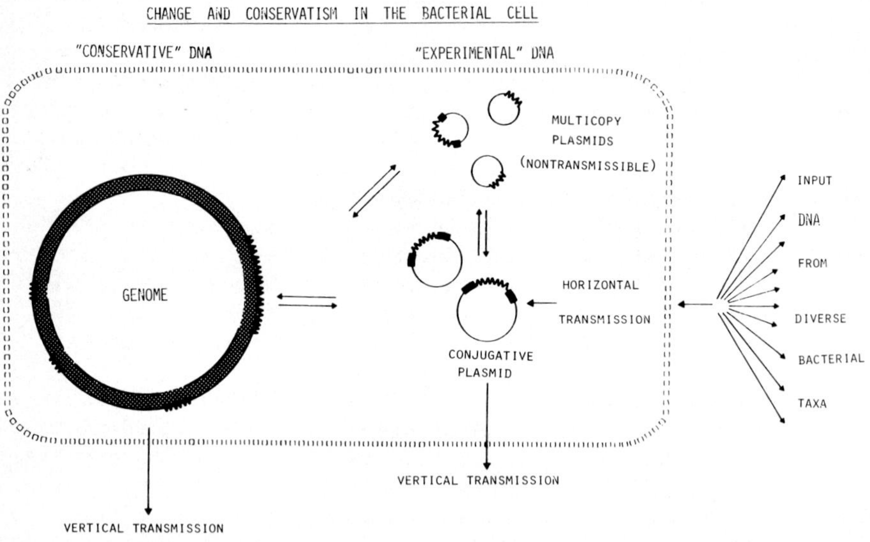

FIGURE 1: the division of genetic labour in the bacterial cell (see text). Solid lines indicate conservative functions needed for the survival of the replicon; in the case of conjugative plasmids for example such functions would include genes for replication and transfer. Sawtooth lines indicate "adventitious" genes; in the case of ECEs these could be genes for drug resistance, in the case of the genome proper genes for prophages and peripheral metabolic functions. The black rectangles represent IS-type sequences occurring at the junctions of essential and nonessential DNA domains.

This shows that prokaryotic DNA is divided into two classes, with a limited degree of overlap (1) *conservative (stationary) DNA:* this can be equated with the cell's chromosome in a conventional sense. For a given taxon the number, type and arrangement of chromosomal genes has normally been optimised by long periods of selection, hence any disruption will tend to be resisted. Since the bulk of cell DNA is chromosomal, tests based on molecular hybridisation or comparisons between orthologous proteins can yield meaningful information about bacterial taxonomy. (2) *experimental (mobile) DNA:* this can be equated with the pool of extra-chromosomal element (ECE) DNA which is such a conspicuous feature of microbial genetics. *Precisely because plasmid DNA is "dispensable", ECEs can engage in a rapid and adventurous evolution without disturbing the background of essential biochemistry directed by the genome.* ECEs are thus an insurance against environmental unpredictability and a laboratory for innovative genetic engineering. The cohabitation of these two DNA classes in the one cell reconciles the conservatism of bacterial phenotypes with the equally remarkable adaptive flexibility displayed by microorganisms under stress.

ECE genetics could almost have been designed by a systems analyst to maximise the role of ECEs as "experimental DNAs". Translocating units move between plasmids with a frequency 10^3 - 10^4 times higher than the corresponding transition from ECE to chromosome (41). Plasmid incompatibility and phage immunity have the consequence that a single cell is barred from accepting similar or isogenic ECEs. This means that heterologous ECEs have a far greater chance of taking up residence in a single cell than homologous. Recombination between such genetically different replicons - a process aided by the modular construction of many ECEs and the enzymes they often encode - creates an extraordinary potential for natural genetic engineering (52, 53).

Conjugal exchange of conservative DNA is confined to closely related taxa because of the requirements for extensive sequence homology between donor and recipient genomes. Such exchanges are exclusive, i.e. one DNA *replaces* its homologue. *By contrast the genetic autonomy of experimental DNAs uncouples ECE exchange processes from taxonomy.* Theoretically plasmids can be accepted by recipient cells which are not detectably related to donors: such exchanges often permit the stable coexistence of both replicons in one cell.

ECE Host Ranges. How far do host range data support the contention that ECEs can enter cells of diverse evolutionary origin? The answer is mixed. Plasmids of the P group can be accepted by members of the enteric taxa, by root symbionts such as *Rhizobia* and by a variety of soil bacteria including strains of *Agrobacterium, Azotobacter* and *Pseudomonas* (53). On the other hand IncFII plasmids seem confined to enteric organisms (33).

To date, almost all plasmid work has been carried out with R factors isolated from human or animal pathogens. This medical bias is reflected in the conditions commonly used to detect recombination; thus "standard" mating techniques are carried out for rather short time intervals, in broth, at 37^{0}C. To set the matter in perspective it is necessary to remember that (1) clinically important organisms constitute a minute fraction of the terrestrial microflora (2) bacteria in a ubiquitous habitat like soil rarely encounter temperatures over about 27^{0}C and (3) in nature much of the bacterial biomass is concentrated at particle/liquid interfaces.

Some hint of the restricted vision we obtain by looking at plasmids through a clinical window has come from two recent studies in New Zealand. In a suggestive survey Cooke (14) studied R$^+$ coliform bacteria taken, not from the cecum or feces but from environmental sources such as filter-feeding shellfish. For several plasmids a *drop* in mating temperature from 35^0 to 15^{0}C - such as might occur on going from intestine to environment - *increased* transfer frequency to levels characteristic of derepressed mutants (14). See Table 1.

TABLE 1

TRANSFER FREQUENCY OF pMC150 AT DIFFERENT TEMPERATURES

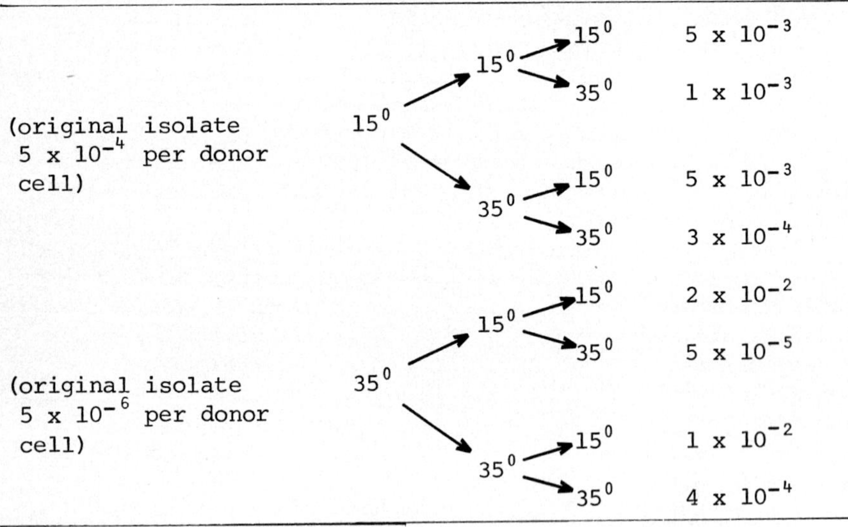

(original isolate 5 x 10^{-4} per donor cell)	15^0	15^0	15^0	5 x 10^{-3}
			35^0	1 x 10^{-3}
		35^0	15^0	5 x 10^{-3}
			35^0	3 x 10^{-4}
(original isolate 5 x 10^{-6} per donor cell)	35^0	15^0	15^0	2 x 10^{-2}
			35^0	5 x 10^{-5}
		35^0	15^0	1 x 10^{-2}
			35^0	4 x 10^{-4}

Transfer efficiencies (per donor cell) of the original isolate to *E. coli* J6-2 lac$^-$ were 5 x 10^{-4}at 15^0 and 5 x 10^{-6} at 35^0. At least 6 R$^+$ lac$^-$ recombinants were purified from each temperature selection, and pMC150 again transferred at 15^0 and 35^0 to *E. coli* 58-161F$^-$lac$^+$. R$^+$ recombinants were again purified and mated with *E. coli* J6-2 lac$^-$ at 15^0 and 35^0; pMC150 transferred with frequencies shown in the final column of the Table. The classical R factor, R1, transferred from 58-161F$^-$ to J6-2 with frequencies of 4 x 10^{-6} at 15^0 and 1 x 10^{-3} at 35^0 under the same conditions (see 14).

TABLE 2

EXTENDED HOST RANGE OF SEVERAL CONJUGATIVE PLASMIDS AT "LOW" TEMPERATURES

DONOR	RECIPIENT	FREQUENCY		
		15^0	26^0	37^0
E. coli J6 3272 pRAS1$^+$	*E. coli* J6-2		8×10^{-3}	
	Erwinia herbicolor 1414		2×10^{-4}	2×10^{-4}
	Pseudomonas aeruginosa OT 109		2×10^{-3}	4×10^{-2}
	P. fluorescens PF015		4×10^{-2}	
	P. putida 3511		4×10^{-3}	
	Pseudomonas spp		5×10^{-6}	
	Achromobacter/ Alcaligenes group		1×10^{-4}	
H_2S^+ Coliform pWK4$^+$	*E. coli* J6-2		4×10^{-2}	
	Erwinia herbicolor 1414		1×10^{-2}	0
Enterobacter Spp pWK5$^+$	*E. coli* J6-2	3×10^{-2}	5×10^{-1}	2×10^{-3}
	Erwinia herbicolor 1414		6×10^{-2}	7×10^{-6}

Matings were carried out for 20 hours under static conditions at the temperatures shown. Recipient cells were present in tenfold excess. The "coliform" organism and the uncoded strains of *Enterobacter, Pseudomonas* and *Achromobacter* were isolated from rhizosphere soil. Plasmid pRAS1 was first detected in *Klebsiella pneumoniae*. This plasmid has been tentatively assigned to IncC because of incompatibility with R40a. The three donor organisms cannot be visibly lysed by the male-specific phages PRR1, PRD1 and PR4; this constitutes weak evidence that the plasmids listed do not belong to the wide host range groups P, N and W (33). Frequency has been calculated as no. of transconjugants/no. of input donor cells. Because of the length of time allowed for mating a variable degree of multiplication of donors and transconjugants may occur. The assay system has been designed to allow such an amplification of transconjugant cells in an attempt to detect low levels of recombination (for details see 54).

I believe that all three sets of noncoding RNA in Table 3 have their origin in a common process. This is the mechanism by which the primary transcriptional units in higher cell nuclei are processed into the smaller derivatives which eventually become mRNAs on the one hand and regulatory signals on the other. Several models have been proposed to explain the inferred processing mechanism(s) (62), although to the best of my knowledge these models have restricted themselves to the spliced gene phenomenon and have not been extended to cover other noncoding sequences in eukaryotic RNA. The most plausible model seems to be Klessig's "cut and splice" mechanism (40). The key feature of the model is the suggestion that mosaic mRNAs are produced by looping out the noncoding portion and covalent rejoining of the remaining coding pieces.

At first sight mosaic mRNAs seem bizarre. I wish to show that their evolution can be rationalised by making one simple assumption. This is that *the particular set of mosaics found in existing cells represents the optimised result of selection among multiple topological precursor possibilities*. That is, a variety of loopouts and excisions have occurred during evolution, generating a large number of mosaic molecules only a few of which have survived. The unexpected feature of this thesis is the prediction that *in higher cells, a great deal of heritable evolution occurs at the level of RNA processing*.

My argument can be developed by seeking out those constraints imposed on any loopout/excision process by established data. Any loopout/excision model has to invoke *ligation* to restore physical continuity to a partly degraded RNA molecule. Ligation will occur most readily when the sensitive HO and PO_4 functions are appropriately juxtaposed by H bonding between strands of opposed polarity (36). This requirement could be met when exonucleolytic chewback of exposed 3' and 5' ends approaches double helical stem regions. An attractive alternative is that nucleases selectively recognise such stem regions and introduce (staggered?) nicks on either side. Examination of the cleavage sites of RNase P (43) and RNase III (56) shows that the cleavage site almost always occurs at or near the point where a base-paired helix gives way to a single-strand domain. Cleavage results in the production of a short single-strand tail. Joining may occur by a process analogous to the RNA ligase catalysed sealing of two H bonded $tRNA^{phe}$ fragments (37).

The above enzymes process the precursors of tRNA and rRNA. These molecules, especially rRNA, are the most strongly conserved gene products in cells and they are needed in huge quantities because of their role in translation. When one comes to those putative enzymes which process mRNA precursors I suggest that the substrate requirement for a stem structure must be retained (to allow ligation) but that the need for specific sequences of substantial length is relaxed. *mRNAs can have variable length untranslated leaders at their 5' ends and variable length untranslated tails at their 3' end (Table 3) without affecting the protein product whose length is specified by the internal positioning of START and STOP triplets (Fig. 2)*

SOME BASIC MOLECULAR FEATURES OF THE LOOP-OUT/EXCISION MODEL

(1) transcribed RNA folds into a minimum free energy structure with multiple stems and loops

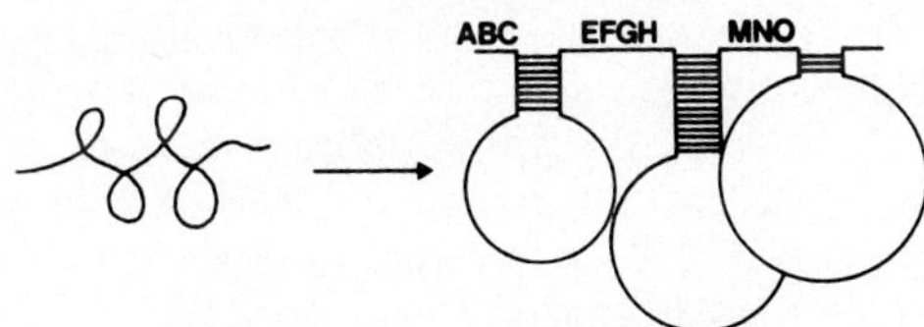

(2) resident endonucleases scan the folded RNA for recognition signals near or within regions of self complementarity; non base paired RNA is excised

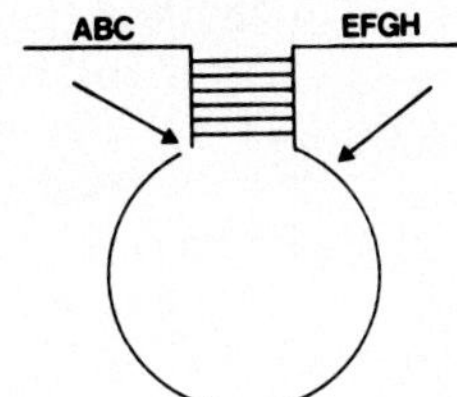

(3) the 5' and 3' functions are covalently sealed by RNA ligase

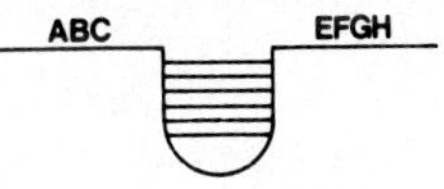

(4) a mosaic molecule is created

5' ABC EFGH MNO 3'

A Model. A possible sequence of molecular events required for a loopout/excision process is schematized in Fig. 2. This shows the production of an unique mosaic. But think about the evolution of such a pathway: it is accepted that evolution normally moves in the direction of increased specificity, not the reverse. Thus likely progenitors for the highly selective nucleases required at step 2 would be molecules of lowered sequence discrimination. An automatic consequence of lowered specificity is an expansion in the number of potential cleavage sites i.e. an increased number of mosaic possibilities.

Further it seems likely that some of the nucleases that process contemporary nuclear RNAs must be only moderately selective with respect to *sequence per se*. If a different, selective nuclease were needed to generate *each* coding mRNA in a higher cell, then the coordinated expression of multiple genes would become almost impossible to negotiate as the nucleases themselves would have to be produced via a mosaic mechanism which, in its turn, would need regulation at a higher level etc. It seems much more likely that all cells in a higher system retain a background level of moderately sequence-specific processing enzymes; in this situation the preferred maturation pathway becomes a function chiefly of a particular transcript topology and of the hierachy of successive RNA - enzyme interactions which is presumed to occur before an acceptable target site is reached (59). Put another way, new pathways can be produced by changing substrate phenotypes, not enzyme sequence specifities.

An examination of nuclease specifities in bacteria supports the view that enzymes of moderate specificity are more widespread than those with unique target sites. While enzymes capable of highly selective interactions and generating staggered nicks do exist e.g. the λ *ter* and integrase enzymes, they are not a general feature of the bacterial cell, as they are confined to the tiny subsection susceptible to λ infection. By contrast site-specific endonucleases, inappropriately called "restriction" nucleases (12), may be ubiquitous in microorganisms (55). The substrates for these enzymes are short palindromes that occur commonly in DNA molecules.

All these points can be condensed into a specific prediction. When substrate specifities are low and/or when various processing enzymes act on a common set of transcripts, then a multiplicity of *different* mosaics can be run off from essentially the *same* template. (See Fig. 3). The loopout/excision model, as defined above, *thus constitutes a novel source of variation in evolution;* it follows that a sizeable fraction of heterogenous nuclear RNA (HnRNA) in eukaryotic cells represents 'experimental' RNA, loosely analogous to the adventitious "experimental" DNA in bacteria (Fig 1).

The major problem posed by the loopout/excision model as a source of variation in evolution is the question "how are adjustments to the morphology of RNA molecules directed by DNA?" The answer is in the fact inherent in the model itself; all sequence changes in DNA can be accommodated within the topologically monotonous structure of the double helix; by contrast, sequence changes in RNA can be expressed as changes in 3D molecular geometry, like proteins. Sometimes many base

RNA SPLICING; A NEW SOURCE OF VARIATION IN EVOLUTION

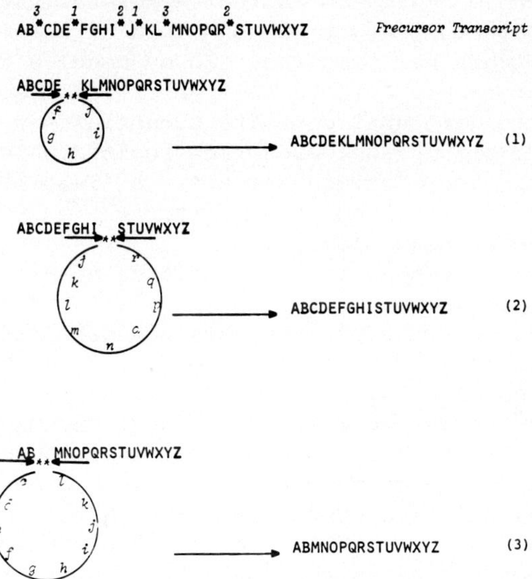

FIGURE 3. The letters of the alphabet schematize a set of information in DNA. Asterisks indicate short nucleotide sequences which, when paired with complementary sequences (indicated by common numbers) acquire the secondary structure needed to generate cleavage sites for processing enzymes. Although all RNAs are copied from a *common* template, the sequences of the mRNAs *differ* depending on the site of action of the cleavage enzyme. The three mosaic molecules in this example could be produced by three separate processing enzymes or by a single enzyme with low substrate selectivity (text).

changes in DNA will have little effect on product phenotype but in other cases the preferred precursor topology may be delicately balanced between several outcomes; often then a few nucleotide substitutions in DNA may suffice to (a) stabilise kinetically rare target sites for cleavage enzymes (b) generate new sites or (c) abolish existing sites. In the case of RNase P for example a single point mutation in the tRNA precursor remote from the cleavage site can drastically reduce the rate of enzyme cleavage (1). *A key effect of the loopout model is therefore an exaggeration of the otherwise trivial adaptive potential of genetic noise.*

This magnifying effect is seen much more clearly when the genetic event is a large-scale rearrangement such as a translocation. In conventional genetics translocation of a DNA block from one site to another generates three new DNA sequence domains: transcribed RNA is a direct molecular photocopy of this new template. In my model a translocated DNA has pleiotropic effects on transcribed RNA. If this transposed "leader" unit contains potential sequence substrates for excision nucleases then regions which base pair with it may become cleavage targets. A "leader" may thus modulate and in a limited sense *direct* the maturation pathways of adjacent sequences. Since the presence of some common coding sequences in the resulting mosaics is likely (Fig. 3) the leader may select mosaics which represent "permutations on a theme". In this way gene batteries whose products have related metabolic properties may be built up from experimental RNAs.

I suggest that the most fundamental consequence of the loopout model is its ability to facilitate the emergence of *new* gene functions grouped under a *common* modulating element. Several late adenoviral mRNAs have a common leader 150-200 bases long (3, 57). The fact that this leader is itself a mosaic suggests that many topological reshufflings must sometimes take place in order to bring diverse genes under coordinated control. Whereas the original Britten-Davidson model (7) explained how existing genes could be regulated, the mosaic model hints at how the gene batteries and their modulating controls may be evolved.

Note that excised,looped-out RNA pieces are unlikely to represent simply "garbage RNA". Fig. 4 shows how such excised fragments can be used to develop positive feedback networks with template DNA. Excised fragments are automatically complementary to corresponding sequences in DNA. Theoretically, any excised subsection of a precursor could bind DNA and prime ongoing RNA synthesis; in fact strong constraints are probably placed on feedback interactions by (1) accessibility of sites in chromatin and (2) the phenotype of the RNA. However, once a single primer has been produced in a cell (or introduced from without) *a self-sustaining cycle is set up in which the primer continuously generates further copies of itself*. By this means a permanent switch-on of genes downstream from primer binding sites is achieved. Once the concentration of primer signals is high enough, all progeny cells will inherit the phenotype specified by the product(s) of the activated gene(s). Concepts like this are obviously relevant to the problems of differentiation (20,51) Gene turn-on by this mechanism would be more reliable if the target set of middle repeat units were amplified at DNA level; accordingly Strom and Dorfman (60) have shown that the

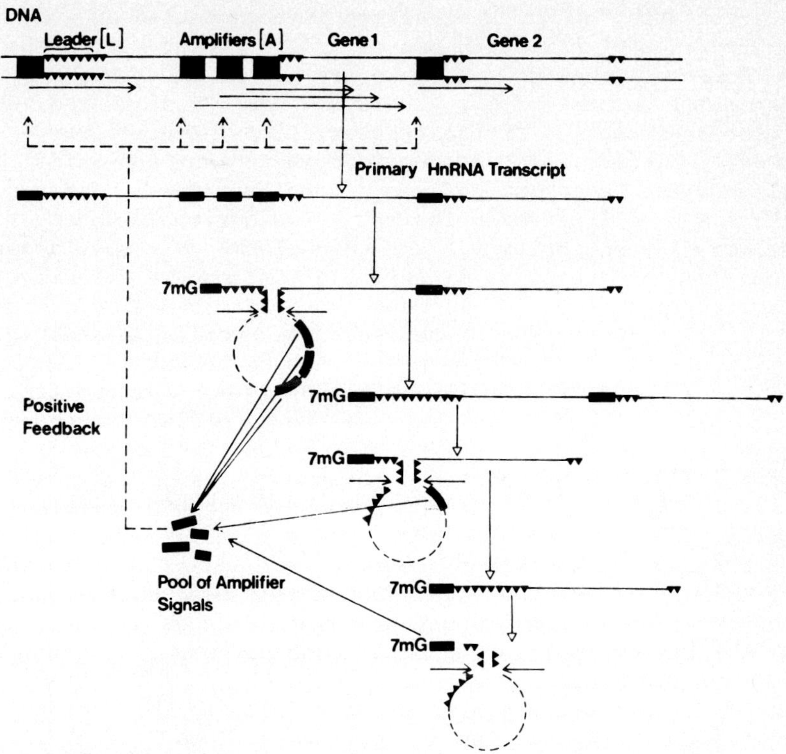

FIGURE 4: The use of excised RNA subsets to switch on distal genes; the folding of the primary HnRNA transcript is dictated largely by the *leader* sequence (sawtooth lines) in interactions with complementary sequences (sawtooth lines). These may or may not be repeated sequences. The primary transcript undergoes successive processings which join the leader sequence first to Gene 1 and then to Gene 2. A set of moderately repetitious DNA sequences (called A for *amplifier* sequences) is interspersed with single copy DNA, as shown by Gulau *et al.* (26). Although not a necessary feature of the scheme, it is attractive to hypothesize that these repeats are *inverted repeats* of the type (190 nucleotides long on average) found by Dott *et al.* (23). On transcription such inverted repeats give rise to snapback RNA regions; the presence of duplex RNA domains in HnRNA is well documented (35). Such fold-back RNA might be expected to be refractory to exonucleolytic degradation. Such signals, on binding to their DNA complements (amplifier sequences), automatically prime further RNA synthesis, thereby producing multiple copies of themselves. Note that the proposed signals have structures and functions similar to those suggested for viroids (22, 52).

differentiation of chick cartilege is accompanied by a limited magnification of a certain set of middle repeat sequences.

Some Implications. When we are almost embarrassed by the plethora of mechanisms for generating new DNA sequences, why now invoke evolution at the transcriptional level? The answer, I believe, lies in the fact that plant and animal cells, unlike bacteria, undergo a complex process of morphogenesis and cell differentiation. Two heritable modes are involved (1) continuity via the germ plasm: this must depend on DNA (2) continuity of differentiation: once a cell is committed to a given developmental step its clonal descendents must inherit the marker phenotype. This heritable somatic commitment need not involve major changes in DNA (27).

Why RNA not DNA? A likely reason is not hard to find. Evolution in higher organisms seems to be mediated more by changes in regulatory networks than by the production of new proteins (63). Most genetic changes e.g. random DNA translocations would tend to scramble such integrated circuits. *However, precisely because RNA is "dispensable" (the analogy with ECE DNA in bacteria presents itself again), experimental reprogamming at the RNA level can be tolerated by the system.* A new function or network may arise at first as a kinetically rare variant of an existing maturation pathway. If this variation improves the fitness of the cell an increasing fraction of precursor RNA in progeny cells may be directed into the new pathway. The new pathway will be "fixed" into the system once a corresponding change in the specificity of a cleavage enzyme has established clear distinction between cleavage targets for pathway A versus pathway B. *In this way new patterns can be generated without disturbing old.*

Until recently experimental evidence in support of epigenetic roles for RNA has been patchy and unconvincing. However, Deshpande and Siddiqui (19) have isolated a noncoding RNA about 200 nucleotides long from the hearts of 16 day chick embryos. This RNA appears to be a genuine inducer of cardiac differentiation. The size of this "cardiac" RNA is similar to that of viroids and the 5' viral leader sequences in Table 3. It is interesting that this size range is close to the modal value for moderately repetitious DNA (26).

The title of this session was "Viroids, Insertion Sequences, Naked Genomes: Do they have Counterparts in Animals and Man?". Since this title was phrased as a question I have designed my talk in the form of an answer. The chief impression to emerge from the model presented is of a flexible RNA subsystem within which substantial evolution can occur. In this context it is not difficult to envisage an origin for many classes of RNA virus from processed nuclear RNAs (47). If the ideas I have presented are broadly correct one might expect that in bacteria (where DNA is the "experimental" polynucleotide) ECEs would consist chiefly of duplex DNA while in higher plants and animals (where RNA is posited to be the prime "experimental" polynucleotide) ECEs should consist of RNA. This expectation is in accord with viral taxonomy (50, 52, 47). Table 4.

TABLE 4

EXPERIMENTAL POLYNUCLEOTIDES AND EXTRACHROMOSOMAL ELEMENTS; A POSSIBLE CORRELATION

PROKARYOTES

"Experimental" Polynucleotide	Extrachromosomal Elements
DNA	Plasmids (all duplex DNA) tailed phages (all duplex DNA) i.e. the vast majority of bacterial ECEs consist of DNA

EUKARYOTES

"Experimental" Polynucleotide	Extrachromosomal Elements
RNA	98% of plant viruses have RNA genomes Approx 70-78% of animal viruses have RNA genomes All known viroids (mal-functioning regulatory signals?) consist of RNA

The model gives regulatory RNAs a key role in the programmed release of epigenetic information. It seems logical to assume that mutations in the DNA specifying such regulatory units will sometimes produce aberrant molecular products. I predict therefore that malfunctioning regulatory polynucleotides will prove to be responsible for multiple diseases in higher plants and animals. As has been noted elsewhere (24) the unusual pathobiology of such degenerative diseases of the mammal nervous systems as kuru, scrapie, and the Kreutzfeldt-Jacob syndrome is consistent with a viroid-like etiology.

Whatever detailed model or models for mosaic mRNA production turns out to be right, one thing is very clear. The classical image of polynucleotides as static, conservative structures is no longer tenable. Instead we are confronted with a dynamic system which makes the evolutionary *conservation* of useful sequence organisations rather harder to explain than their *origins*.

Note added in proof: After this paper had been written and critically assessed by various colleagues, a theory similar to that proposed in the second half of this paper was presented by W. Gilbert in a short commentary article ("News and Views", Nature *271:* 501 (1978).

REFERENCES

1. Altman, S., Bothwell, A., and Stark, B. (1974). Processing of *E. coli* $tRNA^{tyr}$ precursor RNA *in vitro*. Brookhaven Symp. Biol., *26:* 12.
2. Baker, R. F. and Thomas, C.A. (1978). In Ref. 16 p. 463.
3. Berget, S., Moore, C., and Sharp, P. (1977). Proc. Natl. Acad. Sci. (USA), *74:* 3174.
4. Botstein, D., and Herskowitz, I. (1974). Nature *251:* 584.
5. Brack, C., and Tonegawa, S. (1977). Proc. Natl. Acad. Sci. (USA), *74:* 5652.
6. Breathnach, R., Mandel, J., and Chambon, P. (1977). Nature, *270:* 314.
7. Britten, R., and Davidson, E. (1969). Science, *165:* 349.
8. Britten, R., and Davidson, E. (1976). Fed. Proc., *35:* 2151.
9. Brownlee, G.G., and Cartwright, E.M. (1977). J. Molec Biol., *114:* 93.
10. Bukhari, A. (1976). Annu. Rev. Genet., *10:* 389.
11. Chang, J., Temple, G., Poon, R., Neumann, K., and Kan Yuet Wai (1977). Proc. Natl. Acad. Sci. (USA), *74:* 5145.
12. Chang, S., and Cohen, S.N. (1977). Proc. Natl. Acad. Sci. (USA), *74:* 4811.
13. Cohen, S.N. (1976). Nature, *263:* 731.
14. Cooke, M.D. (1978). Symp. Microbiol Ecology *in* Proceedings in the Life Sciences. Springer Verlag, Berlin, In Press, and N.Z. Jl. Marine and Freshwater Res., (1976). *10:* 391.
15. Dasgupta, R., and Kaesberg, P. (1977). Proc. Natl. Acad. Sci. (USA), *74:* 4900.
16. "DNA Insertion Elements, Plasmids and Episomes". (A. Bukhari, J. Shapiro, and S. Adhya, eds.), Cold Spring Harbour Lab., New York.
17. Dawid, I., and Botchan, P. (1977). Proc. Natl. Acad. Sci. (USA), *74:* 4233.
18. Dayhoff, M.O. (1972). Atlas of Protein Sequence and Structure, *5* National Biomedical Research Foundation. Silver Spring Md.

19. Deshpande, A., and Siddiqui. M. (1977). Devl. Biol. *58:* 230.
20: Dickson, E., and Robertson, H. (1976). Cancer Res., *36:* 3387.
21. Diener, T.O. (1971). Virology *45:* 411.
22. Diener, T.O. (1977). In Brookhaven Sym. Biol., *29:* In Press.
23. Dott, P., Chuang, C., and Saunders, G. (1976). Biochemistry, *15:* 4120.
24. Gajdusek, D. (1977). Science, *197:* 943.
25. Gilbert, W., cited in Robertson, M. (1977). Nature *269:* 648.
26. Gulau, G., Chamberlin, M., Hough, B., Britten, R., and Davidson, E. (1976). In *Molecular Evolution*, J. Ayala (ed.), Sunderland Mass; Sinauer Press, p. 200.
27. Gurdon, J. (1978). In Brookhaven Symp. Biol., *29:* In Press.
28. Haywood, G., Jacob, R., Wadsworth, S., and Roizman, B. (1975). Proc. Natl. Acad. Sci. (USA), *72:* 4243.
29. Heffron, F., Bedinger, P., Champoux, J., and Falkow, S. (1977). In Ref. 16, p. 161.
30. Heffron, F., Bedinger, P. Champoux, J., and Falkow, S. (1977). Proc. Natl. Acad. Sci. (USA), *74:* 702.
31. Hsu Ming-Ta, and Ford, J. (1977). Proc. Natl. Acad. Sci. (USA), *74:* 4982.
32. Hyman, R., Brunovskis, I., and Summers, W.C. (1974). Virology, *57:* 189.
33. Jacob, A.E. (1977). In Ref. 16, p. 607.
34. Jeffreys, A., and Flavell, R. (1977). Cell, *12:* In Press.
35. Jelinek, W., and Darnell, J.E. (1972). Proc. Natl. Acad. Sci. (USA) *69:* 2537.
36. Kaufmann, G., and Kallenback, N.C. (1975). Nature, *254:* 452.
37. Kaufmann, G., and Littauer, U.Z. (1974). Proc. Natl. Acad. Sci. (USA), *71:* 3741.
38. Kitchingham, G., Sing-Ping, Lai, and Westphal, H. (1977). Proc. Natl. Acad. Sci. (USA), *74:* 4392.
39. Kleckner, N. (1977). Cell, *11:* 11.
40. Klessig, D.F. (1977). Cell, *12:* 9.
41. Kretschmer, P., and Cohen, S.N. (1977). J. Bacteriol, *130:* 888.
42. Landy, A., and Ross, W. (1977). Science, *197:* 1147.
43. McClain, W., and Seidman, J. (1975). Nature, *257:* 106.
44. McClintock, B. (1967). Devel. Biol., *1:* (suppl.) 84.
45. Marotta, C., Wilson, J., Forget, B., and Weissman, S. (1977). J. Biol. Chem., *252:* 5040.
46. Mellon, P., and Duesberg, P. (1977). Nature, *270:* 631.

47. Nahmias, A., and Reanney, D. (1977). Annu. Rev. Ecol. Syst., *8:* 29.
48. Nevers, P., and Saedler, H. (1977). Nature, *268:* 109.
49. Ohno, S. (1970). Evolution by Gene Duplication. George Allen and Unwin, London.
50. Reanney, D. (1974). Int. Rev. Cytol., *37:* 21.
51. Reanney, D. (1975). J. Theor. Biol., *49:* 461.
52. Reanney, D. (1976). Bacteriol Rev., *40:* 552.
53. Reanney, D. (1978). In Brookhaven Symposia in Biology, No. 29. In Press.
54. Reanney, D., and Kelly, W.J. (1978). J. Bacteriol. In preparation for submission.
55. Roberts, R.J. (1976). Crit. Rev. Biochem., *4:* 123.
56. Robertson, H, Dickson, E., and Dunn, J. (1977). Proc. Natl. Acad. Sci. (USA), *74:* 822.
57. Sambrook, J. (1977). Nature, *268:* 101.
58. Simon, M, Davis, R., and Davidson, N. (1971). In "The Bacteriophage Lambda" (A.D. Hershey, ed.), p. 313. Cold Spring Harbour Lab., New York.
59. Sogin, M., Pace, B., and Pace, R. (1977). J. Biol. Chem., *252:* 1350.
60. Strom, C., and Dorfman, A. (1976). Proc. Natl. Acad. Sci. (USA), *73:* 3428.
61. Tonegawa, S. (1977). Cited in Breathnach *et al.* (1977).
62. Williamson, B. (1977). Nature, *270:* 295.
63. Wilson, A.C., Maxson, L.R., and Sarich, V.M.C. (1974). Proc. Natl. Acad. Sci. (USA), *71:* 2843.
64. Wilson, D., and Thomas, C.A. (1974). J. Mol. Biol., *84:* 115.

COMPETENT AND DEFECTIVE RETROVIRUSES IN THE TRANSFORMATION OF LYMPHOID CELLS[1]

David Baltimore[2], Owen N. Witte[3], Anthony Shields, Naomi Rosenberg*, and Edward J. Siden[4]

Department of Biology and
Center for Cancer Research
Massachusetts Institute of Technology
Cambridge, Massachusetts 02139

*Department of Pathology
and Cancer Research Center
Tufts University School of Medicine
Boston, Massachusetts 02111

ABSTRACT. Two kinds of murine retroviruses have been distinguished, replication competent ones and defective, transforming ones. The competent virus makes 3 polyproteins that form a budding structure at the cell surface. The two internal polyproteins are only cleaved at or soon after completion of the budding process. These cleavages activate reverse transcriptase and make the virus able to infect new cells. The defective, transforming virus, as typified by Abelson mouse leukemia virus (A-MuLV), has only parts of the competent viral genomes but also contains new information. For Abelson MuLV, a 120,000 molecular weight protein is the only viral product we can identify; it has determinants of MuLV p15, p12 and p30 but also has about 70,000 daltons of non-MuLV protein. Abelson MuLV may be a prototype of a general class of defective, transducing viruses. For Abelson virus to induce fibroblast transformation, any competent MuLV appears able to act as helper but for lymphoid cell transformation, only an oncogenic, competent virus can be helper. The helper role seems transient, however, and its expression

[1]This work was supported by grants from the American Cancer Society and the National Cancer Institute and a contract from the Virus Cancer Program of the National Cancer Institute

[2]American Cancer Society Research Professor
[3]Postdoctoral fellow of Helen Hay Whitney Foundation
[4]Postdoctoral fellow of the National Institutes of Health

ISBN 0-12-668350-6

as a complete virion is rapidly lost. This suggests a high mutability of virus-cell complexes, a suggestion that is supported by other data.

In many cases of viral transformation of mammalian cells two different viruses are involved: a replication competent virus and a defective virus. The replication competent virus is considered as a helper virus and provides many if not all of the proteins that make up the virion in which the defective viral genome is contained. Studies on murine sarcoma viruses have indicated that the helper virus genome does not play a crucial role in the transformation process and the defective viral genome by itself is able to transform cells (1, 2). Recent studies in this laboratory on the transformation of lymphoid cells by the Abelson mouse leukemia virus (A-MuLV) have helped to elucidate the roles of the defective virus and the replication competent virus in this transformation system.

THE COMPETENT RETROVIRUS

In a transformation system the competent retrovirus is one that can make all of the necessary components for itself. Three polyproteins represent the total known polypeptide products of the competent retrovirus genome (3). These polyproteins are: 1) Pr^{gag}, the protein that upon cleavage gives rise to the major internal antigens of the virus (p15, p12, p30 and p10 located in that order, 5'-3', on the murine viral genome); 2) $Pr180^{gag\text{-}pol}$, a polyprotein consisting of Pr^{gag} plus the reverse transcriptase polypeptide and, in the case of murine viruses, some extra information; 3) Pr^{env}, the precursor of the envelope glycoproteins that is cleaved to give rise to a disulfide bonded complex on the virion surface (VGP).

The competent retrovirus synthesizes its three polyproteins utilizing two size classes of mRNA. One of these size classes is approximately the size of virion RNA, about 9,000 nucleotides long, and the other is about 3,000 nucleotides long (4, 5). The large mRNA class synthesizes both Pr^{gag} and $Pr^{gag\text{-}pol}$ and it is possible to suppress a UAG codon found at the junction between the *gag* and *pol* genes to generate $Pr^{gag\text{-}pol}$ in an *in vitro* system (6). The lower molecular weight class of mRNA synthesizes only Pr^{env}, as far as is known, and is manufactured by a joining of up to 500 nucleotides from the 5' end of the viral genome to about 2500 nucleotides derived from the 3' end ("splicing") (7).

Recent work with temperature sensitive (*ts*) mutants of murine leukemia virus led us to realize that the viral proteins function in two quite separate contexts. During the synthesis of infectious virions, the two proteins that make up the internal structure of the virion, Pr^{gag} and $Pr^{gag\text{-}pol}$, come together on the cytoplasmic side of the plasma membrane to form a "budding structure." Presumably, the viral glycoprotein, either as Pr^{env} or VGP, is found on the exterior surface of the budding structure. In two temperature sensitive mutants--*ts*24 of Rauscher virus (8) and *ts*3 of Moloney virus (9)--such budding structures represent stable end points of viral maturation at the non-permissive temperature (10-12). Little or no cleavage of either Pr^{gag} or $Pr^{gag\text{-}pol}$ occurs at the non-permissive temperature. This is evident in Figure 1 which shows that when mutant-infected cells are labeled for 2 hours at non-permissive temperature--a time that for wild type virus-infected cells would allow extensive cleavage of the polyproteins--the major products recognized by a series of antisera are the precursor polyproteins: $Pr65^{gag}$, $Pr80^{env}$ or $Pr90^{env}$, and $Pr180^{gag\text{-}pol}$ (12).

When cells that have accumulated budding structures at the non-permissive temperature are shifted to the permissive temperature, budding is rapidly finished and new virions are released. Coincident with the release of virions, the cleavage process begins, most of the cleavages taking place only inside the virions (12). One cleavage in particular, the cleavage of the $Pr^{gag\text{-}pol}$ to generate p85 (the reverse transcriptase), does not take place detectably inside the cell and is in all probability an obligate process of the mature virion. This is seen in Figure 2 where *ts*24-infected cells were labeled at non-permissive temperature and then shifted to permissive temperature in the presence of cycloheximide. Both the released and intracellular viral proteins were then analyzed by immunoprecipitation and electrophoresis. An anti-reverse transcriptase antiserum precipitated mature enzyme (p85) from virions but not from the infected cell (Fig. 2, lanes 1). More than 20 minutes were required after shift down to begin accumulation of p85 even though p30 (derived by cleavage of $Pr65^{gag}$) was evident within less than 20 minutes. At non-permissive temperature, $Pr180^{gag\text{-}pol}$ accumulated in cell membranes but very little reverse transcriptase enzymatic activity could be detected in the membranes. After shift down, enzymatic activity became evident in virions coincident with the appearance of p85.

Mature virions, therefore, contain quite a different set of proteins from budding structures: these are cleavage products of the proteins that carry out budding and they become altered by cleavage to carry out the processes involved during the initial stages of infection. Virion maturation at the chemical level is presumably coincidental with the morphological change from an "immature" to a "mature" virion

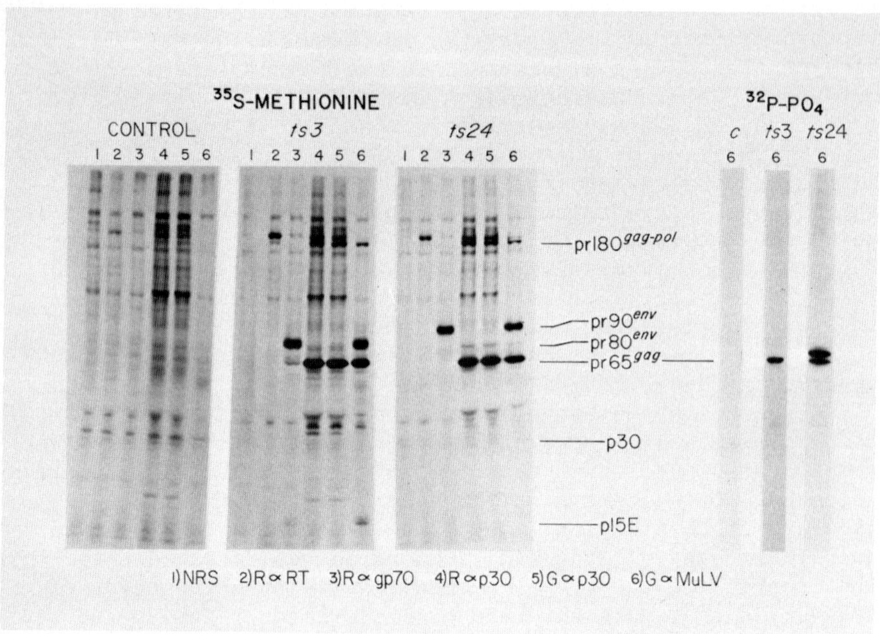

Figure 1. Production of only viral polyproteins by *ts*3 and *ts*24 incubated at non-permissive temperature. Six antisera (noted at the bottom of the Figure) were used to precipitate ^{35}S-methionine-labeled proteins from extracts of NIH/3T3 cells that were either not infected or infected by one of the *ts* mutants and were incubated at 39°. Labeling was for 2 hours. In parallel, cells were labeled with $^{32}PO_4$ for 2 hours and immunoprecipitated with a general anti-MuLV serum to show that Pr65 is a phosphoprotein. Autoradiograms are shown of polyacrylamide gels used for electrophoretic separation of the precipitated proteins in the presence of sodium dodecyl sulfate. Reprinted from reference 12.

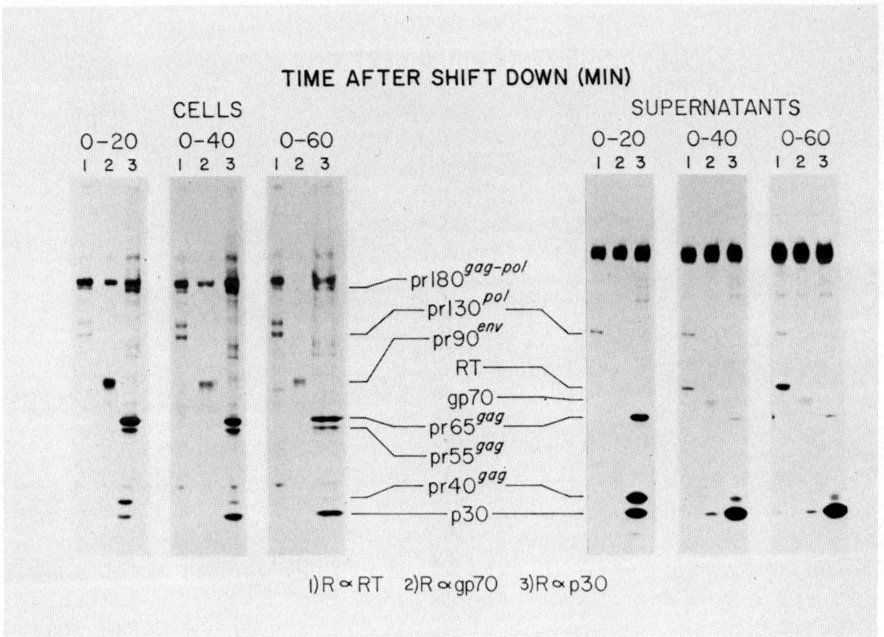

Figure 2. Cleavage of polyproteins after shift down to permissive temperature. The indicated antisera were used to precipitate proteins of *ts*24-infected NIH/3T3 cells that had been labeled at 39° for 2 hours with ^{35}S-methionine and then shifted to 32.5° in the presence of 50 μg/ml cycloheximide at time 0. Harvests of cells and supernatants containing virions were made at 20, 40 and 60 minutes later and extracts were immunoprecipitated. Reprinted from reference 12.

that was defined many years ago. This is made especially evident by the work of Yoshinaka and Luftig (13) who showed that immature particles contain an uncleaved Pr^{gag}.

DEFECTIVE RETROVIRUSES

A-MuLV is a murine retrovirus that arose in a corticosteroid-treated BALB/c mouse inoculated with Moloney MuLV (14). It has the ability to transform continuous lines of fibroblastic cells (15) as well as to transform lymphoid cells (16, 17). In its ability to transform fibroblasts it is very similar to a sarcoma virus except that sarcoma viruses are unable to carry out lymphoid cell transformation. A number of years ago we developed a quantitative *in vitro* transformation assay for Abelson virus using as target cells bone marrow cells from mice (17). The assay involves exposing cells to the virus, plating the cells in agarose, and then picking colonies of transformed cells that arise in the agarose. These colonies develop at a frequency of about 50 per 10^5 bone marrow cells and almost all colonies give rise to continuous lines of transformed cells that are able to cause tumors in syngeneic animals.

The A-MuLV transformed cell lines are probably heterogeneous but many make immunoglobulin heavy and/or light chains (18, 19 and Siden *et al.*, unpublished results). Many also have stem cell antigens as well as the enzyme terminal deoxynucleotidyl transferase, an enzyme of immature lymphoid cells (20). We therefore believe that the A-MuLV-transformed cells are immature lymphoid cells probably of the B-lymphocyte pathway.

We have recently been analyzing transformed fibroblasts and transformed lymphoid cells for their content of viral proteins. Some of the cells are producers of infectious virus and others are non-producers. We have observed that all Abelson virus transformed cells contain a protein that is not the product of the helper virus genome and has a molecular weight of 120,000 (P120) (21). This protein is evident in fibroblastic transformants, lymphoid transformants, producer cells and non-producer cells (Table 1). The producer cells, of course, also contain the standard three viral polyproteins as a consequence of the activity of the helper virus genome. Cells transformed by other means lack P120. P120 is a composite protein containing both standard MuLV antigenic determinants as well as a long segment that has no antigenic determinants related to the helper virus. Antisera against p15, p12 and some antisera against p30, are able to immunoprecipitate P120. Antisera against p10, the reverse transcriptase or gp70 are unable to immunoprecipitate P120. We therefore believe that P120 represents the 5' portion of Pr^{gag}

fused to genetic information of an unknown origin. It is unclear whether P120 represents the total translation ability of the A-MuLV genome.

TABLE 1

OCCURRENCE OF P120 IS CORRELATED WITH TRANSFORMATION BY ABELSON MuLV

Cell line origin	P120 present
A-MuLV + helper (Moloney, Friend, Gross or Rauscher) *in vitro* transformed bone marrow [producer or not]	+ (11 lines)
A-MuLV + Moloney MuLV *in vitro* transformed fibroblast [producer or not]	+ (mouse or rat)
A-MuLV + Moloney MuLV *in vivo* tumor	+
X-ray, carcinogen or Moloney MuLV *in vivo* thymoma or lymphoma	- (5 lines)

Data is summarized from ref. 21.

We have shown that P120 is the product of the Abelson virus genome by demonstrating that the RNA from a mixture of A-MuLV and helper virions directs synthesis of P120 while the RNA from helper virus alone is unable to direct the synthesis of P120 (Fig. 3). Stocks of A-MuLV in which the A-MuLV genome is in excess contain a 30S RNA that we believe to be the A-MuLV genome (Shields *et al.*, unpublished results).

We believe that A-MuLV should be considered a prototype of a general class of retroviruses that we wish to call defective, transducing viruses (DT viruses). Characteristics of DT viruses are that they have one or, more likely, both ends of the standard viral genome but are extensively deleted internally and the deleted information is replaced by information of extrinsic origin. By maintaining the ends of a viral genome, the signals necessary for reverse transcription, integration, processing, etc., may be provided by the parent virus while much of the information that is translated from

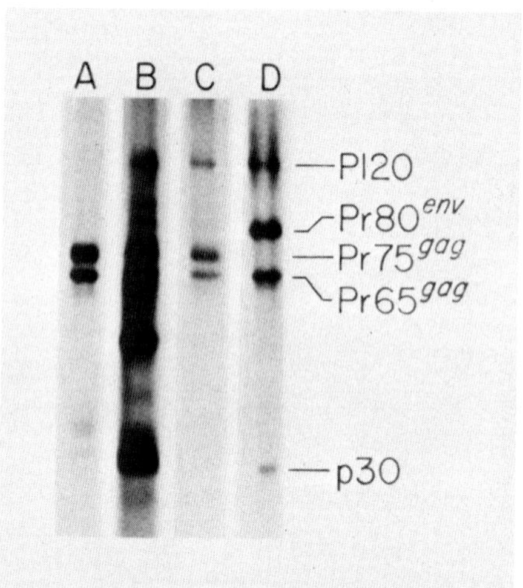

Figure 3. Translation of Moloney MuLV RNA and Abelson/Moloney MuLV RNA. Translation of RNA's in a nuclease-pretreated rabbit reticulocyte system (28) was programmed by (A) Moloney MuLV 38S or (B) total denatured virion 50-70S RNA from Abelson MuLV grown with a Moloney MuLV. (C) Immunoprecipitate of the sample in lane B with an antiserum that recognizes the Moloney MuLV p15. (D) Intracellular standards.

the viral genome is not intrinsic to the viral genome. Such a particle is a classic specialized transducing agent; it is a virus able to carry genes from one cell to another but is sufficiently defective that it requires a helper virus to form a virion. In the case of DT retroviruses, at least as typified by A-MuLV, the DT virus may encode no functional viral proteins and be totally dependent for all of its functions on the helper virus. It is a corollary of the ability to form DT retroviruses that no retrovirus protein can be cis-

REFERENCES

1. Aaronson, S.A. and Rowe, W.P. (1970). *Virology 42*, 9.
2. Bassin, R.H., Tuttle, N. and Fischinger, P.J. (1970). *Int. J. Cancer 6*, 95.
3. Eisenman, R.N. and Vogt, V.M. (1978). *Biochim. Biophys. Acta*, in press.
4. Stacey, D.W., Allfrey, V.G. and Hanafusa, H. (1977). *Proc. Natl. Acad. Sci. U.S.A. 74*, 1614.
5. Van Zaane, D., Gielkens, A.L.J., Hesselink, W.B. and Bloemers, H.P.J. (1977). *Proc. Natl. Acad. Sci. U.S.A. 74*, 1855.
6. Philipson, L., Andersson, P., Olshevsky, U., Weinberg, R., Baltimore, D. and Gesteland, R. (1978). *Cell 13*, 189.
7. Rothenberg, E., Donoghue, D.J. and Baltimore, D. (1978). *Cell*, in press.
8. Stephenson, J.R. and Aaronson, S.A. (1973). *Virology 54*, 53.
9. Wong, P.K.Y., Russ, L.J. and McCarter, J.A. (1973). *Virology 51*, 424.
10. Wong, P.K.Y. and MacLeod, R. (1975). *J. Virol. 16*, 434.
11. Yeger, H., Kalnins, V.I. and Stephenson, J.R. (1976). *Virology 74*, 459.
12. Witte, O.N. and Baltimore, D. (1978). *J. Virol.*, in press.
13. Yoshinaka, Y. and Luftig, R.B. (1977). *Proc. Natl. Acad. Sci. U.S.A. 74*, 3446.
14. Abelson, H.T. and Rabstein, L.S. (1970). *Cancer Res. 30*, 2213.
15. Scher, C.D. and Siegler, R. (1975). *Nature 253*, 729.
16. Rosenberg, N., Baltimore, D. and Scher, C.D. (1975). *Proc. Natl. Acad. Sci. U.S.A. 72*, 1932.
17. Rosenberg, N. and Baltimore, D. (1976). *J. Exp. Med. 143*, 1953.
18. Premkumar, E., Potter, M., Singer, P.A. and Sklar, M.D. (1975). *Cell 6*, 149.
19. Pratt, D.M., Strominger, J., Parkman, R., Kaplan, D., Schwaber, J., Rosenberg, N. and Scher, C.D. (1977). *Cell 12*, 683.
20. Silverstone, A.E., Rosenberg, N., Sato, V.L., Scheid, M.P., Boyse, E.A. and Baltimore, D. (1978). in *Hematopoietic Mechanisms*, Cold Spring Harbor, in press.
21. Witte, O.N., Rosenberg, N., Paskind, M., Shields, A. and Baltimore, D. (1978). *Proc. Natl. Acad. Sci. U.S.A.*, in press.
22. Bister, K., Hayman, M.J. and Vogt, P.K. (1977). *Virology 82*, 431.
23. Stephenson, J.R., Khan, A.S., Sliski, A.H. and Essex, M.

(1977). *Proc. Natl. Acad. Sci. U.S.A. 74*, 5608.
24. Scolnick, E.M., Rands, E., Williams, D. and Parks, W.P. (1973). *J. Virol. 12*, 458.
25. Scolnick, E.M. and Parks, W.P. (1974). *J. Virol. 13*, 1211.
26. Rosenberg, N. and Baltimore, D. (1978). *J. Exp. Med.*, in press.
27. Scher, C.D. (1978). *J. Exp. Med.*, in press.
28. Pelham, H.R.B. and Jackson, R.J. (1976). *Eur. J. Biochem. 67*, 247.

ANALYSIS OF MURINE LEUKEMIA VIRUS RECOMBINANTS[1]

Nancy Hopkins, Douglas V. Faller, Jean Rommelaere, and John Schindler

Biology Department and Center for Cancer Research
Massachusetts Institute of Technology
Cambridge, Massachusetts 02139

ABSTRACT We have used T1 RNA fingerprinting and oligonucleotide mapping to analyze the genomes of parental and recombinant murine leukemia viruses. These studies have allowed us to identify regions of the genome involved in specifying the major viral envelope glycoprotein, gp70, and a determinant of a viral host range phenotype designated N- or B-tropism.

INTRODUCTION

We have analyzed the genomes of an unusual class of murine leukemia viruses (MuLV), designated dual tropic viruses, that appear to arise by recombination between ecotropic and as yet unidentified xenotropic murine C type viruses. We have also analyzed the genomes and virion proteins of a set of "conventional" MuLV recombinants isolated after coinfection of cultured mouse cells with two cloned viruses differing in appropriate phenotypes. These studies have allowed us to determine approximate map locations for the MuLV envelope glycoprotein, or gp70, coding gene and for a region of the genome associated with the N- or B-tropism of MuLVs. Before discussing and comparing these studies, it is helpful to review the biological origins of the viruses we analyzed.

[1]This work was supported by National Cancer Institute grant CA-19308 to N.H. and National Institutes of Health grant CA-14051 to S. E. Luria. J.R. is a Charge de Recherches du Fonds National de la recherche scientifique de Belgique and Fellow of the Foundation Rose et Jean Hoguet.

ISBN 0-12-668350-6

Origin of dual tropic murine leukemia viruses.

MCF viruses. Rowe and his colleagues have shown that the high leukemic AKR mouse inherits two unlinked loci, designated Akv-1 and Akv-2, that appear to be the integrated DNA proviruses of indistinguishable N-tropic ecotropic viruses, designated Akv-1 and Akv-2, or interchangeably, Akv viruses (1,2). Genetic studies in which the Akv-1 or Akv-2 loci were transferred onto the low leukemic NIH Swiss background, revealed that the inheritance of either locus results in the life long, high ecotropic virus expression characteristic of the AKR mouse, and also in a high incidence of leukemia (3).

While the studies clearly implicated the endogenous Akv virus as the "cause" of AKR leukemias, several observations suggested that this virus might not be sufficient to induce leukemia (4,5,6,7). Thus Hartley et al. (8) looked for additional viruses in the leukemic or preleukemic thymuses of AKR mice. This search led to the isolation of viruses called MCF since they induce foci on mink cells in vitro. MCF viruses grow on both mouse cells like ecotropic viruses and on mink cells like xenotropic viruses. Furthermore, in their interference properties, MCF viruses resemble both ecotropic and xenotropic viruses. Since these host range and interference properties of MuLVs are determined by the major viral glycoprotein, designated gp70 and coded for by the so called env gene, Hartley et al. (8) proposed that MCF viruses arise by recombination between the ecotropic Akv virus and an endogenous xenotropic virus, and that the recombination event that generates them results in giving them an env gene derived partially from Akv and partially from their putative xenotropic parent. That MCF viruses arise during the lifetime of the mouse rather than being inherited in the germ line was suggested by the facts that 1) some MCF isolates differ in their biological properties and 2) MCF viruses can be isolated from NIH Swiss mice that inherit the Akv-1 or Akv-2 locus. These observations could be explained alternatively if the Akv-1 and -2 loci were complex and consisted of an integrated Akv provirus as well as a number of different, closely linked MCF proviruses.

While circumstantial evidence is compelling that the appearance of MCF viruses is associated with the onset of AKR leukemias, the reasons for this remain conjectural (8,9). Although at least some MCF viruses are highly leukemogenic in AKR mice, so far, MCF viruses do not appear to be very leukemogenic when injected into low-leukemic strains of mice (J. W. Hartley and W. P. Rowe, pers. comm.).

tropic virus genome that specifies the xenotropic host range.

3) Limited homology between the parents of dual tropic viruses might restrict the number of regions in which recombination can occur.

4) We have been struck by the similarity of Figures 2 and 3 to oligonucleotide maps of certain subgroup E avian type C virus recombinants, designated RAV 60 viruses, that arise after infection of chf^{+} (viral envelope glycoprotein positive) chicken cells by exogenous viruses of different envelope subgroups. To explain the apparent absence of recombination in the 5' halves of several RAV 60 genomes, Coffin et al. (44) suggested that these viruses may arise by recombination between an exogenous virus and subgenomic viral information, in particular, the mRNA for envelope glycoprotein. Biological studies have also suggested this possibility (see 44). Our analysis of the putative Moloney virus gp70 mRNA, indicating that all of the T1 oligonucleotides of Moloney that are missing in HIX are present on the intracellular Moloney virus 21S RNA, might be consistent with the possibility that the 21S RNA species could serve as a parent of dual tropic viruses.

These considerations and the unusual biological properties and origin of MCF and HIX viruses would seem to provide strong motivation for identifying and analyzing the missing, putative xenotropic parents of dual tropic viruses. This search should be greatly facilitated by the use of type specific antisera to dual tropic virus gp70s that have been developed in several laboratories (W. P. Rowe and J. W. Hartley, pers. comm.; E. Scolnick, pers. comm.).

ACKNOWLEDGMENTS

We thank Peter Fischinger for generously giving us HIX and Moloney IC viruses and Janet Hartley and Wallace Rowe for generously giving us Akv and MCF viruses.

REFERENCES

1. Rowe, W. P. (1972). J. Exp. Med. 136, 1272.
2. Chattopadhyay, S. K., Rowe, W. P., Teich, N. M., and Lowy, D. R. (1975). Proc. Natl. Acad. Sci. USA 72, 906.
3. Rowe, W. P. (1977). Harvey Lecture, 1976. Series 71, 173.
4. Kaplan, H. (1967). Cancer Res. 27, 1325.
5. Nishizuki, Y., and Nakakuki, K. (1968). Int. J. Cancer 3, 203.
6. Rowe, W. P., and Pincus, T. (1972). J. Exp. Med. 135, 429.
7. Kawashima, K., Ikeda, H., Hartley, J. W., Stockert, E.,

Rowe, W. P., and Old, L. J. (1976). Proc. Natl. Acad. Sci. USA 73, 4680.

8. Hartley, J. W., Wolford, N. K., Old, L. J., and Rowe, W. P. (1977). Proc. Natl. Acad. Sci. USA 74, 789.
9. Elder, J. H., Gautsch, J. W., Jensen, F. C., Lerner, R. A., Hartley, J. W., and Rowe, W. P. (1977). Proc. Natl. Acad. Sci. USA 74, 4676.
10. Fischinger, P. J., Nomura, S., and Bolognesi, D. P. (1975). Proc. Natl. Acad. Sci. USA 72, 5150.
11. Shih, T. Y., Weeks, M. O., Troxler, D. H., Coffin, J. M., and Scolnick, E. M. (1978). J. Virol., in press.
12. Rommelaere, J., Faller, D. V., and Hopkins, N. (1978). Proc. Natl. Acad. Sci. USA 75, 495.
13. Hartley, J. W., Rowe, W. P., and Huebner, R. J. (1970). J. Virol. 5, 221.
14. Hartley, J. W., Rowe, W. P., Capps, W. I., and Huebner, R. J. (1969). J. Virol 3, 126.
15. Hopkins, N., and Jolicoeur, P. (1975). J. Virol. 16, 991.
16. Hopkins, N., Traktman, P., and Whalen, K. (1976). J. Virol. 18, 324.
17. Schindler, J., Hynes, R., and Hopkins, N. (1977). J. Virol. 23, 700.
18. Stockert, E., Old, L. J., and Boyse, E. A. (1971). J. Exp. Med. 133, 1334.
19. Tung, J. S., Vitetta, E. S., Fleissner, E., and Boyse, E. A. (1975). J. Exp. Med. 141, 198.
20. O'Donnell, P. V., and Stockert, E. (1976). J. Virol. 20, 545.
21. Hopkins, N., Schindler, J., and Gottlieb, P. D. (1977). J. Virol. 21, 1074.
22. Faller, D. V., and Hopkins, N. (1977). J. Virol. 24, 609.
23. Faller, D. V., and Hopkins, N. (1978). J. Virol., in press.
24. Faller, D. V., and Hopkins, N. (1978). J. Virol., in press.
25. Faller, D. V., and Hopkins, N. (1977). J. Virol. 23, 188.
26. Wang, L., Duesberg, P. H., Kawai, S., and Hanafusa, H. (1976). Proc. Natl. Acad. Sci. USA 75, 447.
27. Coffin, J. M., and Billeter, A. M. (1976). J. Mol. Biol. 100, 293.
28. Jolicoeur, P., and Baltimore, D. (1975). J. Virol. 16, 1593.
29. Fan, H., and Paskind, M. (1974). J. Virol. 14, 421.
30. Rommelaere, J., Faller, D. V., and Hopkins, N. (1978). J. Virol. 24, 690.
31. Okabe, H., Gilden, R. V., and Hatanaka, M. (1973). Proc. Natl. Acad. Sci. USA 70, 3923.

32. Rommelaere, J., Faller, D. V., and Hopkins, N. (1978). In "Cold Spring Harbor Conference on Cell Proliferation," Volume 5, in press.
33. Faller, D. V., Rommelaere, J., and Hopkins, N. (1978). Proc. Natl. Acad. Sci. USA, in press.
34. Joho, R. H., Billeter, M. A., and Weissmann, C. (1975). Proc. Natl. Acad. Sci. USA 72, 4772.
35. Hopkins, N., Schindler, J., and Hynes, P. (1977). J. Virol. 21, 309.
36. Stacey, D. W., Allfrey, V. G., and Hanafusa, H. (1977). Proc. Natl. Acad. Sci. USA 74, 1614.
37. Gielkins, A. L. J., Van Zaane, D., Bloemers, H. P. J., and Bloemendal, H. (1976). Proc. Natl. Acad. Sci. USA, 73, 356.
38. Mellon, P., and Duesberg, P. H. (1977). Nature 270, 631.
39. Weiss, S. R., Varmus, H. E., and Bishop, J. M. (1978). Cell, in press.
40. Rothenberg, E., Donoghue, D. J., and Baltimore, D. (1978). Cell, in press.
41. Fan, H., and Verma, I. (178). J. Virol., in press.
42. Joho, R. H., Stoll, E., Friis, R. R., Billeter, M. A., and Weissmann, C. (1976). In "Proc. ICN-UCLA Symp. Mol. Cell Biol." (D. Baltimore, A. Huang, and C. Fox, eds.), Vol. 4, p. 127. Academic Press, New York.
43. Barbacid, M., Robbins, K. C., Hino, S., and Aaronson, S. A. (1978). Proc. Natl. Acad. Sci. USA 75, 923.
44. Coffin, J. M., Champion, M. A., and Chabot, F. (1977). In "Proc. of ICREW-EMBO Workshop on Avian RNA Tumor Viruses," in press.

NOVEL PROTEINS OF REPLICATION-DEFECTIVE MAMMALIAN ONCOGENIC TYPE-C VIRUSES

Edward M. Scolnick, Thomas Y. Shih, Sandra Ruscetti, David Linemeyer, and David Troxler

Laboratory of Tumor Virus Genetics,
National Cancer Institute, National Institutes of Health,
Bethesda, Maryland 20014

ABSTRACT

Protein products coded for by (i) fibroblast transforming viruses, Ki-MuSV and Ha-MuSV and (ii) a hematopoietic transforming agent, spleen focus-forming virus (SFFV) have been identified. Translation of the 5' end of Ki-MuSV RNA or translation of the RNA of the rat viral progenitor of Ki-MuSV yields a 50,000 dalton protein. The peptide map of Ki-MuSV p50 is distinct from the map of the p22 translated from the 5' end of the Ha-MuSV RNA. Comparison of the size of the Ki-MuSV and Ha-MuSV genomes indicates Ha-MuSV is approximately 6.5 kilobases (Kb) and Ki-MuSV 8.0 Kb. Physical mapping of large RNase T1-resistant oligonucleotides indicates that the 1.5 Kb size difference in these two viruses is due to genetic sequences, derived from the endogenous rat virus, at the 5' end of Ki-MuSV RNA. This portion of the endogenous rat virus RNA is not present in Ha-MuSV RNA. The results strongly suggest that the p50 of Ki-MuSV represents the gene product of the 5' end of the endogenous rat virus genome, although technical limitations do not allow an unambiguous proof that this is the case. In the case of SFFV, to detect the protein product of the xenotropic sequences in the SFFV genome, we have developed an MCF specific radioimmunoassay using ^{125}I-labelled gp70 from a Friend MCF virus and antiserum prepared against an MCF strain of Moloney type-C virus. In the MCF immunoassay, we can detect a cross-reacting gene product coded for by the Friend strain of spleen focus-forming virus in rat fibroblasts infected with SFFV. The results support earlier molecular hybridization studies which indicated that the genome of SFFV contained a <u>env</u> gene similar to that in AKR-MCF and Moloney type-C viruses.

ISBN 0-12-668350-6

INTRODUCTION

For the past 6 years, our laboratory has been studying the genesis of oncogenic variants of murine type-C viruses. Two systems in particular have received most of our attention. The first involves study of the Kirsten and Harvey strains of murine sarcoma viruses and the second involves study of the Friend strain of spleen focus-forming virus. This manuscript will focus on recent work related to identifying proteins coded for by these viruses with the goal to identify the gene product(s) of each responsible for their oncogenic potential.

RESULTS

Map of Ki-MuSV and Ha-MuSV. To understand the origin of the map translational products obtained from Ki-MuSV and Ha-MuSV RNA, we have compared the genetic maps of these two viruses by oligonucleotide fingerprinting techniques.

Recent studies have reported a physical map of the mouse and rat sequences in Ki-MuSV (1). Oligonucleotide fingerprints showed that Ki-MuSV RNA has mouse sequences near its 5' and 3' ends, and approximately 7000 nucleotides of rat genetic information between the mouse information. Importantly, the rat information extends close to the very 5' end of the Ki-MuSV genome (Young *et al*., submitted, 1978). Because of the proximity of the rat information to the 5' end of Ki-MuSV, we have attempted to identify proteins coded for by Ki-MuSV and Ha-MuSV by translation of their respective RNAs. In initial experiments, translation of the genomic RNA of Ha-MuSV has shown to yield a 21,000 dalton protein (2).

The current studies were undertaken to compare the physical maps of Ha-MuSV and Ki-MuSV genomes, and to compare the protein products translated from the 5' end of the genomic RNA of each sarcoma virus.

In order to obtain genomic RNA of Ki-MuSV and Ha-MuSV, it was necessary to take into consideration three factors: (i) The ratio of the replication-defective sarcoma virus and the helper virus used to pseudotype it; (ii) Potential contamination of the exogenously added viruses with replication-defective endogenous viruses. This is a particular problem for the smaller genomes of Ki-MuSV and Ha-MuSV which are close in size to endogenous viral RNA known to be expressed in some mouse cells and rat cells (3). Thus, physical separation of large amounts of purified sarcoma virus RNA might be a problem; (iii) The absolute titer of viruses released by a given culture. A good ratio of defective virus to helper virus frequently could be obtained only in cultures producing low titers of virus. As a result of extensive efforts to isolate virus-releasing cultures meeting these criteria, cell lines and viral RNAs shown in Table 1 were chosen for biochemical studies.

TABLE 1

Sources of Viruses Used for Fingerprinting or Translation

Added Viruses		Cells Grown in	Potential Endogenous Virus	Ratio Viral RNAs		
Sarcoma	Helper			Sarcoma :	Helper :	Endogenous
-	Mo-MuLV	SC-1 (mouse)	Mouse 30S	-	20	1
Ki-MuSV	FeLV	Mink	Mink type-C	20	1	1
Ha-MuSV	Mo-MuLV	FRE (rat)	Rat 30S and RaLV	15	1	< 0.2
-	Mo-MuLV	HTP (rat)	Rat 30S	-	1	8
-	Mo-MuLV	NIH (mouse)	Mouse 30S	-	1	1
-	RaLV	NRK (rat)	Rat 30S	-	1	1

Sources of Viruses. As detailed in Materials and Methods, 60-70S viral RNA was isolated from each culture. Hybridization analyses were carried out with 10,000 fold excess of viral RNA over each [^{3}H]cDNA, utilizing the specific cDNAs indicated in Materials and Methods, and the 1/2 C_rt value determined. The ratios of the indicated RNAs were determined from the 1/2 C_rt values. Each RNA which was over 90% pure had a 1/2 C_rt between 3.0 - 4.0 x 10^{-2} moles·sec·liter^{-1} with its respective cDNA. Each of the cDNAs used had less than 3% homology to the other RNAs in the potential mixture that was analyzed.

The Ki-MuSV (FeLV) culture had a stable 20:1 excess of Ki-MuSV RNA as judged by hybridization kinetics analysis with $cDNA_{Ki-MuSV}$ and cDNA FeLV. In addition, a [^{3}H]cDNA to a mink type-C virus (S. A. Aaronson _et al_., C. J. Sherr _et al_., personal communication, 1978) detected less than 5% contamination of the Ki-MuSV RNA. As further confirmation of the relative purity of this Ki-MuSV RNA, oligonucleotide fingerprinting of ^{32}P-labelled viral RNA from this culture has revealed a set of Ki-MuSV specific spots with a complexity of approximately 2.5×10^6 daltons, consistent with the size of Ki-MuSV genome as determined by gel electrophoresis (4).

For Ha-MuSV, a Mo-MuLV pseudotype grown in a Fisher rat embryo cell, FRE Cl 2, was used. This culture produced a stable 15:1 excess of Ha-MuSV over Mo-MuLV as judged by hybridization with $cDNA_{Ki-MuSV}$ and $cDNA_{Mo-MuLV}$. By hybridizing with $cDNA_{rat}$ 30S or $cDNA_{RaLV}$, less than 2% contamination by endogenous rat virus RNA was detected. These two cultures satisfying the above 3 criteria were used to obtain large amounts of sarcoma virus RNA. For pure Mo-MuLV RNA, Mo-MuLV was propagated in SC-1 cells, where relatively low levels of a 30S RNA of a defective mouse virus is expressed (3).

For preparation of the defective rat viral 30S RNA, Mo-MuLV was propagated in HTP cells (5). This culture produced an 8-fold excess of the RNA of the defective rat virus. Unfortunately, the absolute titer of virus produced was low (approximately $10^{3.0}$ XC PFU per ml) and detailed study on this viral RNA was still not possible. For preparation of the defective mouse viral 30S RNA, Mo-MuLV was propagated in NIH cells. The high molecular weight viral RNA was further purified by velocity sedimentation on sucrose gradients (3) to obtain a preparation in which the 30S endogenous RNA was approximately 90% pure.

To identify common or homologous T1 resistant oligonucleotides between Ha-MuSV and Ki-MuSV RNA, protection against RNase digestion of [^{32}P] labelled viral RNA hybridized with cDNA prepared from the other viral RNA was performed. Most of the large T1 resistant oligonucleotides of Ki-MuSV RNA have been mapped and ordered in the 5' to 3' direction within the Ki-MuSV RNA (1). Identification of Ha-MuSV homologous oligonucleotides enables one to place the location of these oligonucleotides within the Ki-MuSV genome. As it is seen in Fig. 1, none of the contiguous stretch of the eight oligonucleotides starting from #29 to #1 at the 5' portion of Ki-MuSV genome is homologous to Ha-MuSV. The oligonucleotides protected by Ha-MuSV cDNA start from #40 and occur intermittently all the way close to the rat-mouse junction at #33. The mouse sequences from #25 to #22 which are protected by Ki-MuLV cDNA are not protected by Ha-MuSV cDNA. In contrast,

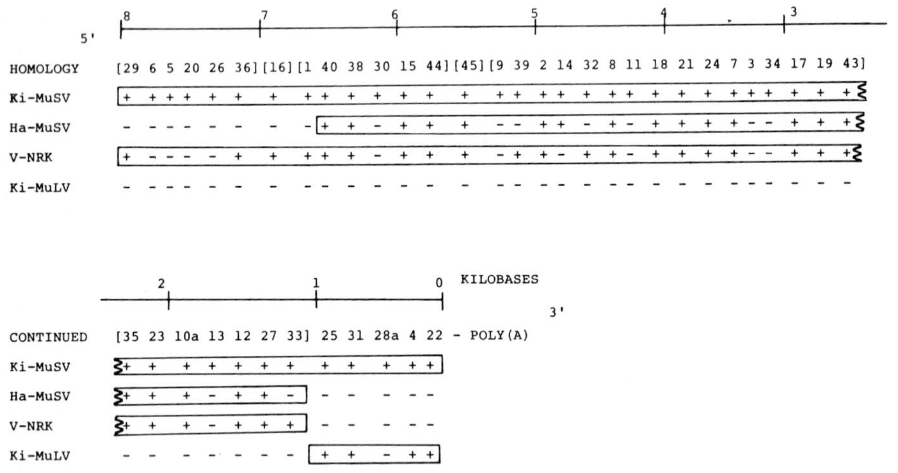

FIGURE 1. The physical map of homologous oligonucleotides among Ki-MuSV, Ha-MuSV, V-NRK and Ki-MuLV genomic RNAs. The homologous oligonucleotides among these viruses were mapped within the Ki-MuSV genome. Numbers indicate those T1 oligonucleotides of Ki-MuSV RNA arranged in the 5' to 3' direction within the viral genome. Ki-MuSV oligonucleotides which are homologous to and were protected by cross hybridization with cDNA prepared from the respective viruses are indicated by "+". Oligonucleotides which appear to represent contiguous segments of the genome are enclosed in blocks. The physical location of oligonucleotides within the Ki-MuSV genome were marked by assuming equal spacings of these marker oligonucleotides. Data on Ki-MuLV cDNA protection were presented in a previous publication (21).

V-NRK prepared from the endogenous rat 30S RNA protects most of the oligonucleotides from the very 5' end to the rat-mouse junction at #33. Although Ki-MuSV retains most of the endogenous rat 30S RNA, the 5' portion represented by the 8 oligonucleotides from #29 to #1, however, has not been incorporated into the Ha-MuSV genome. The Ki-MuSV oligonucleotides from #40 to #33 are intermittently protected by Ha-MuSV cDNA and no clustering of non-protected oligonucleotides is evident in the map. Since the control experiments have indicated that all Ha-MuSV RNA sequences are represented in the Ha-MuSV cDNA preparation, the lack of protection of the 8 oligonucleotides at the 5' portion of Ki-MuSV RNA suggests that this portion of Ki-MuSV genome is not present in Ha-MuSV. On the contrary, 4 oligonucleotides in the group of 8 at the 5' end are protected by V-NRK cDNA if the hybrid is digested with RNases A and T1. In addition, by protecting ^{32}P-labelled Ki-MuSV RNA with V-NRK cDNA and digesting with RNase T1 alone, spot #6 could also be protected. Therefore, 5 out of 8 oligonucleotides at the 5' end of Ki-MuSV can be identified as rat in origin. As reported previously, in the Ki-MuSV genome most of the mouse sequences are at the 3' 1 Kb of the viral RNA, and the rat sequences start from this point and extend very close to the 5' end of the genome.

Since almost all the large T1 resistant oligonucleotides in the size ranges between the two dye markers (ca. 15 to 30 nucleotides long) were resolved in the fingerprints, and were included in the map, it is reasonable to assume that these large T1 resistant oligonucleotides represent random space sampling of the viral genome. The subunit size of Ki-MuSV is 2.5×10^6 daltons as determined by polyacrylamide-formamide gel electrophoresis (4) and is approximately 8 Kb. As it is seen in Fig. 4, 1.4 Kb of the 5' Ki-MuSV genome is not present in the Ha-MuSV RNA. The subunit size of Ha-MuSV is 1.9×10^6 daltons by polyacrylamide-formamide gel electrophoresis (4) or 6.5 Kb by Hg-agarose gel electrophoresis (Chien and Davidson, personal communication, 1978). The 5.5 Kb of rat sequences within the Ha-MuSV as seen in the map is in excellent accord with the subunit size of 6.5 Kb allowing 1 Kb for the mouse sequences homologous to Mo-MuLV.

In Vitro Translation of RNAs. The viral RNAs from Mo-MuLV grown in SC-1 cells, Ki-MuSV grown in mink cells, and Ha-MuSV in FRE cells were translated in a nuclease digested reticulocyte _in vitro_ system (6). Translation of each of these viral RNAs had a magnesium optimum of 1.0mM, and optimum incorporation occurred with 2.0 A_{260} of reticulocyte extract per 0.035 ml reaction mixture. The products of the reaction are shown in lanes 3, 4 and 5 of the slab gel

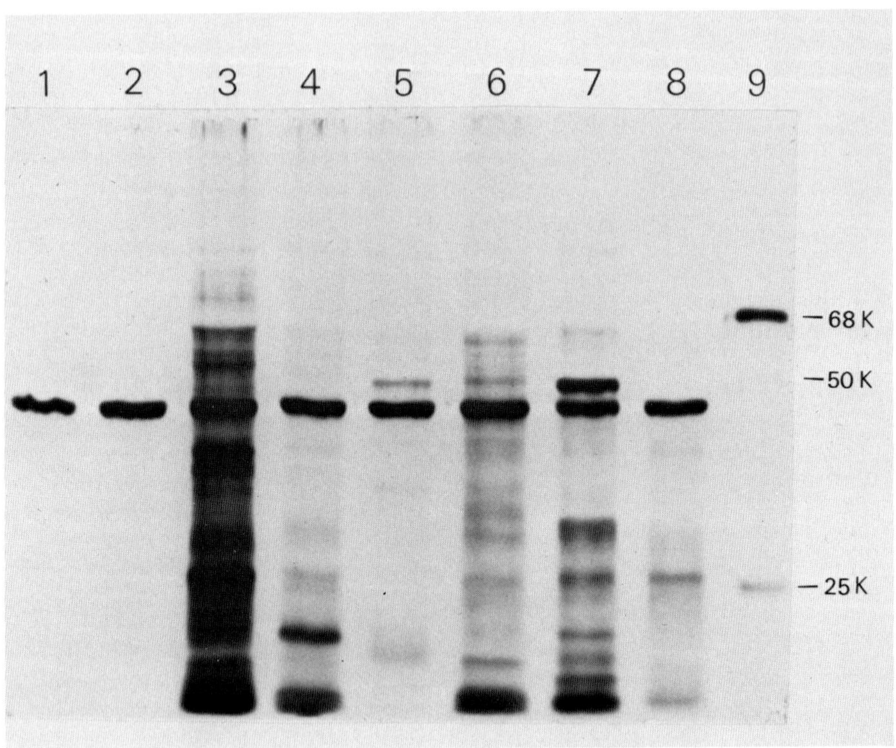

FIGURE 2. Translation of various viral RNAs. Cell free translation and slab gel electrophoresis was carried out as detailed in Materials and Methods.

Lane 1: No added RNA; 1 mM magnesium; $2A_{260}$ extract
Lane 2: No added RNA; 3 mM magnesium; $1A_{260}$ extract
Lane 3: Plus Mo-MuLV RNA; 1 mM magnesium; $2A_{260}$ extract
Lane 4: Plus Ha-MuSV RNA; 1 mM magnesium; $2A_{260}$ extract
Lane 5: Plus Ki-MuSV RNA; 1 mM magnesium; $2A_{260}$ extract
Lane 6: Plus rat endogenous virus RNA; 1 mM magnesium; $2A_{260}$ extract
Lane 7: Plus mouse endogenous virus RNA; 3 mM magnesium; $1A_{260}$ extract
Lane 8: Plus Mo-MuLV RNA; 3 mM magnesium; $1A_{260}$ extract
Lane 9: Iodinated markers bovine serum albumin 68,000, IgG 50,000 and 25,000

autoradiogram in Figure 2. Translation of Mo-MuLV RNA yields major bands at 25,000, 40,000, 65,000, 75,000, 85,000 and 110,000 daltons (7) (lane 3). Translation of Ha-MuSV RNA yields a major band at 22,000 daltons as previously described (2) (lane 4). Translation of Ki-MuSV RNA yields a major band at 50,000 daltons and light bands at 35,000 and 15,000 daltons (lane 5). In earlier studies (2), the 50,000 dalton protein was not resolved from the background band shown in lane 1 and lane 2. In lane 6, is the product of translation of the RNA of the defective endogenous rat virus. The major band is also at 50,000 daltons. In lane 7, is shown the product of the reaction with the RNA of the replication-defective virus recently described in mouse cells. The major band once again is at 50,000 daltons, although this band migrates slightly faster than the product obtained from the Ki-MuSV or rat 30S RNA. Interestingly, the optimum conditions for translation of the mouse 30S RNA were 1.0 A_{260} reticulocyte extract and 3.0mM magnesium. However, in studies not shown, the band was visible at 1.0mM magnesium at about 50% the intensity of that in 3mM magnesium. Thus, the rat 30S RNA (lane 6), mouse 30S RNA (lane 7) and Ki-MuSV RNA (lane 5) each yield a major translation product of approximately 50,000 daltons. Similarly sized product of 50,000 daltons was also observed by translating of Ki-MuSV RNA with extract prepared from NIH 3T3 mouse cells (data not shown). In contrast, Mo-MuLV RNA either at 1mM magnesium and $2A_{260}$ of extract (lane 3) or 3mM magnesium and $1A_{260}$ of extract (lane 8) fails to give a band of this size. Similarly, in other studies not shown, translation of Ki-MuLV RNA also yielded products similar in size to those obtained with Mo-MuLV RNA, and did not give a band of 50,000 daltons.

Tryptic Maps of [^{35}S]Methionine Reaction Products. The bands synthesized from the various templates shown in Fig. 2 were eluted with trypsin and subjected to 2-dimensional fingerprinting on cellulose thin layer plates. In Fig. 3, the p25, p40, p65 and p110 of Mo-MuLV RNA are displayed along with a map of [^{35}S]methionine labelled Mo-MuLV p30 and gp70. It is clear that the p65 contains the spots of the p30, and that the spots of each of the lower molecular weight bands is always included in the bands of increasing molecular weight. In kinetic experiments not shown, the p25 and p40 were synthesized prior to the p65, and were labelled with formylmethionine from the substrate, f-met-$tRNA^F$, suggesting they were N-terminal peptides not degradation products of the higher molecular weight bands. These bands and fingerprints were used as standards for comparison with the translation products of Ki-MuSV, Ha-MuSV, and the endogenous 30S viral

RNAs. In Figure 4A and 4B, the fingerprints of the p50 from Ki-MuSV and the p22 of Ha-MuSV are shown.

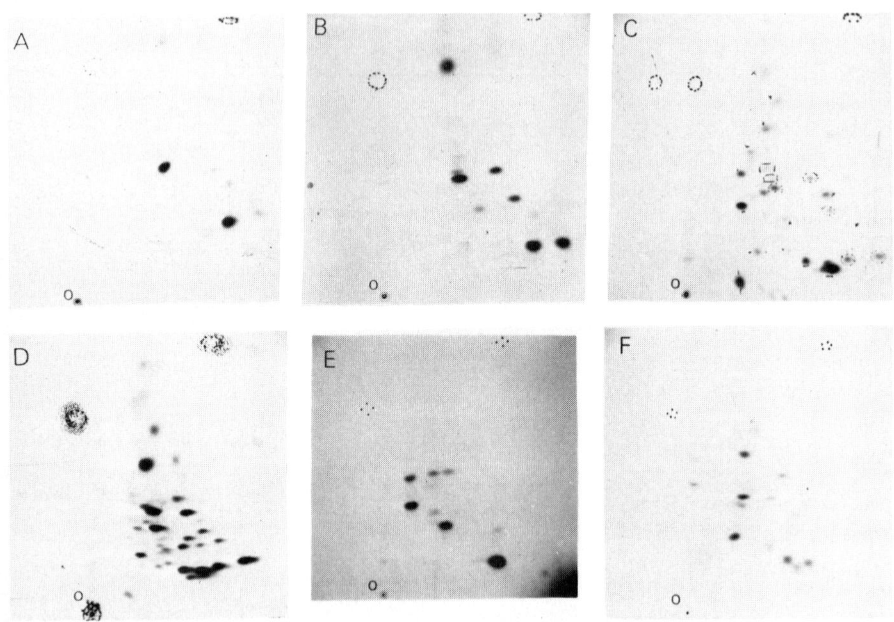

FIGURE 3. Peptide maps of Mo-MuLV structural proteins and *in vitro* products. Peptide mapping was performed as detailed in *Materials and Methods*. Each fingerprint contained approximately 20,000 cpm of [^{35}S]methionine peptides and was developed for 21-28 days. The p30 and gp70 markers were obtained from [^{35}S]methionine labeled Mo-MuLV by purification on phosphocellulose chromatography (10).

In vitro products

A. p25
B. p40
C. p65
D. p110
E. p30
F. gp70

(o) indicates origin, and the dotted circles at the top of the fingerprint are the dye markers crystal violet (top right) and xylene cyanol (top left). Electrophoresis is from left (negative electrode) to right (positive electrode).

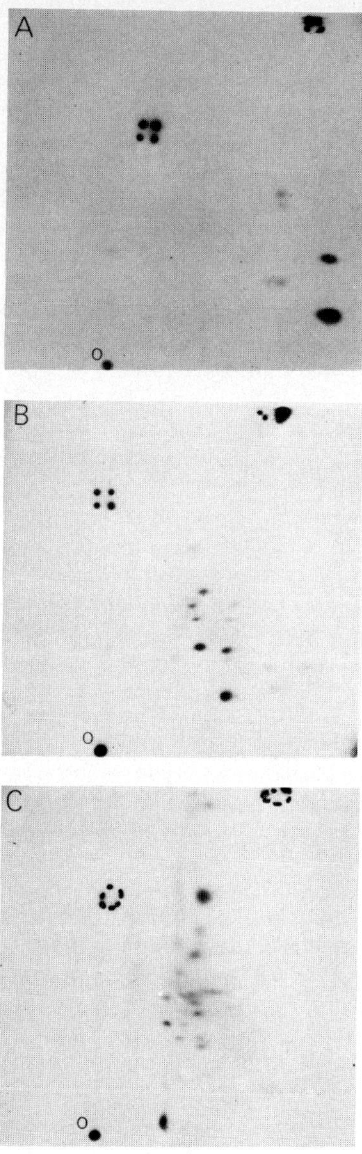

FIGURE 4. Peptide maps of translation products of Ha-MuSV, Ki-MuSV, and mouse 30S RNA. Details are as in the legend to Figure 6.

A. Ki-MuSV p50
B. Ha-MuSV p22
C. Mouse 30S p50

The Ki-MuSV p50 has only 2 clear spots whereas the Ha-MuSV p22 has 3-4 major spots. Comparison of the positions of the peptides indicates no relationship between the Ha-MuSV p22 and the Ki-MuSV p50. By comparison to the fingerprints of either the Mo-MuLV p65, p110 or [^{35}S]methionine labeled gp70, neither the p22 nor the p50 is related. In other studies not shown, the tryptic map of the Ki-MuSV p50 was unrelated to maps of the _gag_ p65 or gp70 of Ki-MuLV or to the _gag_ product translated from FeLV RNA, or to the tryptic map of the major background band (lanes 1 and 2) seen without added RNA.

Unfortunately, we have not been able to obtain sufficient RNA from the defective endogenous rat virus to fingerprint the p50 translated from it. However, we could fingerprint the p50 translated from the endogenous mouse 30S viral RNA. The map is shown in Fig. 4C. Again only 2 major spots are visualized but their positions are clearly distinct from the 2 spots obtained from the Ki-MuSV p50, and the results indicate that the two p50 proteins can be distinguished from each other.

Translation of Different Size Classes of Ki-MuSV and Ha-MuSV RNA. To determine if the p50 of Ki-MuSV and the p22 of the Ha-MuSV were translated from the 5' portion of their respective genomes, different size classes of poly(A) containing Ki-MuSV or Ha-MuSV RNA were prepared and translated with 1.0μg of RNA per assay, an amount which was saturating when utilizing heat denatured high molecular weight viral RNA. The results are shown in Fig. 5A and B. The p50 of Ki-MuSV is obtained with RNA with an average S value of 24-28S. With poly(A) containing RNA sedimenting at 18-22S or 12-16S, the p50 band progressively disappears (Fig. 5A). Similarly, in Fig. 5B, the p22 of Ha-MuSV is translated from the largest cut of poly(A) containing RNA, 23-25S, but not from RNA 18-21S or 12-18S in size. The bands appearing with the 12-18S RNA have not been extensively analyzed; however, the band at approximately 42,000 daltons has a peptide map distinct from any of the peptide maps noted above (Scolnick, unpublished). Preliminary evidence indicates that this band is synthesized from the 3' mouse sequences in Ha-MuSV and is coded for by the so-called common region of Mo-MuLV at its 3' end also.

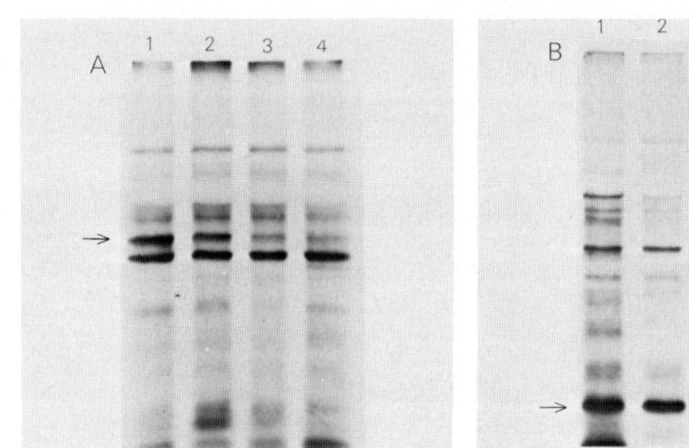

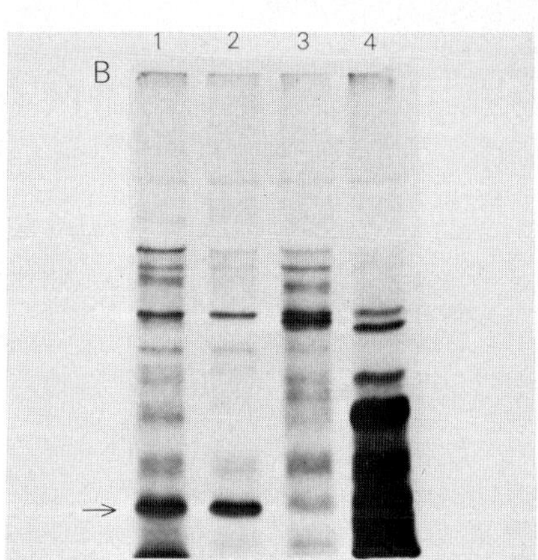

FIGURE 5. Translation of poly(A) containing Ki-MuSV or Ha-MuSV RNA fragments. Ki-MuSV or Ha-MuSV RNA was isolated from the 60-70S viral RNA from the Ki-MuSV (FeLV) or Ha-MuSV (Mo-MuLV) cultures. The RNA was heated from 2 minutes at 80°C in TNE buffer and sedimented at 39,000 rpm in SW41 rotor for 6 hours in a 15-30% sucrose gradient containing TNE as indicated in the legend to Table 2. Size classes of RNAs were pooled with the indicated sedimentation coefficients and purified by one cycle of oligo(dT)$_{12-18}$-cellulose chromatography.

(A) The p50 is indicated by an arrow at the left of the autoradiogram:
Lane 1: Ki-MuSV 70-70 RNA heated 2 minutes, 80°C
Lane 2: Ki-MuSV 24-28S poly(A) containing RNA
Lane 3: Ki-MuSV 18-22S poly(A) containing RNA
Lane 4: Ki-MuSV 12-16S poly(A) containing RNA

(B) The p22 is indicated by an arrow at the left of the autoradiogram:
Lane 1: Ha-MuSV 50-70 RNA heated 2 minutes, 80°C
Lane 2: Ha-MuSV 23-25S poly(A) containing RNA
Lane 3: Ha-MuSV 18-21S poly(A) containing RNA
Lane 4: Ha-MuSV 12-18S poly(A) containing RNA

Competition Radioimmunoassays Specific for the gp70's of Recombinant MCF Viruses. The development of an immunoassay specific for the gp70's of recombinant MCF viruses is of obvious general importance, and was of particular interest to us in analyzing proteins coded for by SFFV (8). We therefore developed a competition radioimmunoassay using a gp70 of a Friend MCF virus (Troxler *et al.*, Ruscetti *et al.*, submitted) and antiserum prepared against Friend MCF virus. Various ecotropic, xenotropic and recombinant viruses were then tested as competing antigens in this assay. As shown in Fig. 6A, Friend MCF virus competed very efficiently in this assay, as did the Moloney MCF virus. AKR MCF virus #247 competed to a lesser degree. A slight cross reaction was seen with a xenotropic NZB virus; Balb:virus-2 or ATS-124 xenotropic virus, competed not at all. Importantly, the ecotropic Friend, Moloney, and AKR viruses did not compete to any degree in this assay. As another negative control a feline virus grown in a Kirsten sarcoma virus transformed mink cell also failed to cross react in this assay. In studies not shown Fr-MCF virus grown in rat cells also reacted fully in this assay.

Since the AKR MCF virus #247 (9, 10), competed poorly in the Friend MCF specific gp70 assay, we sought to broaden the assay in hopes that we would be better able to detect MCF specific determinants on the AKR MCF gp70 without detecting the ecotropic or xenotropic gp70's. We therefore tested a competition radioimmunoassay utilizing Friend MCF gp70 and antiserum to Moloney MCF virus. As shown in Fig. 6B, all three MCF viruses now competed very well and were clearly distinguishable from the xenotropic and ecotropic MuLV's. Importantly, an NZB xenotropic virus competed slightly in this assay while ATS-124 and Balb:virus-2 competed not at all.

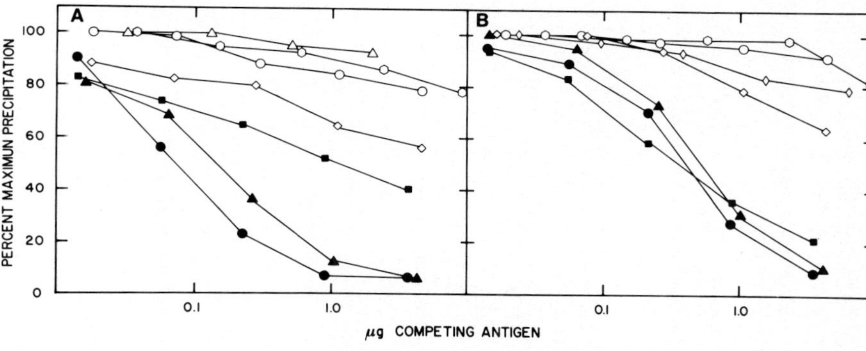

FIGURE 6. Reactivities of mouse type-C viruses in virus-specific MCF assays. Competition radioimmunoassays were carried out as described in Materials and Methods with (A) ^{125}I-labelled Friend MCF gp70 and anti-Friend MCF serum; or (B) ^{125}I-labelled Friend MCF gp70 and anti-Moloney MCF serum. Competing viruses used in these assays were: Friend MCF virus (●); Moloney MCF virus (▲); AKR MCF virus (■); NZB virus (◊); ATS xenotropic virus (◊); Balb xenotropic virus (◊); F/Ki virus (△); and Friend, Moloney or AKR ecotropic viruses (○).

Detection of Cross-reactive Antigen in Cells Nonproductively Infected with the Spleen Focus Forming Virus (SFFV). Using molecular hybridization techniques, the spleen focus forming virus has been shown to be an envelope gene recombinant between ecotropic virus and genes related to the env gene of MCF and xenotropic viruses. The 5' end of the gag gene product, p15 and p12, have been detected in the NRK cells containing SFFV (Barbacid *et al*., in preparation), but gp70 has not been detected in these cells. We therefore utilized several of the immunoassays that we had developed to detect Friend ecotropic or recombinant gp70's to determine if we could detect an antigen cross reactive in any of the assays related to the SFFV genome. In a type-specific ecotropic assay utilizing ^{125}I-labelled Friend MuLV gp70 and either anti-Friend ecotropic MuLV serum (Fig. 7A) or a group-specific assay using anti-xenotropic Balb virus-2 serum (Fig. 7B), and ^{125}I-labelled F-MuLV gp70, Friend MuLV infected cells competed quite efficiently, whereas the NRK cells expressing SFFV competed no better than uninfected NRK cells. However, when Friend MCF gp70 and anti-Moloney MCF serum were utilized in a competition radioimmunoassay, extracts of the SSFV nonproducer NRK cells were clearly competitive compared to extracts of uninfected NRK cells.

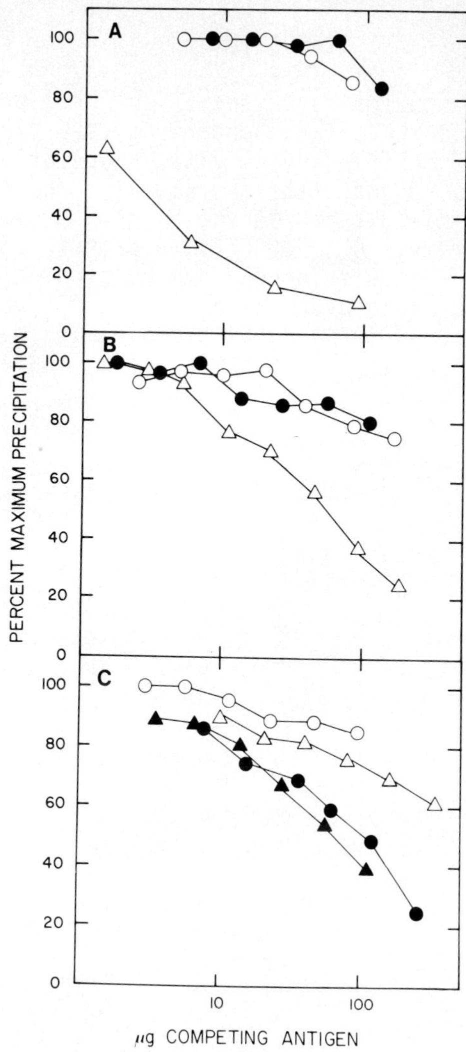

FIGURE 7. Antigen expression in rat cells infected with the spleen focus-forming virus. Competition radioimmunoassays were carried out as described in Materials and Methods utilizing (A) ^{125}I-labelled Friend ecotropic gp70 and anti-Friend ecotropic MuLV serum; (B) ^{125}I-labelled Friend MuLV ecotropic gp70 and anti-Balb virus-2 serum; or (C) ^{125}I-labelled Friend MCF gp70 and anti-Moloney MCF serum. The following cell extracts were used as competing antigens in these assays: SFFV nonproducer NRK cells (●); SR-VNRK cells (○); Friend MuLV infected FRE cells (△); or the FRE cells infected with SFFV and F-MuLV (▲).

DISCUSSION

The current studies have identified novel proteins coded for by Ki-MuSV and SFFV, two oncogenic type-C viruses. The Ki-MuSV p50 was identified by *in vitro* translation of Ki-MuSV RNA. The results suggest that the p50 is coded for by the rat sequences of Ki-MuSV and is the 5' gene product of this class of defective virus. For SFFV, a protein coded for by the xenotropic *env* gene sequences has been detected in a radioimmunoassay. This assay also detects MCF murine leukemia viruses. The role of these novel proteins in the oncogenic potential of each of these viruses is under investigation.

REFERENCES

1. Shih, T. Y., Young, H. A., Coffin, J. M., and Scolnick, E. M. (1978). J. Virol. 25, 238.
2. Parks, W. P., and Scolnick, E. M. (1977). J. Virol. 2, 711.
3. Howk, R. S., Troxler, D. H., Lowy, D., Duesberg, P. H., and Scolnick, E. M. (1978). J. Virol. 25, 115.
4. Maisel, J., Klement, V., Lai, M. C., Ostertag, W., and Duesberg, P. H. (1973). Proc. Nat. Acad. Sci. USA 70, 3536.
5. Scolnick, E. M., Williams, D., Maryak, J., Vass, W., Goldberg, R. J., and Parks, W. P. (1976). J. Virol. 20, 570.
6. Pelham, H. R. B., and Jackson, R. J. (1976). Eur. J. Bioch., 62, 247.
7. Kerr, I. M., Oshevsky, V., Lodish, H., and Baltimore, D. (1976). J. Virol. 18, 627.
8. Troxler, D. H., Lowy, D., Howk, R., Young, H., and Scolnick, E. M. (1977). Proc. Nat. Acad. Sci. USA. 74, 4671.
9. Hartley, J. W., Wolford, N., Old, J., and Rowe, W. P. (1977). Proc. Nat. Acad. Sci. USA. 74, 789.
10. Elder, J. H., Gavtsch, G. W., Jensen, F. C., Lerner, R. A., Hartley, J. W., and Rowe, W. P. (1977). Proc. Nat. Acad. Sci. USA 74, 4676.

CELL-FREE TRANSLATION OF ROUS SARCOMA VIRUS RNA[1]

Karen Beemon and Tony Hunter

Tumor Virology Laboratory, The Salk Institute, San Diego, California 92112

ABSTRACT Rous sarcoma virus (RSV) virion RNA was translated in the messenger-dependent reticulocyte lysate. Three major products, which were synthesized from RNA of non-defective RSV, but not from RNA of transformation-defective deletion mutants, have been characterized. These proteins, which appear to be products of the RSV src gene, were found to have molecular weights of approximately 60,000, 25,000, and 17,000 daltons. Small differences in the electrophoretic mobility of these proteins in SDS-polyacrylamide gels were observed when RNA from different strains of RSV was translated. No significant differences were observed in the translation products of RSV virion RNA from a temperature-sensitive mutant for transformation grown in either transformed or non-transformed cells.

INTRODUCTION

An avian sarcoma virus gene called src is required for both the initiation and maintenance of transformation in cultured fibroblasts and for viral induction of sarcomas in animals. Transformation-defective (td) deletion mutants segregate spontaneously from non-defective (nd) Rous sarcoma virus (RSV) upon passage of the virus. These mutants replicate normally but do not transform cells because of a deletion of 1000-2000 nucleotides in the src gene.

We have identified products of the RSV src gene by cell-free translation of heat-denatured 70S virion RNA (1,2). Three major products synthesized from RNA of nd RSV, but not from RNA of td deletion mutants of RSV, have been observed. The approximate molecular weights of these products are 60,000 (60K), 25,000 (25K), and 17,000 (17K) daltons.

[1]This work was supported by Grant CA 17096 and by National Research Service Award CA 05085, both from the National Cancer Institute.

ISBN 0-12-668350-6

Tryptic peptide analysis of these proteins showed all three to be overlapping, having shared amino acid sequences. However, both similar and distinct peptides were observed when the 60K protein synthesized from RNA of the Prague strain, subgroup B (PR-B), of RSV was compared to the 60K protein synthesized from RNA of the Schmidt-Ruppin strain, subgroup D (SR-D), of RSV, suggesting that these RNAs code for related but different 60K proteins. These proteins were found to be unrelated to any of the RSV virion proteins by tryptic peptide analysis and by immunoprecipitation. The RNAs coding for the 60K, 25K, and 17K proteins were found to be polyadenylated and to sediment in sucrose gradients with peaks at approximately 24S, 20S, and 18S, respectively. It appears likely that these proteins are synthesized from viral RNA because of strain-specific differences in them. The size of the mRNA activity, and the absence of the 60K, 25K, and 17K proteins in the translation products of td RSV RNA, suggest that they are coded for by the RSV *src* gene. Similar cell-free translation products of RSV virion RNA have been observed by Kamine *et al*. (3), and the 60K product has been observed by Purchio *et al*. (4).

Brugge and Erikson (5) have identified a 60K transformation-specific antigen by immunoprecipitation of RSV-transformed cells with sera from rabbits bearing tumors induced by SR-D RSV. Comparison of this 60K antigen with the 60K protein synthesized *in vitro*, by tryptic peptide mapping, has shown them to be largely identical (6).

RESULTS

Strain-Specificity of the Src Gene Products. *Src* gene products synthesized *in vitro* from different strains of RSV have been compared. Heat-denatured 70S RNA from purified virus was translated in the messenger-dependent reticulocyte lysate (7). Comparison of the translation products of RNA from PR-B, PR-C, and SR-D RSV with those from RNA from td deletion mutants of these viruses is shown in Figure 1. In addition, translation products of RNA from RSV(-), grown on quail cells, and from PR-A are shown. The major products synthesized exclusively from the nd RSV RNA had apparent molecular weights of approximately 60K, 25K, and 17K. However, strain-specific differences were observed in the electrophoretic mobilities of these products. The relative molecular weights, determined on the basis of electrophoretic mobility, for the "60K" transformation-specific proteins from several strains of RSV are listed in Table 1. Different preparations of the same strain of virus yielded 60K proteins with identical mobility. Small differences in gel mobilities

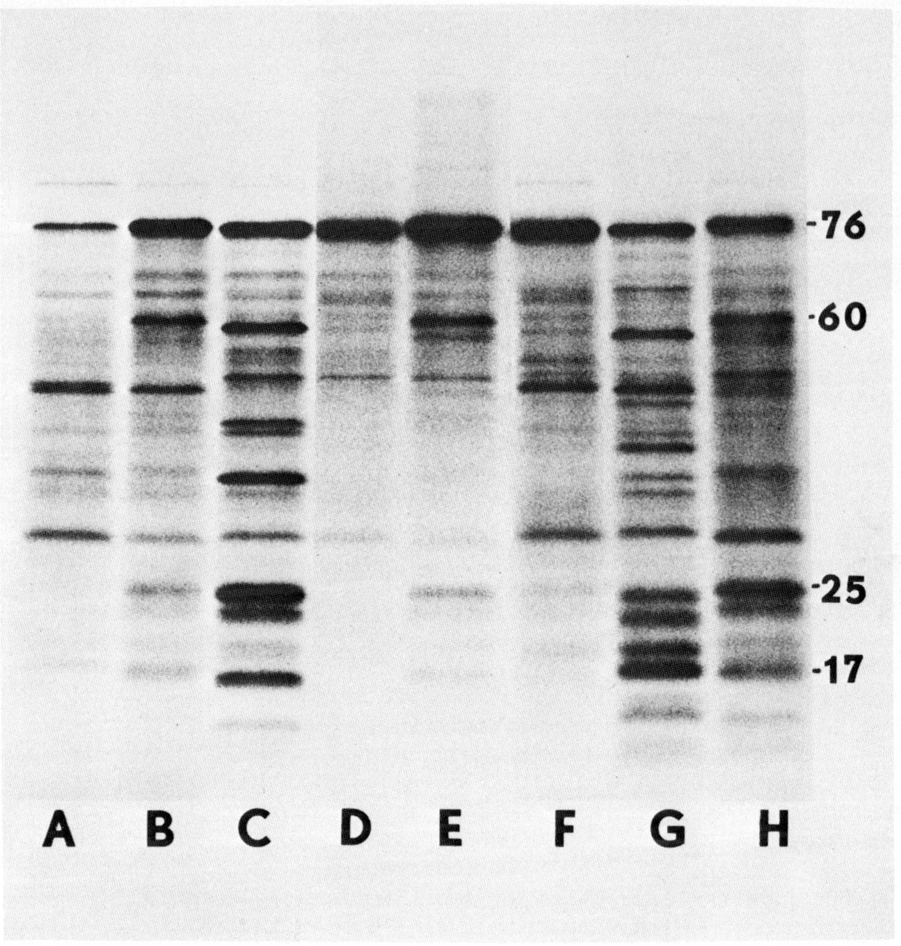

FIGURE 1. Heat-denatured 70S RSV virion RNA was translated in the mRNA-dependent reticulocyte lysate (7) in the presence of ^{35}S-methionine. Products were resolved by SDS-polyacrylamide gel electrophoresis as described (1). Translation products of the following RSV RNAs are shown: (A) td PR-B, (B) PR-B, (C) SR-D, (D) td SR-D, (E) PR-C, (F) td PR-C, (G) RSV(-), (H) PR-A.

TABLE I
STRAIN SPECIFICITY OF SRC PROTEINS

Virus strain and subgroup	Molecular weight of "60K protein"
PR-A	60,000
PR-B	60,000
PR-C	60,500
SR-A	57,000
SR-D	58,000
RSV(-)	56,000

of the 76K precursor to the virion internal structural proteins have also been observed between different strains of RSV.

Strain-specific differences in the 60K, 25K, and 17K proteins from SR-D and PR-B have also been observed by two-dimensional mapping of methionine-containing tryptic peptides (2). This analysis revealed that 7 of the 15 tryptic peptides present in the SR-D 60K protein were also present in the PR-B 60K protein. However, about half of the methionine tryptic peptides in each protein were unique to that strain. Strain-specific antigenic differences in the _src_ proteins have also been observed. Sera were prepared from rabbits bearing SR-D RSV-induced tumors, by the procedure of Brugge and Erikson (5). Sera which immunoprecipitated the 60K protein from the _in vitro_ translation products of RNAs from SR-D, SR-A, and one clone of PR-C, did not cross-react with PR-B, PR-A, B77, or RSV(-) 60K proteins. Similar antigenic specificity was observed _in vivo_ (6). Therefore, strain-specific differences in the 60K protein have been observed by three different techniques: electrophoretic mobility in acrylamide gels, tryptic peptide mapping, and immune precipitation.

Is mRNA Activity for the Src Proteins Present Only in RSV Grown in Transformed Cells? In earlier experiments we compared the _in vitro_ translation products of RNA from nd RSV grown in transformed cells with that from td RSV grown in non-transformed cells. Although the most likely hypothesis for the differences observed in the translation products of RNA from these two viruses is that the td deletion overlaps at least part of the coding sequence for the 60K, 25K, and 17K proteins, it was also formally possible that nucleases specific to transformed cells were necessary to generate mRNA activity for these proteins or that cellular mRNA, coding for these proteins, is only packaged by RSV grown in transformed cells.

To investigate these possibilities, a temperature-sensitive (ts) transformation mutant, LA56, derived from PR-B RSV, was grown at both permissive and non-permissive temperatures for transformation. After the virus was purified, its RNA was extracted, and denatured 70S RNA was translated *in vitro*. The products of translation of RNA from this virus grown at the permissive temperature (36°) and at the non-permissive temperature (42°), together with RNA from the nd parent PR-B, grown at 39°, are shown in Figure 2. No significant differences were observed between products of viral RNA from virus grown in transformed and non-transformed cells, ruling out the second and third hypotheses mentioned above. This result substantiates our earlier data suggesting that these proteins are coded for by the RSV *src* gene.

A difference was apparent in the 17K protein synthesized from RNA from nd PR-B and from the LA56 mutant grown at either temperature. A similar alteration was observed when translation products of SR-A RSV RNA were compared to that from RNA of NY68, another ts transformation mutant. The basis of this alteration is not known. No such differences were observed in products of RNA from a number of other ts transformation mutants derived from PR-A, so it is not a change that is necessary for the expression of a ts transformed phenotype.

DISCUSSION

Proteins having molecular weights of approximately 60K, 25K, and 17K have been synthesized *in vitro* from virion RNA of nd RSV. Several lines of evidence suggest that these proteins are encoded by the RSV *src* gene:

(1) These proteins are not synthesized from virion RNA of td deletion mutants of RSV, which lack 10-20% of the RSV genome and all or part of the *src* gene. However, they are synthesized from RNA of ts mutants for transformation, even when grown at the non-permissive temperature.

(2) By fractionation of virion RNA on sucrose gradients and translation of the poly(A)-containing RNA from each fraction, we have shown that the 60K, 25K, and 17K proteins are synthesized from polyadenylated RNAs sedimenting at 24S, 20S, and 18S, respectively. We have also shown, by oligonucleotide fingerprinting, that polyadenylated fragments of RSV genomic 38S RNA having these sedimentation coefficients contain the *src* gene and the 3' terminal common region. Therefore, it is likely that these proteins are synthesized from sequences within the *src* gene. Furthermore, cells infected with nd RSV contain an RNA species, sedimenting at approximately 21S, which contains sequences of both the *src*

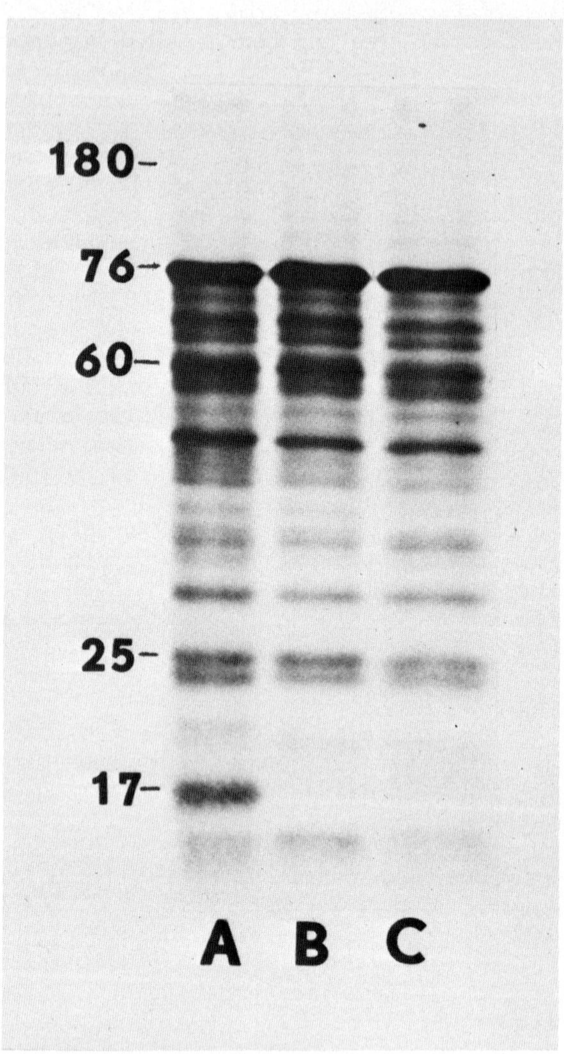

FIGURE 2. SDS-polyacrylamide gel electrophoresis of ^{35}S-methionine labeled cell-free translation products of denatured 70S virion RNA from (A) PR-B RSV grown at 39°, (B) LA56, ts transformation mutant of PR-B, grown at 36°, and (C) LA56 grown at 42°.

gene and the 3' terminal common region (8,9). This 21S RNA is thought to represent _src_ mRNA.

(3) The strain specificity observed for the 60K, 25K, and 17K proteins suggests that these are viral coded proteins. However, these proteins appear to be unrelated to any of the RSV structural proteins as demonstrated both by tryptic peptide mapping and by immune precipitation. This is consistent with these products being coded for by the _src_ gene. These strain-specific physical differences in the 60K, 25K, and 17K proteins may be reflected in strain-specific biological differences observed in transformed cells (10).

(4) Anti-tumor sera from rabbits bearing tumors induced by SR-D RSV precipitates a "60K" protein from cells infected with SR-A, SR-D, and one clone of PR-C, but not from cells transformed by PR-A, PR-B, B77, or another clone of PR-C. In the case of SR-D, tryptic peptide mapping has shown that the protein found _in vivo_ is essentially identical to that made _in vitro_. This identity with a transformation-specific intracellular protein provides positive proof that the 60K, 25K, and 17K proteins made _in vitro_ originate from the _src_ gene.

REFERENCES

1. Beemon, K., and Hunter, T. (1977) Proc. Nat. Acad. Sci. U.S.A. 74, 3302.
2. Beemon, K., and Hunter, T. (1978) Manuscript in preparation.
3. Kamine, J., Burr, J.G., and Buchanan, J.M. (1978) Proc. Nat. Acad. Sci. U.S.A. 75, 366.
4. Purchio, A.F., Erikson, E., and Erikson, R.L. (1977) Proc. Nat. Acad. Sci. 74, 4661.
5. Brugge, J.S., and Erikson, R.L. (1977) Nature 269, 346.
6. Sefton, B., Hunter, T., and Beemon, K. (1978) Manuscript in preparation.
7. Pelham, H.R.B. and Jackson, R.J. (1976) Eur. J. Biochem. 67, 247.
8. Weiss, S.R., Varmus, H.E., and Bishop, J.M. (1977) Cell 12, 983.
9. Hayward, W.S. (1977) J. Virol. 24, 47.
10. Vogt, P.K. (1977) In "Comprehensive Virology", vol. 9 (H. Fraenkel-Conrat and R.R. Wagner, eds.), p. 379. Plenum Press, New York.

GENETICS OF REOVIRUS: ASPECTS RELATED TO VIRULENCE AND VIRAL PERSISTENCE

Bernard N. Fields,[1,2] Howard L. Weiner[1,2]
Robert F. Ramig[1], and Rafi Ahmed[1]

Department of Microbiology and Molecular Genetics, Harvard Medical School ([1]) and Department of Medicine, Peter Bent Brigham Hospital ([2]), Boston, Massachusetts 02115.

ABSTRACT The individual dsRNA genome segments of reovirus types 1, 2 and 3 can be distinguished by differences in their migration on SDS-PAGE. Ts^{+} recombinants have been isolated following mixed infections of ts mutants of reovirus type 3 and clones of type 1 or 2. Utilizing these recombinants we have determined that the S1 dsRNA segment (that codes for the $\sigma 1$ outer capsid polypeptide) is the hemagglutinin as well as the gene coding for the type specific polypeptide (as determined by neutralization tests). Furthermore, using such clones to infect newborn mice, we have found that the $\sigma 1$ polypeptide is the determinant of cell tropism, determining whether the virus will infect ependyma to produce a "non-lethal" disease (type 1 pattern) or infect neurons to produce a highly lethal necrotizing encephalitis (type 3 pattern). The highly specific interaction of the reoviral hemagglutinin ($\sigma 1$) with cell surface receptors, is thus the major determinant of differential neurovirulence associated with the different reovirus serotypes.

INTRODUCTION

Recent advances in the genetic analysis of reovirus types 1, 2 and 3 have enabled us to begin to study the role of specific viral genes in pathogenesis. Reoviruses are animal viruses whose genomes consist of 10 segments of double-stranded RNA (dsRNA) named according to size classes: 3 large segments (L1, L2, L3), 3 medium segments (M1, M2, M3), and 4 small segments (S1, S2, S3, S4). The observation on which the genetic analysis of reovirus is based is the finding that the genome RNAs and the polypeptides of the three serotypes can be distinguished following polyacrylamide-gel electrophoresis (PAGE). The aims of this report are to summarize

ISBN 0-12-668350-6

genetic studies leading to the map of reovirus; to describe biologic functions of specific viral genes and relate these functions to acute reoviral virulence; to document the phenomenon of suppression of the temperature sensitive phenotype of reovirus; and, lastly, to describe recent findings relating to viral persistence.

RESULTS

The dsRNA Genomes of Reovirus types 1, 2 and 3 Contain Species of RNA with Differing Electrophoretic Mobilities. Examination by polyacrylamide gel electrophoresis revealed substantial heterogeneity (Figure 1) (1). Although the exact nature of the variations is not yet defined, these differences have provided very useful markers for assigning temperature-sensitive (ts) lesions to genome segments since efficient reassortment of the genome segments occurs following mixed infections between serotypes.

A Genetic Map of Reovirus (2,3,4). When cells are mixedly infected with a type 3 ts mutant and wild type 1 or 2 and the progeny are analyzed, clones that plaque at the nonpermissive temperature select against the parental type 3 ts mutant, as well as recombinant progeny containing the ts lesion derived from the type 3 parent (2). By analyzing the dsRNA of such progeny clones, it has been possible to map the segments responsible for the ts phenotype (3,4) and to construct a map correlating segments between serotypes (2, Figure 2).

The Role of the S1 Gene. Using recombinant clones derived from mixed infections with type 1 or 2 reovirus and type 3 reovirus, we have found that the S1 dsRNA segment is the reoviral hemagglutinin, is responsible for type specificity (as determined by neutralization testing) and is the determinant of specific cell tropisms in the central nervous system.

Among the various serotypes of mammalian reoviruses hemagglutination is type specific; type 1 agglutinates human erythrocytes, while type 3 agglutinates bovine erythrocytes. There are three outer capsid polypeptides in mammalian reoviruses, $\mu 2$, $\sigma 1$ and $\sigma 3$, the products of the M2, S1 and S4 dsRNA segments respectively (5). Two recombinants that contain M2 and S4 segments from type 1 (204) or type 3 (802) but an S1 segment from the opposite type, were utilized to show that the S1 dsRNA segment is responsible for hemagglutination (Table 1) (Figure 3) (6). In a similar fashion we have found that the S1 gene is also responsible for type specificity as determined by plaque reduction neutralization

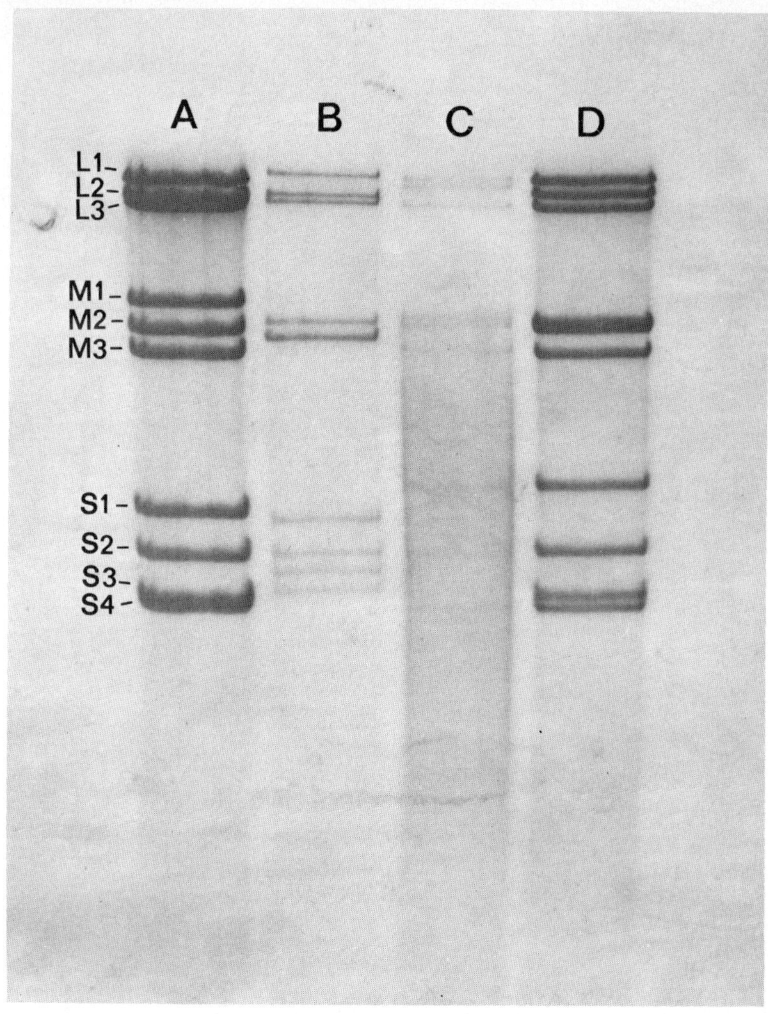

FIGURE 1. Analysis of cytoplasmic dsRNA's extracted from cells infected with the reovirus serotypes and labeled with [^{14}C]uridine. Electrophoresis was carried out on a 10% polyacrylamide slab gel as described by Laemmli (1) for 8 h at 40 mA. Electrophoresis was from top to bottom. A, Type 1 Lang; B, type 2 Jones; C, type 3 Abney; D, type 3 Dearing. Reprinted from ref. (1) with permission.

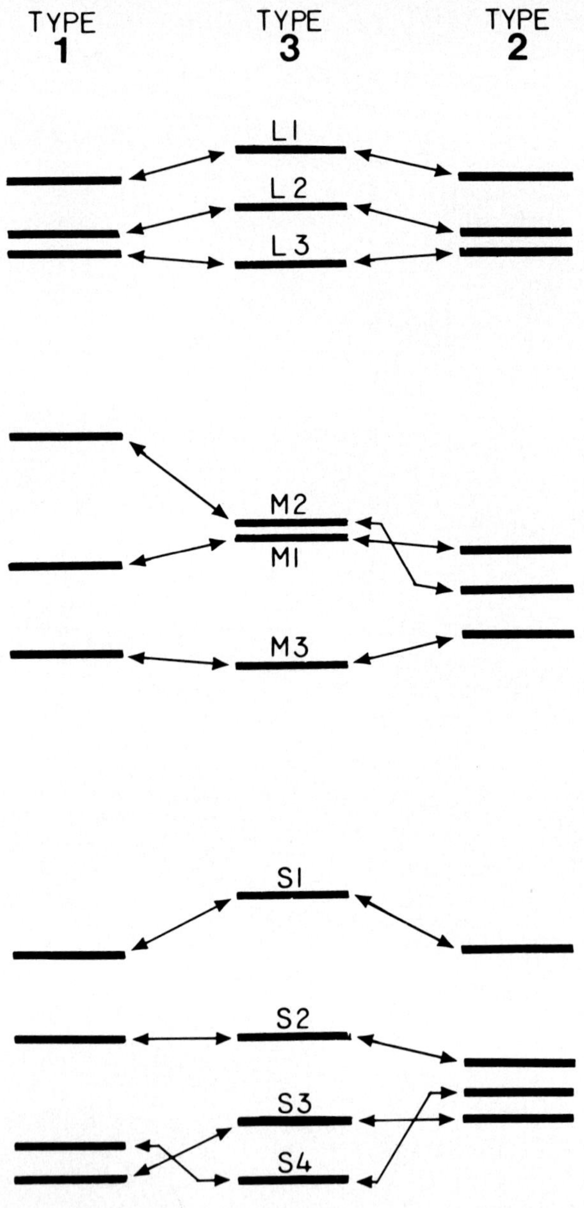

FIGURE 2. The map of the genome segments of reovirus serotypes 1, 2 and 3 as resolved in Tris-glycine buffered gel electrophoresis systems. Band identities between gel systems and correlation of bands between serotypes are as determined

(Figure 4 con't) clones. A suspension of 10^7 mouse cells were mixedly infected with a multiplicity of infection of 10, each with freshly cloned 101 and wild type. Two hours after infection, unadsorbed virus was removed by centrifuging the infected cells and resuspending them in fresh medium. Forty-eight hours after infection, the cells were sonicated to release cell-associated virus and to disrupt viral aggregates. Appropriate dilutions were plated on L cell monolayers and incubated for 13 days at 31°C. The culture plates were overlaid with neutral red agar and, after overnight incubation, plaques were picked. The plaques were passaged twice on L cell monolayers at 31°. The titer and efficiency of plating of second passage virus was determined by plating on L cell monolayers at 39°C and 31°C. Plates at 39°C were overlaid and counted on day 5; plates at 31°C were overlaid and counted on day 13 after infection. Wild-type and clone 101 controls were the same, except that for single infection a multiplicity of infection of 20 was used. The EOP is the ratio of the titer at 39°C to that at 31°C. (A) EOP of wild type control clones. (B) EOP of clone 101 control clones. (C) EOP of clone 101 wild-type progeny clones. Reprinted from ref. (12) with permission.

DISCUSSION

The results presented in this report indicate that it is possible using *in vivo* animal model systems, to assign specific stages of viral pathogenesis to specific viral genes. It has been shown in this manner that the S1 dsRNA segment, the gene that codes for the viral hemagglutinin is responsible for cell tropism in the acute viral infection. In addition, others have shown that once persistent virus infection has been established in cell cultures, the phenotype of the virus changes (in general leading to the generation of ts mutants). In the case of persistent infections due to reovirus, the viral phenotype changes in that the input ts mutant becomes ts^+. We have recently shown that this is due, in part, to second site suppressor mutation(s). In one novel clone, 3 ts mutations were present. Thus viral mutation is occuring coincident with the establishment and/or maintenance of the persistent state. Although the role of specific viral genes associated with, and perhaps responsible for, viral persistence has not yet been established, the ability to map biologic function to specific genes should provide important clues as to the precise mechanism(s) of viral persistence.

ACKNOWLEDGMENTS

This work was supported by grants from the National Institutes of Health (AI 13178), and the Milton Fund of Harvard University. Howard L. Weiner is the recipient of a Teacher-Investigator Award from the National Institute of Neurological and Communicative Disorders and Strokes (#NS1-EA 1 K07 NS 237-01 NSPB). Robert F. Ramig is the recipient of an NIH Research Fellowship Award (#1 F32 CA 05709-011). We would like to thank Roz White and Elaine Freimont for excellent technical assistance.

REFERENCES

1. Ramig, R.F., Cross, R.C., and Fields, B.N. (1976). J. Virol. 22, 726.
2. Sharpe, A.H., Ramig, R.F., Mustoe, T.A., and Fields, B.N. (1978). Virol. 84, 63.
3. Ramig, R.F., Mustoe, T.A., Sharpe, A.H., and Fields, B.N. (1978). Virol. In press.
4. Mustoe, T.A., Ramig, R.F., Sharpe, A.H., and Fields, B.N. (1978). Virol. In press.
5. Joklik, W.K. (1974). Comp. Virol. 2, 231.
6. Weiner, H.L., Ramig, R.F., Mustoe, T.A., and Fields, B.N. (1978). Virol. In press.
7. Weiner, H.L., and Fields, B.N. (1977). J. Exp. Med. 146, 1305.
8. Margolis, G., Kilham, L., and Gonatos, N.K. (1971). Lab. Invest. 24, 101.
9. Raine, C.S., and Fields, B.N. (1973). J. Neuropath. Exp. Neurol. 32, 19.
10. Kilham, L., and Margolis, G. (1969). Lab. Invest. 21, 189.
11. Weiner, H.L., Drayna, D., Averill, D.R., and Fields, B.N. (1977) Proc. Nat. Acad. Sci. (U.S.) 74, 5744.
12. Ramig, R.F., White, R.M., and Fields, B.N. (1977). Science 195, 406.
13. Ramig, R.F., and Fields, B.N. (1977). Virol. 81, 170.
14. Ahmed, R., and Graham, A.G. (1977). J. Virol. 23, 250.
15. Preble, O.T., and Youngner, J.S. (1975). J. Infect. Dis. 131, 467.
16. Fields, B.N. (1971). Virol. 46, 142.

DEFECTIVE INTERFERING PARTICLES: THEIR EFFECT ON GENE EXPRESSION AND REPLICATION OF VESICULAR STOMATITIS VIRUS[1]

Alice S. Huang, Sheila P. Little,[2]
M.B.A. Oldstone and Donald Rao

Department of Microbiology and Molecular Genetics
Harvard Medical School
Boston, Massachusetts 02115

and

Department of Immunopathology
Scripps Clinic and Research Foundation
La Jolla, California 92037

ABSTRACT

Cells are infected either with standard vesicular stomatitis virus (VSV) alone or co-infected with standard VSV and its DI-T particle. With co-infected cells the presence of VSV proteins and the shedding of VSV antigens are largely unaffected by the presence of DI particles; there is, however, an overall reduction in the budding out of extracellular particles as well as a reduced expression of antigen(s) at the surface of co-infected cells when compared to cells infected by standard virus alone. This reduced expression of surface antigen(s) may aid DI particle-infected cells to escape immune surveillance. Virus-specific RNA synthesis is markedly altered by the presence of DI particles. In co-infected cells standard VSV expresses only primary transcription products. The bulk of the RNA is detected as plus and munus strand DI particle-specific RNA. The successful competition of the DI RNA against standard RNA is interpreted to be the result of the generation of two high affinity sites for the replicase on DI particle-specific RNA compared to only one high and one low affinity binding site on the templates responsibile for the replication of standard VSV RNA.

[1]This work was supported by research grants AI10100 and AI09484 from the USPHS and VC-63 from the American Cancer Society.

[2]Sheila P. Little was a Ford Foundation Predoctoral Fellow.

ISBN 0-12-668350-6

INTRODUCTION

Defective interfering (DI) particles are generated by every virus group (1). They have been postulated to be one of the important parameters responsible for the termination of acute viral infections as well as to participate in the establishment and maintenance of persistent infections (2).

Two general properties of DI particles contribute to their role during viral pathogenesis. These are: 1) inhibition of excessive virus growth; and 2) amelioration of cell killing. Although other mutations in animal virus genomes as well as host defense mechanisms may result in similar inhibitory effects of virus growth, the specificity of interference caused by DI particles makes it a unique virus-controlled event which guarantees a limit on the growth of the virus itself.

Since fluctuation in the total amount of virions, both standard and DI, as well as in the ratio of the two types of particles result in a variety of host cell response (3), it is necessary to detail the synthesis and distribution of the viral gene products when DI particles are present as well as when they are absent. In order to study this virus-host interaction, we have utilized the vesicular stomatitis virus (VSV) system. This system has several advantages. Interference of virus growth by DI particles is readily demonstrated in a variety of host cells (4) and VSV DI particles can be separated from the standard infectious virus from which they are derived (5). Therefore, it is possible to study the virus-host interaction when only standard virus or only DI particles are present or when cells are co-infected with different ratios of standard virus and DI particles.

RESULTS

We have chosen to compare two different states of infection: 1) when cells are infected only with standard virus and producing standard virus maximally and 2) when cells are co-infected at equal multiplicities where maximal interference with standard virus growth by DI particles is occurring. Our particular DI particle, known as T particle or DI-T particle, produces no detectable virus-specific products when it infects cells alone (5, 6).

Synthesis of Protein. When virus-specific products in the extracellular environment are examined, the co-infected cells are producing less than 0.001% of the expected yield of standard virus progeny (7). DI particles, however, are produced by these co-infected cells. When extracellular, non-virion-associated viral antigens are examined, both types of infected cells shed them to approximately the same degree (7). This shedding is not due to the degradation of virions or due to the nonspecific lysis of cells.

When the surface of an infected cell is assayed for the expression of viral antigen(s), there is enough expression to give positive immunofluorescence (Fig. 1). Titration of these infected cells indicate a quantitative difference, with co-infected cells showing less surface antigen(s) than cells producing the normal yield of standard virus (Fig. 2). In other on-going studies using L cells, these findings are substantiated. By radiobinding assays and quantitating the number of antibody molecules to VSV, less antigen is expressed on the surface of co-infected cells than on cells infected by standard virus alone. Similarly, the effective numbers of cytotoxic thymus-derived T cells more efficiently lyse standard virus-infected cells than co-infected cells (unpublished observations). Hence, all these observations indicate diminished expression of VSV antigen(s) on the surface of cells co-infected by standard VSV and DI particles when compared to cells infected by standard virus alone. Similar differences in the expression of viral antigen at the cell surface have been seen with the lymphocytic choriomeningitis virus system (8). Such diminished expression of antigen(s) at the cell surface caused by DI particles may allow such cells to escape immune surveillance in the infected host.

Nevertheless, these studies on VSV and biochemical analysis of cell-associated VSV proteins (7) lead to the conclusion that there is considerable virus-specified protein synthesis whether cells are singly infected with standard virus or co-infected with both standard virus and DI particles. There is no significant impairment of the VSV-specified translational apparatus by DI particles and that subsequent distribution of these VSV proteins to the extracellular environment occurs in co-infected cells.

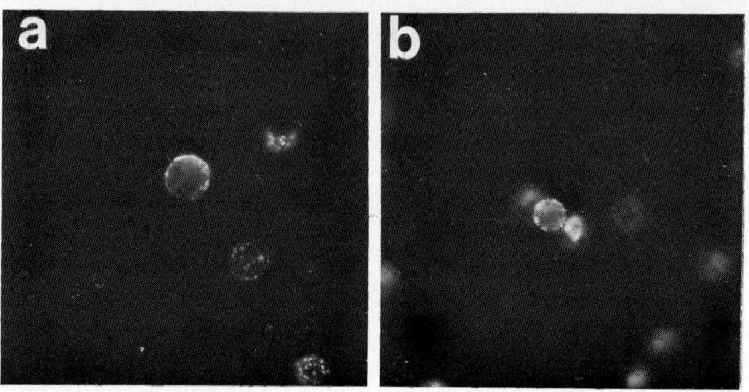

FIGURE 1. Immunofluorescence of Chinese hamster ovary cells infected with standard VSV alone or co-infected with standard virus and DI particles. Cells were infected at 34°C with standard or standard and DI VSV, each at a multiplicity of 40. Cells were treated with mouse-anti-VSV serum at a 1:5 dilution and then exposed to fluorescein-conjugated rabbit-anti-mouse IgG. The difference in size of cells shown is due to the extent of magnification. (a) Standard virus alone; (b) Standard virus and DI particles.

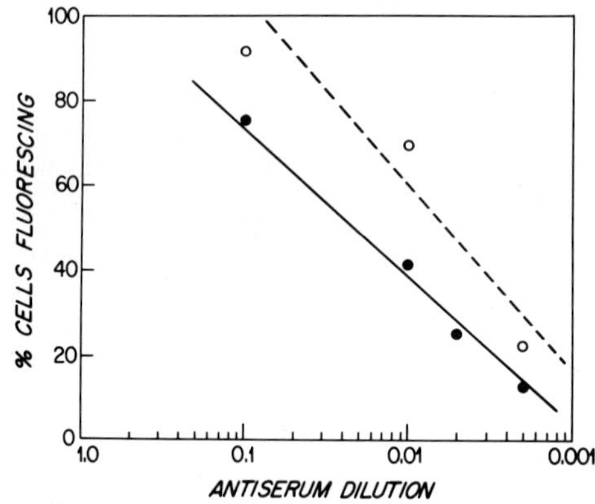

FIGURE 2. Antibody binding by cells infected with standard VSV alone or co-infected with standard virus and DI particles. Chinese hamster ovary cells were infected and stained as shown for Fig. 1, except that the indicated dilutions of mouse-anti-VSV were used. (o---o) standard virus alone; (●——●) standard virus and DI particles.

Synthesis of RNA. On the other hand, VSV-specific RNA synthesis is rather drastically altered by DI particles. Co-infected cells synthesize less than 20% of the VSV RNA made by cells infected with only standard virus (9). Standard virus codes for the synthesis of five small, monocistronic messenger RNAs and the synthesis of large complementary 40S RNA (10, 11). Cells co-infected with standard and DI particles carry on only primary transcription off of the input standard virions (9, 12). No replication of the standard virus genome is detectable.

The majority of the RNA that is found intracellularly is specific to the DI particle (13). Fig. 3, slot B, shows that the major stable RNA species appear as two bands after denaturation in DMSO and separation in an acid-urea containing agarose gel. In comparison, RNA from DI particles appear only as one band (slot A).

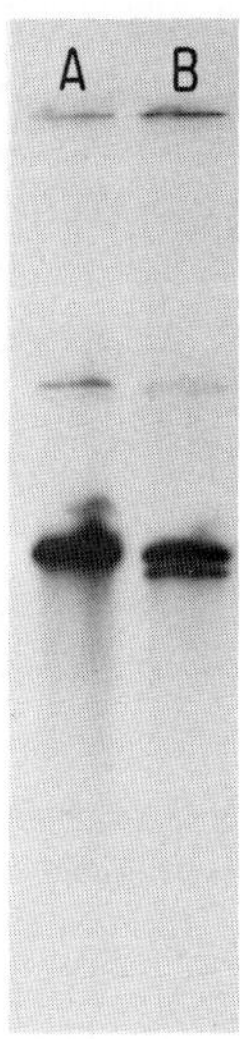

FIGURE 3. Strand separation and identification of a VSV DI particle-specific RNA species. Cells were infected with approximate equal multiplicities of standard and DI particles of VSV and labeled with ^{32}P as previously described (22). DI particles were purified on sucrose gradients (13). Their RNA and total cytoplasmic RNA from the infected cells were extracted by phenol:chloroform and then ethanol precipitated. Prior to electrophoresis the RNA was treated with 90% dimethylsulfoxide. Electrophoresis was on a 1.5% agarose gel containing 6M urea at pH 3.8 (23). (slot A) DI particle RNA; (slot B) cytoplasmic RNA.

Hybridization analysis of RNA from DI particles indicate that DI particles contain mainly minus-stranded RNA covalently linked with no more than 5% complementary RNA (14). Intracellular RNA contains 60% minus-stranded material and 40% plus-stranded material (15). The results in Fig. 3 suggest that plus and minus-stranded RNA of the same size may be separated into two bands, although in sucrose gradients the two species co-sediment as 19S RNA.

To confirm this interpretation and further analyze the major RNAs specified by DI particles, the intracellular RNAs from co-infected cells were examined in their undenatured form on a neutral agarose gel. In addition, lithium chloride fractionation was used to identify fully double-stranded (not precipitated) material from single-stranded RNA (precipitated). For comparison these fractionated RNAs were also denatured in DMSO and treated with glyoxal (16) in order to determine if there were identities in molecular weight.

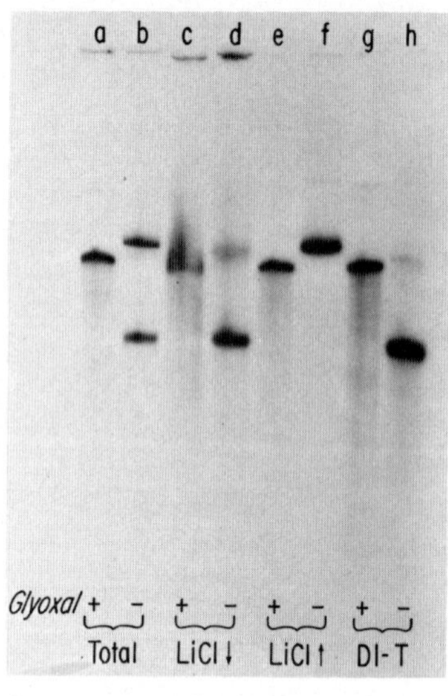

FIGURE 4. Identification of cytoplasmic RNA from cells co-infected with standard and DI particles of VSV. Cells were infected with standard and DI particles of VSV as described for Fig. 3. ^{32}P-labeled total cytoplasmic RNA was fractionated by 2 M LiCl (24) and electrophoresed with and without denaturation. Denaturation consisted of exposure of the RNA to dimethylsulfoxide and glyoxal (16). Then the RNA was electrophoresed in a neutral 1.5% agarose gel containing 6 M urea and "E" buffer (25): (a & b) total cytoplasmic RNA; (b & c) LiCl precipitated RNA; (d & e) LiCl soluble RNA; (f & g) DI particle RNA.

Fig. 4, slot a, shows total undenatured intracellular VSV-specific RNA from cells co-infected with standard and DI particles. There are two major well separated bands. Lithium chloride precipitation indicates that the faster migrating band contains single-stranded RNA (slot d) and the slower migrating band contains double-stranded RNA (slot f). When each of these two species is denatured and treated with glyoxal, the result is one band which co-migrates with RNA from DI particles (slots c, e, and g).

Therefore, VSV RNA synthesis in co-infected cells consists of primary transcription by input standard virions without any subsequent amplification (9). Standard genome replication is inhibited but DI particle-specific RNA synthesis is readily detected.

Mechanism of interference during genome replication. Because the primary effect of interference by DI particles appears to be on genome replication, an important question is how this competition occurs, resulting in the enrichment of DI genomes at the expense of standard virus genomes. Size alone does not appear to offer the DI RNA an advantage (17). Therefore, there must be some sequence alteration or sequence unmasking during the generation of the RNA for DI particles which result in a competitive advantage for the binding of polymerase.

Recent oligonucleotide analyses, self-hybridization experiments and electron microscopy have indicated complementary, extragenomic sequences in a variety of DI particle RNAs (18, 19, 20), as well as the particular DI particle, which are used in these experiments (14,21). We have mapped the complementary extragenomic sequences on the DI RNA to be at or near the 3' end of the RNA (14). Comparison to individual messenger RNAs complementary to the standard virus genome indicates that these additional complementary sequences on DI RNA are probably derived by copying nontranscribed regions of the standard virion RNA. We have already hypothesized that DI RNA can be generated by the formation of hairpin loops during the synthesis of genome RNA (21). If this is so, then the molecular basis for the competition between the replication of DI RNA and standard RNA could be based on the generation of two high affinity sites for polymerase during the replication of the DI genome in contrast to the existence of one high affinity site and one low affinity site during the replication of the genome of standard virus (Fig. 5). Therefore, without further mutations or alterations of the DI genome, a competitive situation can be envisaged with just the acquisition of extragenomic sequences.

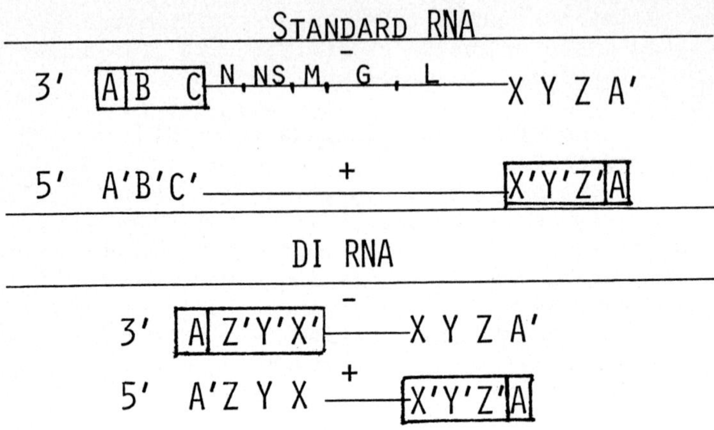

A = POLYMERASE RECOGNITION SITE
BC = LOW AFFINITY BINDING SITE
X′Y′Z′ = HIGH AFFINITY BINDING SITE
A′ = N PROTEIN RECOGNITION SITE

FIGURE 5. Possible explanation for the competition between DI particle RNA and standard virus RNA during in vivo synthesis.

For DI genomes which appear to derive from the 5′ end of standard virion RNA, such a mechanism would explain their competitive advantage for polymerase attachment. For DI genomes derived from the 3′ end of the standard virus RNA, another mechanism may have to be postulated, perhaps involving the initial attachment of the RNA-binding N protein. Obviously, continued analysis of these polymerase attachment sites and the ends of both standard RNA and the RNA from a large number of DI particles derived from the same standard RNA would help us to understand the generation of deletion mutants with selective advantages in replication.

For clarifying the role of DI particles during the persistence in the host, RNA sequence data are only a part of the picture. How cells from different species or even different tissues affect interference, i.e., the replication of DI and standard RNA genomes, remains the challenge for the future.

ACKNOWLEDGMENTS

We thank Gertrude Lanman for excellent technical support and Suzanne Ress for typing.

REFERENCES

1. Huang, A.S., and Baltimore, D. (1977) In "Comprehensive Virology" Vol. 10 (H. Fraenkel-Conrat and R.R. Wagner, eds.), pp. 73-116, Plenum Publishing Corp., New York
2. Huang, A.S., and Baltimore, D. (1970) Nature 226: 325-327.
3. Palma, E.L., and Huang, A.S. (1974) J. Infect. Dis. 129:402-410.
4. Perrault, J., and Holland, J. (1972a) Virology 50: 148-158.
5. Huang, A.S., Greenawalt, J.W., and Wagner, R.R. (1966) Virology 30:161-172.
6. Huang, A.S., and Wagner, R.R. (1966) Virology 30: 173-181.
7. Little, S.P., and Huang, A.S. (1977) Virology 81:37-47.
8. Welsh, R.M., and Oldstone, M.B.A. (1977) J. Expt. Med. 145:1449-1468.
9. Huang, A.S., and Manders, E.K. (1972) J. Virol. 9: 909-916.
10. Soria, M., Little, S.P., and Huang, A.S. (1974) Virology 61:270-280.
11. Knipe, D., Rose, J.K., and Lodish, H.F. (1975) J. Virol. 15:1004-1011.
12. Perrault, J., and Holland, J.J. (1972b) Virology 50: 159-170.
13. Stampfer, M., Baltimore, D., and Huang, A.S. (1969) J. Virol. 4:154-161.
14. Freeman, G.J., Rao, D.D., and Huang, A.S. (1978) In "Negative Strand Viruses and the Host Cell" (B.W.J. Mahy and R.D. Barry, eds), Academic Press, New York
15. Huang, A.S., and Palma, E.L. (1974) In "Mechanisms of Virus Disease" (W.S. Robinson and C.F. Fox, eds.), pp. 87-99, Benjamin, Inc., Menlo Park, California.
16. McMaster, G.K., and Carmichael, G.G. (1977) Proc. Nat. Acad. Sci USA 74:4835-4838.
17. Reichman, M.E., personal communication
18. Perrault, J., and Leavitt, R.S. (1977) J. Gen. Virol. (in press).

19. Kolakofsky, D. (1976) Cell 8:547-555.
20. Leppert, M., Kort, L., and Kolakofsky, D. (1977) Cell 12:539-552.
21. Huang, A.S. (1977) Bacteriol. Rev. 41:811-821.
22. Freeman, G.J., Rose, J.K., Clinton, G.M., and Huang, A.S. (1977) J. Virol. 21:1094-1104.
23. Lehrach, H., Diamond, D., Wozney, J.M., and Boedtker, H. (1977) Biochem. 16:4743-4751.
24. Baltimore, D. (1966) J. Mol. Biol. 18:421-428.
25. Loening, U.E. (1969) Biochem J. 113:131-138.

GENE EXPRESSION BY A DEFECTIVE INTERFERING PARTICLE OF VESICULAR STOMATITIS VIRUS

Leslye D. Johnson and Robert A. Lazzarini

Molecular Virology Section, Laboratory of Molecular Biology, National Institute of Neurological and Communicative Disorders and Stroke
NIH, Bethesda, Maryland 20014

ABSTRACT A 3' end defective interfering particle, DI-LT, of vesicular stomatitis virus directs the synthesis in cultured cells of four viral proteins: nucleocapsid (N), nonstructural (NS), matrix (M) and glycoprotein (G). By several criteria these four proteins are functional. The only viral protein not synthesized in these infected cells is the L protein which may represent the limiting factor in this infection.

INTRODUCTION

The genome of vesicular stomatitis virus (VSV) is a negative strand of RNA (1). It has information for five genes which are arranged as shown in Fig. 1 (2,3). When passaged several times at high multiplicity, VSV produces defective interfering (DI) particles (1). Such defective virions are shorter than the standard virus due to a smaller RNA genome; contain all the standard viral proteins; are not capable of productive infections by themselves and reduce the yield of standard virus in mixed infections (1). Two types of DI particles can be distinguished (4). Those having various portions of the L protein gene (Fig. 1), i.e., 5' end DI particles, are able to transcribe a 2-4S RNA _in vitro_ (5-8). The second type has genetic information from the 3' half of the viral genome. An example of this class is the DI-LT virus derived from the heat resistant strain of VSV_{IND} (9). It has the entire region for the N, NS, M and G proteins (4-6). _In vitro_, DI-LT is transcribed into a leader RNA and four mRNAs (10). These species are the same as those made from the

ISBN 0-12-668350-6

3' half of VSV *in vitro*.

The expression of VSV and DI particles has been evaluated in cultured cells. VSV infected cells synthesize five mRNAs but replicate the viral genome only after proteins are made (11). This new

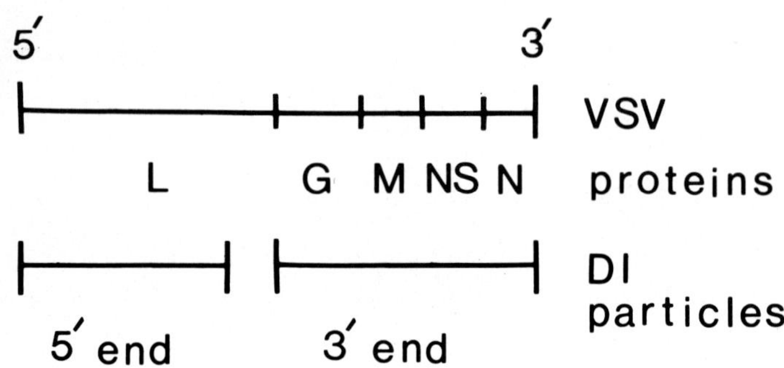

FIGURE 1. Gene order of VSV and genome retention by DI particles.

genomic RNA serves as a template for further transcription and replication. A 5' end DI particle makes no detectable RNA under similar conditions (12). On the other hand, a 3' end DI virion (DI-LT) synthesizes both mRNAs and replicative RNAs (13). The rate of transcription of the input DI-LT template is the same as that observed with VSV. However, the rate of total RNA synthesis by the DI particle is much less than that by VSV suggesting that replication and/or secondary transcription is curtailed.

The synthesis of mRNAs and replicative RNAs by DI-LT led us to examine both the synthesis and biological activity of proteins by cells infected with this DI virion.

METHODS

Suspension BHK-21 cells were infected with purified preparations of either DI-LT or VSV as described earlier (13). When RNA was analyzed, the infections were labeled with ^{3}H-uridine (20 uC/10^{6}

cells) at 20 min. post infection. For protein analysis, ^{35}S-methionine (2 uC/10^6 cells) was added to cells 3 h. post infection. In both cases, cytoplasmic extracts were made after 4 h. of infection (13). Labeled RNAs were analyzed on SDS-sucrose gradients as described earlier (13). The DI ribonucleoprotein (RNP) structure was isolated by centrifugation of cytoplasmic extracts in 5-20% sucrose gradients containing either

Table 1 Association of 28S RNA with a 110S RNP Structure

Sample	^{3}H-Uridine CPM
28S RNA[a]	193,600
28S RNA made from 110S RNP[b]	177,100

[a]Infections and labeling were performed as described in METHODS. Half of the cytoplasmic extract was made 1% SDS and centrifuged on a deproteinizing sucrose gradient. Radioactivity migrating as 28S was determined.

[b]The second half of the cytoplasmic extract was centrifuged on a nondeproteinizing sucrose gradient. RNA was prepared from the 110S RNP and analyzed on sucrose gradients containing 80% DMSO. Radioactivity migrating as 28S was determined.

NEB ("low salt") or NEB-1 M NH_4Cl ("high salt") (14,15). RNP containing replicative RNA was located by its resistance to ribonuclease (13,14). RNAs prepared from it were centrifuged on 5-20% sucrose with 80% dimethylsulfoxide (13). Proteins of cytoplasmic extracts, RNPs, or virions were electrophoresed on SDS-polyacrylamide gels in a Laemmli system (16). Complementation tests were performed as described in Table 2.

RESULTS

In an earlier study we observed the synthesis of both + and - strand replicative RNAs. These data implied that the + strand was present as an adequate template for - strand RNA synthesis. The results of other studies indicate that the template for - strand RNA synthesis is a + strand RNA associated with the N protein (17,18, L.D. Johnson, unpublished results). We thus examined an extract of DI-LT infected cells for encapsidated replicative RNAs (Table 1). A comparison of the radioactivity in total 28S replicative RNA with that of 28S RNA sedimenting in a 110S RNP structure shows that 91.5% of the RNA is encapsidated. The ribonuclease resistance of the RNP indicates that the RNA-protein association is similar to that of the standard and DI virion RNPs and of RNPs found in infected cells (17).

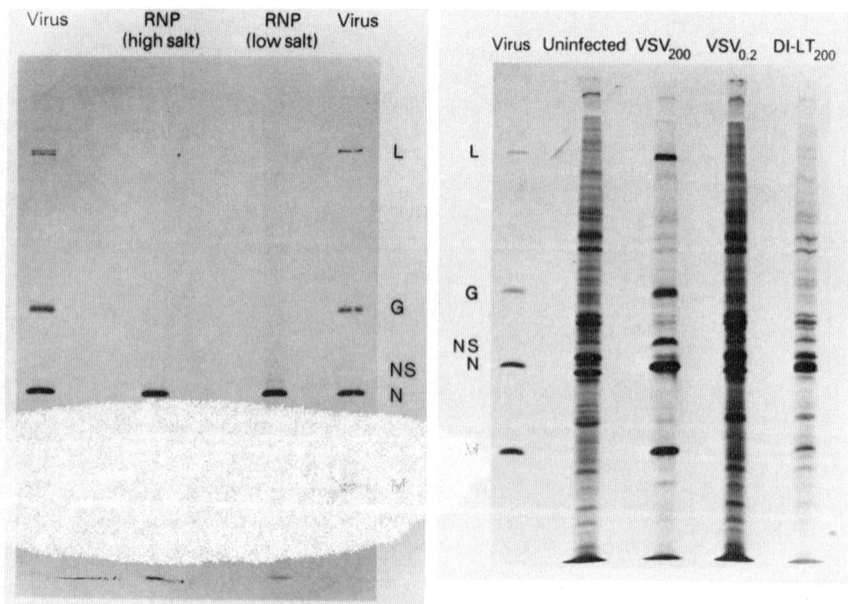

FIGURE 2 (left). Radioautogram of SDS-polyacrylamide gel separations of ^{35}S-methionine labeled proteins present in "high" and "low" salt RNPs of DI-LT infected cells. ^{35}S-methionine labeled proteins of VSV are shown for comparison.

FIGURE 3 (right). Radioautogram of SDS-polyacrylamide gel separations of ^{35}S-methionine labeled proteins of VSV or DI-LT infected cells. The ^{35}S-methionine labeled protein profiles for uninfected cells and VSV particles are also shown.

This RNP was analyzed for the presence of newly synthesized N protein as well as RNA. The proteins of DI infected cells were labeled with ^{35}S-methionine and RNP prepared in 1 M NH_4Cl ("high salt") as in the RNA study above. The proteins present in this fraction were analyzed on SDS-polyacrylamide gels. Lane 2 of Fig. 2 demonstrates that labeled N protein is associated

Table 2 Complementation Studies with VSV Temperature Sensitive Mutants and DI-LT[a]

TS Mutant (Viral Protein)	DI-LT	PFU/ml 32°	PFU/ml 39.5°	CI[b]
	+	$<10^1$	$<10^1$	
G-11(L)	-	$1.0x10^3$	$<10^1$	
	+	$5.5x10^2$	$<10^1$	0.55
G-22(NS)	-	$1.0x10^3$	$<10^2$	
	+	$2.8x10^4$	$4.0x10^2$	28
G-31(M)	-	$2.8x10^3$	$4.1x10^4$	
	+	$1.2x10^5$	$1.7x10^4$	37
G-41(N)	-	$1.5x10^3$	$2.0x10^1$	
	+	$7.2x10^3$	$<10^1$	4.8
0/45(G)	-	$5.0x10^1$	n.d.	
	+	$2.5x10^4$	$8.5x10^2$	480

[a] BHK-21 monolayers were infected with TS mutants (3PFU/cell) with or without DI-LT (3physical particles/cell) at 39.5°. After 7h, the supernatant was harvested and plaqued at 32° and 39.5°.

[b] $$CI = \frac{(TSxDI\text{-}LT)^{32^\circ} - (TSxDI\text{-}LT)^{39.5^\circ}}{(TS)^{32^\circ} + (DI\text{-}LT)^{32^\circ}} = \text{complementation index}$$

with a 110S RNP structure in cells infected with this DI particle. These data indicate that newly synthesized + and - strand replicative RNAs are associated with labeled N protein to form RNP structures. Furthermore, this new + strand RNP functions as a template for the production of - strand RNA.

In addition to N protein, the synthesis of two other viral proteins is demonstrated by ^{35}S-metionine labeling of cells infected with this DI virion. When intracellular structures containing replicative RNA are isolated under conditions of low salt, NS and L proteins remain associated with the RNP structure (15). Using this technique, we have tentatively identified ^{35}S-methionine labeled NS in the RNP structure (Fig. 2, lane 3). Labeled L protein is notably absent from this RNP.

Electrophoretic patterns of labeled proteins in unfractionated cytoplasmic extracts of DI virus or VSV infected cells are shown in Fig. 3. Infection with the same level of VSV as DI particle results in synthesis of all five viral proteins and marked shut off of host protein synthesis. In contrast, VSV at the level contaminating the DI preparation has the same protein pattern as uninfected cells. Cells infected with DI-LT clearly make a protein migrating with M; synthesis of N and NS is obscured by host proteins. Neither labeled L nor G proteins is seen in the gel. The inability to detect newly synthesized L protein is expected since the DI genome does not code for this protein. However, the absence of labeled G protein remains unexplained since the gene for this protein is clearly present (19,20, and below).

The biological activity of the proteins specified by this DI particle is measured by its ability to rescue temperature sensitive (ts) mutants for each of the five complementation groups of VSV. This rescue requires that functional proteins be supplied to the ts virus by the DI particle. As shown in Table 2, addition of the DI particle increases the yield of ts virus substantially for mutants in N, NS, M and G proteins. The complementation index with 0/45 (G) is particularly striking. Values as high as 1200 are seen when the ts virus is used at 30 PFU/cell.

No complementation by DI-LT of a mutant in the L protein is seen consistent with the absence of this gene in the DI-LT RNA. VSV at the level contaminating the DI-LT preparation did not complement any of these ts mutants (data not shown).

DISCUSSION

The data presented here demonstrate the synthesis of four viral proteins--N, NS, M and G--in cells infected with a 3' end DI particle (DI-LT). We conclude that the polyadenylylated RNAs sedimenting at 12-17S described previously

(13) are truly mRNAs and contain all four of the species characteristic of this size class.

Several lines of evidence indicate that these four proteins are functional. In the case of N protein it associates with replicative RNA to form a RNP structure. This RNP is resistant to ribonuclease like those present in VSV and DI virions and in infected cells. The synthesis of - strand RNA suggests that newly synthesized + strand RNP is a functional template.

The sedimentation of NS protein with this RNP might suggest that it too is functional. The competence of N and NS is indicated by the ability of DI-LT to kill cells (21).

NS and L proteins along with - strand RNP are required for *in vitro* transcription (22). The inability to detect newly synthesized L protein in this infection would appear to be the only factor limiting transcription of new RNP structures. However, neither the ability of purified L protein to stimulate total RNA synthesis in these cells nor the ability of this incomplete RNP to be transcribed outside the cells when supplied with L protein has been tested.

The biological activity of the proteins specified by DI-LT is clearly demonstrated by the fact that new ts VSV particles are produced from infections with N, NS, M and G protein ts mutants and DI-LT. The detection of G protein synthesis by complementation and not by ^{35}S-methionine labeling remains unexplained but may reflect different sensitivities for detection in the two methods.

REFERENCES

1. Huang, A.S., and Baltimore, D. (1977). In "Comprehensive Virology" (H. Fraenkel-Conrat and R.R. Wagner, eds.), pp. 73-116. Plenum Press, New York.
2. Ball, L.A., and White, C.N. (1976). Proc. Nat. Acad. Sci. USA. 73, 442.
3. Abraham, G., and Banerjee, A.K. (1976). Proc. Nat. Acad. Sci. USA. 73, 1504.
4. Johnson, L.D., and Lazzarini, R.A. (1978). In "Rhabdoviruses" (D.H.L. Bishop, ed.), CRC Press, in press.

5. Leamnson, R.N., and Reichmann, M.E. (1974). J. Mol. Biol. 85, 551.
6. Stamminger, G., and Lazzarini, R.A. (1974). Cell. 3, 85.
7. Emerson, S.U., Dierks, P.M., and Parsons, J.T. (1977). J. Virol. 23, 708.
8. Perrault, J., Semler, B.L., Leavitt, R.A., and Holland, J.J. (1978). In "The Negative Stand Viruses" (B.W. Mahy and R.D. Barry, eds.), Academic Press, London, in press.
9. Petric, M., and Prevec, L. (1970). Virol. 41, 615.
10. Colonno, R.A., Lazzarini, R.A., Keene, J.D., and Banerjee, A.K. (1977). Proc. Nat. Acad. Sci. USA. 74, 1884.
11. Perlman, S.M., and Huang, A.S. (1973). J. Virol. 12, 1395.
12. Stampfer, M., Baltimore, D. and Huang, A.S. (1969). J. Virol. 1, 154.
13. Johnson, L.D., and Lazzarini, R.A. (1977). Proc. Nat. Acad. Sci. USA. 74, 4387.
14. Johnson, L.D., and Lazzarini, R.A. (1978). In "The Negative Strand Viruses" (B.W. Mahy and R.D. Barry, eds.), Academic Press, London, in press.
15. Emerson, S.U., and Wagner, R.R. (1972). J. Virol. 10, 297.
16. Laemmli, U.K. (1970). Nature 222, 680.
17. Soria, M., Little, S.P., and Huang, A.S. (1974). Virol. 61, 270.
18. Moyer, S.A., Holmes, K.S., Hamilton, D.H., and Moyer, R.W. (1977). In "Abstracts of 1977 American Society for Microbiology"
19. Schnitzlein, W.M., and Reichmann, M.E. (1976). J. Mol. Biol. 101, 307.
20. Chow, J.M., Schnitzlein, W.M., and Reichmann, M.E. (1977). Virol. 77, 579.
21. Marcus, P.I., Sekellick, M.J., Johnson, L.D., and Lazzarini, R.A. (1977). Virol. 82, 242.
22. Emerson, S.U., and Yu, Y-H. (1975). J. Virol. 15, 1348.

EVOLUTION OF VIRUS POPULATIONS IN PERSISTENT INFECTIONS OF L CELLS WITH VESICULAR STOMATITIS VIRUS[1]

Julius S. Youngner, Olivia T. Preble, Elaine V. Jones, and Richard S. Creager

Department of Microbiology, University of Pittsburgh, School of Medicine, Pittsburgh, Pennsylvania 15261

ABSTRACT We have previously reported (1) that non-cytocidal persistent infection can be established with wild-type (wt) VSV in mouse L cells at 37C, and that a rapid selection of RNA$^-$, group I ts mutants consistently occurs in this system. In order to assess the selective advantage of the RNA$^-$ ts phenotype, we studied the evolution of the virus population when persistent VSV infections were initiated in L cells using VSV ts 0 23 and ts 0 45, RNA$^+$ ts mutants belonging to complementation groups III and V, respectively. Fluids from the persistently infected cells were harvested at intervals and the virus population present was characterized by clonal analysis and compared to clones of parental ts 0 23 and ts 0 45.

In L cell cultures persistently infected with ts 0 23, the frequency of ts RNA$^-$ virus clones rose from 10% at 2 days after initiation to 81% by 198 days. Similar results were also obtained when persistent infections were initiated with 6-times cloned ts 0 23; by 211 days, 96% of the virus clones had an RNA$^-$ phenotype. In contrast, 29 of 30 parental ts 0 23 clones tested were RNA$^+$.

VSV ts 0 45 has another marker in addition to reduced virus yield at 39.5C: a defective protein which renders virions heat-sensitive (hs) at 50C. Virus clones isolated very early after infection with 6-times cloned ts 0 45 had properties identical to those of parental ts 0 45. Between 2 and 3 weeks after infection, however, most of the virus population consisted of ts$^+$, heat-resistant (hr), apparently wild-type virus. This ts$^+$ hr virus population was then rapidly replaced by ts hr virus. Although most of the virus clones were

[1]This work was supported by research grant AI-06264 from the National Institute of Allergy and Infectious Diseases, NIH.

ISBN 0-12-668350-6

leaky or had a high reversion frequency which made complementation analyses difficult, 22 out of 26 (85%) of the stable ts hr clones isolated between 16 and 235 days after infection were RNA$^-$. In contrast, 53 of 56 parental ts 0 45 clones tested were RNA$^+$.

These two examples of persistent VSV infections in L cells suggest that the ts RNA$^-$ phenotype may have a unique selective advantage in this system. Other mechanisms were explored which also might be operative in the maintenance of persistent VSV infections of L cells, in addition to selection of ts RNA$^-$ mutants. These included i) mediation by DI particles; (ii) mediation by interferon, and (iii) the role of a mutation in the viral function (P) required for inhibition of host cell protein synthesis. Of these 3 alternatives evidence was obtained that interferon may play a role in the regulation of persistent infections of L cells with VSV.

INTRODUCTION

In 1976 we reported (1) that when noncytocidal persistent infections were established in mouse L cells (L_{VSV}) at 37C with infective B particles of vesicular stomatitis virus (VSV) in the presence of large numbers of DI particles, there was a rapid spontaneous selection of temperature-sensitive (ts) virus. A significant rise in the frequency (73%) of ts mutants in the virus population was noted by 11 days and by 63 days 100% of the clones isolated were ts at 39.5C, the non-permissive temperature used. The ts clones isolated had an RNA$^-$ phenotype and belonged to VSV complementation group I. It was also determined that ts mutants of VSV are conditionally defective particles which can interfere with the replication of wild-type VSV (2). The selection of ts mutants, the dominance of the replication of ts mutants at 37C, and the apparent absence of significant numbers of DI particles in the culture fluids from L_{VSV} cells, led us to speculate about the role of ts mutants in both the establishment and the maintenance of the persistently infected state.

The present report describes the evolution which occurs in virus populations when persistent VSV infections of L cells are established with phenotypically RNA$^+$ mutants belonging to complementation groups III and V. In addition, an assessment is made of the current status of the various mechanisms which may operate in the maintenance of persistent infections by VSV in L cells.

METHODS

Detailed descriptions of the cells, viruses, and methods employed are given in the articles referenced below.

RESULTS AND DISCUSSION

EVOLUTION OF POPULATIONS IN PERSISTENT INFECTIONS OF L CELLS INITIATED WITH PHENOTYPICALLY RNA$^+$ ts MUTANTS OF VSV

The persistent infections which had been initiated in L cells with wild-type VSV in the presence of large numbers of DI particles showed a rapid selection of ts viruses with RNA$^-$ phenotypes belonging to VSV complementation group I (1). This finding was so consistent that the possible selective advantage of the RNA$^-$ phenotype in persistent infection had to be considered. An alternate explanation stemmed from the finding by Flamand (3) that 82% of spontaneous mutants present in VSV (Indiana) populations were group I mutants. If the ts mutants that maintain the persistently infected state were selected from those spontaneously present in the wild-type inoculum, there is a high probability that they would belong to complementation group I.

To investigate the possible selective advantage of RNA$^-$ VSV mutants in persistence, infections were initiated with RNA$^+$ mutants belonging to complementation group III. Persistence was established in L cells infected with a low multiplicity (0.4) of ts 0 23 (III) grown in BHK-21 cells under conditions designed to minimize the presence of DI particles. Fluids were collected from the persistently-infected line at the time of each cell passage. At selected intervals after initiation of persistence, the virus in harvested fluids was cloned in primary chick embryo (CE) cells by isolating plaques at terminal dilutions at 33C, the permissive temperature. After confirming the ts phenotype of each clone, its RNA phenotype at the non-permissive temperature in BHK-21 cells was determined. In addition, the parental virus (ts 0 23) was cloned to determine its homogeneity in regard to the RNA$^+$ phenotype. The results of these determinations are displayed in Table I.

Clonal analysis revealed that in the case of the ts 0 23 parental population, 29 of 30 clones (97%) showed an RNA$^+$ phenotype; one clone synthesized RNA at the non-permissive temperature with an effectiveness only 10% of that of wild-type virus. By 2 days after initiation of the infection the frequency of RNA$^-$ mutants present in the fluids from persistently infected cells had risen to 10%, an increase not

TABLE I
APPEARANCE OF RNA^- PHENOTYPE IN PERSISTENT INFECTIONS OF L CELLS INITIATED WITH A GROUP III (RNA^+) ts MUTANT OF VSV (ts 0 23)

Cell passage	Days after initiation	No. clones ts / No. isolated	No. of clones with RNA^- phenotype[a]
Parental virus (ts 0 23)	——	30/30	1 (3%)
P-0	2	30/30	3 (10%)
P-8	37	28/28	19 (68%)
P-18	75	29/29	21 (72%)
P-44	198	26/26	21 (81%)

[a]RNA^- phenotype = 39.5/33C ratio of RNA synthesis in BHK-21 cells is less than 20% of the ratio of wild-type virus.

statistically significant. However, at 37 days the frequency of RNA^- mutants rose to 68%; at 75 and 198 days after initiation the frequency of RNA^- mutants increased to 72% and 81%, respectively. The criterion for classifying a clone as RNA^- was its ability to synthesize RNA with an efficiency less than 20% of the 39.5C/33C ratio of RNA synthesis by wild-type VSV.

To confirm these results another persistent infection was started in L cells using a subclone (F-3) of ts 0 23 which had been plaque-purified 6 times in CE cells. The sixth clonal passage was amplified at low multiplicity of infection (0.01) in MDBK cells which are poor producers of DI particles. When clonal analysis techniques were applied to fluids harvested at different times after initiation of the persistence, the pattern observed was similar to that described above with parental ts 0 23. At 8 days after initiation, all 30 clones tested were ts and 29 (97%) were RNA^+ at the non-permissive temperature (39.5C); by 43 days, all 29 clones isolated still were ts but of these, 23 (79%) were RNA^-. At 211 days after initiation, of 28 ts clones isolated, 27 (96%) had an RNA^- phenotype.

Attempts were made to carry out complementation analysis of the RNA^- clones isolated from the persistent infections initiated with complementation group III RNA^+ mutants. These efforts were unsuccessful. In addition to problems of the leakiness of these mutants when used at high multiplicities in complementation analyses, the RNA^- clones also appear to have multiple mutations.

Persistent infection was also begun by infecting L cells with VSV ts 0 45 (V), an RNA^+ mutant. In addition to reduced viral replication at 39.5C, ts 0 45 has an additional phenotypic marker: the virion G protein is defective and renders the virion heat-sensitive (hs) at 50C. This marker enabled further characterization of the evolution of virus in persistent VSV infections.

The virus used for infection had been serially cloned 6 times in CE cells, then amplified under conditions designed to minimize production of DI particles. Persistent infection was established using an input moi of 0.01. Samples of the culture fluids were harvested at various intervals after initiation, the efficiency of plaquing (e.o.p.) at 39.5C was determined, and individual virus clones were isolated from terminal dilution plates incubated at 33C. The properties of virus clones isolated at intervals from 7 to 235 days after initiation are summarized in Table II. Although infectivity titers were barely detectable in fluids harvested during the first two passages, titers subsequently fluctuated between 10^3 and 10^5 PFU/ml.

No correlation was apparent between viral titer, extent of CPE in the cultures, and the properties of the individual isolated virus clones. However, at 16 and 21 days after infection, the high e.o.p. at 39.5C of the uncloned culture fluids was indicative of the presence of a high proportion of non-ts (ts^+), heat-resistant (hr) apparently wt revertant clones in the population. The ts^+ clones replicated to high titers in CE and BHK cells at both 33C and 39.5C. The reasons for the appearance of wt revertants at this time and for their subsequent elimination from the virus population are unclear. A second persistent infection in L cells established with 6-times cloned ts 0 45 is currently under study to determine if the same pattern of reversion and reselection is observed.

Concurrent with the appearance and disappearance of ts virus in the population, the heat stability of the ts virus clones changed dramatically (Table III). At 7 days after initiation of infection, all 14 virus clones isolated were ts and hs, with heat inactivation ratios ranging from 6×10^{-2} to 2.4×10^{-3} relative to wild-type virus. This range was very similar to that found for 56 clones of

TABLE II

EVOLUTION OF VIRUS IN L CELLS PERSISTENTLY INFECTED WITH VSV ts 0 45 (V)

Cell passage	Days after initiation	e.o.p. (39.5/33C)	No. clones ts / No. isolated	% ts
P-1	7	ND[a]	15/15	100
P-3	16	.27	11/30	37
P-4	21	.18	5/30	17
P-7	36	$<1.3 \times 10^{-3}$	23/27	85
P-9	49	$<6.6 \times 10^{-4}$	30/30	100
P-11	79	$<3.1 \times 10^{-3}$	30/30	100
P-16	118	9.1×10^{-3}	28/28	100
P-44	235	$<9.0 \times 10^{-5}$	30/30	100

[a]Not determined.

TABLE III

HEAT SENSITIVITY AND RAN PHENOTYPE OF VIRUS CLONES ISOLATED FROM PERSISTENT INFECTION INITIATED WITH ts 0 45

Days after initiation	ts hs Clones		ts hr Clones	
	No.	RNA^+/RNA^- [a]	No.	RNA^+/RNA^- [a]
Parental virus (ts 0 45)	56	53/3	0	——
7	14	13/1	0	——
16	6	5/0	5	1/3
21	0	——	7	ML[b]
36	1	ML[b]	24	0/1
49	0	——	30	ML[b]
79	0	——	30	2/4
118	0	——	28	0/8
235	0	——	29	1/6

[a]RNA phenotype of stable ts pi clones only.

[b]ML = multiplicity leak.

parental VSV ts 0 45 (1×10^{-2} to 2×10^{-3}). The 11 ts clones isolated 16 days after initiation of the cultures fell into two groups: six clones (ts hs) were similar to ts 0 45, while five clones (ts hr) were 3.5- to 10-fold more heat-resistant than wild-type VSV. All ts virus clones isolated from subsequent passages of the persistently infected cells, with only one exception, were ts hr.

Ts virus clones from 16, 79 and 118 days failed to complement any of the 5 known genetic groups of VSV. However, most of the virus clones isolated 21 days or later after initiation produced unacceptably high yields under the conditions of infection required for either complementation analyses, or cumulative RNA synthesis measurements at 39.5C. When clonal fluids of 21- or 36-day ts pi clones were tested in CE, L, or BHK cells, all clones were uniformly ts, but the 33C yield from this first amplification passage in either CE or BHK cells had high replication efficiency at 39.5C. This effect was much more pronounced at an moi of 5 than at an moi of 0.05 and was termed multiplicity leak (ML). However, of 26 stable ts hr clones isolated between 16 and 235 days, 22, or 84.6%, were RNA^-, as summarized in Table III.

Therefore, during the course of establishment and maintenance of this persistent VSV infection initiated with a group V mutant in L cells, the ts hs RNA^+ virus population present 7 days after infection evolved by 2 to 3 weeks into a mixture in which apparently wt virus predominated. The wt virus (ts^+ hr RNA^+) was then rapidly replaced by ts hr RNA^- virus, suggesting that the pressures within the persistent infection may select specifically for certain types of ts mutants most suited for maintaining the carrier state.

In summary, when persistent infections are initiated in L cells with RNA^+ VSV mutants representing complementation groups III and V, the viruses which evolve are quite different than the parental virus. In addition to the long-term maintenance of the ts phenotype, there is a selection of viruses with RNA^- phenotypes. Since clonal analysis of the parental ts viruses revealed the presence of RNA^- viruses at a frequency ranging from 3% to 6%, it is likely that the RNA^- mutants which predominate in the persistent infection were selected from those present in the parental population. In the case of the group V mutant, the heat sensitivity marker due to a defective G virion protein was lost rather soon after the establishment of the carrier infection. Detailed analyses are being carried out of the virion proteins and other properties of the mutants selected during these persistent infections.

MECHANISMS WHICH MAY MEDIATE THE MAINTENANCE OF PERSISTENT VSV INFECTION IN L CELLS

A number of mechanisms have been suggested which may operate in the maintenance of persistent infections. The following sections attempt to summarize their status in regard to the specific system we have been studying intensively, namely, VSV persistence in mouse L cells.

Mediation by Defective Interfering (DI) Particles. We have reported previously that noncytocidal persistent infections of L cells with infective B particles of VSV could be established only in the presence of large numbers of DI particles (1). Although essential for initiation, DI particles were not considered crucial to the maintenance of this persistent infection on the basis of two findings. First, we failed to demonstrate significant numbers of DI particles in L_{VSV} fluids ($<$ 1 DI/PFU) by an interference assay. Second, low multiplicities of 6-times cloned VSV ts mutants belonging to RNA^+ complementation groups III and V (see above), as well as a clonal isolate from L_{VSV} (ts pi 364) were capable of establishing persistent infection in L cells (1). The virus inocula for these experiments were prepared using conditions to minimize the number of DI particles present. Other indirect evidence has been cited which reduces the likelihood that DI particles are required for the maintenance of the carrier state in L_{VSV} cells. When the persistently infected cells maintained at 37C are shifted down to 32C, there is an increased replication of virus and complete cell destruction occurs within 3 days (1). Holland and Villarreal (4) also reported a rapid cytocidal effect accompanied by virus replication when their line of BHK-21 cells persistently infected with VSV was shifted down from 37C to 33C. These findings are difficult to reconcile with maintenance of the carrier state by the presence of DI particles in L cells. There is no evidence in the literature and no *a priori* reason to believe that DI particles show a temperature-dependent interference effect (5).

Further efforts were made to quantitate the DI particles present in culture fluids by using the highly sensitive amplification technique of Holland and Villarreal (6). Employing BHK-21 cells for amplification, this method showed that there was less than one DI particle for each 10 PFU in the culture fluids from L_{VSV} cell cultures. In addition, serial undiluted passages in BHK-21 cells of fluids from L_{VSV} cell cultures showed that multiple passages were necessary before interference and DI bands were detected in the harvested fluids, again indicating the low concentration of DI particles in the fluids from L_{VSV} cell cultures. We have also found that, compared to BHK-21 cells, L cells are poor hosts for the replication of DI particles (unpublished data). The deficient replication of VSV DI particles in L cells has also been noted by others (7,8).

As Holland _et al_. (9) have pointed out, it is practically impossible to rule out the presence of small numbers of DI particles in VSV replicating systems. In the case of L_{VSV} cells, the paucity of demonstrable DI particles in the culture fluids and the inherent defectiveness of the host L cells in regard to DI particle replication make it seem unlikely that in the system we are studying DI particles play a crucial role in maintenance of the carrier state.

Temperature-Sensitive Mutants as Conditionally Defective Interfering Particles. It has been reported that ts mutants of NDV (10), Sindbis virus (11), and VSV (2) are capable of interfering with the replication of homologous wild-type viruses at permissive and nonpermissive temperatures. We showed with VSV (2) that the inhibition of wild-type virus replication at the nonpermissive temperature (39.5C) was accompanied by a marked enhancement of the yields of co-infecting ts mutant. Our original experiments dealing with the "dominance" of ts mutants over wild-type replication were carried out with mutants belonging to complementation groups I, II, and IV which have RNA^- phenotypes (2). We have extended these experiments to include mutants of complementation groups III and V which are phenotypically RNA^+. We found that members of these complementation groups also behave as conditionally defective interfering particles (unpublished data).

The dominance of the replication of ts virus over that of the wild-type at 37C must be considered a rationale for the spontaneous selection and maintenance of ts mutants in persistently infected cell lines. The ts mutants present in a persistent infection can inhibit the replication of wild-type virus, thus preventing revertants from providing a significant portion of the virus population. A ts virus population selected during a persistent infection could, therefore, maintain its mutant character. Still unclear is the precise mechanism by which ts mutants are able to establish themselves in persistent infections.

Mediation by Interferon. In our original studies with L_{VSV} cell cultures we reported that although the cells were resistant to challenge with wild-type VSV and a heterologous virus, pseudorabies, no interferon was detected in the culture fluids (1). Even 100-fold concentration of the fluids did not result in detection of a viral inhibitor.

More recently, Ramseur and Friedman reported that prolonged VSV infections of L cells, established with the aid of exogenous interferon, may be maintained by endogenous interferon production as well as by the selection of ts mutants (8). In their experiments, endogenous interferon was implicated not by the detection of the inhibitor in fluids from the infected cells (12) but by the enhancement of virus replication and cell destruction after treatment of the cell cultures with anti-mouse interferon antibody. This approach had previously been used by Inglot et al. (13) to elucidate the role of endogenous interferon in persistent infections of L cells with Sindbis virus.

We carried out experiments with several of our L_{VSV} cell lines to determine the effects of treatment with anti-mouse interferon antibodies. Persistently infected cell lines were treated with a 1:100 dilution of rabbit anti-mouse interferon IgG, normal rabbit IgG, or with growth medium alone. The results obtained with two L_{VSV} cell lines show that in both instances continued presence of anti-interferon antibody in the medium increased virus titers 34- to 180-fold in fluids harvested 3 and 6 days after treatment was begun (Table IV). Correlated with the increased virus yield there was also a significant increase in cell destruction in both cell lines when treated with anti-interferon antibody. These results, which are quite similar to those obtained by Ramseur and Friedman (8), point to the possible role of endogenous interferon production in the maintenance of persistent VSV infections in L cells. There are still some unresolved questions these observations raise. How is this endogenous interferon production related to the selection of VSV ts mutants in persistently infected L cells? Are VSV ts mutants exceptionally effective inducers of the interferon system in L cells? Why do persistently infected cells maintained at 37C show high virus yields and undergo cell destruction when shifted down to 32 or 33C (1,4)? Why is it impossible to establish prolonged VSV infections in L cells pretreated with interferon at 37C and then infected and shifted down to 32C (8)?

In addition to the studies which have been summarized briefly above, Marcus and Sekellick (personal communication) have proposed that activation of the interferon system by the [±] RNA class of DI particles may provide the means by which persistently infected cells survive in the presence of otherwise lethal virus. However, it is difficult to invoke the interferon mechanism in the case of cell lines which are defective interferon producers; despite this deficiency, these cell lines (Vero or BHK-21) have been persistently infected with a variety of viruses (14,15,16).

TABLE IV

INFLUENCE OF ANTI-MOUSE INTERFERON ANTIBODY ON VIRUS YIELD AND CELL DESTRUCTION IN L_{VSV} CELL LINES

L_{VSV} cell line	Duration of treatment	Infectivity of culture fluid (PFU/ml) from cells treated with		
			Rabbit IgG (1:100 dil.)	
		MEM	Normal	Anti-mouse IF
402 (P-17, 70 days)	3 days	1.1×10^4 (0)[a]	1.6×10^4 (0)	3.0×10^6 (++++)
312 (P-115, 547 days)	3 days (Cells split 1:2)	1.0×10^4 (±)		4.5×10^5 (±)
	6 days	5.3×10^4 (±)		1.8×10^6 (++++)

[a]( ) = degree of cell destruction

Role of a Mutation in Viral Function (P) for Inhibition of Protein Synthesis. Stanners *et al.* reported that a group I ts mutant of VSV (T 1026), derived from the HR strain, contains a second non-ts mutation in addition to its ts transcriptase mutation (17). The second mutation is in a viral function termed "P" which is responsible for the inhibition of total protein synthesis in infected cells. Whereas wild-type VSV is P^+ and effectively shuts off protein synthesis in infected cells, T 1026 and its ts^+ revertants are phenotypically P^-. Rates of total protein synthesis in cells infected with these P^- viruses are equal to or greater than in uninfected cells. Stanners and his colleagues suggested that loss of P function may be a necessary condition for establishment of persistence, since interference with cellular translational efficiency would prevent cell division and could lead to cell death.

The results of Stanners *et al.* in regard to the P^+ function of wild-type VSV and the P^- mutation in T 1026 were duplicated in L cells by their techniques (17) in our laboratory. We then examined the P function of a variety of ts pi clones isolated from different lines of persistently infected L cells which had been established with either wild-type VSV, ts 0 23 (III), or ts 0 45 (V). A total of 17 ts clones which had been recovered from fluids harvested as early as 10 days or as late as 211 days after initiation were tested for their P phenotype. All 17 ts pi clones tested were unequivocally P^+ and were able to inhibit protein synthesis of infected L cells as well as, or better than, wild-type VSV. On the basis of these results, we conclude that ts mutants recovered from persistent infection of L cells with VSV do not have a mutation in the P function and that the loss of the P function does not appear to be necessary for establishment or maintenance of persistence in L cells.

ACKNOWLEDGMENTS

The authors thank Marion E. Kelly and Gregory S. Buzard for their valuable assistance.

REFERENCES

1. Youngner, J. S., Dubovi, E. J., Quagliana, D. O., Kelly, M., and Preble, O. T. (1976). J. Virol. 19, 90.
2. Youngner, J. S., and Quagliana, D. O. (1976). J. Virol. 19, 102.
3. Flamand, A. (1970). J. gen. Virol. 8, 187.
4. Holland, J. J., and Villarreal, L. P. (1974). Proc. Nat. Acad. Sci. USA 45, 385.
5. Huang, A. S. (1973). Ann. Rev. Microbiol. 27, 101.
6. Holland, J. J., and Villarreal, L. P. (1975). Virology 67, 438.
7. Potter, K. N., and Stewart, R. B. (1976). Can. J. Microbiol. 22, 1458.
8. Ramseur, J. M., and Friedman, R. M. (1977). J. gen. Virol. 37, 523.
9. Holland, J. J., Villarreal, L. P., and Breindl, M. (1976). J. Virol. 17, 805.
10. Preble, O. T., and Youngner, J. S. (1973). J. Virol. 12, 472.
11. Stollar, V., Peleg, J., and Shenk, T. E. (1974). Intervirology 2, 337.
12. Nishiyama, Y. (1977). J. gen. Virol. 35, 265.
13. Inglot, A. D., Albin, M., and Chudzio, T. (1973). J. gen. Virol. 20, 105.
14. Stanwick, T. L., and Hallum, J. V. (1974). Infect. and Immun. 10, 810.
15. Youngner, J. S., and Quagliana, D. O. (1975). J. Virol. 16, 1332.

16. ter Meulen, V., and Martin, S. J. (1976). J. gen. Virol. 32, 431.
17. Stanners, C. P., Francoeur, A. M., and Lam, T. (1977). Cell 3, 273.

RNA SPLICING IN THE EARLY mRNAs OF SIMIAN VIRUS 40 AND ADENOVIRUS 2[1]

Arnold J. Berk[2] and Phillip A. Sharp[3]

Department of Biology and Center for Cancer Research
Massachusetts Institute of Technology
Cambridge, Massachusetts 02139

ABSTRACT The structure of the early mRNAs of Simian Virus 40 (SV40) and of Adenovirus 2 (Ad2) are deduced by analyzing the S_1 and Exonuclease VII digestion products of hybrids between cytoplasmic RNA and DNA restriction fragments. Two major early SV40 mRNAs are detected and both are spliced. Eight major early Ad2 mRNAs are observed and each of these is spliced. The early region of Ad2 located at the left end of the genetic map is known to encode functions required for transformation by Ad2. The splicing pattern of the mRNAs encoded in this region is similar to the splicing pattern of the early SV40 mRNAs. Two mRNAs occur which have the same 5' and 3' termini but have different intervening sequences deleted by splicing. DNA sequencing has shown that in one of the two mRNAs, the deleted sequence contains termination codons in all three reading frames. Thus in this mRNA, translation can proceed through the point at which the RNA is spliced. In the case of the early SV40 mRNAs, the translation products of the two closely related mRNAs have identical N-terminal but different C-terminal amino acid sequences.

INTRODUCTION

Simian Virus 40 (SV40) and Adenovirus 2 (Ad2) are well studied examples of DNA viruses which can transform cells in tissue culture. Cells which are transformed by SV40 in tissue culture or tumor cells which arise after SV40 injection into animals have common nuclear antigens and contain SV40 DNA integrated into cellular chromosomal DNA (1,2,3,). The

[1]This work was supported by grants from NIH and ACS

[2]Supported by a postdoctoral fellowship from the Helen Hay Whitney Foundation.

[3]Recipient of an American Cancer Society Career Development Award.

ISBN 0-12-668350-6

integration of the circular genome occurs in such a way as to leave the early region of the genome intact (4). SV40 RNA detected in the cytoplasm of transformed cells (5,6) is similar to that expressed during the early phase of productive infection. Two early SV40 gene products, small and large T-antigens, are found in transformed cells (7,8). A class of viral temperature-sensitive mutants (ts-a) are defective for transformation at the non-permissive temperature (9,10), and some types of transformed cells produced at the permissive temperature by these mutants are temperature sensitive for the expression of the transformed phenotype (11,12,13,14). These mutants map in the early region of the genome (15) and produce a temperature-sensitive large T-antigen (16). SV40 deletion mutants (17) which induce an altered small T-antigen (18) are defective for transformation as assayed by colony formation in soft agar (Bouck, N., Beules, N., di Mayorca, G., Shenk, T., and Berg, P.; Fluck, M., and Benjamin, T., personal communications). These data demonstrate that expression of early SV40 gene products is required for induction and maintenance of the transformed state in SV40 transformed cells.

A similar situation occurs in cells transformed by Ad2. Four separate regions of the 35kb viral genome are expressed during the early phase of productive Ad2 infection (19,20). One of these early regions which is encoded at the left end of the viral genome (early region 1) is invariably found integrated into host cell DNA in transformed cells (21) and Ad2 specific RNA transcribed from this region is found in the cytoplasm of transformed cells (19). Graham et al. have also shown that a 7% fragment of Ad2 DNA adjacent to the left terminus of the genome can be used to transform rat cells in culture. These transformed cell lines are "T" antigen positive and appear to be identical to virus-transformed cell lines (22). Viral mutants in early functions mapping in the left end of the genome are defective for transforming activity under non-permissive conditions (23,24). These data strongly suggest that early Ad2 gene products encoded in early region 1 are required for transformation by Ad2. Thus, in both SV40 and Ad2 transformed cells, it appears that early viral gene functions are required to induce and maintain the transformed state.

In this paper, we present data defining the map positions of viral genome sequences which comprise the early SV40 and Ad2 mRNAs. These mRNAs have a spliced structure (25), i.e., single covalently continuous RNAs are composed of sequences which map in separated regions of the viral genome. An intriguing observation which resulted from these studies is that the pattern in which genome sequences are spliced to produce the mRNAs encoded in the oncogenic region of Ad2 (early region 1) closely resembles the splicing pattern observed in the early SV40 mRNAs. The pattern of splicing directly affects the translation of these SV40 mRNAs (see below) and, therefore, the expression of early SV40 DNA sequences. We postulate (see below) that genetic information in early region 1 of Ad2 is organized in an analogous manner to the arrangement of genetic information in the early region of SV40, and, therefore, that the oncogenic functions encoded by these two viruses may have similar activities.

MATERIALS AND METHODS

Strategy and terminology: A spliced RNA is defined as a covalently continuous RNA molecule composed of sequences which are encoded in separate regions on a DNA genome. The RNA sequence which is colinear with DNA sequence is referred to as a colinear transcript. Thus a spliced RNA is composed of two or more colinear transcripts. The point in the RNA sequence at which the 3' end of one colinear transcript and the 5' end of the neighboring colinear transcript are joined is referred to as a splice point. Intervening DNA sequence refers to DNA sequence in the genome between DNA sequences which are transcribed into RNA colinear transcripts. Therefore, a hybrid between genome DNA and a spliced RNA is composed of RNA-DNA duplex over the colinearly transcribed sequences with loops of intervening single stranded DNA occurring at splice points in the RNA (25, Figure 1). When the products of this S_1 digestion are analyzed by electrophoresis in a neutral agarose gel followed by autoradiography, a labeled band is observed migrating at the rate of a duplex DNA molecule equal in length to the mRNA. When the products of S_1 digestion are denatured and analyzed by electrophoresis in a denaturing gel system followed by autoradiography, labeled bands are observed which migrate as expected for single stranded DNA molecules equal to the lengths of the colinear transcripts from which the RNA is composed. When a spliced RNA genome DNA hybrid is digested by Exo VII, single stranded DNA is removed from the ends of the RNA-DNA hybrid, but the single stranded loops of intervening DNA at the splice points are left intact. Following denaturation of the Exo VII products, electrophoresis on a denaturing gel system and autoradiography, a band is observed migrating as a single stranded DNA molecule equal in length to the lengths of the colinear transcripts plus the length of intervening DNA (length a+b+c in Figure 1). By performing this type of analysis with different restriction fragments having ends within the colinearly transcribed or

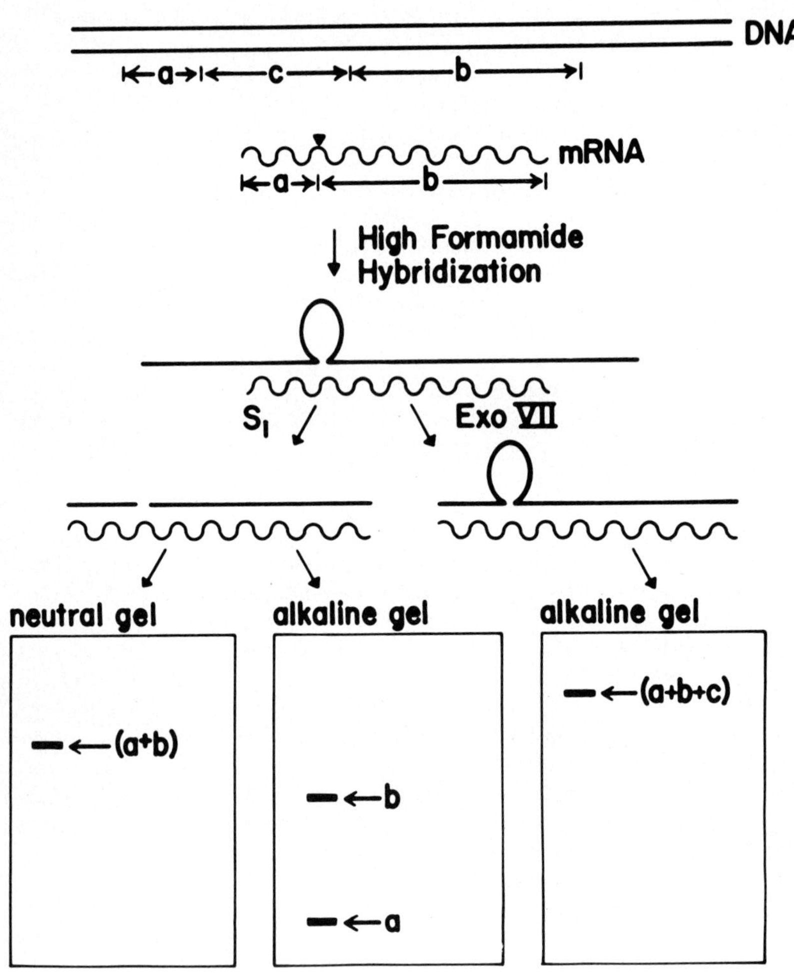

FIGURE 1. Strategy for the analysis of spliced mRNA structure by gel electrophoresis of endonuclease S_1 and exonuclease VII digested RNA-DNA hybrids.

intervening DNA sequences it is possible to map these regions on the DNA genome.

Cytoplasmic RNA was prepared from infected cells cultured in the presence of 20 μg/ml cytosine arabinoside 18 hours (SV40) or 8 hours (Ad2) post infection as described (28). Cytoplasmic RNA was hybridized to ^{32}P-labeled DNA restriction fragments for 3 hours in 80% formamide (29) at 49°C (SV40) or 60°C (Ad2) as described (29). S_1 digestion of products was as described (28) and Exo VII digestion as described (30). Neutral and alkaline agarose gels were as described (28) and 8M urea gradient polyacrylamide gels were as described (30). Preparation of ^{32}P-labeled DNA and isolation of restriction fragments were as described (28,30).

S_1 nuclease is an endoribonuclease as well as an endodeoxyribonuclease (27). In experiments with SV40 RNA, there often was cutting to some extent by S_1 in the RNA at a splice point in RNA-DNA hybrids. This occurred less frequently in experiments with Ad2 RNA and may be due to the A-U richness of SV40 RNA (∿60%) compared to Ad2 RNA (∿40%).

RESULTS

Analysis of early SV40 mRNA: Early SV40 RNA is transcribed from about one half of the viral DNA sequence. Transcription is in the counter-clockwise direction on the standard circular map beginning at approximately the origin of DNA replication at 67 unit lengths from the single *Eco* RI site (5,6). To determine the full length of early SV40 mRNAs, ^{32}P-labeled SV40 DNA cut one time in the late region of the genome at 0/100 (*Eco* RI) was hybridized to early cytoplasmic RNA, and digested with S_1. The products were resolved by electrophoresis on a neutral agarose gel (Figure 2a). Bands migrating at lengths of 2500, 2200, and 1900 nucleotides were reproducibly observed. The relative intensities of the 2200 and 2500 nucleotide bands were approximately 4 to 1 in all experiments, while the relative intensity of the 1900 nucleotide band was variable. The 1900 nucleotide band was found to be due to S_1 cutting at the splice point in the early mRNA-DNA hybrids (see below).

To determine the lengths of the co-transcripts of these mRNAs, the 2500 and 2200 nucleotide RNA-DNA hybrids were cut from a preparative neutral gel, denatured, and electrophoresed in a denaturing gel system (Figure 3). The 2500 nucleotide RNA-DNA band gave rise to two DNA bands migrating at 630 and 1900 nucleotides, while the 2200 nucleotide RNA-DNA hybrid gave rise to two DNA bands migrating at 330 and 1900 nucleotides. From these results we conclude that there are two early SV40 mRNAs composed of two colinear transcripts each. One is 2500 nucleotides long and is composed of 630 and 1900 nucleotide colinear transcripts. The other is 2200 nucleotides long and is composed of 330 and 1900 nucleotides.

To determine the mapping positions of the 1900 nucleotide colinear transcripts, early RNA was hybridized to SV40 DNA cut with various restriction enzymes, digested with S_1, and analyzed by electrophoresis on alkaline agarose gels. A single prominent 1900 nucleotide band was observed when the DNA was cut at 0/100 (Eco RI), 67 (Bgl I), 14 (Bam I), or at 57 (Taq I) (Figure 2b). The 1900 nucleotide band was also generated when the isolated restriction fragment mapping from 57 to 14 (Taq I - Bam I B) was used in the hybridization (Figure 2b). When the DNA in the hybridization was originally cut at 27 and 4 (Pst I) major bands produced by S_1 digestion were observed migrating at 1350 and 650 nucleotides (Figure 2c). From this data it is concluded that the two early SV40 mRNAs contain the same 1900 nucleotide colinear transcript mapping between 54 and 14 (1350 and 650 nucleotides from the Pst I site at 27, respectively).

To determine the mapping position on the genome of the 5' and 3' sequences of these spliced mRNAs, early RNA was hybridized to SV40 DNA cut with various restriction enzymes and the hybrids were digested with Exo VII. When the DNA was initially cut in the late region with Eco RI, a single prominent band migrating at 2600 nucleotides was observed on an alkaline agarose gel (Figure 2c). This same band was observed when the DNA was cut at 67 (Bgl I) and 14 (Bam I). When the DNA was initially cut at 27 and 4 (Pst I) bands migrating at 1950 and 650 nucleotides were observed. From these results we conclude that the 5' sequences of both the 2500 and 2200 nucleotide mRNAs map at 67 (1950 nucleotides clockwise from the Pst I site at 27), and the 3' sequences of these mRNAs map at 14 (650 nucleotides counter-clockwise from this Pst I site).

The structure of the two early SV40 mRNAs deduced from these data are diagrammed in Figure 4. Both mRNAs contain 5' sequence which is transcribed from the SV40 DNA sequence mapping at 67. The 2500 nucleotide mRNA has a 5' colinear transcript 630 nucleotides in length, while the 2200 nucleotide mRNA has a 5' colinear transcript 330 nucleotides in length. The 5' colinear transcripts of both of these mRNAs are spliced to a 1900 nucleotide colinear transcript mapping from 54 to 14. Data from experiments with other SV40 DNA restriction fragments are consistent with these proposed structures for the early SV40 mRNAs (30).

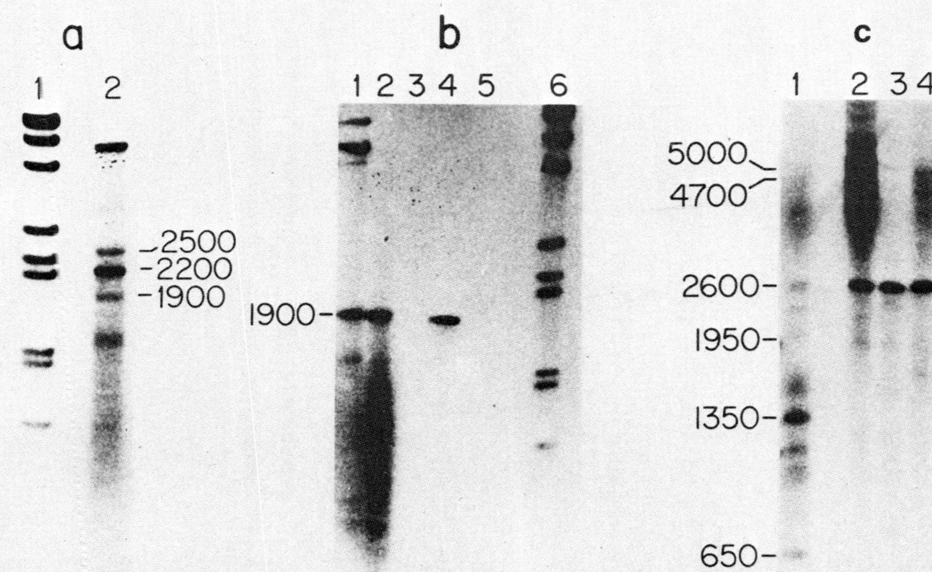

FIGURE 2. Autoradiograms of S_1 and Exo VII digested SV40 early RNA-SV40 DNA hybrids. (a) Neutral agarose gel of hybrid to DNA cut with Eco RI (track 2). Marker Sma I digest of Ad2 DNA (track 1). (b) Alkaline agarose gel of hybrids to: DNA cut at 0 with Eco RI (track 1); DNA cut at 14, 57, and 67 with Bam I, Taq I, and Bgl I (track 2); the fragments from 67 to 14 (track 3); from 14 to 57 (track 4); and from 57 to 67 (track 5). Marker Sma I digest of Ad2 DNA (track 6). (c) Alkaline agarose gel of: S_1 digestion products of hybrid to SV40 cut at 4 and 27 with Pst I (track 1); Exo VII digestion products of hybrids to SV40 DNA cut at: 0 with Eco RI (track 2), 0 and 14 with Eco RI and Bam I (track 3), and 67 with Bgl I (track 4).

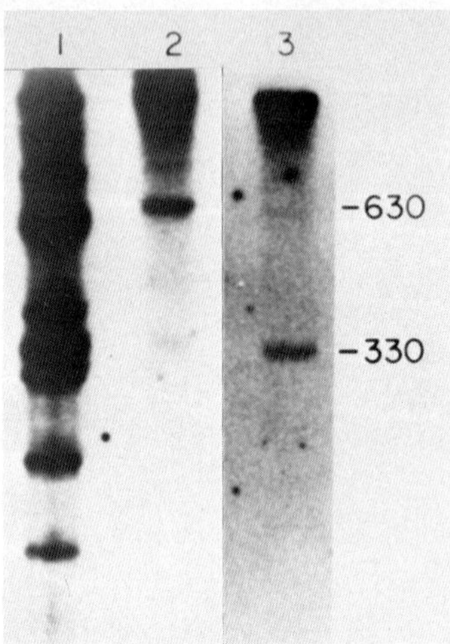

FIGURE 3. Autoradiogram of 8M urea gradient acrylamide gel. The 2500 (track 2) and 2200 (track 3) nucleotide RNA-DNA hybrids prepared as in Figure 2a were cut from an agarose gel, denatured with alkalai and electrophoresed. Marker *Hae* III digest of SV40 DNA (track 1).

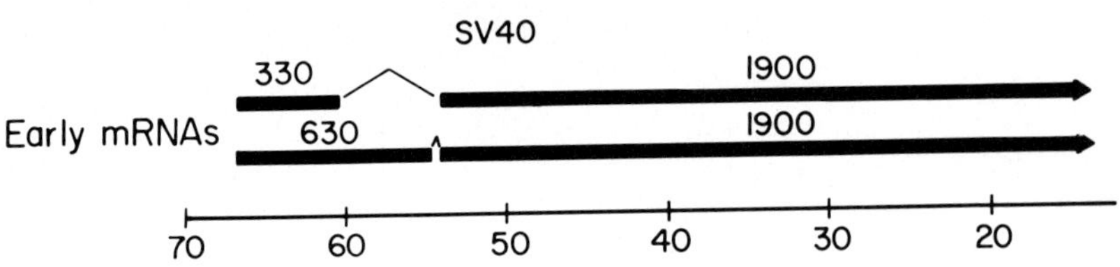

FIGURE 4. Structure of the early SV40 mRNAs. The SV40 genome is represented by a narrow line marked-off in percentage map units. Heavy lines above the genome map represent genome sequences included in the mRNAs. The caret symbol indicates that sequences are covalently joined in the mRNA molecule by a 3'-5' phosphodiester bond. Numbers above the heavy lines indicate lengths in nucleotides, and arrowheads indicate the 3' direction.

Analysis of early Ad2 mRNA: Similar experiments were performed on cytoplasmic RNA isolated from cells early in the course of infection with Ad2. Figure 5 shows the products of S_1 digests of Ad2 early RNA-DNA hybrids resolved by electrophoresis on a neutral agarose gel. This analysis exhibits bands corresponding to the full lengths of spliced RNAs. The full complement of early RNAs is observed following hybridization to full length Ad2 DNA, and mRNA transcribed from each of the four early regions (20,21) is analyzed separately by hybridizing RNA to isolated DNA restriction fragments which entirely encompass each of the early regions. Analysis of the splicing pattern of each of the detected early RNAs required S_1 and Exo VII experiments utilizing many different restriction fragments. These data will be presented elsewhere. However, the deduced structures of the abundant RNAs encoded in each of the four early regions are depicted in Figure 6.

Three abundant mRNAs transcribed in the rightward direction are observed in early region 1 as depicted in Figure 6. Two of these have the same 5' and 3' sequences, but differ from each other in the pattern of genome sequences deleted by splicing. A third mRNA with a splice towards its 3' end maps in this region immediately to the right of these two mRNAs.

The two abundant mRNAs observed in early region 2, are transcribed in the leftward direction. These mRNAs are composed of three colinear transcripts which map in a region encompasing 13% of the genome. They are identical to each other except in the region of the 3'-most colinear transcript where one mRNA contains approximately 100 more nucleotides of genome sequence than the other.

Early region 3 encodes two abundant mRNAs. These are transcribed to the right and are each composed of two colinear transcripts. They differ only at their 3' ends where the longer mRNA includes approximately 900 nucleotides more of the genome sequence.

Early region 4 is transcribed to the left and encodes one abundant mRNA. It includes a 5' colinear transcript <50 nucleotides long spliced to a 1900 nucleotide colinear transcript.

DISCUSSION

Translation of the early SV40 mRNAs: Several observations bear on the problem of how the early SV40 mRNAs are translated: 1. The sequence of the early strand of SV40 DNA contains two sets of translation termination codons in each of all three possible reading frames near position 54 on the

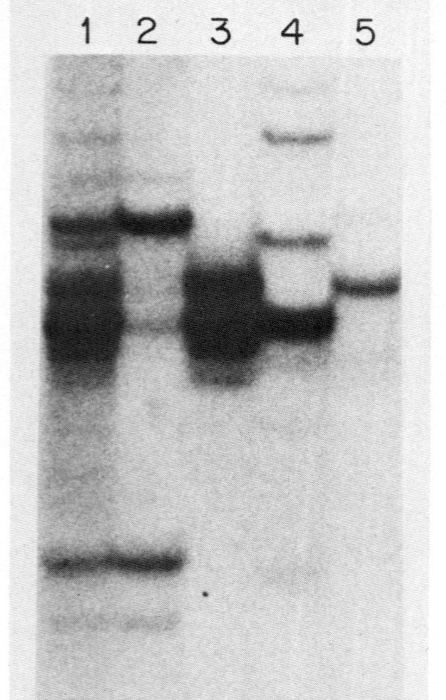

FIGURE 5. Autoradiogram of a neutral agarose gel of Ad2 early cytoplasmic RNA hybridized to total Ad2 DNA (track 1), or restriction fragments <u>Bam</u> I B (early region 1, track 2), <u>Sma</u> I A (early region 2, track 3), <u>Sma</u> I C (early region 3, track 4), <u>Eco</u> RI C (early region 4, track 5).

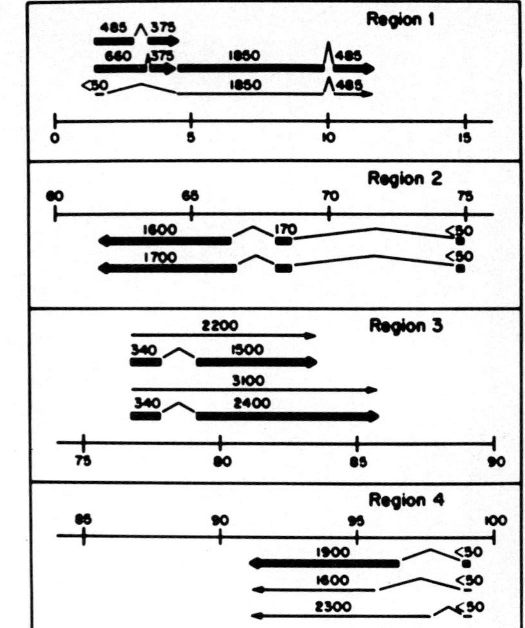

FIGURE 6. Structure of the early Ad2 mRNAs represented by narrow lines are less abundant than those represented by heavy lines.

genome (31). 2. Two SV40-specific proteins have been detected in productively infected cells during the early phase of infection and similarly in transformed cells. These are known as large and small T-antigens and have molecular weights estimated by SDS-polyacrylamide gel electrophoresis to be 94,000 and 17,000 daltons, respectively (7,8). The mobility of large T-antigen may be anomalously slow in this gel system, leading to an overestimate of its molecular weight which may in fact be closer to 80,000 daltons (32). 3. Mutants of SV40 with deletions on the genome in the interval between 54 and 59 (17) induce a wild-type sized large T-antigen, but either fail to produce any detectable small T-antigen, or induce an altered protein of decreased molecular weight (18, Sleigh, M. J., Topp, W. C., Hanich, R., and Sambrook, J., personal communication). 4. The large and small T-antigens have amino acid sequence in common. Seven of nine total methionine-containing tryptic peptides of small T-antigen co-chromatograph on ion-exchange columns with methionine-containing tryptic peptides of large T-antigen (33). 5. Large and small T-antigen can be translated _in vitro_ from viral mRNAs which sediments at ∿19S (34). The message for small T-antigen sediments slightly faster than the large T-antigen mRNA. 6. The amino-terminal sequences of large and small T-antigen synthesized _in vitro_ are identical and correspond to DNA sequences of the early strand beginning with an ATG at approximately 65.5 map units (A. E. Smith, E. Paucha, personal communication).

These data are consistent with the following model for translation of the early SV40 mRNAs: translation of both the 2200 and 2500 nucleotide mRNAs initiates at the AUG transcribed from DNA sequence at map position 65.5. Translation of the 2500 nucleotide mRNA is terminated at termination codons transcribed from the DNA sequence at position 54. The 2500 nucleotide mRNA, therefore, is proposed to be the small T-antigen message. Translation of the 2200 nucleotide mRNA continues through the splice point in this mRNA to termination codons encoded in the DNA sequence at 17 (35). This is possible because DNA sequence near position 54, which contains termination codons in all three reading frames, is not present in the 2200 nucleotide mRNA because of the pattern in which genome sequences are spliced to form this messenger (Figure 4). The 2200 nucleotide mRNA is, therefore, proposed to be the messenger for large T-antigen. This model for the translation of the early SV40 mRNAs accounts for the relative sizes and shared amino acid sequence of the large and small T-antigens, the phenotype of SV40 mutants with deletions mapping between 60 and 54, the relative sizes of the large and small T-antigen mRNAs, and the occurrence of termination codons in the DNA sequence at map position 54.

A similarity between the splicing pattern of mRNAs from the oncogenic region of Ad2 and the early SV40 mRNAs: One of the intriguing results of these studies was the finding that the splicing pattern of two of the mRNAs in the oncogenic region of Ad2 (early region 1) mapping from 1.5 to 4.5 is similar to the splicing pattern of the early SV40 mRNAs. The 860 and 1035 nucleotide Ad2 mRNAs which map in early region 1 between 1.5 and 4.5 map units (Figure 6) contain the same 5'-nucleotide sequence, and the 5' colinear transcripts of these mRNAs are spliced to the same colinear transcript which maps immediately adjacent to the 5' colinear transcript of the 1035 nucleotide mRNA. Furhtermore, DNA sequence data from another laboratory (J. Maat and H. van Ormondt, personal communication) indicates that translation termination codons exist in the long reading frames of the DNA sequence at map positions $\sim$3.5 and $\sim$4.5.

This similarity in the pattern of splicing of these mRNAs and the occurrence of termination codons to the 5' side of the splice point in the longer SV40 and Ad2 mRNAs suggests that the organization of translation of these Ad2 mRNAs may be analogous to that of the SV40 mRNAs. That is these two mRNAs may encode polypeptides which have identical N-terminal amino acid sequence corresponding to DNA sequence between map positions 1.5 and 3.0 map units. The polypeptide encoded by the 860 nucleotide mRNA may have C-terminal amino acid sequence corresponding to DNA sequence between 3.0 and 3.5 map units. The polypeptide encoded by the 1035 nucleotide mRNA may have C-terminal sequence encoded by DNA sequence mapping between 3.5 and 4.5 map units.

Whether this similarity in organization of RNA segments and DNA sequence in the oncogenic regions of SV40 and Ad2 is reflected in a similarity in the activities of proteins encoded by these mRNAs remains to be determined.

Possible control of early SV40 gene expression at the level of RNA splicing: Studies on the transcription and processing of late Ad2 mRNAs (36,37,38) suggest that these messengers are produced by "post-transcriptional splicing." According to this model an initial transcript which is to be processed into an mRNA is synthesized as a single colinear transcript which contains intervening sequences as well as RNA sequences which will form the colinear transcripts of the mature mRNA. This initial transcript is then processed into a mature mRNA by removal of intervening sequences and religation of colinear transcripts at splice points. If this is the mechanism of synthesis of the early SV40 and Ad2 mRNAs, then the post-transcriptional splicing events will determine the fraction of initial transcripts which are matured into specific mRNAs. It is possible that the postulated post-transcriptional splicing events are subject to control allowing regulation of expression of the early SV40 and Ad2 genes at the level of RNA processing.

REFERENCES

1. Black, P. H., Rowe, W. P., Turner, H. C., and Huenber, R. J. (1963). Proc. Nat. Acad. Sci. USA 50, 1148.
2. Pop, R. J. and Rowe, W. P. (1964). J. Exp. Med. 120, 121.
3. Gelb, L. D., Kohne, D. E., and Martin, M. A. (1971). J. Mol. Biol. 57, 129.
4. Botchan, M., Topp, W., and Sambrook, J. (1976). Cell 9, 269.
5. Khoury, G., Martin, M. A., Lee, T. N. H., Danna, K. J., and Nathans, D. (1973). J. Mol. Biol. 78, 377.
6. Sambrook, J., Sugden, B., Keller, W., and Sharp, P. A. (1973). Proc. Nat. Acad. Sci. USA 70, 3711.
7. Rundell, K., Collins, J. K., Tegtmeyer, P., Ozer, H. L., Lai, C. J., and Nathans, D. (1977). J. Virol. 21, 636.
8. Prives, C., Gilboa, E., Revel, M., and Winocour, E. (1977). Proc. Nat. Acad. Sci. USA 74, 457.
9. Kimura, G., and Dulbecco, R. (1972). Virology 52, 529.
10. Tegtmeyer, P. (1972). J. Virol. 10, 591.
11. Tegtmeyer, P. (1975). J. Virol. 15, 613.
12. Martin, R. G., and Chou, J. Y. (1975). J. Virol. 15, 599.
13. Osborn, M., and Weber, K. (1975). J. Virol. 15, 636.
14. Brugge, J. S., and Butel, J. S. (1975). J. Virol. 15, 619.
15. Lai, C.-J., and Nathans, D. (1974). Cold Spring Harbor Symp. Quant. Biol. 39, 53.
16. Alwine, J. C., Reed, S. I., Ferguson, J., and Stark, G. R. (1975). Cell 6, 529.
17. Shenk, T. E., Carbon, J., and Berg, P. (1976). J. Virol. 18, 664.
18. Crawford, L. V., Cole, C. N., Smith, A. E., Paucha, E., Tegtmeyer, P., Rundell, K., and Berg, P. (1978). Proc. Nat. Acad. Sci. USA 75, 117.
19. Sharp, P. A., Gallimore, P. H., and Flint, S. J. (1974). Cold Spring Harbor Symp. Quant. Biol. 39, 457.
20. Pettersson, U., Tibbetts, C., and Philipson, L. (1976). J. Mol. Biol. 101, 479.
21. Sambrook, J., Botchan, M., Gallimore, P., Ozanne, B., Pettersson, U., Williams, J., and Sharp, P. A. (1974). Cold Spring Harbor Symp. Quant. Biol. 39, 615.

22. Graham, F. L., Abrahams, P. J., Mulder, C., Heijneker, H. L., Warnaar, S. O., de Vries, F. A. J., Fiers, W., and van der Eb, A. J. (1974). Cold Spring Harbor Symp. Quant. Biol. 39, 637.
23. Harrison, T., Graham, F., and Williams, J. (1977). Virology 77, 319.
24. Graham, F. L., Harrison, T., and Williams, J. (1978). Virology, in press.
25. Berget, S. M., Moore, C., and Sharp, P. A. (1977). Proc. Nat. Acad. Sci. USA 74, 3171.
26. Vogt, V. M. (1973). Eur. J. Biochem. 33, 192.
27. Chase, J. W., and Richardson, C. C. (1974). J. Biol. Chem. 249, 4545.
28. Berk, A. J., and Sharp, P. A. (1977). Cell 12, 721.
29. Casey, J., and Davidson, N. (1977). Nucleic Acid Res. 4, 1539.
30. Berk, A. J., and Sharp, P. A. (1978). Proc. Nat. Acad. Sci. USA 75, in press.
31. Thimmapaya, B., and Weissman, S. M. (1977). Cell 11, 837.
32. Light, S., Griffin, J. D., and Livingston, D. M. (1978). Proc. Nat. Acad. Sci. USA, submitted.
33. Simons, D. T., and Martin, M. A. (1978). Proc. Nat. Acad. Sci. USA 75, in press.
34. Paucha, E., Harvey, R., Smith, R., and Smith, A. E. (1977). INSERM, Collog. 49, 189.
35. Dhar, R., Zain, B. S., Weissman, S. M., Pan, J., and Subramanian, K. N. (1974). Proc. Nat. Acad. Sci. USA 71, 371.
36. Bachenheimer, S., and Darnell, J. E. (1975). Proc. Nat. Acad. Sci. USA 72, 4445.
37. Berget, S. M., Berk, A. J., Harrison, T., and Sharp, P. A. (1977). Cold Spring Harbor Symp. Quant. Biol. 42, in press.
38. Goldberg, S., Weber, J., and Darnell, J. E. (1977). Cell 10, 617.

MONKEY DNA SEQUENCES IN DEFECTIVE SIMIAN VIRUS 40 VARIANTS

M. F. Singer, M. Rosenberg, H. Rosenberg[1], T. McCutchan, T. Wakamiya and S. Segal[2]

Laboratory of Biochemistry and Molecular Biology, National Cancer Institute, Bethesda, Maryland 20014

ABSTRACT Nucleotide sequences have been determined in two defective variants of SV40 that contain covalently linked monkey and viral DNA sequences. In each variant, part of the monkey sequences are derived from the highly reiterated portion of the cellular genome called α-component. However each variant contains a different member of the set of closely related, 172 base pair long monomer units contained in α-component. Each variant also contains sequences tentatively identified as low reiteration frequency monkey sequences: the sequences in the two variants are different. Sequences around the joints between the SV40 and monkey DNA have been determined. These studies afford information on the nature of the recombinational events involved in the evolution of the variants.

INTRODUCTION

Defective variants of the papova viruses polyoma and simian virus 40 (SV40) that are recombinants between the DNA of the virus and the DNA of permissive host cells arise upon multiple serial passage of wild type virus at high multiplicity (1,2,3). With SV40, the cloned and characterized recombinant variants are typically closed circular duplex molecules somewhat shorter than the wild type genome [about 5200 base pairs (bp)]. They contain several tandem repeats of DNA segments comprised of both monkey sequences and portions of the wild type SV40 sequences including the origin of replication. Thus, both recombinations and reiterations occurred during the evolution of these variants.

[1]Present address: Israel Institute for Biological Research, Ness-Ziona, Israel

[2]Present address: Laboratory of DNA Tumor Viruses, National Cancer Institute, National Institutes of Health, Bethesda, Maryland 20014

ISBN 0-12-668350-6

We report here studies of the fine structure of two such SV40 variants. These studies have yielded information on the nature of the recombinational events involved in the evolution of the variants as well as on the nature of the monkey sequences that interacted with the viral DNA.

STRUCTURE OF TWO DEFECTIVE SV40 VARIANTS

One variant, called CVP8/1/P2 (EcoRI res) (abbreviated here as CVP8/1/P2) was obtained after passage of wild type SV40 strain 777 on the BSC-1 line of African green monkey kidney cells and was previously characterized as to repeating unit, some restriction endonuclease cleavage sites, and the presence of both highly reiterated and infrequently reiterated monkey sequences (4,5,6,7). The primary sequence of the highly reiterated monkey DNA in CVP8/1/P2 has also been reported (7). CVP8/1/P2 was isolated from a stock of mixed defectives initially obtained in the laboratory of Ernest Winocour (8). An independently cloned variant isolated from the same stock and called F161F appears to be very similar to CVP8/1/P2 (9). Our current knowledge of the structure of CVP8/1/P2 is summarized in Figure 1. The basic repeating unit of approximately 1200 bp is reiterated four times: one out of every four repeats is altered by an internal duplication of about 400 bp including the region of the endoR·BamHI site and its neighboring endoR·HindII site (not shown on Figure 1) (5).

The second cloned variant to be discussed is called 1103 and was obtained in the laboratory of Daniel Nathans after passage of wild type SV40 strain 776 on BSC-1 cells (10). Figure 2 summarizes our present data concerning the structure of variant 1103. The repeating unit is 440 base pairs long and is usually reiterated about nine times in the defective genome (10).

The two variants have certain common features. Both are defective and require the presence of wild type virus as helper for productive infection. Depending on conditions, the yield of the variant genome from such mixed infections is usually over half of the total viral DNA and can be much higher. Each repeating unit of both variants contains a region derived from the origin of replication of the wild type SV40 genome, a region homologous to the highly reiterated monkey DNA called α-component, and a region tentatively identified as originating from infrequently reiterated monkey sequences. The orientation of the α-component sequences and the SV40 origin sequences, relative to one another, is the same in both variants (Figures 1 and 2).

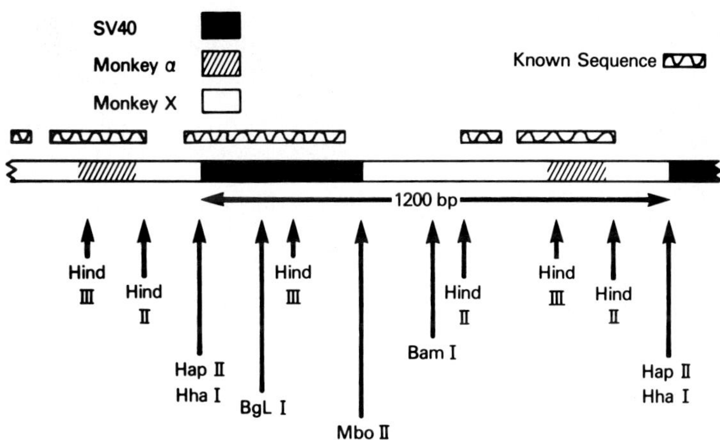

FIGURE 1. The structure of variant CVP8/1/P2. The variant genome is composed of four reiterations of the basic 1200 bp repeat unit shown here (with certain modifications as explained in the text). The filled areas indicate sequences derived from wild type SV40 sequences; cross hatched areas represent sequences derived from the highly reiterated portion of Cercopithecus aethiops (African green monkey) DNA known as α-component; open areas (Monkey X) indicate sequences derived from infrequently reiterated monkey DNA sequences. The characterization of the various segments is based on hybridization experiments (5,6,7) and nucleotide sequence analysis (7,11, and unpublished experiments of T. Wakamiya). The wavy bars along the top indicate those regions for which primary nucleotide sequence data have been obtained. The nucleotide sequence of the α-component segment in CVP8/1/P2 has been described (7). The orientation of the α-component segment, from left to right, corresponds to the sequence shown in Figure 3. The other sequences have been determined (T. Wakamiya, unpublished results) by the method of Maxam and Gilbert (12). The

(Figure 1, legend, continued)
wild type SV40 sequences were identified by comparison with published sequences (13): the SV40 segment corresponds to sequences between map positions 0.62 and 0.73 on the SV40 genome, reading from right to left on this Figure: within this region a segment corresponding to 185 bp of wild type SV40 is deleted in the defective. The characterization of the remaining segments as being derived from low reiteration frequency components of the monkey genome is tentative and is based on the following considerations. Restriction endonuclease fragments containing those segments do not hybridize to monkey DNA under conditions which permit detection of only highly reiterated sequences nor do they hybridize to SV40 DNA. Hybridization experiments under conditions designed to detect homology to low reiteration frequency monkey components demonstrated that the analogous segments in variant F161F, which appears very similar to and related to CVP8/1/P2 (see text), do indeed hybridize with monkey DNA (9).

The drawing is to scale: the cross-hatched area is equivalent to 157 bp.

Wild type SV40 sequences in variants CVP8/1/P2 and 1103. Both variants contain only short regions of wild type SV40 sequences including the origin of replication. Variant 1103 (Figure 2) has 155 bp of wild type sequences and these are from the region between wild type map positions 0.65 and 0.68. Variant CVP8/1/P2 (Figure 1) contains about 425 bp of wild type SV40 sequence from the region between 0.62 and 0.73 map units. A stretch of about 185 bp of the wild type SV40 sequences in this region are deleted in the variant. The deletion is between map positions 0.70 and 0.73.

HIGHLY REITERATED MONKEY DNA IN THE DEFECTIVE GENOMES

Structure of α-component DNA. The most abundant family of repetitive DNA sequences in the African green monkey (*Cercopithecus aethiops*) accounts for between 10 and 20 percent of the total DNA of the organism (11,14,15) and has been termed α-component (15). Treatment of α-component DNA or total African green monkey DNA with endoR·HindIII results in the conversion of the high molecular weight α-component to a series of multimeric fragments (7,11,14,16,17). The monomeric fragment is 172 bp in length and represents about 7 to 10 percent of the total monkey DNA (11). The bulk of the monomeric repeat units are present in long tandem arrays in α-component (11). Sequences homologous to α-component are

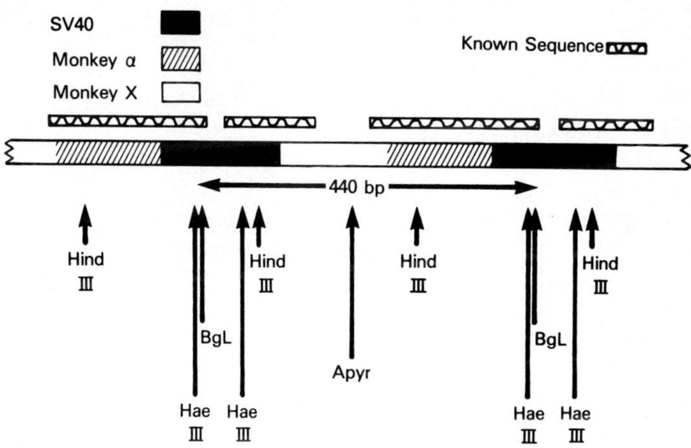

FIGURE 2. The structure of variant 1103. The variant genome is composed of about 9 reiterations of the basic 440 bp repeat unit shown here. The symbols are as for Figure 1. The scale is such that the cross-hatched area is 143 bp long. The characterization of the various segments is based on hybridization data (10,7) and nucleotide sequence analysis by the method of Maxam and Gilbert (12) as described for Figure 1 (unpublished experiments of T. McCutchan). The SV40 segment corresponds to sequences between map positions 0.65 and 0.68 on the SV40 genome reading from right to left on this figure. The orientation of the α-component segment, from left to right, corresponds to the sequence shown in Figure 3. As with variant CVP8/1/P2 (Figure 1) the characterization of the sequences indicated as derived from low reiteration frequency monkey sequences is tentative.

localized in the centromeres of African green monkey chromosomes and also occur in some chromosome arms (6, 18).

A unique nucleotide sequence representing the most abundant residue at each of the 172 positions in the monomeric unit has been determined (11) and is shown in Figure 3. However evidence was obtained indicating that the isolated monomeric unit (defined by the 172 bp distance between endoR·HindIII sites) is in fact a set of closely related sequences the members of which differ from one another at one or a few residues: the sequence divergence appears to be nonrandom (11). For example, some members of the set were shown to contain base pair variations at residues 31 and 64, and these are indicated in Figure 3. Thus the determined nucleotide sequence (Figure 3) actually represents the most abundant residue at each of the 172 positions for the set as a whole but not necessarily the structure of any particular member of the set. Neither the number of members of the set nor their relative abundances are known. The data (11) showed that at least 90 percent of the molecules in the isolated set contain the most abundant nucleotide at each position but that some positions probably contain the most abundant nucleotide in more than 99 percent of the molecules.

α-Component sequences in variants CVP8/1/P2 and 1103. As indicated in Figures 1 and 2, both CVP8/1/P2 and 1103 contain sequences derived from α-component DNA. The portions of the α-component repeat unit that appear in CVP8/1/P2 and 1103 are indicated by the bars along the top and bottom, respectively, of the sequence shown in Figure 3. At intervals along the bars sequence variations found in the two variants, as compared to the sequence determined for the α-component monomeric unit, are indicated.

CVP8/1/P2 contains 157 of the 172 residues of the monomer, beginning with residue 141 and going through to residue 124. In two positions, residues 155 and 169, transitions have occurred, compared to the sequence determined for the α-component monomer.

Variant 1103 contains 143 of the 172 residues of the monomer, beginning with residue 133 and going through residue 102. In six positions, residues 150,17,45,76,80 and 94, five transversions and one transition yield a sequence that is different from the sequence determined for the α-component monomer.

These data suggest that each of the two independently derived SV40 variants contains a different member of the set of sequences comprising the α-component repeating unit. In contrast to the uncloned fragments obtained directly from monkey DNA, the fragments derived from the defective variants

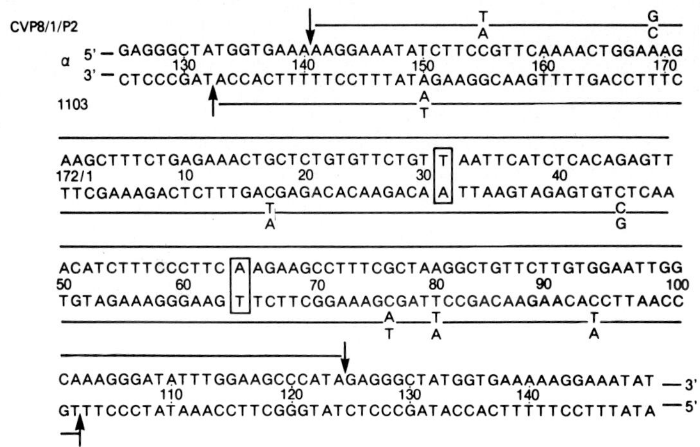

FIGURE 3. African green monkey α-component sequences in variants 1103 and CVP8/1/P2. The sequence shown is that of the most abundant residue at each of the 172 positions in the monomeric repeating unit of α-component (11). Both strands are shown. Position 1 corresponds to the single endoR·HindIII cleavage site in the repeating unit. As written the sequence starts with residue 125, goes through the entire 172 base pairs and then shows residues 125 through 150 again, in order to emphasize that the sequences occur most often in tandem repeats within the monkey genome. The boxed residues at positions 31 and 64 indicate known variations within the set of sequences that comprise the α-component repeat unit (11). The bar along the top, starting at residue 141 and going through residue 124, shows the segment that occurs in variant CVP8/1/P2 (7, 11 and unpublished experiments of T. Wakamiya). The observed base changes from the most abundant residue in α-component monomer are indicated at residues 155 and 169. The alteration at residue 155 was not previously noted (7, 11): the reason for the discrepancy is not known, but may be related to the new stock of CVP8/1/P2 used in these experiments. The bar along the bottom, starting at residue 133 and going through residue 102 shows the segment that occurs in variant 1103 (unpublished data of T. McCutchan). The observed base changes from the most abundant residue in α-component monomer are indicated at residues 150,17,45, 76,80 and 94.

The orientation of the sequence (left to right) is the same as that shown in Figures 1 and 2.

are cloned segments. Data obtained in the course of the sequence determination of the α-component segment in CVP8/1/P2 were consistent with it being a cloned fragment: no sequence divergence was detected under conditions that readily revealed the divergence in the monomer isolated directly from monkey DNA (11). Analogous data are not available for the α-component segment in variant 1103.

It is possible that the base pair alterations in the α-component sequences in 1103 and CVP8/1/P2 reflect mutational events after the defectives were formed and during their propagation, rather than the cloning of different members of the original set. We consider this unlikely for several reasons. First, the wild type SV40 sequences in variant 1103 show no divergence from the sequence found in wild type SV40 DNA of strain 776 (13) and a similar number of base pairs have been examined. Variant CVP8/1/P2 does show variation from the published wild type strain 776 sequence (13) at one position (T. Wakamiya, unpublisned experiments), but this could reflect differences between strains 777 and 776. Second, if the base substitutions occurred after formation of the recombinants it would have to have occurred before the extensive reiterations and replications involved in propagating the DNA, thus permitting amplification of the unique divergent sequences. In that case, however, additional divergences might be expected during propagation and the final pool of variant DNA molecules would not be expected to give a unique sequence. It therefore appears that different members of the set of sequences comprising α-component were incorporated into the two defectives and that there is no single member of the set that specifically recombines with SV40.

As indicated in Figure 3 different, but overlapping, regions of the α-component sequence appear in the two defectives. A completely random recombination between the SV40 genome and the α-component sequences (and a lack of selective pressure after recombination) might have resulted in very different lengths of the α-component monomer sequence being incorporated into the two defectives. In particular, multimeric lengths would be expected since the bulk of the monomeric units are arranged in long tandem arrays within α-component. Further, random recombination (and lack of selective pressure) might be expected to result in recombinants in which the α-component segment started and stopped at any one of the 172 possible residues. Therefore it is striking that the beginning and end of the α-component sequence in each of the two defectives are close to one another. There are a total of 8 bp, residues 125 through 132, that do not appear in either defective. Even allowing for secondary elimination of α-component sequences after the initial recombinational events, this suggests that some specificity operated during the initial joining. In this regard it may be of interest to point out that inverted repeat sequences occur at positions 117-121 and 127-131, 120-125 and 130-135, and 107-113 and 145-151 in the α-component monomer, relatively close to the beginning and end of the segment in the two variants.

DNA sequences flanking the α-component sequences in the two defectives. In variant CVP8/1/P2 the sequences flanking the α-component segment have been tentatively identified as infrequently reiterated monkey DNA sequences (Figure 1). In variant 1103 the α-component segment is joined directly to SV40 sequences on one side and to sequences tentatively identified as low reiteration frequency monkey sequence on the other (Figure 2). CVP8/1/P2 has a total of about 750 bp of such low reiteration frequency monkey sequences while 1103 has only about 140. The two variants do not share any common sequences of this class at least insofar as they would be detectable by cross hybridization of relevant fragments (Table I)(9).

Both variants then contain contiguous stretches of monkey DNA much longer than a single α-component repeat unit. But, as pointed out above, each variant contains less than a single complete repeat unit of the α-component monomer even though the bulk of α-component sequences occur in long tandem repeats within the monkey genome. There are several possible explanations for this finding. First, if the sequences in the variants reflect the initial recombination events between virus and host, then rare copies of the α-component sequence were involved in the recombination. In the monkey genome these rare copies might be found at the ends of tandem repeats where α-component sequences join other DNA or else might represent an α-component segment that is interspersed between other DNA sequences. Since each of the defectives contains a different member of the set of α-component sequences, and different flanking low reiteration frequency sequences as well, a different such rare segment would have been involved in the formation of each of the two defectives. A second possible explanation is that the structures of the purified defectives represent multiple recombinational events and therefore do not indicate anything about the relation of the α-component sequences to the flanking sequences within the monkey genome. In the case of CVP8/1/P2 there is some evidence to suggest that the sequences closely flanking the α-component segment have not been markedly altered in the course of multiple amplifying passages. Thus, in the very first high multiplicity passage of

TABLE I

HYBRIDIZATION OF FRAGMENTS FROM CVP8/1/P2 AND 1103 DNA TO THE DEFECTIVE DNAs ON FILTERS

DNA on Filter	% ^{32}P-DNA Fragment Hybridized	
	1103 fragment	CVP8/1/P2 fragment
Wild type SV40	3.5	1.3
1103	80	< 0.1
CVP8/1/P2 variant	1.7	34

Hybridization to unlabeled DNA immobilized on nitrocellulose filters was carried out as previously described (7). The filters contained 2.5 μg of plaque-purified wild type strain 777 SV40 DNA or 0.5 μg of either variant 1103 or CVP8/1/P2 DNA freed of wild type SV40 DNA by treatment with endoR·EcoRI and subsequent separation of circular DNA (variant) from linear DNA (wild type) by centrifugation (5). The radioactive DNA in solution was, in the case of 1103, the shorter (210 bp) of the two endoR·HindIII fragments obtained from this defective (see Figure 2 and reference 10) labeled with ^{32}P at both 5' termini. Previous experiments indicated that this fragment contained low reiteration frequency monkey sequences (10). In the case of CVP8/1/P2 the radioactive DNA was the 235 bp long fragment generated by combined cleavage with endoR·HindII/HindIII (see Figure 1 and reference 5) labeled with ^{32}P at the 5'-HindIII terminus. Previous experiments indicated that this fragment contained low reiteration frequency monkey sequences (5,8,9). In each case the input to the hybridization was about 4000 cpm.

plaque purified strain 777 in the passage series leading to CVP8/1/P2, a restriction fragment containing highly reiterated monkey sequences and corresponding in size to the analogous fragment obtained from CVP8/1/P2 (see Figure 1) was generated by endoR·HindII/HindIII (5,8,7). Although it has been suggested (1,3) that defective variants of SV40 containing host sequences originate late in a series of high multiplicity passages, the history of CVP8/1/P2 suggests that they may also arise very early in the course of the passaging. Experiments designed to elucidate the relation between the flanking sequences and α-component sequences within the monkey genome are currently being carried out.

JOINTS BETWEEN SV40 AND MONKEY DNA SEQUENCES

Nucleotide sequence analysis of variant 1103 demonstrates joints between wild type SV40 sequences and either α-component sequences or putative low reiteration frequency monkey sequences. The data are summarized in Figure 4. The nucleotide sequences of the neighboring portions of either α-component or SV40 that are contiguous with the sequences in the variant are shown for comparison. It is evident that there are no long stretches of homologous sequences between the wild type SV40 segment and the monkey segment in the region of the joints. In the joint between α-component and SV40, where both complete sequences on either side are known (upper portion of Figure 4), there are no short homologies surrounding the joint itself. As shown in Figure 4, the two joined segments have only a single AT base pair in common, right at the joint (making precise definition of the joint impossible). Thus, regardless of whether the joint represents the primary recombinational event, or is the result of deletions subsequent to the primary event, homologous interactions do not appear to have played a role in the formation of the joint. The sequence 5'-GCCCAT occurs 15 bp from the joint in the SV40 sequences and also, in the same orientation, 17 bp from the joint within the SV40 sequences that were lost in the recombination. This hexanucleotide is part of a series of multiple repeating units within this portion of the wild type genome (13). The sequence, 5'-GCCCAT, also occurs within the α-component sequence that was lost during generation of the variant, starting 16 bp from the joint (Figure 4). Within the α-component monomer, portions of this hexanucleotide are contained within two different short palindromic sequences (see Figure 3, residues 117-121 and 127-131, residues 120-125 and 130-135).

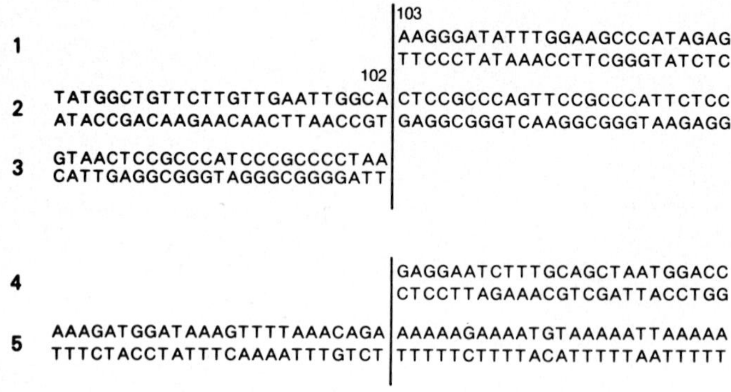

FIGURE 4. The joints between wild type SV40 sequences and monkey sequences in variant 1103. The sequence determinations were made with appropriate restriction endonuclease fragments by the procedure of Maxam and Gilbert (12) (unpublished experiments by T. McCutchan). The upper portion shows (line 2) the nucleotide sequences around the joint between wild type SV40 (at map position approximately 0.68) (13) and α-component sequences (at residue 102 as shown in Figure 3). The vertical bar represents one possible position of the joint: it could also be one base pair to the left. α-component sequences are to the left of the bar: SV40 sequences are to the right. The sequence on line 1 is the continuation of the α-component segment that is "missing" in the variant (see Figure 3). The sequence on line 3 is the continuation of the SV40 segment in the wild type genome (13) that is missing in the variant (13).

The lower portion shows (line 5) the nucleotide sequences around the joint between wild type SV40 (at map position approximately 0.65) (13) and putative low reiteration frequency monkey sequences. The vertical bar represents the end of identifiable SV40 sequences. SV40 sequences are to the left of the bar: monkey sequences to the right. The sequence on line 4 is the continuation of the SV40 segment in the wild type genome (13) that is missing in the variant.

In variant CVP8/1/P2 the SV40 sequences are joined to infrequently reiterated monkey DNA: there is no joint between SV40 and α-component sequences (Figure 1) as there is in variant 1103.

The joint between SV40 sequences and the putative low reiteration frequency monkey sequences in variant 1103 are shown in the bottom section of Figure 4. In this case, the "missing" monkey sequences that are adjacent to these monkey sequences in the cellular genome are not known. The sequences surrounding the joint are notably rich in AT base pairs. The monkey DNA sequences are part of a stretch of 140 bp between this joint and the joint with α-component (see Figure 2). Approximately 60 of these base pairs are known (T. McCutchan, unpublished experiments) and 83 percent of them are AT pairs. The region of wild type SV40 shown in Figure 4 is itself 72 percent AT pairs.

Earlier experiments (11) indicated that at one of the joints between α-component sequences and infrequently reiterated monkey sequences in CVP8/1/P2 (the joint to the right of the α-component segment in Figure 1) the infrequently reiterated sequences are also AT rich. Furthermore, in both CVP8/1/P2 and 1103 one joint between α-component sequences and infrequently reiterated sequences (the joint to the left of the α-component segment in Figures 1, 2, and 3) occurs within or near an AT rich region of α-component. It is possible that these AT rich regions play a role in the recombinational events leading to the variants.

DISCUSSION

The isolation and characterization of defective variants of SV40 containing monkey DNA segments covalently linked to SV40 DNA sequences indicated that recombination between the monkey genome and the viral genome can occur during lytic infection. The ease with which such variants can be detected in lysates from high multiplicity serial passages (reviewed in 1, 2 and 3; see also 8, 9, 10, 19, 20) suggests either that recombination is not uncommon or that the recombinants replicate efficiently, compared to the wild type genome, or both. The only SV40 sequences that appear to be required to assure replication of the recombinants is the origin of replication (1, 2, 3). Recent evidence (T. Lee and D. Nathans, unpublished experiments) supports earlier hypotheses that the variants are indeed replicated preferentially and that preferential replication can be correlated with the multiple SV40 replication origins within the variants.

Several variants have been and are being studied in some detail. Nevertheless, the mechanism by which recombination occurs remains obscure.

It has been recognized for some time that recombinant variants might reflect the integration of SV40 DNA into the genome of the host during lytic infection and subsequent faulty excision. However, it is not known whether SV40 does indeed integrate into the monkey genome during lytic infection. Therefore it is also possible that recombination occurs between independent SV40 genomes and fragments of monkey DNA present within the infected cell. Those recombinants that have been characterized are also reiteration variants, containing multiple repeats of segments consisting of SV40 sequences and covalently linked monkey sequences. The mechanism of reiteration is not known, but it has been shown that infection of monkey cells with short segments containing either SV40 DNA alone or SV40 and monkey sequences leads to reiteration and replication of the segments: there is a strong selection for the reiterated forms because of a minimum size required for encapsidation (21, 22). It is possible that the mechanism of reiteration is related to the replication process. In the case of the naturally occurring reiterated recombinant variants such as CVP8/1/P2 and 1103 it is possible that the product of the initial recombination was a longer segment than the characterized repeat unit, containing either additional SV40 or monkey sequences. Subsequent modifications may have yielded and then reiterated the short segment. Alternatively, the characterized short repeat units may have been formed directly and then reiterated.

Hybridization experiments previously indicated that while different monkey DNA segments were present in a group of independently derived defective SV40 variants, certain monkey sequences could be detected in common in several such variants. Among these, α-component sequences were frequently detected (7, 19, and M. F. Singer, unpublished experiments) as well as some infrequently reiterated monkey sequences (19). The demonstration here that 1103 and CVP8/1/P2 contain different members of the set of closely related sequences found (11) in α-component indicates that even with regard to the highly reiterated monkey segments, recombination is not completely specific. The insertion of SV40 DNA segments into the genome of transformed cells is not highly specific either and can occur at different places in the cellular genome (23, 24). Furthermore, the putative low reiteration frequency monkey sequences in 1103 and CVP8/1/P2 are also different. Preliminary sequence data on those segments in

CVP8/1/P2 (7, and T. Wakamiya and M. Rosenberg, unpublished experiments) confirm the hybridization experiments presented here, and have indicated no marked similarities between 1103 and CVP8/1/P2 in this regard.

Study of other SV40 variants will indicate whether such variants generally contain less than a single copy of the α-component monomer and, if so, whether the missing region always involves the same sequences as observed in CVP8/1/P2 and 1103. Study of the relative positions, in the monkey genome, of the infrequently reiterated monkey sequences and the α-component sequences found in variants will permit additional insights into the recombinational events between SV40 and monkey DNA.

For the present we conclude that recombination mechanisms other than those involving homologous crossing-over also occur in eukaryotic cells and were involved in the formation of variants CVP8/1/P2 and 1103. In the SV40-monkey system our data give no indication that mechanisms analogous either to bacterial IS sequence insertion (25) or to the site-specific mechanism involved in integration of bacteriophage lambda (26, 27, 28) are operative. Recombination may thus depend on unknown or unsuspected mechanisms.

ACKNOWLEDGMENTS

We are grateful to Dr. Daniel Nathans for a stock of variant 1103 and to Dr. Ernest Winocour for providing results from his laboratory prior to publication.

REFERENCES

1. Fareed, G. C., and Davoli, D. (1977). Annu. Rev. Biochem. 46, 471-522.
2. Fried, M., and Griffin, B. E. (1977). Adv. Cancer Res. 24, 67-113.
3. Kelly, T. J., Jr., and Nathans, D. (1977). Adv. Virus Res. 21, 85-173.
4. Rao, G. R. K., and Singer, M. F. (1977). J. Biol. Chem. 252, 5115-5123.
5. Rao, G. R. K., and Singer, M. F. (1977). J. Biol. Chem. 252, 5124-5134.
6. Segal, S., Garner, M., Singer, M. F., and Rosenberg, M. (1976). Cell 9, 247-257.
7. Rosenberg, M., Segal, S., Kuff, E. L., and Singer, M. F. (1977). Cell 11, 845-857.

8. Rozenblatt, S., Lavi,S., Singer, M. F., and Winocour, E. (1973). J. Virol. 12, 501-510.
9. Oren, M., Lavi, S., and Winocour, E. (1978). Virology, in press.
10. Lee, T. N. H., Brockman, W. W., and Nathans, D. (1975). Virology 66, 53-69.
11. Rosenberg, H., Singer, M. F., and Rosenberg, M. (1978). Science, in press.
12. Maxam, A., and Gilbert, W. (1977). Proc. Natl. Acad. Sci. USA 74, 560-564.
13. Dhar, R., Subramanian, K. N., Pan, J., and Weissman, S. M. (1977). Proc. Natl. Acad. Sci. USA 74, 827-831.
14. Fittler, F. (1977). Eur. J. Biochem. 74, 343-352.
15. Maio, J. J. (1971). J. Mol. Biol. 56, 579-595.
16. Grüss, P., and Sauer, G. (1975). FEBS Lett. 60, 85-88.
17. Brown, F. L., Musich, P.R., and Maio, J. J. (1978). J. Mol. Biol., in press.
18. Kurnit, D. M., and Maio, J. J. (1973). Chromosoma 42, 23-36.
19. Oren, M., Kuff, E. L., and Winocour, E. (1976). Virology 73, 419-430.
20. Winocour, E., Oren, M., Lavi, S., Vogel, T., and Gluzman, Y. (1977). In "Genetic Manipulation as It Affects the Cancer Problem", Miami Winter Symposia, Vol. 14 (J. Schultz and Z. Brada, eds.), pp. 181-194. Academic Press, New York.
21. Ganem, D., Nussbaum, A. L., Davoli, D., and Fareed, G. C. (1976). J. Mol. Biol. 101, 57-83.
22. Shenk, T. E., and Berg, P. (1976). Proc. Natl. Acad. Sci. USA 73, 1513-1517.
23. Kettner, G., and Kelly, T. J. (1976). Proc. Natl. Acad. Sci. USA 73, 1102-1106.
24. Botchan, M., Topp, W., and Sambrook, J. (1976). Cell 9, 269-287.
25. Bukhari, A. I., Shapiro, J., and Adhya, S. (1977). "DNA Insertion Elements, Plasmids and Episomes." Cold Spring Harbor Laboratory, New York.
26. Weisberg, R. A., Gottesman, S., and Gottesman, M.E. (1977). In "Comprehensive Virology " (H. Fraenkel-Conrat and R. R. Wagner, eds.) Vol. 8, pp. 197-258.
27. Landy, A., and Ross, W., (1977). Science 197, 1147.
28. Davies, R. W., Schreier, P. H., and DeBuchel. (1977). Nature 270, 757-759.

ADENOVIRUS 2 CYTOPLASMIC RNAs FROM THE LEFT 11% OF THE GENOME[1]

D. Spector, M. McGrogan[2], and H.J. Raskas

Department of Pathology and Department of Microbiology and Immunology, Washington University, School of Medicine, St. Louis, Missouri 63110

ABSTRACT Five cytoplasmic RNA species are identified as transcripts of the left end of the adenovirus 2 genome. A 22S RNA contains sequences from coordinates 4.4-11.0. One 13S RNA maps within 1.2-4.4 while a second 13S ($13S_f$) is a fusion of sequences from within 5.0-6.5 and 9.7-11.0. Two 9S RNAs map from 2.9-4.4 and 9.7-11.0 ($9S_{IX}$) respectively. Each of these RNAs has a characteristic rate of accumulation during the course of productive infection. The 22S and 13S RNAs are synthesized in relatively constant amounts. $13S_f$ is made in small quantities early and stimulated more than 20 fold at late times. The 9S RNAs appear after viral DNA synthesis begins. Each approaches the rate of labeling of the 13S RNA with which it shares sequences. Altered control of the transition from early to late rates of synthesis may be related to abortive infection and to transformation.

INTRODUCTION

The gene products coded for by the left 11% of the adenovirus 2 (Ad2) genome are of particular interest, for it is these sequences which are present as a minimum in cells transformed by the virus (1,2). Early in productive infection,

[1]Supported by grant CA16007 from the National Cancer Institute (NCI) and American Cancer Society grant VC-94D. D.S. was supported by PHS training grant CA09129 and PHS postdoctoral fellowship CA05815 from the NCI. Cell culture media were prepared in a cancer center facility funded by the NCI. This study was also supported by Brown & Williamson Tobacco Corp.; Larus and Brother Co., Inc.; Liggett & Myers, Inc.; Lorillard, a Division of Loews Theatres, Inc.; Philip Morris, Inc.; R.J. Reynolds Tobacco Co.; and Tobacco Associates, Inc.

[2]Present address: Department of Biological Sciences, Stanford University, Stanford, CA 94305

ISBN 0-12-668350-6

before the onset of viral DNA synthesis, poly(A)-containing cytoplasmic viral RNA is transcribed from four distinct genome regions, including map positions 0 to 11 which specify rightward transcripts (3-5). A definitive transcription map of the region has not been described. Two major size classes of early RNA, 19-24S and 13-14S, were detected by DNA fragment hybridization of labeled RNA fractionated by size (6,7). The 13S RNA has been thought to consist of two distinct species (6-10). R-loop mapping of DNA-RNA hybrids in the electron microscope (11,12) and the examination of the products of S1 nuclease treated hybrids (13) both suggest that some of these RNAs are spliced molecules.

There is late transcription of some early RNA sequences (9,10) including the mRNA for viral structural polypeptide IX encoded by the 9.7-11.0 region (14). It is not known whether other left end sequences are transcribed at late times.

We have mapped five distinct cytoplasmic RNAs specified by the left end. The data are consistent with both R-loop and S1 nuclease studies and previous filter hybridization

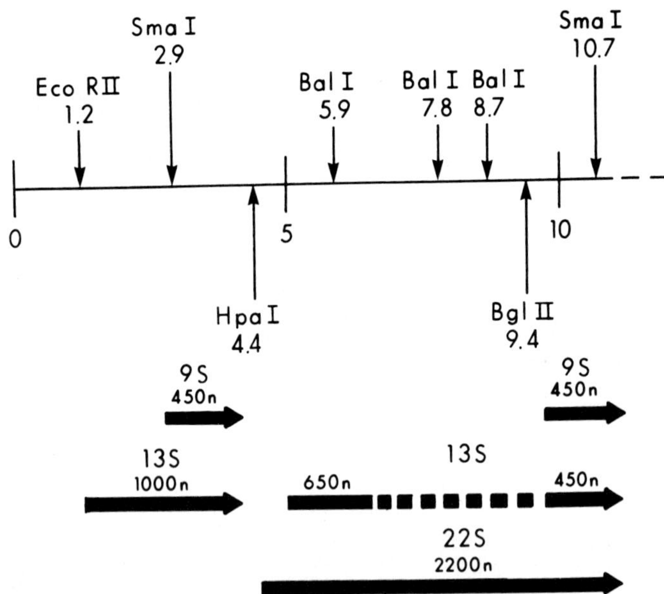

FIGURE 1: Transcription map of the left end of the adenovirus 2 genome. Approximate map positions and numbers of nucleotides (n) are taken from best agreement of our filter hybridization data with R-loop and S1 nuclease studies (11-13, 15). Restriction endonuclease cleavage sites have been previously reported (7,8,16) or were obtained from R. Gelinas and M. Zabeau (personal communication).

analyses. Each RNA has been identified by its sequence content and size. Moreover, each RNA species has a characteristic time course of appearance in the cytoplasm of infected cell, suggesting ways in which alteration of normal regulation may play a part in abortive infection and transformation.

RESULTS

The transcription map shown in Fig. 1 is based on studies of viral RNA synthesized at early and late times. As a first step, direct hybridization identified regions of the genome coding for a size class of RNA (Fig. 2). For example, early in infection all fragments in the region 4.4-10.7 hybridized to 22S RNA, and all fragments except 0-1.2 and 7.8-8.7 hybridized to 13S RNA (8.7-9.4 DNA was not tested). However, this procedure did not reveal precisely how many discreet species were present in each size class.

A selection and rehybridization procedure (Fig. 2) separated RNA species and demonstrated the existence of a spliced

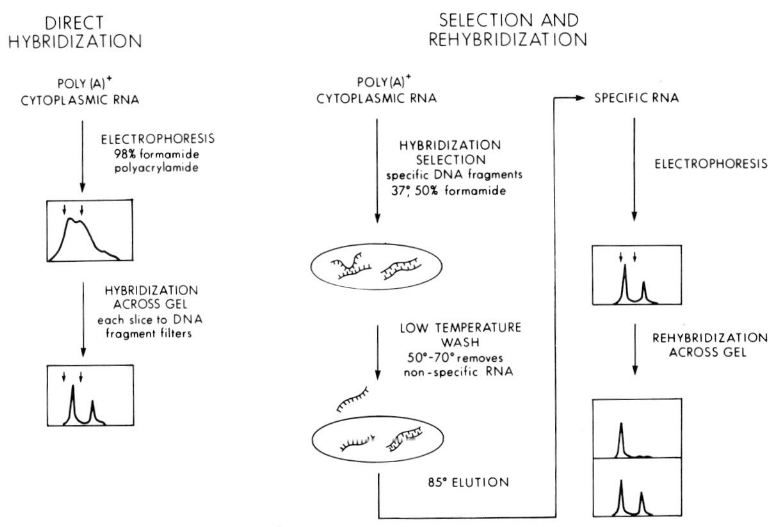

FIGURE 2: Methods for analysis of adenovirus 2 RNAs. Hybridization-selection was according to Buttner *et al.* (17) except that mismatched RNA was removed at a temperature between 50° and 70° depending on the GC content of the fragment (McGrogan, Spector, Goldenberg and Raskas, in preparation). All other procedures have been previously described (16).

RNA (Fig. 3). Selection confirmed that 13S RNA does not hybridize to the 7.8-8.7 fragment. Rehybridization of 8.7-10.7 selected RNA identified a "fusion" RNA product ($13S_f$), seen originally in R-loops (11,12). This selected RNA does not hybridize to 2.9-4.4 DNA (not shown) which encodes much of the second 13S species.

The 9S RNAs shown in Fig. 1 were identified by a similar analysis and can be separated slightly in our gel system. The faster migrating species ($9S_{IX}$) maps to the 8.7-10.7 region and is most likely the polypeptide IX mRNA (14).

<u>Time Course of Appearance of Left End Cytoplasmic RNAs.</u>
To analyze the accumulation of left end RNAs during productive infection, samples were labeled with ^{3}H-uridine for two hour intervals. Poly(A)-containing cytoplasmic RNAs were fractionated by size and then hybridized to appropriate DNA fragments under DNA excess conditions. Probes for 13S and 9S RNA were chosen to exclude by sequence the co-migrating species.

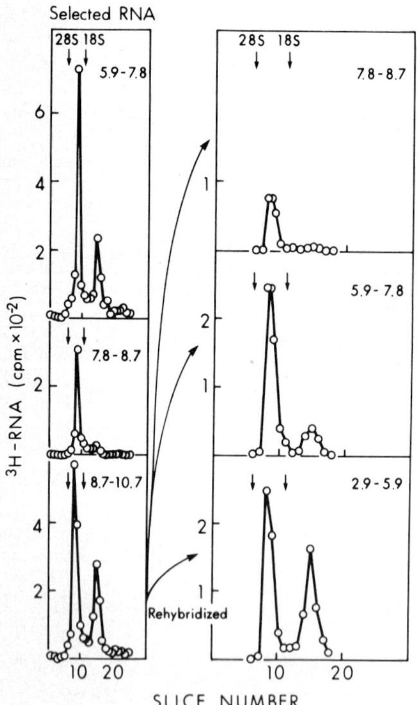

FIGURE 3: Selection and rehybridization of 5.9-10.7-specific RNA. Left panels: Polyadenylated cytoplasmic RNA was labeled from 3-6 h after infection with ^{3}H-uridine in the presence of 20 μg/ml of cytosine arabinoside. Samples were hybridized at 37° for 16 h to DNA filters containing 100 μg equivalents (the amount derived from 100 μg of whole adenovirus 2 DNA) of the indicated fragment. Filters were rinsed as previously described (17) and washed at 55° with 0.002 M EDTA containing 10 μg/ml tRNA and specific RNA was eluted at 85° in the same buffer. After electrophoresis RNA was eluted from gel slices and sampled as described previously (16). Right panels: Samples of the 8.7-10.7 selected RNA gel fractions were hybridized to filters containing ∿1 μg equivalent of the fragment indicated.

The gel profiles showed 22S and both 13S RNAs in early samples and all five RNAs in late samples. With the DNA fragments used we have detected no sequence differences in RNAs of the same size made both early and late (i.e., 22S and both 13S). For each time point the radioactivity in each peak was summed and normalized to sample size. To estimate relative molar amounts of each RNA produced, values were corrected for approximate length. The amount of the 13S RNA encoded by 0-4.4 was further corrected since only half of its sequences were detected with the 0-2.9 fragment used as a probe.

Analysis of the data (Fig. 4) revealed two different patterns of RNA accumulation. Assuming no significant changes in precursor pool specific activities, the synthesis of one 13S RNA remained essentially constant throughout and the 22S increased no more than two fold. The other RNAs either appear (the 9S RNAs) or increase sharply ($13S_f$) when viral DNA replication accelerates (8-12 h). Note the high level of production of $13S_f$ and $9S_{IX}$ RNA compared to the 22S species which contains common sequences. Likewise the second 9S RNA is stimulated in the absence of any substantial change in the overlapping 13S.

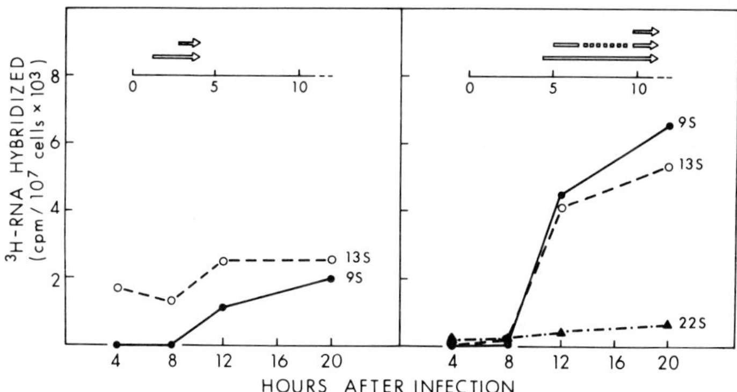

FIGURE 4: Relative rates of accumulation of left-end cytoplasmic RNAs during productive infection. Monolayers of KB cells were infected and labeled with ^{3}H-uridine at 4-6, 8-10, 12-14 and 20-22 h after infection. Poly(A)-containing RNA was prepared and fractionated by size as previously described (16). Eluted gel samples were hybridized to 1 μg equivalent DNA fragment filters containing 4.4-10.7 DNA (right panel), 0-2.9 DNA (13S, left panel), or 2.9-5.9 DNA (9S, left panel). The data were treated as described in the text. Results are expressed as relative molar amounts of RNA species labeled in two hours per 10^7 cells.

INTRODUCTION

The adenovirus and SV40 tumor antigens are detected and defined by antibodies present in tumor bearing animals (1,2). These antibodies have been used to immunoprecipitate radioactively labeled, viral induced proteins synthesized at early times (prior to initiation of viral DNA replication) after productive infection. In this manner, two SV40 tumor antigens of 94,000 and 17,000 apparent molecular weights have been detected (3,4). Similarly, adenovirus proteins of 72,000, 58,000 and 10,000 MW have been observed (5,6,7). Antibodies to the 58,000 MW protein, and possibly the 10,000 MW protein, are present in all tumor sera so far examined but antibodies to the 72,000 MW protein are found in only some adenovirus tumor sera. Because of this situation, the adenovirus tumor antigens were defined as those viral induced proteins (antigenic determinants) detected by antibodies common to all tumor sera (i.e. the 58,000 MW and possibly the 10,000 MW proteins) (6). The study of these tumor antigens are of some interest because they may be involved in viral transformation (8-13) and they appear to be similar or identical to the early viral proteins synthesized during productive infection or translated from early viral mRNA in vitro (3,4,14,15). To determine whether additional adenovirus induced early proteins could be detected and to see if additional adenovirus tumor sera contain antibodies to only a subset of these viral induced proteins, sera from hamsters bearing tumors derived from six different adenovirus transformed cell lines were examined for the presence of antibodies to adenovirus induced proteins synthesized in productively infected cells. By examining which portions of the adenovirus genome were present, and transcribed into cytoplasmic mRNA in these transformed cell lines, it might further be possible to locate the map position of the adenovirus structural gene that coded for these proteins. This analysis detected seven adenovirus induced proteins: antibodies to two of these proteins, 58,000 and 10,000 MW were present in all six antisera. The sites of the structural genes for five of these proteins could be located on the adenovirus physical map.

MATERIALS AND METHODS

Viruses, cell culture and experimental procedures. SV40 and adenoviruses were propagated as described previously in monkey (16) or human cells (17). Virus infected cells were labeled with ^{35}S-methionine (7). Tumor sera were prepared by injection of newborn hamsters with adenovirus transformed hamster cell lines (1-5x10^6 cells): 1) Ad2-HK-

A2325, (Ad2), 2) ND1-HK-A1704, (ND1), 3) ND4-HK-A1678, (ND4), (each from A. Lewis) and 4) Ad5-297-C43 (from A. van der Eb). Hamsters were bled several months after the appearance of a tumor. Normal hamster sera was obtained from animals in the same colony not carrying tumors. Antiserum from hamsters bearing Ad1-SV40 tumors (18) and HT14B tumors (19) were kind gifts of Dr. R. Gilden and Dr. J. Williams respectively. Immunoprecipitates were analysed by electrophoresis on SDS-polyacrylamide gels by procedures similar to those described elsewhere (7). The analysis of adenovirus and SV40 DNA segments and transcripts present in the transformed cells was carried out as described by Flint *et al.* (19).

RESULTS AND DISCUSSION

I. The Adenovirus Tumor Antigens. In order to identify adenovirus tumor antigens, KB cells were infected with type 5 adenovirus and the proteins synthesized in these cells were labeled with ^{35}S-methionine at 8-16 hours after infection. Soluble protein extracts were prepared from infected cells and incubated with serum from hamsters not bearing tumors (normal) or sera from hamsters bearing tumors derived from an injection of one of six different transformed cell lines (Ad2, ND1, ND4, Ad1-SV40, 14B, 293-C43). The antigen antibody complexes were mixed with formalin-treated *S. aureus* bacteria and centrifuged out of suspension. Such immunoprecipitates were then analysed for viral specific proteins by SDS-polyacrylamide gel electrophoresis and the labeled proteins detected by autoradiography (7). Figure 1 is an autoradiogram of ^{35}S-methionine labeled adenovirus induced proteins immunoprecipitated by the addition of sera from normal hamsters or animals bearing ND4, ND1, Ad2 or Ad1-SV40 induced tumors. The ND4(1) and ND4(2) serum samples were obtained from two different hamsters carrying tumors derived from the same ND4 transformed cell line. Sera from different hamsters bearing tumors induced by the same adenovirus cell line were analysed separately (not pooled) to examine the possibilities a) that individual animals might respond differently (make antibodies) to the same tumor or b) that a particular cloned transformed cell line might express different antigens when it initiates a tumor in different animals. The four adenovirus tumor sera, taken together, detected seven different adenovirus induced proteins in productively infected cells (Figure 1). These proteins had molecular weights of 72,000, 58,000, 49,000, 19,000, 15,500, 15,000 and 10,000.

Table 1 summarizes the results of a large number of experiments where six, independently-derived, adenovirus tumor sera were employed to identify these seven viral induced proteins. All six antisera contained antibodies

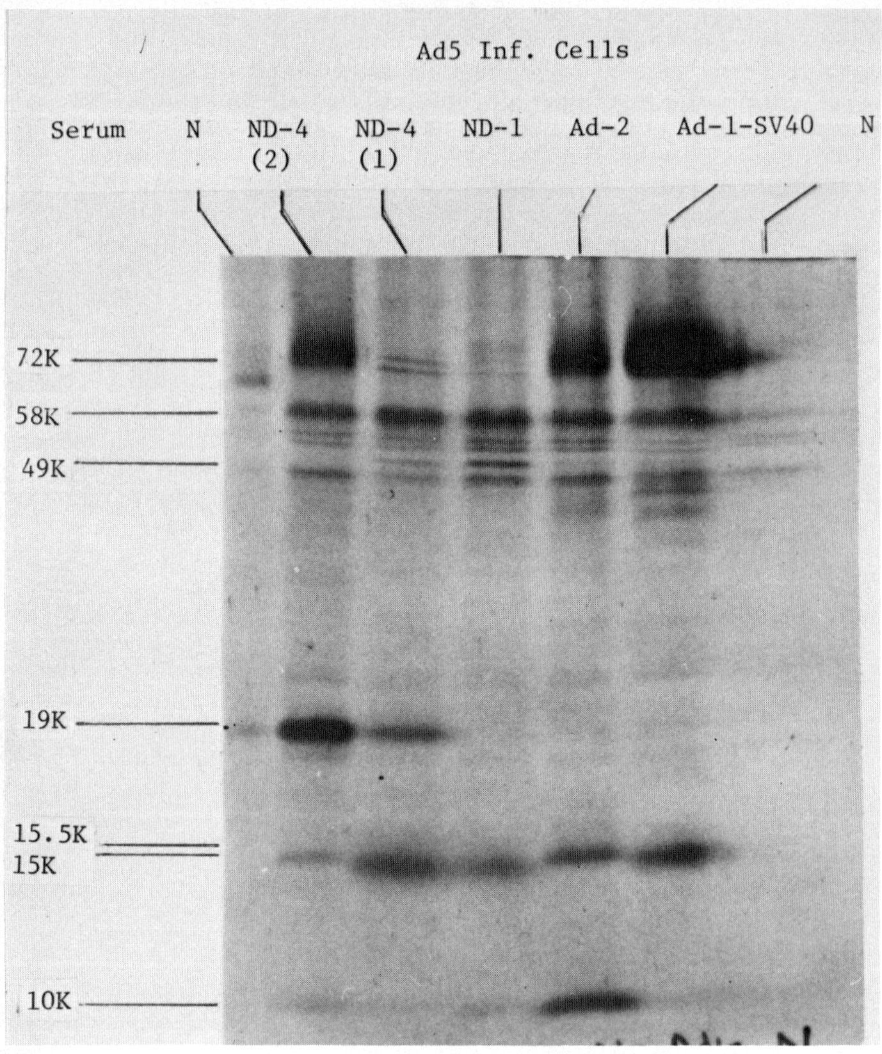

Figure 1. SDS-polyacrylamide gel autoradiogram of ^{35}S-methionine labeled proteins immunoprecipitated from adenovirus infected cells by five different tumor sera (ND4(1), ND4(2), ND1, Ad2, Ad1-SV40) and normal (N) sera.

TABLE 1
DISTRIBUTION OF ANTIBODIES IN TUMOR BEARING HAMSTERS TO ADENOVIRUS INDUCED PROTEINS FROM INFECTED CELLS

proteins	SERA Ad2	ND1	ND4(1)*	Ad1-SV40	14b	297-C43
72K	+	-	+	+	-	-
58K	+	+	+	+	+	+
49K	+	+	+	-	+	-
19K	-	-	+	-	-	-
15.5K	+	-	+	-	-	-
15K	-	+	+	+	-	-
10K	+	+	+	+	+	+

The pattern of proteins immunoprecipitated by ND4 sera is the sum of eleven separate tumor sera (see Table 2).

to the 58,000 and 10,000 MW proteins while the other five proteins were detected by some, but not all, antisera. This analysis demonstrates that the 72,000, 49,000, 19,000, 15,500 and 15,000 MW proteins are all immunologically distinct entities. For example, antibodies to the 72,000 MW protein do not cross react with the 15,000 MW protein, because the Ad2 serum contains antibodies to the 72,000 MW but not the 15,000 MW antigen, whereas ND1 serum has anti-15,000 MW activity and no detectable antibody against the 72,000 MW protein. Similar arguments can be made in the case of the 72,000, 49,000, 19,000, 15,500 and 15,000 MW proteins but it remains possible that the 58,000 and 10,000 MW proteins do cross react antigenically because antibodies to these two proteins are present in all six tumor sera analysed. Figure 1 also demonstrates that two sera obtained from two different hamsters bearing tumors derived from the same cloned transformed cell line, ND4(1) and ND4(2), contain antibodies to different adenovirus induced proteins. Thus ND4(1) serum contains antibodies to the 58,000, 49,000, 19,000, 15,000 and 10,000 MW proteins while ND4(2) serum (from a different hamster) contains detectable antibody to the 72,000, 58,000, 19,000, 15,500 and 10,000 MW proteins. Table 2 summarizes the results of a survey of eleven sera from 11 different hamsters each bearing an ND4 transformed cell line induced tumor. Four different sets of antibody combination were found, all of which exhibited antibody activity to the 58,000, 19,000 and 10,000 MW proteins. Most of the animals (8/11) produced antibody of the ND4(1) type. Thus there can be some heterogeneity in the tumor sera obtained from different animals in response to a tumor induced by the same cloned cell line.

To determine if these seven adenovirus induced proteins were synthesized at early times, adenovirus infected cells were labeled with ^{35}S-methionine at 4-9 hrs, 9-13 hrs and 13-17 hrs after viral infection (under the conditions of infection employed here viral DNA synthesis began at about 10 hours after infection). The soluble proteins were extracted and immunoprecipitated with either ND4 serum or Ad2 serum so that all seven proteins could be examined. Figure 2 presents an autoradiogram of ^{35}S-methionine-labeled proteins separated on an SDS-polyacrylamide slab gel. By 4-9 hours after infection the 72,000, 58,000, 19,000, 15,500 and 10,000 MW proteins were detected (a 44,000 MW protein which is the proteolytic breakdown product of the 72,000 MW protein (20) can also be seen). All of these proteins were also synthesized at late times after infection when the 49,000 MW protein (and 15,000 MW protein) were first detected (9-13 hrs postinfection). Normal hamster serum did not react with any of these seven proteins (Figure 2) and none of these proteins are detected in uninfected cells treated with immune sera

TABLE 2

ANTIBODIES DETECTED IN SERA FROM DIFFERENT HAMSTERS BEARING ND4 INDUCED TUMORS

# of Sera	8	1	1	1
proteins	ND4(1)	ND4(2)	ND4(3)	ND4(4)
72K	-	+	+	+
58K	+	+	+	+
49K	+	-	+	+
19K	+	+	+	+
15.5K	-	+	+	-
15K	+	-	+	+
10K	+	+	+	+

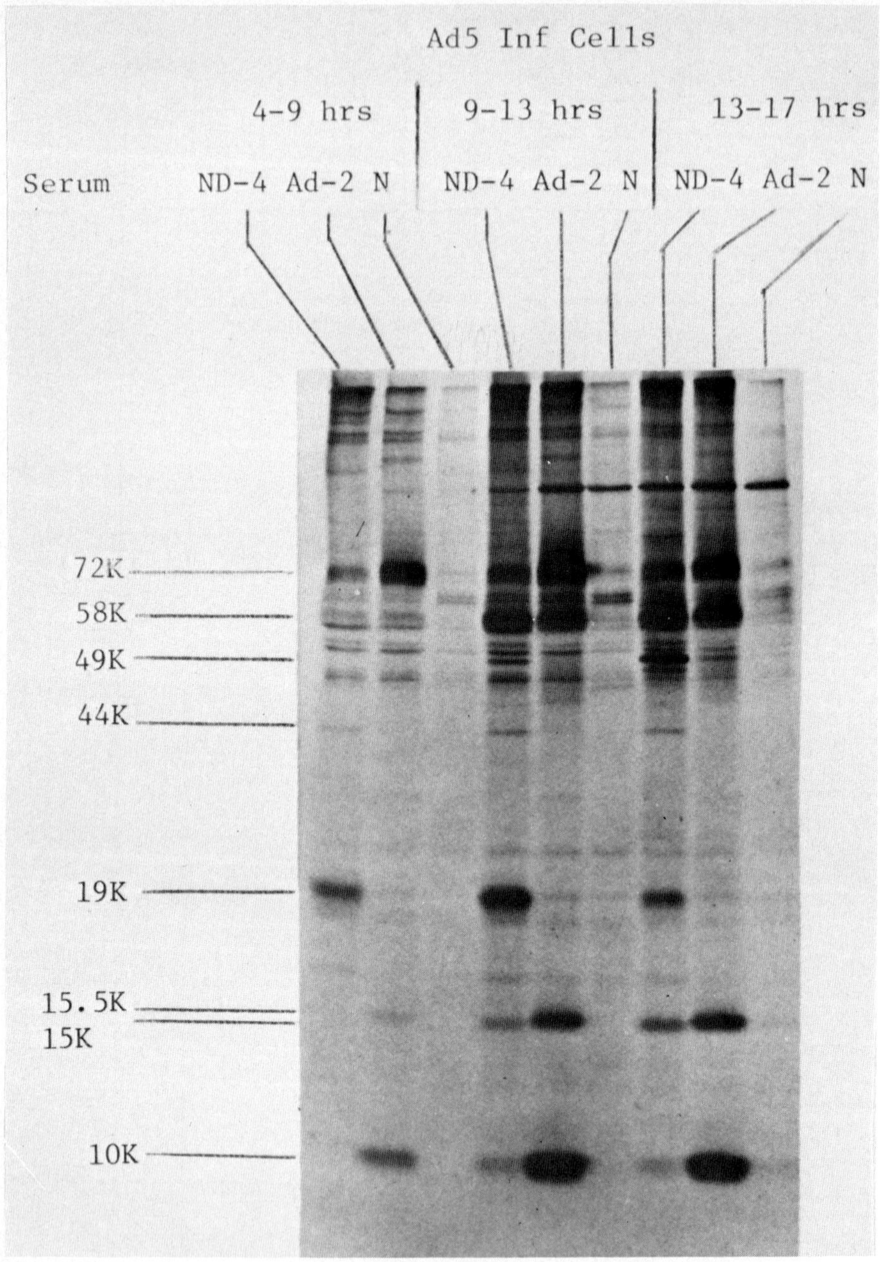

Figure 2. SDS-polyacrylamide gel autoradiogram of ^{35}S-methionine labeled proteins immunoprecipitated from adenovirus infected cells (labeled at 4-9 hrs, 9-13 hrs, 13-17 hrs after infection) by two different tumor sera (ND4 and Ad2) and normal sera (N).

(results not shown). In good agreement with these results, the 72,000, 58,000, 19,000, 15,500, 15,000 and 10,000 MW proteins were all synthesized in adenovirus type 5 infected KB cells treated with cytosine arabinoside to block viral DNA replication (results not shown). Thus at least six of these adenovirus induced proteins can be classified as early viral proteins.

II. The SV40 Tumor Antigens. Three of the antisera employed in this study were derived from hamsters bearing tumors made with transformed cell lines that could carry SV40 genetic information (ND1, ND4 and Ad1-SV40). To determine if these sera contained antibodies to the 94,000 MW SV40 tumor antigen, CV-1 cells were infected with SV40 or mock infected and the proteins were labeled with ^{35}S-methionine at 37-43 hours after infection. Soluble labeled protein was extracted from infected and mock infected cells and either normal serum or serum from ND1, ND4 or SV-T (a hamster carrying a SV40 virus induced tumor) were added to see if they would immunoprecipitate the SV40 T-antigen. The SV40 tumor antigen was analysed on SDS-polyacrylamide gels. Figure 3 shows the autoradiogram of this gel. It can be seen that ND4 and SV-T sera both contained antibodies to the 94,000 MW SV40 T-antigen (in infected cells only) whereas ND1 and normal serum contained no detectable antibody to the SV40 94,000 MW protein. The unexpected result with ND1 sera (no anti-SV40 94,000 MW activity) is explained by the fact that this ND1 transformed hamster cell line does not contain any detectable SV40 DNA (see next section). The SV40 large T-antigen in ND4 transformed cells was next examined by labeling such cells with ^{35}S-methionine and immunoprecipitating this antigen with SV40 anti-T serum (SV-T). Figure 4 presents an autoradiogram of an SDS-polyacrylamide gel of the SV40 large T-antigen labeled and immunoprecipitated from either SV80 (an SV40 transformed human cell line) or ND4 transformed cells. The ND4 transformed cell large SV40 T-antigen was found to be 92,000 MW, which is slightly smaller than the SV40 large T-antigen (94,000 MW from SV80 cells or productively infected monkey cells) (results not shown). Whereas SV40 tumor sera (SV-T) also immunoprecipitates a low molecular weight viral t-antigen (15-17,000 MW) the ND4 sera does not appear to detect this antigen. Whether this is a quantitative problem, or reflects the absence of antibody to little t antigenic determinants in ND4 sera is presently under investigation.

III. Adenovirus and SV40 specific DNA and RNA in viral transformed cell lines employed in this study. All group C adenovirus transformed cells so far examined, contain at least the left hand 12% of the viral genome and some transformants contain additional viral DNA segments (19,21): no

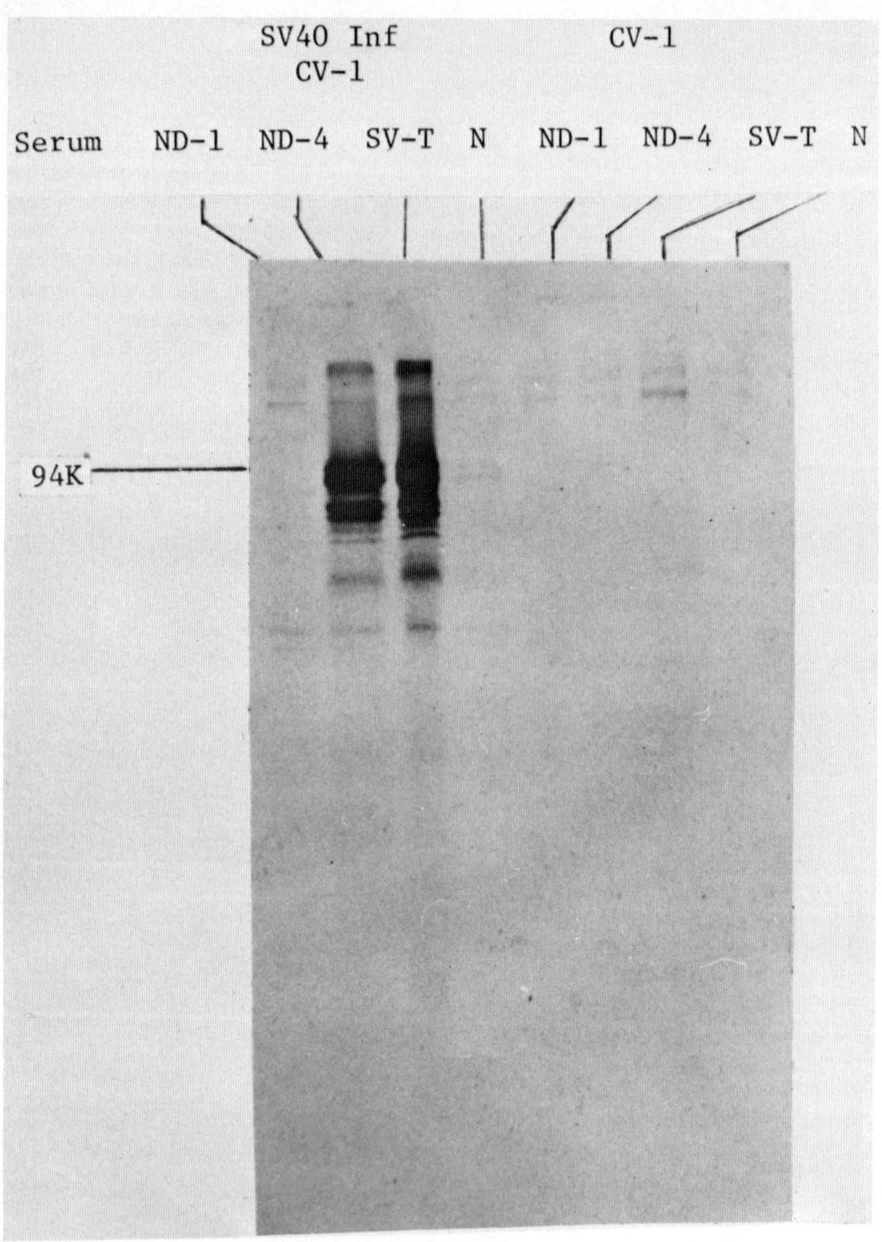

Figure 3. SDS-polyacrylamide gel autoradiogram of ^{35}S-methionine labeled proteins immunoprecipitated from SV40 infected or mock infected CV-1 cells by ND1, ND4, SV-T or normal (N) sera.

Figure 4. SDS-polyacrylamide gel autoradiogram of ^{35}S-methionine labeled T-antigen immunoprecipitated from SV40 transformed cell lines SV80 (94,000 MW) and ND4 (92,000 MW) by sera from tumor bearing animals (SV) or normal sera (N).

transformed cell line, however, contains the entire adenovirus genome. These results suggest that it might be possible to correlate the presence or absence of a segment of adenovirus or SV40 DNA and its expression as cytoplasmic viral RNA sequences in a transformed cell line with the presence or absence of a specific antibody in serum from hamsters bearing tumors produced by that transformed cell line. In this way it should be possible to assign the production of a particular protein (antigenic determinant) to a localized region of the viral genome, as is done in deletion mapping of a gene and gene product. The regions of the adenovirus and SV40 genomes and cytoplasmic RNA sequences present in Ad2, ND1 and ND4 transformed cells were therefore determined by solution hybridization to selected, labeled SmaI, BamHI, and EcoRI restriction enzyme fragments of viral DNA (19). Figure 5 presents a summary of these results, along with those for the cell line HT14B which has been analysed previously (19). All four cell lines contain the left hand 13% of the adenovirus genome and have cytoplasmic RNA complementary to 11-12% of this left hand end region. An examination of the matrix of adenovirus proteins and antibodies present in sera from hamsters bearing tumors derived from these transformed cell lines (Table 1) indicates that the 58,000 MW, 10,000 MW and 49,000 MW viral induced proteins are coded for by this region of the genome (Figure 5). Analogously, the position of the structural gene for the 72,000 MW protein can be located between 58.5-70.7% of the genome which is in good agreement with other results (15,22,23) which place this structural gene at 60-68% of the genome. The genetic information for the 15,500 MW protein can be placed in the region of 80% from the left hand end of the genome. Two ambiguities remain in assigning the adenovirus-induced proteins to locations on the genome. The 15,000 MW protein detected by antibodies in ND1 sera could well be derived from the low levels of mRNA corresponding to the 90-100% region of the genome (Figure 5, Table 1). Antibodies to this protein are not detected in Ad2 and HT14B sera and no RNA from this region of the genome is present in these transformed cell lines. Based upon the assignment of five adenovirus induced proteins (72,000, 58,000, 49,000, 15,500 and 10,000 MW) the pattern of mRNA production in ND4 transformed cells most closely resembles the expression of proteins detected by ND4(2) sera (Table 2). The ND4(2) sera does not contain antibodies to the 15,000 MW protein and no mRNA from the 90-100% region of the genome was observed (Figure 5). Thus the best assignment of this 15,000 MW protein, which remains tentative, would be to a structural gene in the 90-100 unit region of the genome. Antibodies to the 19,000 MW protein are found only in the ND4 sera. The only cytoplasmic RNA

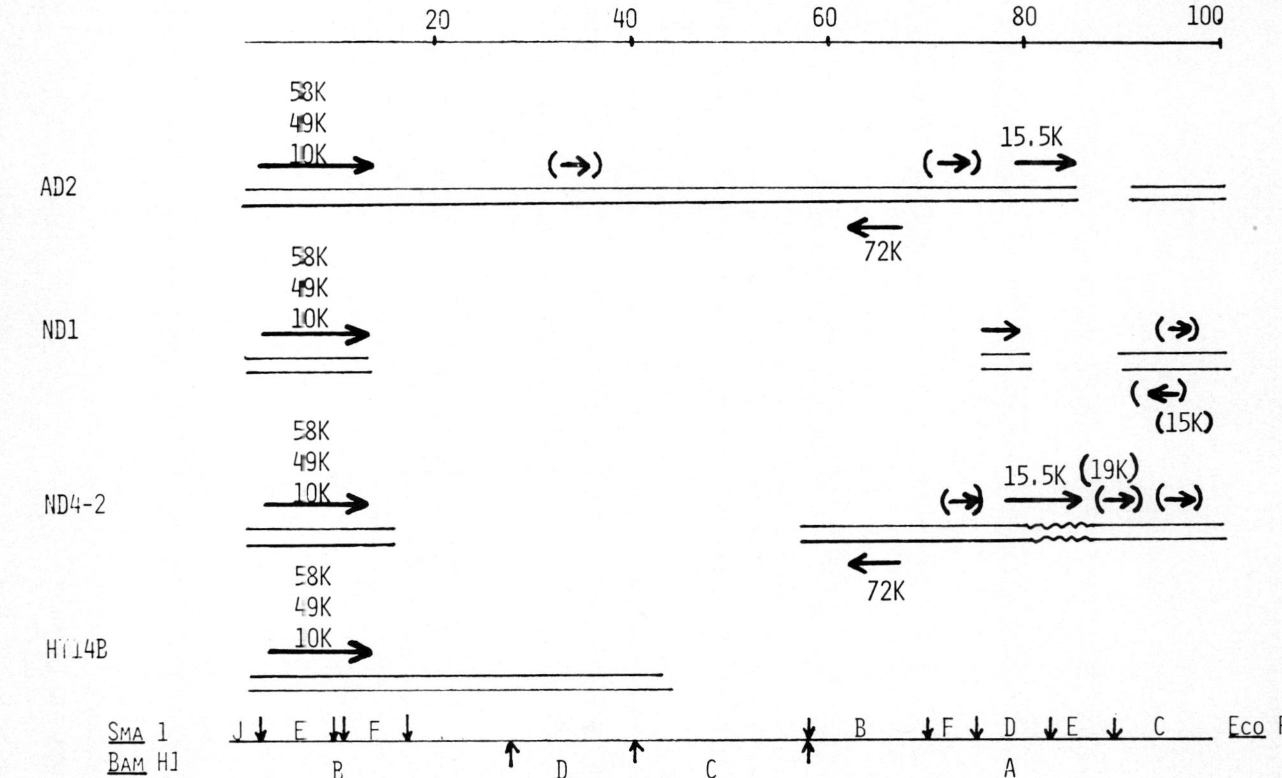

Figure 5. A Diagramatic representation of the Adenovirus specific DNA ===== ; cytoplasmic RNA, ---- ; and proteins, MW of protein; present in adenovirus transformed cell lines Ad2, ND1, ND4 and HT14B.

corresponding to an early gene region and uniquely found in the ND4 transformed cell line is from the 72-90% region of the genome. These data tentatively place the structural information for the 19,000 MW protein in this region of the adenovirus genome (Figure 5).

ND1 virus infected cells produce an antigen that cross reacts with the 94,000 MW large SV40 T-antigen (at the C-terminal end) (24). Sera from hamsters bearing the ND1 transformed hamster cell induced tumors fail to produce antibody to the SV40 large T-antigen (Figure 3). However, the transformed cell line employed in this study does not contain any SV40 DNA sequences (Figure 5), which must have been lost either during integration or subsequent events in the establishment or passage of this cell line. The ND4 transformed cell lines do contain SV40 sequences homologous to the region comprising 0.11-0.59 units of the SV40 genome (24): these cells produce a large SV40 T-antigen that is about 2,000 MW smaller than wild type SV40 T-antigen from transformed cells (SV80) or lytic infection. Peptide map analysis should reveal whether more extensive differences exist between SV40 and ND4 induced SV40 T-antigens. A portion of the SV40 large T-antigen (and small t antigen) are coded for by the region of the viral genome comprising 0.67-0.59 units and so it will be interesting to analyse the SV40 large and small (if present) tumor antigens in these ND4 transformed cells.

Whereas SV40 appears to code for two early viral proteins, both of which are expressed in viral transformed cells (3,4), the group C adenoviruses can express up to six distinct virus induced early proteins in different virus transformed cell lines (see Table 1, Figure 1). In addition, some adenovirus tumor sera contained antibodies that react with (or cross react with) a viral induced protein of 49,000 MW that is synthesized predominantly at late times after infection. The regions of the adenovirus genome that encode the information for the production of these seven proteins have been determined by an experimental procedure analogous to deletion mapping and the correlation of cytoplasmic transcripts with the presence of individual proteins (see Figure 5). The 58,000 MW and 10,000 MW early viral induced proteins that map in the left hand 0-12% of the genome are the adenovirus tumor antigens by virtue of the fact that they are present in all transformed cell lines examined to date. The antigenic determinants for the 49,000 MW protein map in this region as well. The 72,000 MW protein (mapping at 58.5-70% of the genome) is the adenovirus single strand specific DNA binding protein (7,23) required for viral DNA replication (23). Nothing is known about the functions of the 15,500 MW protein (located at 80% of the genome) or the two additional proteins of 19,000 and 15,000 MW detected in these studies.

The availability of an antisera bank of the type described here, should permit the purification of these adenovirus induced proteins and an analysis of their function.

ACKNOWLEDGMENTS

The authors thank A. Teresky, N. Tick and C. McIver for excellent technical assistance. This research was supported by grants from the American Cancer Society, VC-57F and NP-239, and the National Cancer Institute, CA11049-10.

REFERENCES

1. Black, P.H., Rowe, W.P., Turner, H.C. and Huebner, R.J. (1963). Proc. Nat. Acad. Sci. U.S. 50, 1148.
2. Huebner, R.J., Rowe, W.P., Turner, H.C., Lane, W.T. (1963). Proc. Nat. Acad. Sci. U.S. 50, 379.
3. Tegtmeyer, P., Schwartz, M., Collins, J.K. and Rundell, K. (1975). J. Virol. 16, 168.
4. Crawford, L.V., Cole, C.N., Smith, A.E., Tegtmeyer, P., Rundell, K. and Berg, P. (1978). Proc. Nat. Acad. Sci. U.S. 75, 117.
5. Gilead, Z., Jeng, Y.H., Wold, W.S., Sugawara, K., Rho, H.W., Harter, M.L. and Green, M. (1976). Nature 264, 263.
6. Levinson, A.D. and Levine, A.J. (1977). Virology 76, 1.
7. Levinson, A.D. and Levine, A.J. (1977). Cell 11, 871.
8. Kimura, G. and Itagki, A. (1975). Proc. Nat. Acad. Sci. U.S. 72, 673.
9. Brigge, J.S.and Butel, J.S. (1975). J. Virol. 15, 619.
10. Osborn, M. and Weber, K. (1975). J. Virol. 15, 636.
11. Tegtmeyer, P. (1975). J. Virol. 15, 613.
12. Martin, R. and Chou, J.Y. (1975). J. Virol. 15, 599.
13. Williams, J.F., Young, H. and Austin, P. (1974). Cold Spr. Hbr. Symp. Quant. Biol. 39, 427.
14. Saborio, J., Oberg, G. and Phillipson, L. (1975). INSEM Symp. (Haenne, A.L. and Beard, G., eds.) 47, pp. 325-330.
15. Lewis, J.B., Atkins, J.F., Baum, P.R., Salem, R. and Gesteland, R.F. (1976). Cell 7, 141.
16. Levine, A.J., Jang, H.S. and Billheimer, F.E. (1970). J. Mol. Biol. 50, 549.
17. van der Vliet, P.C. and Sussenbach, J. (1972). Eur. J. Biochem. 30, 584.
18. Black, P.H., Lewis, A.M., Blacklow, N.R., Austin, J.B. and Rowe, W.P. (1967). Proc. Nat. Acad. Sci. U.S. 57, 1324.
19. Flint, S.J., Sambrook, J., Williams, J.F. and Sharp, P. (1976). Virology 72, 456.
20. Rosenwirth, B., Anderson, C. and Levine, A.J. (1976). Virology 69, 617.
21. Gallimore, P.H., Sharp, P.A. and Sambrook, J. (1974). J. Mol. Biol. 89, 49.

frequent exposure to virus experienced by members of these groups. The carrier rates in most Western European countries are similar to those in the U.S. (reviewed in ref. 7) but in other areas of the world where sanitary conditions are poor such as in many countries in Africa, Asia and Oceania the rate may exceed 50 or 100 per 1000 (7,9). It has been estimated that there are more than 100 million people in the world today who are chronically infected with HBV (10). Although chronic antigen carriers may become HBsAg negative at any time, almost all remain infected for many years (11) and infection for life is probably the rule.

As in humans, the chronic carrier state develops in a significant number of HBV infected chimpanzees providing a useful animal model to study persistent infection by this virus.

2. Hepatitis B Viral Forms in the Blood and Liver During Persistent Infection. Several structures bearing virus specified antigens have been found in patients with persistent HBV infection. Sera containing infectious virus almost always contain detectable hepatitis B surface antigen (HBsAg). The most numerous structures carrying HBsAg determinants are roughly spherical and have an average diameter of 20 nm. (12). Concentrations of these particles appear to exceed 10^{13} per ml. or 200/ug per ml. in the sera of many HBsAg carriers (13). Highly purified 20 nm. forms contain no detectable nucleic acid and are thus probably not complete virus but incomplete virus coat particles.

It is well established that the surface of the HBsAg bearing particles in blood is antigenically complex (reviewed in ref. 14). Different sera may contain different antigenic specificities and more than one antigenic determinant is present on individual HBsAg bearing particles. A group specific determinant designated a is shared by all HBsAg positive sera. In addition, particles generally carry two type specific determinants, either d or y and either w or r.

Larger and more complex structures which share surface antigens with the 20 nm. particles have been observed as occasional components in some antigen-positive sera. These are long, filamentous structures of variable length and spherical 42 nm. diameter structures which have 28 nm. electron-dense cores (15). The former are thought to be variant forms of the 20 nm. particles. The latter which are uniform in appearance and have a virus-like morphology have been termed Dane particles after their discoverer.

Twenty-eight nm. core structures prepared by detergent treatment of Dane particles (which removes the outer HBsAg layer of the Dane particle) have been shown to contain an antigen (BHcAg) which is distinct from HBsAg (16). HBcAg containing particles with an electron microscopic appearance similar to that of Dane particle core structures have been isolated from liver tissue of patients infected with HBV (17).

Dane particles have been shown to contain within their inner core a circular DNA molecule with a molecular weight of 2×10^6 daltons and a DNA polymerase that uses that small circular DNA as a template (18,19,20). Two thirds of the circular DNA molecule appears to be double stranded and one third single stranded. The DNA polymerase uses the single stranded DNA as template to make the single stranded regions double stranded during the reaction. The biological significance of the unique DNA structure and DNA polymerase activity in Dane particles is not known. The presence of DNA in the Dane particle suggests that it may be the complete virus of hepatitis B although its infectivity has not been directly shown. Other findings consistent with the Dane particle being HBV include the protection against HBV infection resulting from active immunization with HBsAg (21) (the antigen on the Dane particle surface) and passive immunization with anti-HBs (22). The concentrations of Dane particles known to occur in sera and their size are also consistent with the limited information available on HBV infectivity titers and the size of the infectious virus. Dane particle concentrations are variable and have been estimated to be between 10^5 and 10^9 ml. in different sera (23). A single serum from an HBsAg carrier has been shown to infect human volunteers at a 10^{-7} dilution but not at 10^{-8} (24). Infectious virus has been shown to pass through a filter with a pore size of 52 nm. (25). If the Dane particle does represent the infectious form of HBV, its ultrastructure, DNA size and structure, antigenic composition and presence of a DNA polymerase indicate that it is not a member of any known group of viruses but is a unique virus. The size of the DNA of Dane particles is smaller than that of any known virus and it could specify only about 120,000 daltons of protein.

E antigen (HBeAg) which is physically and antigenically distinct from HBsAg and HBcAg, has been found exclusively in patients infected with hepatitis B virus and is present in the blood of some chronically infected patients (26-28). It is thought to consist of protein in a form which has a sedimentation coefficient of approximately 12 S. HBeAg is commonly found in the serum of carriers with high concentrations of Dane particles and anti-HBe in sera with low or undetectable levels of Dane particles (27). Whether there is a physical relationship between HBeAg and Dane particle (or HBV) is not clear. Recent studies have claimed that HBeAg is immunoglobulin IgG4 with idiotypic determinants to which anti-HBe is directed (28.a) or associated with the lactic dehydrogenase isoenzyme 5 (28 b). Further research is needed to confirm or refute these findings and resolve the discrepancies.

HBsAg detected by immunofluorescence is observed in the cytoplasm and surface of between 0.5 and 100% of hepatocytes in persistently infected patients (29). HBcAg is present in the nucleus of a smaller number of hepatocytes (29). Particles with the appearance of Dane particle cores are also observed in some hepatocyte nuclei. (30).

3. <u>Diseases Associated with Persistent Infection</u>. Chronic HBV infection may rarely be associated with a histologically normal liver and normal liver functions. More often, the syndrome of chronic persistent hepatitis occurs; it is not associated with progressive liver disease although mild histologic and liver function abnormalities are present. Chronic HBV infection may also be associated with chronic active hepatitis (CAH) which can progress to cirrhosis, hepatic failure and death. The exact role of HBV infection in the pathogenesis of CAH is unknown. In a recent series of patients who became chronically infected after hospitalization with acute hepatitis B, 30% developed associated chronic active hepatitis and 70% chronic persistent hepatitis (6).

In some parts of the world where HBsAg carrier rates are very high, more than 80% of patients with hepatocellular carcinoma have HBsAg in their blood and most have postnecrotic cirrhosis raising the possibility of a causal relationship between HBV infection and some cases of hepatocellular carcinoma (31).

Certain extrahepatic manifestations of persistent HBV infection appear to be related to circulating immune complexes consisting of HBsAg and anti-HBs (32). Polyarteritis nodosa and chronic HBV infection may coexist (32). For example, in one series of 21 patients with biopsy proven polyarteritis nodosa, 9 were HBsAg carriers. These patients have low serum complement levels and circulating HBsAg-antibody complexes; immune complexes and complement components were detected in diseased blood vessels by immunofluorescent staining.

Membranous glomerulonephritis has been associated with chronic active hepatitis and persistent HBV infection through observation of deposited immune complexes along the subepithelial surface of glomerular basement membranes (32). Nodular deposits of HBsAg, immune globulin, and C_3 in glommeruli have been demonstrated by immunofluorescent staining in these cases.

More than 85% of patients with infantile papular acrodermatitis in Mediterranean countries (33) and the Far East (34) have been found to be HBsAg positive suggesting HBV infection may be causal in these cases.

4. Spread of Infection by Carriers. Chronic HBsAg carriers transmit infection by blood transfusions (7) and by other routes (26,35-40). Although blood and blood products are the best documented vehicles of infection, HBsAg has also been found in feces, urine, bile, tears, saliva, semen, breast milk, vaginal secretions, cerebro-spinal fluid, and cord blood. Often the concentrations of antigen in these liquids are lower than in the serum of infected patients and sometimes antigen can be demonstrated only after concentration. If infectious virus is also present, its concentration may be lower than in serum. Infectious HBV has also been shown to be present in blood without detectable HBsAg so that the absence of the antigen does not exclude the presence of infectious virus.

All routes of transmission of HBV from carriers may not be recognized at this time. The most clearly documented route of transmission is by overt parenteral inoculation such as by blood transfusion or puncture with contaminated needles. It is possible that non-parenteral transmission occurs commonly. In underdeveloped areas of the world where HBV infection rates are much higher than in the United States and the opportunity for virus transmission by parenteral routes may be less, it is likely that non-parenteral transmission is the rule. In the United States it appears that crowded or close living conditions, such as occur among family members of the same household and among institutionalized children, may favor transmission, although the exact mode of transmission in such circumstances is not known.

Transmission between heterosexual as well as male homosexual partners may occur commonly but whether the route of transmission is venereal or non-venereal is not proven.

Neonatal transmission has been clearly documented although whether the mechanism involves transplacental infection, infection at delivery or infection post-delivery via maternal milk or other routes has not been shown. The presence of significant Dane particle concentrations and HBeAg in the blood of chronic carriers has recently been shown to correlate with transmission of infection from mothers to neonates (26) and from carriers to normal health-care personnel accidentally inoculated with contaminated needles (27, 27a).

The dramatic transmission of HBV infection by some chronic HBsAg carrier health care personnel to multiple patient contacts (36,37) and the frequent transmission to neonates from chronic carrier mothers (26,40) emphasize the hazard some carriers represent for the spread of infection. The frequent transmission to neonates and the large number of carriers in the world suggest that elimination of this virus from the population by vaccination is unlikely in the foreseeable future.

5. Host Factors Associated with Persistent HBV Infection

The incidence of the HBsAg carrier state is related not only to the incidence of primary infection in the population but clearly also to host and possibly viral factors which appear to promote persistent infection. A possible genetic predisposition for persistent infection has been suggested by family studies in which the chronic carrier state appears to segregate as an autosomal recessive trait (41-43). The results of others have failed to confirm this (44) and there has been no association found between the chronic carrier state and HLA or blood group antigens. Thus the question of a genetically determined predisposition remains unsettled.

It is possible that the very high carrier rates in some underdeveloped countries (see Frequency of persistent infection above) are not only related to poor sanitary conditions and increased exposure at very early ages but also to differences in predisposition for chronic infection on a genetic, nutritional or other basis.

The severity of initial disease and age also appear to influence the probability for developing persistent infection. Persistent infection is probably more frequent after initial anicteric hepatitis compared with initial icteric disease (45, 46) and survivors of fulminant hepatitis rarely become persistently infected (6). Persistent infection almost invariably follows anicteric neonatal hepatitis (47) compared with 10% of adults with clinically apparent initial infection who become chronic carriers (6). The HBsAg carrier state appears to be most common in young age groups with a peak around 15 to 19 years and the frequency in males is several times that in females in this age group (reviewed in ref. 9 and 48). Anecdotal cases suggest that immunosuppression may be associated with milder initial disease and more frequent persistent infection than occurs in immunological normals. Finally the HBsAg carrier state appears to be more common in patients with certain diseases such as Down's syndrome, lepromatous leprosy and chronic lymphocytic leukemia than in the general population (reviewed in ref. 9).

Future research must determine whether the increased frequency of persistent HBV infection under each of the conditions described above occurs merely because of increased exposure to the virus and opportunity for infection or because of a common underlying mechanism such as altered immune response to infection which promotes persistent infection. Altered immune response could be common to young age, acquired immunosuppression, genetically determined altered immune response, and the diseases such as Down's syndrome, lepromatous leprosy and chronic lymphocytic leukemia.

6. Viral Factors Associated with Persistent HBV Infection. Viral factors associated with persistent HBV infection are less well documented than the host factors described above. The infecting virus dose, however, has been clearly shown to be correlated with the probability for developing persistent infection. Experimental infections with different dilutions of infectious serum have shown that lower doses of virus result more often in mild initial disease and subsequent chronic infection compared with larger virus doses (45).

The dose of HBV required to infect human subjects by mouth appears to be greater than needed by parenteral injection (49). For cases of accidental transmission the usual virus dose received by mouth or by percutaneous puncture with contaminated needles is probably much smaller than the dose received by transfusion of large blood volumes containing virus. Thus virus dose undoubtedly varies with route of transmission and in this way the probability for persistent infection may be influenced by the route of virus transmission.

A striking and unique feature of persistent HBV infections is the presence of incomplete virus coat particles (nucleic acid free 20 nm. HBsAg particles). These particles may exceed in number Dane particles by between 10^4 and 10^8 to one in different sera. Recently it has been postulated that defective-interfering (DI) particles may play a role in maintaining persistent non-cytocidal infection with vesicular stomatitis virus (VSV) (50) and lymphcytic choriomeningitis virus (LCM) (51,52) in tissue culture and VSV *in vivo* (53,54). Although viral forms with the features of true DI particles (i.e. containing DNA with large deletions) have not been described for hepatitis B, the remarkable persistent HBV infections associated with very high concentrations of incomplete particles makes it important to consider a possible role for defective virus in future investigations of persistent HBV infection.

There is no evidence at this time that hepatitis B virus strains (e.g. HBsAg sub-types) have different virulence or propensity for persistent infection.

7. Immunological Aspects of Persistent Infection. Humoral antibody against both known viral antigens (HBsAg and HBcAg) is regularly present in persistently infected patients (55). Anti-HBc is present in high titer in all persistently infected patients and appears early in the course of acute infection, commonly before the onset of acute hepatitis (56). The anti-HBs response occurs even in the presence of large

quantities of circulating HBsAg. It is not known whether the response is quantitatively diminished, but the regular presence of this antibody shows that immunological tolerance to HBsAg does not account for persistent infection. The role of humoral antibody or immune complexes in maintaining persistent infection or influencing viral antigen expression is not known.

Antibodies against the normal hepatocyte surface and other hepatocellular antigens are present in persistently infected patients (55) but the role of such antibody in disease and in maintaining persistent infection is not known.

Cellular immune responses detected by HBsAg induced lymphocyte stimulation and leukocyte migration inhibition are said to show cellular immunity in convalescence from hepatitis B and not during acute hepatits (55,57). Variable results have been obtained in persistently infected patients. Some with chronic active hepatitis and liver injury are said to have partial responses. Others without liver injury as well as patients with chronic persistent hepatitis show no response by these tests. The problem of evaluating cellular immunity in the presence of antigenemia suggests that these studies do not necessarily imply an inadequate cellular immune response during persistent infection or differences in patients with and without associated liver disease. Cellular immune responses to HBcAg have not been studied.

A difference in the pattern of immunofluorescent staining for HBsAg is said to occur in hepatocytes of persistently infected patients with and without associated liver disease. HBsAg has been observed predominantly in the cytoplasm of patients without liver disease and on the cell surface as well as in the cytoplasm of patients with chronic active hepatitis (29). How this difference may be related to differences in immune response in the two kinds of patients or to development of hepatic disease is unclear at this time.

There is evidence suggesting that cellular immunity to hepatic antigens occurs in chronic hepatitis B (55). Lymphocyte mediated hepatolysis *in vitro* has been demonstrated with lymphocytes from patients with chronic active hepatitis B suggesting that liver injury in such patients may be mediated at least in part by the cellular immune response.

Edgington (55) has described a plasma factor in patients with acute hepatitis B which is capable of inhibiting E-rosette formation by T-lymphocytes. This factor is said to persist in patients who develop chronic hepatitis. The role of the humoral factor which suppresses T-lymphocyte function in the initiation or maintenance of persistent infection, or in altering immune mediated disease is not known.

8. Response of Persistent HBV Infection to Antiviral Agents. An intriguing feature of persistent HBV infection has been the regular response to treatment with two antiviral agents. In recent studies, human leukocyte interferons (HLI) (58,59) and adenine arabinoside (ara A) (59,60) have been shown to inhibit Dane particle production in all persistently infected patients treated. To date seven persistently infected patients with high Dane particle concentration in the blood and with chronic hepatitis have completed treatment with HLI and 4 with ara A, and Dane particle markers including Dane particle-associated DNA polymerase and HBcAg in serum were rapidly reduced in concentration in all. Three types of response were recognized. First, in 4 patients treated with HLI and 2 with ara A, reduction in Dane particle markers was maintained only during treatment and returned to pretreatment levels immediately after stopping treatment. No change in HBsAg titer was observed. Second, in 1 patient treated with HLI and 2 with ara A, Dane particle markers and HBeAg were permanently reduced to below the level of detection (i.e. the effects were maintained after stopping treatment) and a significant fall in HBsAg titer was observed in all. Third, in 2 patients treated with HLI Dane particle markers (HBeAg and HBsAg were permanently reduced to below the level of detection (i.e. the effects were maintained after stopping treatment). Liver biopsies after treatment were remarkable for the absence of detectable cells staining for HBsAg or HBcAg, whereas most hepatocytes in pretreatment biopsies revealed staining for these antigens. Thus in 5 our of 11 patients treated with either antiviral, Dane particles and HBeAg were permanently reduced in blood to below the level of detection, and in 2 of the 5 HBsAg similarily disappeared from the blood, and HBsAg and HBcAg from the liver (i.e. all markers of HBV eliminated from blood and liver). Two interesting features of the responses were observed. First, Dane particles responded to treatment much more rapidly (and regularly) than total HBsAg (mostly 20 nm particles) suggesting that both antivirals tested have a significantly different effect on production of the two HBV forms. Second, females appeared to respond more readily than males. Three out of four females and only 2 out of 7 males had type 2 or 3 (permanent) responses. The more favorable response of females to antiviral treatment correlates with their greater tendency to eliminate the virus after acute infection (males become persistently infected two or more times as frequently as females) and indicates that the response to HBV infection differs in females and males. Further studies will be required to determine the frequency and extent of the responses and whether complete eradication of infectious virus and resolution of liver disease can be achieved with these antiviral agents.

No other treatment has been shown to effect the course of persistent hepatitis B infection or the associated liver disease.

REFERENCES

1. Blumberg, B.S., Alter, H.J., and Visnich, S. (1965). J. Am. Med. Assoc. 191, 541.
2. Blumberg, B.S., Gerstley, B.J.S., Hungerford, D.A. (1967). Ann. Int. Med. 66, 924.
3. Okachi, K., and Murakiam, S. (1968). Vox Sang. 15, 374.
4. Prince, A.M. (1968). Proc. Natl. Acad. Sci. USA 60, 814.
5. Barker, L.F., Maynard, J.E., Purchell, R.H., Hoofnagle, J.H., Berquist, H.R., and London, W.T. (1970). Am. J. Med. Sci. 276, 189.
6. Redeker, A.G. (1975). Am. J. Med. Sci. 270, 9.
7. Shulman, N.R. (1970). Am. J. Med. 49, 669-692.
8. Szmuness, W., Much, W.M., Prince, A.M., Hoofnagle, J.H., Cherubin, C.E., Harley, E.J., and Block, G.H. (1975). Am. Int. Med. 83, 489.
9. Blumberg, B.S. In "Hepatitis and Blood Transfusion" (G.N. Vyas, H.A. Perkins and R. Schmid, eds.), p. 63. Grune and Stratton, New York (1972).
10. Gerin, J.L. (1976). Fractions 1, 1-9.
11. Zuckerman, A.J., and Taylor, P.E. (1969). Nature 223, 81.
12. Bayer, M.E., Blumberg, B.S., and Werner, B. (1968). Nature 218, 1057.
13. Kim, C.Y., and Tilles, J.G. (1973). J. Clin. Invest. 52, 1176.
14. Le Bouvier, G.L. (1973). Ann. Int. Med. 79, 894.
15. Dane, D.S., Cameron, C.H., and Briggs, M. (1970). Lancet 1, 695.
16. Almeida, J.D., Rubenstein, D., and Stott, E.H. (1971). Lancet 2, 1225.
17. Barker, L.F., Almedia, J.D., Hoofnagle, J.H., Gerety, R.J., Jackson, D.R., and McGrath, P.P. (1974). J. Virol. 14, 1552.
18. Kaplan, P.M., Greenman, R.L., Gerin, J.L., Purchell, R., and Robinson, W.S. (1973). J. Virol. 12, 995.
19. Greenman, R.L., and Robinson, W.S. (1974). J. Virol. 13, 1231.
20. Robinson, W.S., Clayton, D.A., and Greenman, R.L. (1974). J. Virol. 14, 384.
21. Purcell, R.H., and Gerin, J.L. (1975). Am. J. Med. Sci. 270, 395.
22. Alter, H.J., Barker, L.F., and Holland, P.V. (1975). New Engl. J. Med. 293, 1093.
23. Almeida, J.D. (1972). Am. J. Dis. Child. 123, 303.
24. Barker, L.F., and Murray, R. (1972). Am. J. Med. Sci. 263, 27.
25. McCollum, R.W. (1952). Proc. Soc. Exp. Biol. Med. 81, 157.

26. Okada, K., Kamiyama, I., Inomata M., et al. (1976). N. Engl. J. Med. 294, 746-749.
27. Alter, H.J., et al. (1976). N. Engl. J. Med. 295, 909.
27a. Grady, G.F., et al. (1976). Lancet 2, 492.
28. Magnius, L.O. and Epsmark, J.A. (1972). J. Immunol. 109, 1017; Magnius, L.O., et al. (1975). JAMA 231, 356.
28a. Neurath, A.R. and Strick, N. (1977). Proc. Natl. Acad. Sci. USA 74, 1702-1706.
28b. Vyas, G.N., Peterson, D.L., Townsend, R.M. (1977). Science 198, 1068-1070.
29. Barker, L.F., et al. (1973). J. Inf. Dis. 127, 648.; Gudat, F., et al. (1975). Lab. Invest. 32, 1; Ray, M.B., et al. (1976) Gastroenterol. 71, 462.
30. Almeida, J.D., et al. (1970). Microbios 2, 145; Huang, S.N., (1971). Am. J. Path. 64, 483.; Caramia, F., et al., (1972). Am. J. Dis. Child. 123, 309.; and Hoofnagle, et al. (1973). Lancet 2, 869.
31. Williams, A.O. (1975). Am. J. Med. Sci. 270, 53.
32. Goche, D.J. (1975). Am. J. Med. Sci. 270, 49.
33. Ishimaru, Y., et al. (1976). Lancel 1, 707.
34. Gianotti, F. (1973). Arch. Dis. Child. 48, 794.
35. Schalm, W.S., Ammon, H.V., Summerskill, W.H.J. (1976). Ann. Clin. Res. 8, 221.; Schalm, W.S., Summerskill, W.H.J. et al. (1976). Gastroenterol 70, 947.; Summerskill, W.H.J., et al. (1976). Gut
36. Levin, M.L., Addrey, W.C., Wands, J.R. et al. (1974). J. Am. Med. Assoc. 228, 1139-1140.
37. Rimland, D., Parkin, W.E. (1974). Eastroenterology 67, 822.
38. Wright, R.A. (1975). J. Amer. Med. Assoc. 232, 717-721.
39. Szmuness, W., Much, I.M., Brince, A.M., et al. (1975). Ann. Int. Med. 83, 489-495.
40. Stevens, C.E., Beasley, R.P. Tsui J., et al. (1975). New Engl. J. Med. 292, 771-774.
41. Blumberg, B.S., Friendlander, J.S., Woodside, A., Sutnick, A.I. and London, W.I. (1969). Proc. Natl. Acad. Sci. USA, 62, 1108.
42. Ceppellini, R., Bedarida, G., and Carbonara, A.O. (1970). Minerva Medica Torino, p. 53.
43. Grossman, R.A., Benenson, M.W., Scott, R.M., Sutbhan, R., Top, F.H., and Pantuwatana, S. (1975). Am. J. Epidem. 101, 144.
44. Stevens, C.E., and Beasley, R.P. (1976). Nature 260, 715.
45. Barker, L.F., and Murray, R. (1971). J. Am. Med. Assoc. 216, 1970.
46. Krugman, S. In "Hepatitis and Blood Transfusion" (G.N. Vyas, H.A. Perkins and R. Schmid, eds.), p. 349 (1972).

47. Schweitzer, I.L., Dunn, A., Peterson, R., and Spears, R. (1973). Am. J. Med. 55, 762.
48. Mosely, J.W. (1972). In "Hepatitis and Blood Transfusion" (G.N. Vyas, H.A. Perkins and R. Schmid, eds.) p. 23. Grune and Stratton, New York.
49. Krugman, S., and Giles, J.P. (1970). J. Am. Med. Assoc. 212, 1019.
50. Holland, J.J., and Villarreal, L.P. (1974). Proc. Natl. Acad. Sci. USA. 71, 2956.
51. Lehman-Grube, F. (1971). Virol. Monograph 10, Sprinter-Verlag, New York.
52. Welsch, R.M., O'Connel, C.M., and Fair, C.J. (1972). J. Gen. Virol. 17, 355.
53. Doyle, M., and Holland, J.J. (1973). Proc. Natl. Acad. Sci. USA. 70, 2105.
54. Holland, J.H., Villarreal, L.P., and Etchinson, J.R. (1974). In "Mechanisms of Virus Disease" (W.S. Robinson and C.F. Fox, eds.), p. 131. W.A. Benjamin, Menlo Park, Calif.
55. Edgington, T.S., and Chisari, F.V. (1975). Am. J. Med. Sci. 270, 218.
56. Krugman, S., Hoofnagle, J.H., Gerety, R.J., Kaplan, P.M., and Gerin, J.L. (1974). New Engl. J. Med. 290, 1331; Hoofnagle, J.H., Gerety, R.J., and Barker, L.F. (1975). Am. J. Med. Sci. 270, 179.
57. Tong, M.J., et al. (1975). New Engl. J. Med. 293, 318.
58. Greenberg, H.P. Pollard, R.B., Lutwick, L.I., et al. (1976). New Engl. J. Med. 295, 517.
59. Merigan, T.C., Robinson, W.S., et al. Clin. Res., in press.
60. Pollard, R.B., Smith, J.L., Neal, E.A. Gregory, P.B., Merigan, T.C., and Robinson, W.S. J. Am. Med. Assoc., in press.

AN INTEGRATING IMMUNOREGULATORY HYPOTHESIS FOR THE IMMUNOPATHOGENESIS OF LIVER DISEASE ASSOCIATED WITH HEPATITIS B VIRUS INFECTION[1]

Francis V. Chisari and Thomas S. Edgington

Department of Molecular Immunology, Research Institute of Scripps Clinic, La Jolla, California 92037

ABSTRACT Hepatitis B virus infection as associated with highly divergent forms of hepatic injury. No rigorous association between patterns of viral synthesis and disease has been identified; whereas a number of correlations between the immune response, immune attack systems, and immunoregulatory events are observed. An integrating hypothesis has been developed to incorporate the character and regulation of the immune attack systems. This hypothesis accommodates current observations regarding the immune responses to viral and autoantigens in evolution of disease, and serves as a model for more specific analysis of the encounter between host and virus.

INTRODUCTION

The molecular biology of the hepatitis B virus (HBV) has reached an advanced state of understanding. In contrast, the biology of this agent in man has presented a major investigative challenge.

The pathogenesis of hepatic injury is central to the biology of this agent. First, it is untenable to adopt a direct cytopathic role for HBV. This is based on evidence that: 1) infection of hepatocytes is typically diffuse throughout the liver although injury is focal; 2) hepatocellular injury is an elective event and not a requisite consequence of hepatocellular infection; 3) hepatocellular injury can continue in the absence of demonstrable viral synthesis; 4) hepatocellular injury is diminished in immunosuppressed patients. If hepatic disease is a consequence of hepatocellular infection it must either: a) reflect a rather selective and specific pattern of aberrant cell function

[1]This work was supported by NIH grants CA-14346, AI-13393 and Research Career Development Award 5 K04 AI-00174 (FVC).

ISBN 0-12-668350-6

induced by the virus; or b) the host response to the virus infected and modified cell. In the absence of substantive evidence to support the former and a sizeable body of data to support the latter, a significant role for the host immune response in the genesis and occasional perpetuation of hepatic injury secondary to HBV infection has been implicated.

Immune responses of two specificities have been implicated: 1) the response to HBV structural or induced antigens that are expressed at the hepatocyte surface; 2) the response to hepatocyte-specific autoantigens, specifically a liver specific lipoprotein antigen designated LSP. It is notable that the elective forms of disease associated with HBV infection, namely: 1) acute transient hepatitis, 2) acute slowly resolving hepatitis, 3) chronic persistent hepatitis, a mild form of chronic hepatitic injury, 4) chronic active hepatitis, a more severe form of chronic hepatic injury, or 5) the silent chronic carrier state can not be clearly distinguished only by respect to the type of immune response that is induced but rather by reference to the regulation of the two types of immune responses, anti-viral and autoimmune. Indeed the key to understanding the immunopathogenesis of these diseases seems embodied in the immunoregulatory features of these diseases, which have now been incorporated in an integrated hypothesis that accommodates the more critical available data.

LYMPHOCYTE DISTRIBUTIONS

Distribution of T and B Lymphocytes. Thymus-derived (T) and bone marrow derived (B) lymphocytes are, respectively, the primary effectors of cellular and humoral immune responsiveness. Based on differences in cell surface membrane markers, i.e., surface immunoglobulin (SIg) for B and rosette formation with sheep erythrocytes (E) for T cells, quantitation of these lymphocytes in a variety of tissues is possible if the cells remain phenotypically normal. When applied to study of patients with viral hepatitis (Table 1), all investigators have observed a significant decrease in the relative and absolute numbers of E rosette forming T lymphocytes in the peripheral blood of patients with HBV related AVH (1-4) and CAH (1-7). Most studies reveal little if any change in the B lymphocyte populations. The appearance of increased "null" lymphocytes which lack these characteristic surface markers of T or B cells occurs in most cases.

LYMPHOCYTE DYSFUNCTION

Mechanisms. Because of the apparent decrease in peripheral T cells in hepatitis patients and a suggested T cell predominance in the hepatic infiltrates some investigators

TABLE 1
DISTRIBUTION OF LYMPHOCYTES IN PATIENTS WITH HBV ASSOCIATED LIVER DISEASE[a]

	# Patients	Peripheral blood			Reference
		T	B	Null	
AVH[b]	27	↓	↑	↑	1
	67	↓	↔	↑	2,3
	4	↓	↔	↑	4
CAH[b]	5	↓	↑	↑	1
	5	↓	↔	↑	2,3
	9	↓	↔	↑	4
	11	↓	↑	↔	5
	5	↓	↔	↑	6
	15	↓	↔	↑	7

[a]From references 1-7.
[b]AVH, acute viral hepatitis; CAH, chronic active hepatitis

have postulated increased homing of these cells to inflammatory foci within the liver with depletion from the peripheral vascular compartment. This hypothesis is unsupportable since: 1) simple T lymphocytopenia would be accompanied by a compensatory relative B lymphocytosis. However, in all but two studies (1,5) the B cell distribution remained normal and a third, "null" cell population appeared which accounted for the decrease in phenotypically normal T cells. 2) The assay used to identify T cells depends on the binding by T cells of sheep red blood cells with the formation of rosettes. Rosette (ER) formation is known to be a receptor-mediated (9,10), energy dependent process (11), dependent on intact protein and nucleic acid synthetic pathways (12) and modulated by intracellular cyclic nucleotide levels (13,14). Variable expression of ER function may be observed under a variety of conditions; and a decrease of ER forming lymphocytes is not equivalent to a decrease in T lymphocytes. 3) The number of peripheral T lymphocytes, identified by rosetting with neuraminidase treated sheep red blood cells (N-SRBC) is normal in acute and chronic hepatitis B (2,15) when ER forming cells are reduced. N-SRBC rosette formation is a specific T cell marker (16) but more sensitive. Thus, reduced ER positive T cells in HBV infection does not indicate peripheral T lymphocytopenia, but rather a phenotypic alteration reflecting a functional alteration of T cell metabolism.

DEFECTIVE E ROSETTE FUNCTION

Utilizing *in vitro* regeneration (1,2), two independent mechanisms have been implicated in defective ER function. The first (extrinsic) mechanism is readily reversible upon cultivation of defective lymphocytes in the absence, but not in the presence, of autologous serum. Furthermore, such serum inhibits the ER function of normal T cells in a metabolic fashion and the active constituent, termed Rosette Inhibitory Factor (RIF), has been identified as a unique plasma lipoprotein (17). The second (intrinsic) mechanism is not readily reversible *in vitro* and is not associated with an autologous serum factor which induces the defect in normal T lymphocytes.

During the active phase of acute hepatitis most if not all patients exhibit the extrinsic defect when evaluated at sufficiently frequent intervals. About half exhibit an extrinsic defect at single sampling, 30% exhibit an intrinsic defect and 20% are normal (Table 2). There is no association between the type of defect present and: a) the severity of injury, b) cholestasis, or c) HBV serologic markers. The type of defect present at <4 weeks has no predictive value relative to evolution of disease. When examined 12-16 weeks after onset, the extrinsic defect is uniformly absent from those patients with resolution but is universally present in those with continued hepatocellular injury, referred to as unresolved hepatitis (Table 2). Nearly 90% of patients with HBV related chronic active hepatitis exhibited the extrinsic defect; and prolonged persistence of the extrinsic defect in E rosette function of T cells attributable to RIF is associated with prolonged liver disease.

The extrinsic defect, characterized by the presence of the serum regulatory lipoprotein RIF, also occurs in hepatitis A virus infection, an RNA virus, and in viral hepatitis of the "non-A, non-B" variety. However, it does not occur in a

TABLE 2
EXTRINSIC AND INTRINSIC DEFECTS IN E ROSETTE FUNCTION OF PERIPHERAL T LYMPHOCYTES IN HEPATITIS B VIRUS INDUCED LIVER DISEASE*

	Number	Extrinsic	Intrinsic	None
Acute‡	30	16	8	6
Resolved†	9	0	8	1
Unresolved†	10	10	0	0
Chronic Active§	8	7	1	0

*From references 2,3. ‡<4 weeks.
†12-16 weeks. §>3 years.

variety of other non-hepatocytopathic viral disease, non-viral hepatopathies of CMV or EBV mononucleosis associated with hepatitis. Thus, there appears to be some poorly understood specificity of the extrinsic mechanism for the typical hepatotrophic viral infections of man.

Mitogen Responsiveness. Stimulation of cellular DNA synthesis by polyclonal activators such as phytohemagglutinin (PHA), a parameter of T lymphocyte function, is frequently diminished in a number of viral (18), neoplastic (19), and hepatic diseases including acute and chronic hepatitis B (20). In uncomplicated acute hepatitis B response to PHA is attenuated during the first few weeks of illness and thereafter returns to normal (21,22). In contrast, response to PHA is persistently suppressed in patients with chronic active hepatitis B but generally remains normal in the silent carrier state. These observations have prompted the suggestion (23) that the evolution of disease in HBV infection is dependent upon the functional integrity of the T lymphocyte. Since the bulk of available evidence implicates immunologic effector mechanisms in the mediation of hepatocellular injury and because such systems are modulated by assorted molecular and cellular immunoregulatory events the functional integrity of not only effector cell populations but also regulatory cells may determine the course of HBV related disease. Indeed the existence of a spectrum of HBV associated liver diseases is the anticipated reflection of a biological continum of immune reactivity to relevant intrinsic and induced hepatocyte antigens.

SENSITIZATION TO VIRAL AND HEPATOCELLULAR ANTIGENS

Viral Antigens. Despite deficits in nonspecific lymphocyte activation and cell surface membrane receptor expression, patients with viral hepatitis are quite capable of specific cellular immune reactivity to relevant antigens. Utilizing leukocyte migration inhibition and lymphocyte transformation techniques (21,24-33), a high incidence of sensitization to HBsAg has been demonstrated (Table 3). Such studies, though differing in specific techniques and results do in composite indicate a high frequency (59%) of sensitization to HBsAg during acute injury. Sensitization persists at higher frequency during resolution (87%), and is also observed in chronic active hepatitis (80% of untreated cases) but not in those who become silent carriers.

The virtual absence of sensitization among silent $HBsAg^+$ carriers suggests that cellular immunity to this antigen may play a role in the pathogenesis of hepatocellular injury; however, the high incidence of cellular sensitization in

TABLE 3
CELLULAR IMMUNE SENSITIZATION TO HBsAg AND TO LIVER SPECIFIC LIPOPROTEIN ANTIGEN (LSP)*

Antigen	Acute viral hepatitis		Chronic active hepatitis		Silent carrier
	Acute	Convalescent	Untreated	Total	
HBsAg	59% (87)	87% (35)	80% (10)	50% (88)	6% (66)
LSP	50% (16)	0% (16)	71% (118)	56% (177)	--

*From references 21, 24-33, 37-40.

convalescence indicates that in the absence of an appropriate HBV target antigen at the hepatocyte surface such sensitization is pathogenetically ineffective. The absence of detectable sensitization in some cases of acute and chronic HBV-related hepatitis suggests that this pathway is not requisite for hepatocellular injury, but may be only one of the responses involved in the immune attack.

Based on evidence derived from other viral infections and the parallel with HBV infection we have suggested (15) that the antibody response to HBsAg may suppress the surface expression of HBV, subsequently the cellular synthesis and perhaps even the HBV genome itself. Suppression of surface expression of HBsAg or other relevant HBV antigens (perhaps HBcAg and HBeAg) would then eliminate susceptibility of hepatocytes to HBV specific cellular attack systems. Suppression of viral synthesis would determine the duration of antigenemia and the development of the chronic carrier state. Thus, the humoral immune response to at least one hepatitis B viral antigen (HBsAg) may play an important but indirect role in the termination of hepatocellular injury. Modulation of the magnitude and duration of the antibody response by immunoregulatory molecules adds an important additional dimension to the complexity of pathogenesis. It is unlikely, however, that anti-viral antibody is a requisite hepatocytotoxic mechanism since extensive hepatocellular injury has been observed during viral hepatitis in agammaglobulinemic individuals (36).

Liver Specific Antigens. Cellular sensitization to liver specific proteins in viral hepatitis and other liver diseases (Table 3) has been demonstrated by several laboratories (27, 28, 37-42). Sensitization to a liver specific hepatocyte surface membrane lipoprotein (LSP) is detectable in half of patients with acute hepatitis, is transient, and disappears upon recovery (28,43). In contrast, sensitization persists with chronic active hepatitis, particularly untreated patients

(71%) and is found with equal frequency in chronic active hepatitis of both HBsAg-positive and -negative types. Thus, sensitization to LSP is better correlated with hepatocellular injury than is sensitization to HBsAg.

Sensitization to LSP is lost with clinical recovery. Absolute disappearance of antigen specific cytolytic T lymphocytes (T killer cells) suggests diversion to the memory cell pool. Suppression of differentiation to T killer cells specific for autoantigen is a much more likely possibility especially in a disease characterized by similar temporal expression of molecular immunoregulatory events. For example, Rosette Inhibitory Factor (RIF) is known to occur in approximately 50% of patients with acute hepatitis B, to disappear with recovery and to persist in chronic active hepatitis. Since LSP is an antigen on the surface of cells of a vital organ one might anticipate that the normal unresponsive state may be maintained at least in part by active suppression, one of three general pathways of tolerance. If RIF inhibits active cellular suppression, the presence of RIF would facilitate clonal expansion and differentiation of killer T cells specific for a variety of autoantigens, responses that are suppressed by normal control mechanisms when RIF is absent. This hypothesis fits current data.

While the induction of cytolytic cells specific for LSP may be initiated by other events such as HBV infection, perpetuation of the response may require only the attenuation of mechanisms responsible for maintenance of the tolerant state. If sensitization to LSP is a pathogenetic determinant of virus initiated hepatocellular injury such injury could persist despite the clearance of HBV. Indeed the progression of liver disease from HBsAg-positive acute to HBsAg-negative chronic active hepatitis has been amply documented (44).

THE IMMUNE RESPONSE TO LIVER SPECIFIC LIPOPROTEIN ANTIGEN

Experimental Autoimmune Hepatitis. Meyer zum Buschenfelde and colleagues (45-48) have induced chronic active hepatitis in rabbits immunized with LSP. Histopathologic lesions followed the development of demonstrated humoral and cellular immune responses to LSP. Hepatocyte membrane bound immunoglobulin was detectable in most affected animals. Whether liver disease is due to humoral or cellular immune mechanisms acting alone or in concert was not resolved. The studies also demonstrated the existence of mechanisms capable of suppressing the immune response and preventing hepatocellular injury, suggesting a role for immunoregulation in the evolution of certain types of hepatic disease.

IMMUNE RESPONSES TO LIVER SPECIFIC LIPOPROTEIN

Antibody Initiated Reactions. Hopf, et al. (34,49) have demonstrated the presence of anti-LSP antibodies in the serum of patients with HBsAg-negative chronic active hepatitis but not in a variety of other liver diseases. These patients also exhibited linear deposits of IgG on the surface of viable hepatocytes dissociated from liver biopsies. Neither serum anti-LSP antibody nor linear hepatocyte surface IgG was found in patients with HBs-Ag positive acute or chronic active hepatitis. These patients did, however, display granular hepatocyte surface IgG deposits, a hallmark of local immune complex formation, the antigenic specificity of which is unproven but suggested to be specific for hepatocyte membrane associated HBsAg. Thus, in at least one subgroup of patients with chronic active hepatitis both the afferent (sensitization) and the efferent (specific antibody production) limbs of the humoral immune response to LSP can be initiated and sustained. Similar observations have been described in LSP immunized rabbits that developed a chronic form of experimental autoimmune hepatitis that appeared histologically similar to chronic active hepatitis in man (46). These results support autoimmunity to LSP as a candidate pathogenetic pathway in certain types of liver disease.

Cell Mediated Hepatocytotoxicity. The cytotoxic capacity of peripheral blood lymphocytes from patients with a variety of liver diseases has been analyzed using a variety of hepatocyte related targets. Autologous human hepatocytes, isolated rabbit hepatocytes, and human liver-derived Chang cells have been used. The validity of the Chang cell is debatable (50, 51); and since these cells appear to lack cell surface LSP (52) this assay most likely is responsive to "natural" killer (NK) lymphocyte cells, not necessarily a liver specific attack system.

In the other studies (53-59), autologous hepatocytes or rabbit hepatocytes have been used with rather more definitive results (Table 4). Both cell types express antigenically similar LSP on their surface membranes (47). A reproducible pattern was observed: 1) nearly all patients had significant cytotoxic effector cell activity during the acute phase of hepatitis; 2) cytotoxicity disappeared with recovery, and 3) cytotoxicity persisted in most individuals with untreated chronic active hepatitis (Table 4). There was no difference in cytotoxic activity between serum $HBsAg^+$ and $HBsAg^-$ groups. Notably hepatocytotoxicity was not restricted only to viral disease but was also found in some cases of alcoholic hepatitis (60, neonatal hepatitis (59), hepatitis with α_1-antitrypsin deficiency (59).

TABLE 4
LYMPHOCYTE MEDIATED HEPATOCYTOTOXICITY
ASSOCIATED WITH VIRAL HEPATITIS*

Disease	Number	Percent Positive
Acute Viral Hepatitis B		
Acute	15	93
Resolved	7	0
Chronic Active Hepatitis		
Untreated	41	85
Total	118	63

*From references 53-59.

The antigenic specificity of this induced hepatocytotoxicity was analyzed in acute and chronic active hepatitis (56,59), and was found to be neutralized by either purified LSP or antibody to LSP suggesting that LSP is the target antigen. Blocking of cytotoxicity by aggregated IgG suggested dependence of the reaction. The effector cells appeared devoid of ER function (T cells) and had complement receptors (55,56). It is not known whether they possess surface membrane immunoglobulin. This cytotoxic cell appears phenotypically to be E^-, Fc^+, C^+, $Ig^?$; and on the basis of its behavior in blocking experiments, the reaction appears to be relatively specific for hepatocyte surface LSP.

Further implication of LSP as the relevant hepatocyte target follows comparison of the relatively similar incidence of sensitization to LSP (Table 3) and cellular hepatocytotoxicity (Table 4). Correlation is especially good in resolved acute hepatitis and chronic hepatitis. Thus, it appears, at least with respect to LSP, that both the capacity to mount this suspected NK cell mediated response is functionally intact in viral hepatitis. Clear identification of cytolytic T lymphocyte responses, the potential role of activated macrophages and armed macrophages is awaited. Considerable evidence suggests profound defects in various nonspecific aspects of cellular immunity in this disease (1-3,20-24); and at least one group of investigators has suggested that the degree of cell mediated immunologic dysfunction in a given individual may play a determinant role in the evolution of infection and hepatic disease (24).

These two sets of observations are indeed quite compatible and can be synthesized into a unitary hypothesis in which the primary antigen reactive cells are functionally normal while antigen-nonspecific immunoregulation is susceptible to varying degrees of functional disequilibrium. The net effect would be: a) variable expression of antigen specific immune reactivity, and b) heterogeneity in character and severity of hepatocellular injury.

IMMUNOREGULATORY DYSFUNCTION

Cellular Immunoregulatory Mechanisms. Both cellular and humoral immune responses are dependent upon and regulated by helper and suppressor lymphoid cells (61-63). Excessive suppressor cells can be demonstrated in some *in vitro* systems and have been implicated in the immunodeficiency seen in common variable hypogammaglobulinemia (64), multiple myeloma (65), Hodgkin's disease (66), etc. Suppressor cell activity has been identified within the T lymphocyte and also the adherent cell populations (63).

Relative suppression of such hepatocytotoxic immune responses should favor survival of the host and perpetuation of the species, and thus could enjoy selectional pressures in evolution. Circumstantial evidence for suppression exists in the difficulty of inducing in rabbits an immune response to LSP with the subsequent development of chronic active hepatitis (45). Such suppression of the immune response to hepatocellular surface membrane autoantigens would serve to prevent the induction of an autoimmune pathogenesis, and functional suppressor cell defects could hypothetically account for acute and chronic active hepatitis. Indeed, Hodgson, et al. (67) and we (unpublished observations) have observed two distinct types of suppressor cell defects in association with HBV infection and hepatic injury.

In Hodgson's study (67) mitogen-induced T suppressor cell activity was found to be deficient in chronic active hepatitis and the deficit was accentuated with exacerbations of hepatic injury. Serial studies in patients with acute viral hepatitis have demonstrated diminished suppressor cell activity which returned to normal during recovery.

In an ongoing study, spontaneous suppressor activity is monitored using a system which detects suppressor monocytes capable of suppressing a one way allogeneic lymphocyte response (68). Deficient suppressor activity was observed in 5/7 patients during the acute hepatocellular injury during the first two weeks after onset of jaundice. By four weeks only 3/7 exhibited the defect; and normal suppressor activity was observed in all patients by six weeks after onset coincident with biochemical evidence of healing. Two individuals with moderate clinical exacerbations at 8 and 12 weeks had deficient suppressor activity but reverted to normal with remission. Thus, the phenotypic features of deficient suppressor cell function occurs during the acute phase of HBV associated injury, returns to normal with recovery and remains deficient in chronic active hepatitis. These observations are compatible with the suggestion of Dudley, et al. (24) that defective cell mediated immunity is associated with development of chronic hepatocellular injury in viral hepatitis. They also support the evidence that antigen-specific sensitization and effector mechanisms are not only intact but may be pathogenetically responsible for the disease.

Humoral Immunoregulatory Mechanisms. A number of antigen non-specific immunoregulatory molecules (17,69-75) have been identified in human plasma (Table 5). Most are synthesized or modified by the liver. Most of these molecules suppress lymphocyte activation; some of them inhibit initiation of the immune response and some also suppress the primary induction of cytotoxic effector cells (76). The hepatitis virus induced lipoprotein RIF is different from the rest of these molecules in that in preliminary experiments it has demonstrated facilitating rather than suppressive properties.

It is not surprising that quantitative and qualitative abnormalities of several of these substances occur in association with hepatic disease. Whereas RIF is induced only in viral hepatitis (2,3); various structural abnormalities in the plasma lipoproteins have been reported in viral hepatitis and other liver diseases (17,77-79). Alpha fetoprotein levels increase in association with hepatocyte regeneration (80). It is postulated that certain types of lymphocyte dysfunction observed in association with viral hepatitis might follow from

TABLE 5
IDENTIFIED IMMUNOREGULATORY MOLECULES IN PLASMA

Alpha Fetoprotein (AFP)*
C-Reactive Protein (CRP)
Immunoregulatory Peptide (IRP)†
Fibrinogen cleavage fragments*
Hepatocyte derived regulatory protein*
Immunoregulatory Lipoproteins
Rosette Inhibitory Factor (RIF)*
Chylomicrons*
Very Low Density Lipoproteins (VLDL)*
Intermediate Density Lipoproteins (IDL)*
Low Density Lipoprotein Inhibitor (LDL-In)*
Low Density Lipoproteins (LDL)*
High Density Lipoproteins (HDL)*

*Definite or probable liver dependent metabolism.
†Peptide of previously described immunoregulatory alpha globulin.

either hepatic induction of regulatory molecules such as RIF, or to quantitative and qualitative abnormalities in the hepatic metabolism of other immunoregulatory molecules. The net phenotypic expression of altered immunologic reactivity would thus depend on the character of specific abnormalities in the biosynthesis of individual immunoregulatory molecules and would be expected to vary between individual hosts.

Finally, an immunosuppressive molecule has been extracted directly from normal human liver (81). We have characterized this inhibitory liver extract (LEx) as a protein of molecular weight approximately 65,000 (82). It suppresses activation of lymphocytes by mitogens and allogeneic cells, mitogen-activated and basal protein synthesis of lymphocytes and immunoglobulin synthesis by pokeweed mitogen stimulated lymphocytes. Anticipating that LEx is probably released as a consequence of hepatocellular injury, it may modulate local lymphocyte effector function in the vicinity of adjacent hepatocytes and thus limit the magnitude and extent of injury in accord with local concentration. The yield of LEx from different livers purified under identical conditions differed by 10 fold, suggesting that intracellular concentrations of LEx may vary from individual to individual. If true, the variable severity and the focal nature of hepatocellular necrosis may relate to differences in local concentration of LEx released during the initial phase of hepatocellular injury.

By virtue of its role in the metabolism of humoral immunoregulatory molecules, hepatic function can be implicated in the regulation of immunological function. Dysfunctional metabolism of these substances may result in overt immunologic abnormalities; indeed both hypergammaglobulinemia with autoantibody formation on the one hand and anergy on the other accompany certain types of chronic active liver disease (83) and viral hepatitis (84) respectively.

RIF has an unusual apolipopeptide composition and serves as a prototype of dysfunctional metabolism of immunoregulatory molecules. If it is postulated that RIF alone or in concert with other factors can attenuate suppressor cell activity, sufficient data is available to formulate an association between the pattern of immunoregulation, cellular immune reactivity to LSP, and the evolution of viral hepatitis (Table 6). RIF is present in a pattern concordant with the frequent presence of phenotypically decreased suppressor activity. This association is also concordant with evidence of sensitization and cytotoxicity towards a hepatocellular autoantigen (LSP). A key observation is that with diminishing injury there is: a) loss of autoantigen reactivity in clinically resolved hepatitis, b) normal suppressor cell activity, and c) disappearance of RIF.

TABLE 6
CORRELATION OF THE INCIDENCE OF A PROTOTYPE IMMUNOREGULATORY MOLECULE (RIF) WITH AN INDEX OF SUPPRESSOR CELL FUNCTION AND EVIDENCE OF CELLULAR IMMUNE REACTIVITY TO LSP IN VIRAL HEPATITIS

	Acute viral hepatitis B		Untreated chronic
	Acute phase*	Resolved*	active hepatitis*
RIF	52.9% (70)	0% (9)	87.5% (8)
Suppressor cell activity	↓	N	↓
LSP specific sensitization	50% (16)	0% (16)	71% (118)
LSP specific cytotoxicity	93% (15)	0% (7)	85% (41)

* % (number tested)

INTEGRATING HYPOTHESIS

The remarkable, strong and consistent association between these variables has led us to postulate a hypothesis which accommodates the diverse and complex biology of HBV and the associated hepatic injury. This hypothesis is based on the premises outlined in Table 7.

Based on these premises the pathogenesis of hepatocellular injury in viral hepatitis is viewed as illustrated in Figure 1. Incorporation of the HBV genome into infected hepatocytes not only a) leads to HBV synthesis, stimulation of immuneresponses, and hepatocyte surface expression of target antigens, but b) induces a state of disordered hepatocellular metabolism resulting in the induction of unique or quantitatively or qualitatively abnormal immunoregulatory molecules. Dependent on type, quality and magnitude of these alterations complex disturbances of the immunologic function of the host is produced. These would include inhibition of normal active suppressor cell function, effector cell function and antibody synthesis. Inhibition of normal suppressor cell activity may result in antibody responses to autologous molecules such as actin and hepatocyte surface membrane antigens, LSP. Sensitization of the cellular immune response with differentiation to cytolytically active immune effector cells would be permitted by the suppressor cell deficit.

TABLE 7
PREMISES UNDERLYING AN INTEGRATING HYPOTHESIS OF THE IMMUNOPATHOGENESIS OF VIRAL HEPATITIS

1. Hepatocellular injury is mediated by cellular effector systems specific for hepatocyte autoantigens (LSP) and viral antigens (HBsAg).
2. Cellular immune effector pathways (especially for LSP) are actively suppressed in the normal tolerant state.
3. Humoral immunoregulatory mechanisms normally modulate cellular and humoral immune responses.
4. HBV infection induces significant disorders of humoral immunoregulation.
5. Cell mediated anti-viral attack mechanisms can be expressed only in the presence of hepatocyte surface membrane viral antigens.
6. The anti-viral antibody response modulates the expression of viral antigens at the hepatocyte surface and suppresses viral genome synthesis.
7. The local release of immunosuppressive molecules from injured hepatocytes influences the local distribution and severity of immunologically mediated hepatocellular necrosis.
8. Correction of the immunoregulatory disorder leads to recovery of normal active suppressor mechanisms with abrogation of autoimmune responses and recovery.

With hepatocyte injury intracellular constituents including LEx would appear in the local extracellular micro-environment but variable concentrations in different individuals, and LEx would directly inhibit local cytotoxic effector cell function. Suppression of cytotoxic effector cell function *in situ* would explain the focal nature of injury in the face of diffuse infection, a characteristic of acute viral hepatitis. Selective modulation of these events would influence the evolution of disease, permitting either clinical recovery or development of the silent carrier state.

Cellular attack mechanisms specific for hepatocyte surface membrane expressed viral antigens (i.e., HBsAg) and other viral coded antigens may also be important. Operationally, such systems depend on the coexistent expression of the antigen at the hepatocyte surface. As indicated in Figure 1 and discussed in a previous report (15) anti-viral antibody may suppress viral genome expression by infected hepatocytes.

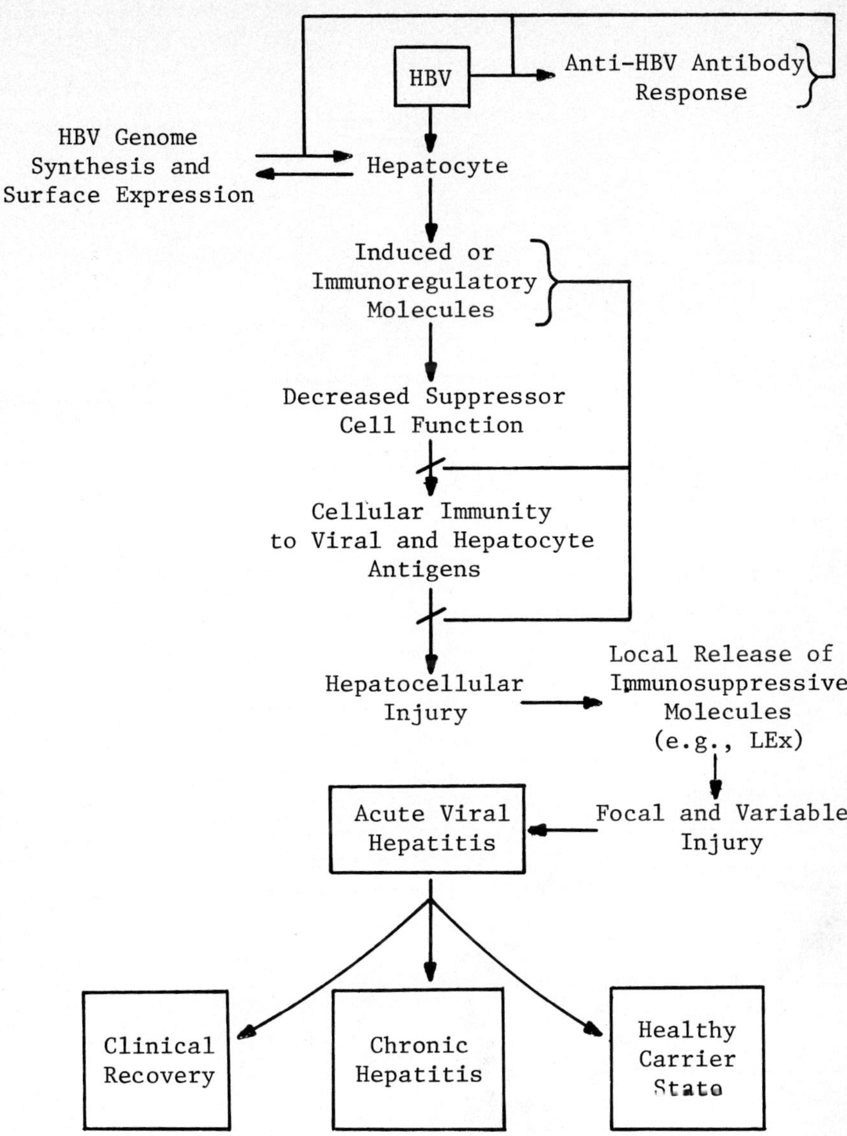

FIGURE 1. Sequence of events in the pathogenesis and evolution of disease associated with HBV infection.

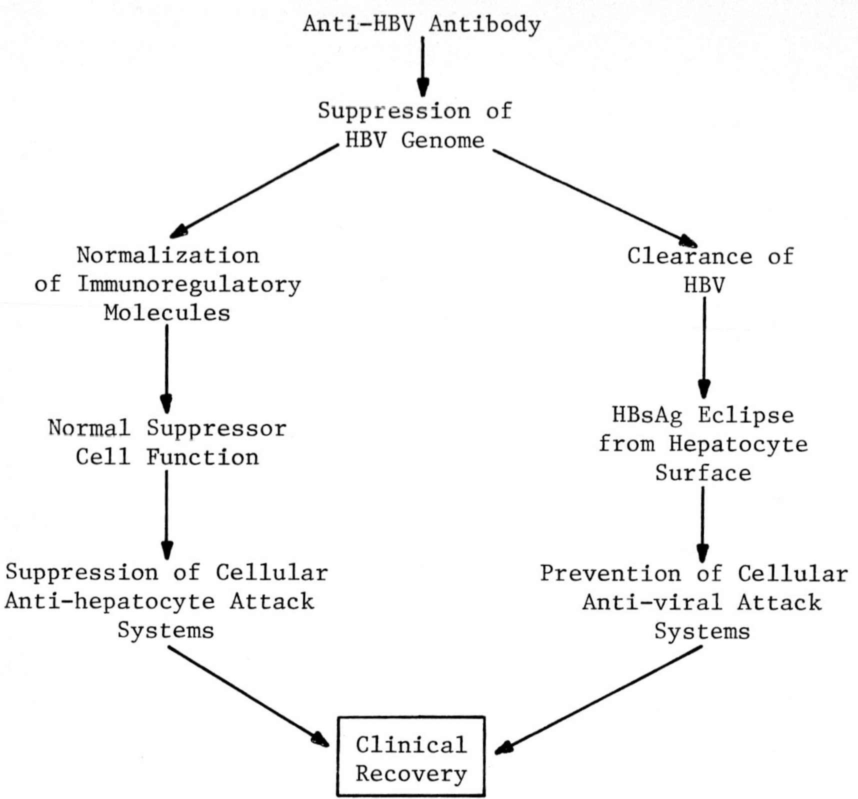

FIGURE 2. Hypothesis sequence of events in the evolution of disease associated with recovery from HBV infection.

This event has two logical consequences both of which would promote evolution to the recovered state (Figure 2). First, HBsAg would be capped-off resulting eclipse from the hepatocyte surface, eliminating the target for cellular anti-viral attack. Second, normal hepatocellular metabolism of immunoregulatory molecules may resume thus permitting the resumption of normal suppressor cell function with active suppression of the autoimmune cellular attack specific for LSP or other hepatocyte surface autoantigens.

Anti-viral antibody may also be operative as a direct pathogenetic determinant of hepatocellular injury by binding to hepatocyte surface viral antigens and providing a target for Fc receptor mediated K cell cytotoxicity. Although it may contribute to the pathogenesis as an elective mechanism, this is unlikely as a pathway in view of reported cases of acute viral hepatitis in patients with agammaglobulinemia (36).

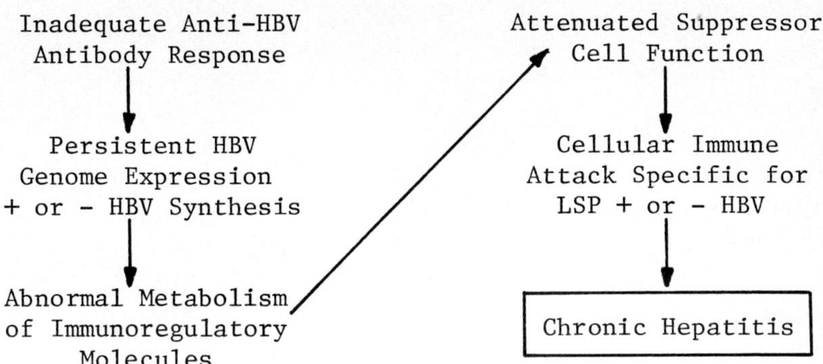

Figure 3. Hypothesized sequence of events in pathogenesis of chronic hepatitis.

The abnormal synthesis of immunoregulatory molecules induced by hepatocellular HBV infection could also attenuate or otherwise modify the character of the anti-viral antibody response. This could permit persistent hepatocellular HBV genome expression with or without associated cell surface HBsAg eclipse (Figure 3). Persistent genome with disturbance of hepatic biosynthesis would result in modified metabolism of immunoregulatory molecules and dysequilibrium of the immune network through one or more pathways allowing progressive cellular immune attack of hepatocyte autoantigens and perhaps also of viral antigens, the latter dependent on the status of hepatocyte surface HBV expression. The end result of this process would be the progressive hepatocellular injury of chronic hepatitis.

There is sufficient latitude in the current hypothesis to permit hepatocellular HBV infection to result in selective disorders of molecular immunoregulation (Figure 4). One type of selective abnormality would involve an attenuated antiviral antibody response without effect on suppressor cell function. This can also be accounted for on the basis of two other mechanisms: 1) antigen induced clonal abortion of those B cells capable of participating in the response as described by Nossal and Pike (85); or 2) antigen induced inactivation of specific B cells by a variety of mechanisms including receptor eclipse (86), antigen-antibody complex inactivation (87), and de-suppression of anti-idiotype responses (88). All would result in an attenuated antibody response to a high concentration antigen such as HBsAg. The net effect of this situation would be continued viral synthesis without hepatocellular injury, i.e., the healthy HBsAg carrier state.

28. Lee, W. M., Reed, W. D., Osman, C. G., Vahrman, J., Zuckerman, A. L. W. F., and Williams, R. (1977). *Gut 18*, 250.
29. DeSaules, M., Frei, P. C., Libanska, J., and Lovillert, B. (1976). *J. Inf. Dis. 134*, 505.
30. Erard, P. (1974). *Clin. exp. Immunol. 18*, 439.
31. Irwin, G. R., Jr., Hierholzer, W. J., Jr., Cimis, R., and McCollum, R. W . (1974). *J. Inf. Dis. 130*, 580.
32. Tong, M. J., Wallace, A. M., Peters, R. L., and Reynolds, T. B. (1975). *N. Engl. J. Med. 293*, 318.
33. Ibrahim, A. B., Vyas, G. N., and Perkins, H. A. (1975). *Infect. and Immun. 11*, 137.
34. Hopf, U., Meyer zum Buschenfelde, K. H., and Arnold, W. (1976). *N. Engl. J. Med. 294*, 578.
35. Alberti, A., Realdi, G., Tremoladi, F., and Spiva, G. P. (1976). *Clin. exp. Immunol. 25*, 396.
36. Good, R. A., and Page, A. R. (1960). *Am. J. Med. 29*, 804.
37. Miller, J., Smith, M. G. M., Mitchell, C. G., Reed, W. D. Eddleston, A. L. W. F., and Williams. R. (1972). *Lancet 2*, 296.
38. Vergani, C., Oldoni, T., and Dioguardi, N. (1974). *J. Clin. Path. 27*, 772.
39. Meyer zum Buschenfelde, K. H., Knolle, J., and Berger, J. (1974). *Klin. Wschr. 52*, 246.
40. Meyer zum Buschenfelde, K. H., Alberti, A., Arnold, W., and Freudenberg, J. (1975). *Klin. Wschr. 53*, 1061.
41. Sorrell, M. F., and Leevy, C. M. (1972). *Gastroenterology 63*, 1020.
42. Mihas, A. A., Bull, D. M., Davidson, C. S. (1975). *Lancet 1*, 951.
43. Moussouros, A., Cochrane, A. M. G., Thomson, A. D., Eddleston, A. L. W. F., and Williams, R. (1975). *Gut 16*, 835.
44. Nielsen, J. O., Dietrichson, O., Elling, P., and Christofferson, P. (1971). *N. Engl. J. Med. 285*, 1157.
45. Meyer zum Buschenfelde, K. H., and Hopf, U. (1974). *Brit. J. Exp. Pathol. 55*, 498.
46. Hopf, U., and Meyer zum Buschenfelde, K. H. (1974). *Brit. J. Exp. Pathol. 55*, 509.
47. Hopf, U., Meyer zum Buschenfelde, K. H., and Freudenberg, J. (1974). *Clin. exp. Immunol. 16*, 117.
48. Meyer zum Buschenfelde, K. H., and Miescher, P.A. (1972). *Clin. exp. Immunol. 10*, 89.
49. Hopf, U., Arnold, W., and Meyer zum Buschenfelde, K. H. (1975). *Clin. exp. Immunol. 22*, 1.
50. Wands, J. R., Perrotto, J. L., Alpert, E., and Isselbacher, K. J. (1975). *J. Clin. Invest. 55*, 921.
51. Vierling, J. M., Nelson, D. L., Strober, W., Bundy, B. M., and Jones, E. A. (1977). *J. Clin. Invest. 60*, 1116.

52. Mutchnik, M. G., Kawanishi, H., and Hopf, U. (1976). *Gastroenterology 70,* 989.
53. Wands, J. R., and Isselbacher, F. J. (1975). *Proc. Natl. Acad. Sci. 72,* 1301.
54. Paronetto, F., and Vernace, S. (1975). *Clin. exp. Immunol. 19,* 99.
55. Eddleston, A. L. W. F. (1975). In *Chronic Hepatitis* pp. 117-122. Karger, Basel.
56. Cochrane, A. M. B., Moussouros, A., Thomson, A. D., Eddleston, A. L. W. F., and Williams, R. (1976). *Lancet 1,* 441.
57. Thomson, A. D., Cochrane, A. M. G., McFarlane, I. G., Eddleston, A. L. W. F., and Williams, R. (1974). *Nature (Lond.) 252,* 721.
58. Geubel, A. P., Keller, R. H., Summerskill, W. H. J., Dickson, E., Tomasi, T. R., and Shorter, R. G. (1976). *Gastroenterology 71,* 450.
59. Smith, A. L., Cochrane, A. M. G., Mowat, A. P., Eddleston, A. L. W. F., and Williams, R., (1977). *J. Pediatrics 91,* 584.
60. Kakumu, S., and Leevy, C. M. (1977). *Gastroenterology 72,* 594.
61. Gershon, R. K. (1974). *Comtemp. Top. Immunobiol. 3,* 1.
62. Waldmann, T. A., and Broder, S. (1977). In *Progress in Clinical Immunology* (R. Schwartz, ed.), pp. 155-199. Grune and Stratton, New York.
63. Waldmann, T. A. (1977). *Ann. Intern. Med. 88,* 226.
64. Waldmann, T. A., Broder, S., Blaese, R. M., Durm, M., Blackman, M., and Strober, W. (1974). *Lancet 2,* 609.
65. Broder, S., Humphrey, R., Dunn, M., Blackman, M., Meade, B., Goldman, C., Strober, W., and Waldmann, T. A. (1975). *N. Engl. J. Med. 293,* 887.
66. Twomey, J. J., Laughter, A. H., Farrow, S., and Douglass, C. C. (1976). *J. Clin. Invest. 56,* 467.
67. Hodgson, H. J. F., Wands, J. R., and Isselbacher, K. J. (1977). *Gastroenterology 72,* 1070.
68. Laughter, A. H., and Twomey, J. J. (1977). *J. Clin. Invest. 119,* 173.
69. Occhino, J. C., Glasgow, A. H., Cooperband, S. R., Mannick, J. A., and Schmid, K. (1973). *J. Immunol. 110,* 685.
70. Girmann, G., Pees, H., and Scheurlen, P. G. (1975). *Nature 259,* 399.
71. Murgita, R. A., and Tomasi, T. B. (1975). *J. Exp. Med. 141,* 440.
72. Mortensen, R. J., Osmand, A. P., and Gewurz, H. (1975). *J. Exp. Med. 141,* 821.
73. Curtiss, L. K., and Edgington, T. S. (1976). *J. Immunol. 116,* 1452.

74. Chisari, F. V. (1977). *J. Immunol. 119*, 2129.
75. Morse, J. H., Witte, L. D., and Goodman, D. S. (1977). *J. Exp. Med. 146*, 1791.
76. Edgington, T. S., Henney, C. S., and Curtiss, L. K. (1976). In *Regulatory Mechanisms in Lymphocyte Activation* (D. O. Lucas, ed.), pp. 736-738. Academic Press, New York.
77. Seidel, D., Alaupovic, P., and Furman, R. H. (1969). *J. Clin. Invest. 48*, 1211.
78. Magnani, H. H. (1976). *Biochem. Biophys. Acat 450*, 390.
79. Blomhoff, J. P. (1976). *Scand. J. Gastroent. 11*, 753.
80. Sell, S., Skelly, H., Leffert, H. L., Muller-Eberhard, U., and Kida, S. (1975). *Ann. N. Y. Acad. Sci. 259*, 45.
81. Schumacher, K., Maerker-Alzer, G., and Wehmer, V. (1974). *Nature 251*, 655.
82. Chisari, F. V. (1978). *Fed Proc.* in press.
83. Wright, R. (1975). In *Clinical Aspects of Immunology* (P. G. H. Gell, R. R. A. Coombs, and P. J. Lachman, ed.), pp. 1269-1300. Blackwell Scientific Publications, London.
84. Mella, B., and Taswell, H. F. (1970). *Am. J. Clin. Path. 53*, 141.
85. Nossal, G. J. V., and Pike, B. L. (1974). In *Immunological Tolerance: Mechanisms and Potential Therapeutic Applications* (D. H. Katz, and B. Benacerraf, eds.), p. 351. Academic Press, New York.
86. Sidman, C. L., and Unanue, E.R. (1977). *J. Exp. Med. 144*, 882.
87. Uhr, J. W., and Moller, G. (1968). *Adv. Immunol. 8*, 81.
88. Frischknecht, H., Binz, H., and Wigzell, H. (1978). *J. Exp. Med. 147*, 500.

EVIDENCE FOR THE ASSOCIATION OF A CHRONIC DISEASE (LIVER CANCER) WITH PERSISTENT HEPATITIS VIRUS INFECTION

Joseph L. Melnick

Department of Virology and Epidemiology,
Baylor College of Medicine, Houston, Texas 77030

ABSTRACT. Patients with primary carcinoma of the liver (hepatoma) have a high prevalence of persistent infection with hepatitis B virus (HBV). This association occurs in areas of the world where hepatoma is common, as in parts of Asia and Africa, as well as in areas where hepatoma is an uncommon disease, as in the United States. Several possible explanations have been offered for this association of hepatitis B virus and liver cancer. One explanation is that patients with liver cancer or cirrhosis are particularly susceptible to hepatitis B infection or to becoming chronic carriers of the virus. Another is that the infection leads to cirrhosis, which is the actual precursor to the cancer. A final possibility is that the hepatitis B virus may transform the liver cells it invades, in other words, that it is a tumor-causing virus influenced, perhaps, by genetic, environmental, and other factors. If either of the last two interpretations is correct, the control of hepatitis B by vaccination should reduce the incidence of liver cancer.

INTRODUCTION

Primary liver cell carcinoma, or hepatoma as it will be called in this article, varies in the frequency of its occurrence in different parts of the world. It has a high prevalence in many parts of Africa and Asia and a low prevalence in Western Europe, the United States, and Australia (1-5). In some areas, hepatoma is one of the most frequent types of cancer seen in the male population. It may even be the most frequent -- as in Taiwan in young males. In many of these patients, the carcinoma occurs in livers also exhibiting cirrhosis, particularly of the macronodular type, but in many others there is no evidence of an underlying cirrhosis. Cirrhosis and hepatoma may arise from the same cause in some

ISBN 0-12-668350-6

instances, but in others they may arise from completely different causes.

With the recognition of the hepatitis B viral (HBV) antigens, serological tests have been conducted in many areas of the globe, and there has been a striking co-incidence between areas of high frequency of hepatoma and high frequency of HBV infection. Even in areas where hepatoma is not a common cancer, the prevalence of persistent HBV infection in hepatoma patients is more than 40 times higher than in the population at large. For example, 8 years ago when the first evidence was obtained of the striking association of HBV with hepatoma in the U.S. (6), it had already been known for some time that the prevalence of the disease in the U.S. was very low in contrast to a reported prevalence over one hundred times higher in some parts of Africa (7). Nevertheless, among hepatoma patients themselves, regardless of whether they develop their disease in the U.S. or elsewhere, there is a similar high frequency of persistent infection with HBV. It has been suggested by several investigators that following infection with HBV, there may be progression to chronic active hepatitis, to cirrhosis, and to hepatoma (5, 8, 9). What has recently become apparent is that a high proportion of hepatoma patients manifest persistent infection with HBV.

The regional differences in prevalence of hepatoma appear to be due to environmental rather than genetic or ethnic factors. Thus, prevalence in both blacks and whites in the U.S. is relatively low, in contrast to the high prevalence among blacks in Africa. Similarly, the prevalence is high in China, and in those who moved from China to Singapore, but is much lower among Chinese born and raised in Singapore, or in the U.S. Europeans who move to Africa or Southeast Asia continue to have a low prevalence even though the native population is at high risk. These observations suggest that early infection before the immune system has matured, perhaps at birth from a virus-carrier mother, may be more important than infections in later years, as far as causing hepatoma is concerned (4, 10).

Recent Epidemiologic Studies. The early studies on hepatoma patients gave varying results, in large measure because tests for antigenemia were often carried out by insensitive methods. With the development and use of highly sensitive methods, the picture has become clearer. As shown in Table 1, radioimmunoassay is a thousandfold more sensitive than the original agar gel diffusion method that Blumberg used when he discovered the antigen. Thus the reported prevalence of antigenemia is not only related to age, sex, and socioeconomic status, but also to the sensitivity of the

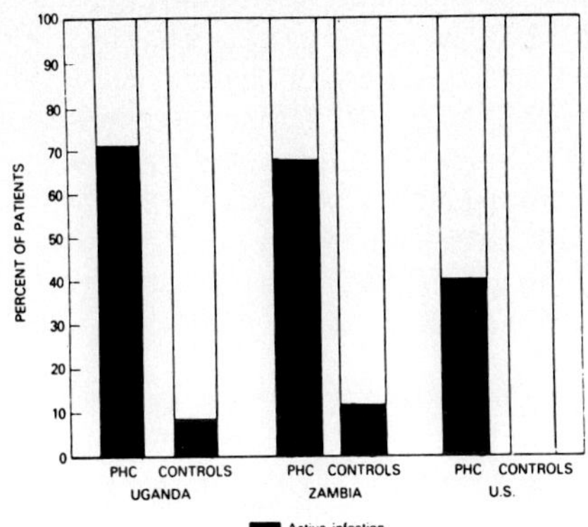

Figure 4. Evidence of active HBV infection in patients with primary hepatocellular carcinoma (PHC). Active infection is indicated by the presence of HBsAg (with or without anti-HBs) or of anti-HBc (without anti-HBs). U.S. controls represent the prevalence of active HBV infection in healthy U.S. adult populations, as reported in the literature (≤ 1%). From Tabor et al. (14).

TABLE 2

HBV INFECTIONS IN SERA OF HEPATOMA PATIENTS AND CONTROLS[a]

Area	Percent positive			
	Persistent active infection[b]		Persistent active[b] and past[c] infections combined	
	Hepatoma patients' sera	Control sera	Hepatoma patients' sera	Control sera
Uganda	72	8	94	76
Zambia	68	12	100	63
USA	41	0.1-0.5	74	4-25

[a] Data from Tabor et al. (14).

[b] Persistent active infection as indicated by positive test for HBsAg or for anti-HBc without anti-HBs.

[c] Past infection as indicated by positive test for anti-HBs (without HBsAg).

In the hepatoma patients, HBeAg and anti-HBe tests by agar gel diffusion were not helpful, in that they did not detect any cases of HBV infection other than those detected by other tests. Actually, the number of positive results for HBeAg or its antibody in hepatoma patients (5, 14, 17) is too low to show meaningful association. At this time, studies of HBeAg and anti-HBe involve the use of the relatively insensitive agar gel diffusion assay. However, with the development of more sensitive tests the picture may change.

Tabor et al. (14) point out that an etiologic role for HBV in the development of hepatoma is usually discussed in the context of altered immunologic surveillance or of chronic liver damage by HBV in concert with aflatoxin or other carcinogens. The possible interplay of HBV and host immune responses in the development of hepatoma, and the possible role of T-lymphocytes in the clearance of HBsAg, have led to the suggestion that a patient who cannot respond adequately to HBV might also be tolerant of altered liver cell antigens during malignant transformation as a result of HBV infection. Persistent HBV infection can result in chronic hepatitis and cirrhosis, and some studies have shown an association between cirrhosis (18-20) or liver cell dysplasia (21) and hepatoma. Potential carcinogens that have been associated with hepatoma include aflatoxin, which is present in the staple diet in Uganda and elsewhere (22) and which has been shown to be carcinogenic for the livers of experimental animals, and several alkaloids found in medicinal plants used in East Africa during childhood and pregnancy.

Cocarcinogenesis Studies. A number of chemicals are carcinogenic, and indeed the nitrosamines induce primary liver cancer in monkeys (23-25). Aflatoxins, which cause hepatomas in several species, occur in many of the same regions of the globe that have a high prevalence of hepatoma. However, as Higginson (2) points out, if man has a susceptibility similar to that of animals the human carcinogenic dose would have to be several kilograms per day. Nonetheless, lower doses of chemical carcinogens may act in concert with a virus to induce the cancer. We have attempted to obtain an answer in the laboratory (25), asking whether HBV could act as a cocarcinogen and enhance the effect of diethylnitrosamine in macaque monkeys. Because the appearance of nitrosamine-induced tumors in monkeys was known to require a number of years, the experiment was set up to determine whether the virus would materially shorten the induction time or influence the type of tumor induced. The results failed to demonstrate, over a 3-year period, a carcinogenic potential for HBV. However, HBV given to juvenile monkeys before nitrosamine treatment resulted in subsequent gross and microscopic alterations consistent with mild chronic hepatitis and postnecrotic cirrhosis. Multifocal liver carcinoma apparently developed within these cirrhotic nodules.

Hepatoma as a World Problem. There are large differences in the prevalence of antigenemia throughout the world. Szmuness (5) has compiled the global data based on a large number of studies and has estimated that in 1970, when the world population was 3.5 billion people, there were about 176 million carriers (see Table 3).

Prevalence rates as high as 10% are found in Central and South Africa and Southeast Asia. The second highest rates, between 2 and 5%, exist in other regions of Asia, in North Africa, and in the Middle East. Rates of 1-2% are present in South America, South and East Europe, and European Russia. The lowest rates, of 0.2 to 0.5%, occur in West Europe, North America, Australia, and New Zealand.

It is striking that the prevalence of hepatoma follows the same geographic pattern of distribution as that of persistent HBV infection.

TABLE 3

AN ESTIMATE OF HBsAg PREVALENCES AND NUMBERS OF PERSISTENT CARRIERS IN THE WORLD[a]

Geographic region	Population: 1970 est. (millions)	Average prevalence (%)	Number of carriers (thousands)
USA and Canada	275	0.25	688
Central and South America	232	1.5	3,481
North and West Europe	232	0.25	580
South and East Europe	266	1.5	3,990
USSR -- European	100	1.5	1,500
USSR -- Asian	143	5.0	7,150
Middle East	70	5.0	3,510
South Asia	691	5.0	34,580
Southeast Asia	300	10.0	30,020
China	800	7.5	60,000
Japan	103	1.5	1,551
Australia and New Zealand	19	0.25	47
North Africa	70	5.0	3,505
Central and South Africa	256	10.0	25,680
World total	3,560	--	176,284

[a] From Szmuness (5).

Antigenic Subtypes. The particles containing HBsAg are antigenically complex. Each contains a group-specific antigen, a, in addition to two pairs of mutually exclusive subdeterminants, d/y and w/r. Thus, 4 phenotypes of HBsAg have been observed: adw, ayw, adr, and ayr. Even though there is a high correlation between the geographic distribution of persistent HBV infection and that of hepatoma, the correlation does not extend to subtypes. In areas of high hepatoma prevalence, almost all of the different subtypes may be present, but with one predominating. For example, adw and ayw predominate in different parts of Africa, adr in Southeast Asia, and ayw in South and East Europe and the Middle East (Fig. 5).

The situation is noteworthy in Taiwan, where the antigenemia prevalence is 82% in hepatoma patients, in contrast to 13% in the population at large. In a recent study (27), the predominant subtype among carriers who were born in Taiwan was found to be adw (91%). Among Chinese antigen-carriers born on the mainland south of the Yangtze River who had moved to Taiwan, the same subtype, adw, was predominant (76%). However, for Chinese carriers born north of the Yangtze River and now living in Taiwan, a different subtype, adr, predominated (78%).

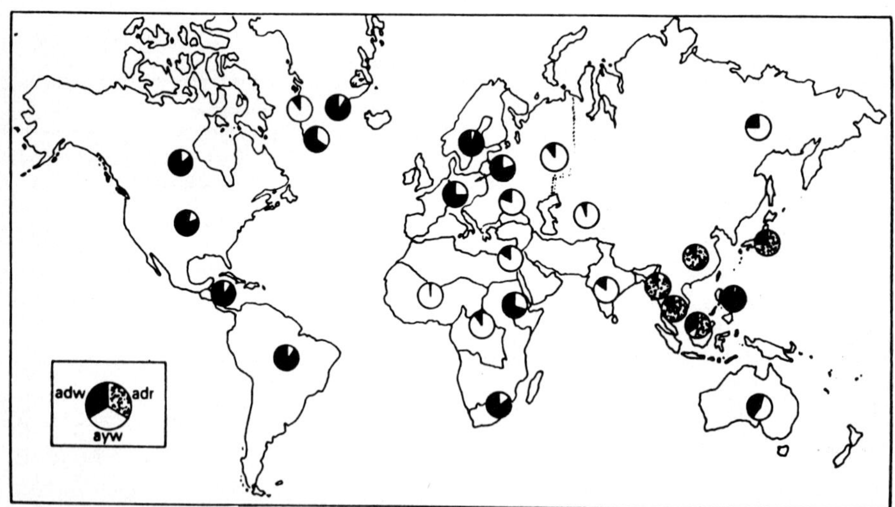

Figure 5. Relative prevalence of major HBsAg subtypes (adw, ayw, adr) in the world. The ayr subtype is very rare. From Advances in Viral Hepatitis (26).

Age Factor. Although there are millions of HBV carriers in the world, hepatoma occurs in very few. This situation is not unusual in viral infections; wild poliovirus produces paralysis in 1% of those infected, and some strains are known in which only a single paralytic case occurs for over a million persons exposed. What risk factors other than HBV are associated with development of hepatoma is not known, but age of infection and perhaps the accompanying dose of virus are suggestive.

Blumberg's group (28) has reported a study from Senegal in which the mothers of hepatoma patients were four times more apt to have a persistent HBV infection than the fathers. In contrast, the mothers and fathers of matched controls showed no differences, and had the same relatively low rates as the fathers of the hepatoma patients (see Table 4). When infection occurs at birth from a viremic mother, the exposure of the infant to the mother's infectious blood must be high indeed.

The above data suggest that mothers with persistent HBV infections are more likely to produce offspring at high risk of developing hepatomas. This observation is in keeping with the vertical transmission of HBV from mothers to their infants that commonly occurs in geographic areas of high hepatoma prevalence (29-31).

As Szmuness (5) has pointed out, "Should these observations concerning the role of vertical transmission be confirmed, then interruption of [neonatal] transmission, postponing the time of primary infection, and shifting transmission from vertical to horizontal by means of passive or active immunization programs, would be the most rational approach to the control of cirrhosis and hepatocellular carcinoma."

"Assuming that a substantial proportion of hepatoma patients actually acquired HBV neonatally, the age-distribution of patients with hepatoma would reflect the distribution of incubation periods and would indicate a mean of 35

TABLE 4
PERSISTENT HBV INFECTION AMONG PARENTS OF HEPATOMA PATIENTS IN SENEGAL[a]

	HBsAg: Per cent positive	
	Mothers	Fathers
Hepatoma patients	71.4	18.5
Matched controls	14.3	18.5

[a] Data from Larouze, Blumberg et al. (28).

years. By the same token, since hepatoma patients in Western countries are on the average older than patients in high endemic areas, we may speculate that in Western countries fewer hepatoma patients acquire HBV during birth or childhood."

Virus Markers in Hepatoma Cells. HBsAg may be found in the cytoplasm of hepatocytes in liver tissue of hepatoma patients (32, 33). In addition, a human hepatoma cell line has been established which produces HBsAg of an *ad* subtype. The antigen can be found as typical 20-nm spherical particles in the cell culture fluids (34). The antigenic activity was blocked by specific antibody. The frequency with which hepatitis B markers can be detected in hepatoma cells remains to be determined.

Under investigation is the search for DNA in hepatoma cells that can hybridize with labeled HBV-DNA probes. Even though the virus has not yet been grown in cell culture, a specific probe can be made (35). HBV-DNA is double-stranded and circular, with one full circle containing 3200 nucleotides and with a shorter open strand. The open strand serves as the primer for the viral DNA polymerase. Under proper conditions, the *in vitro* reaction can produce labeled DNA containing the full 3200 nucleotide pairs which can then be used to detect viral DNA by reassociation kinetics. The original viral DNA and viral polymerase for making the labeled probe can be purified from viral products in the blood of carriers or from liver tissue of hepatitis B patients.

SUMMARY

Patients with primary carcinoma of the liver (hepatoma) have a high prevalence of persistent infection with hepatitis B virus (HBV).

Three interpretations have been offered for this association of HBV infection and liver cancer.

(1) HBV infects patients with liver cancer or with cirrhosis (a precursor condition), who have a high susceptibility to infection and the development of the chronic carrier state. However, in areas of high hepatoma prevalence, HBV carrier infections occur most frequently in childhood, and it seems clear that the virus carrier state occurs before the tumor and not after.

(2) The virus carrier state is a cause of cirrhosis, and the hepatoma arises from regenerative nodules by mechanisms in which HBV is not involved. This view is supported by

finding an increase of hepatoma among alcoholic cirrhosis patients. However, hepatoma is more apt to develop in cirrhosis patients who are persistent carriers of HBV than in such patients who are not infected by the virus. Furthermore, many cases -- perhaps the majority -- of hepatoma associated with HBV develop in persons without cirrhosis.

(3) HBV is an oncogenic virus transforming the liver cells it invades. Other cocarcinogenic influences may be necessary for the induction of the cancer; these might be genetic, hormonal, immunologic, or environmental.

Since persons living in areas where there is a high prevalence of both HBV and hepatoma are at high risk, they are being studied for the evaluation of HBV vaccines. If these trials are successful, a viral vaccine would become available which may decrease not only the incidence of viral hepatitis but also that of primary carcinoma of the liver.

REFERENCES

1. Steiner, P. E. (1960). Cancer 13, 1085.
2. Higginson, J. (1970). In "Tumours of the Liver" (G. T. Pack and A. H. Islami, eds.), p. 38. Springer-Verlag, Berlin.
3. Williams, A. O. (1975). Am. J. Med. Sci. 270, 53.
4. Zuckerman, A. J. (1975). "Human Viral Hepatitis." North Holland, Amsterdam.
5. Szmuness, W. (1978). Prog. in Med. Virol. 24, in press.
6. Hersh, T., Hollinger, F. B., Goyal, R. K., Grubb, M.N., and Melnick, J. L. (1971). Internatl. J. Cancer 8, 259.
7. Higginson, J. (1963). Cancer Res. 23, 1625.
8. Blumberg, B. S., Larouze, B., London, W. T., Werner, B., Hesser, J. E., Millman, I., Saimot, G., and Payet, M. (1975). Am. J. Pathol. 81, 669.
9. Prince, A. M., Szmuness, W., Michon, J., Demaille, J., Diebolt, G., Linhard, J., Quenum, C., and Sankale, M. (1975). Internatl. J. Cancer 16, 376.
10. Ohbayashi, A., Okochi, K., and Mayumi, M. (1972). Gastroenterology 62, 618.
11. Primack, A., Vogel, C. L., and Barker, L. F. (1973). Brit. Med. J. i, 16.
12. Szmuness, W., Prince, A. M., Diebolt, G., Leblanc, L., Baylet, R., Massayeff, R., and Linhard, J. (1973). Am. J. Epidemiol. 98, 104.
13. Nishioka, K., Levin, A. G., and Simons, M. J. (1975). Bull. World Hlth. Organiz. 52, 293.

14. Tabor, E., Gerety, R. J., Vogel, C. L., Bayley, A. C., Anthony, P. P., Chan, C. H., and Barker, L. F. (1977). J. Natl. Cancer Inst. 58, 1197.
15. Proceedings of a Symposium on Viral Hepatitis, National Academy of Sciences. (1975). Am. J. Med. Sci. 270.
16. Melnick, J. L., Dreesman, G. R., and Hollinger, F. B. (1977). Scientific Amer. 237, 44.
17. Maupas, P., Werner, B., Larouze, B., Millman, I., London, W. T., O'Connell, A., and Blumberg, B. (1975). Lancet ii, 9.
18. Vogel, C. L., Anthony, P. P., Mody, N., and Barker, L. F. (1970). Lancet ii, 621.
19. Dudley, F. J., Scheuer, P. J., and Sherlock, S. (1972). Lancet ii, 1388.
20. Chainuvati, T., and Viranuvatti, V. (1975). Gastroenterology 68, 1261.
21. Anthony, P. P., Vogel, C. L., and Barker, L. F. (1973). J. Clin. Pathol. 26, 217.
22. Alpert, M. E., Wogan, G., and Davidson, C. S. (1968). Gastroenterology 54, 149.
23. Kelly, M. G., O'Gara, R. W., Adamson, R. H., Gadekar, K., Botkin, C. C., Reese, W. H., Jr., and Kerber, W. T. (1966). J. Natl. Cancer Inst. 36, 323.
24. Ruebner, R. H., Michas, C., Kanayama, R., and Bannasch, P. (1976). J. Natl. Cancer Inst. 57, 1261.
25. Gyorkey, F., Melnick, J. L., Mirkovic, R., Cabral, G. A., Gyorkey, P., and Hollinger, F. B. (1977). J. Natl. Cancer Inst. 59, 1451.
26. WHO Expert Committee on Viral Hepatitis: Advances in viral hepatitis (1977). WHO Tech. Rept. Series, No. 602.
27. Chen, D-S.,and Sung, J-L. (1977). New Engl. J. Med. 297, 668.
28. Larouze, B., London, W. T., Saimot, G., Werner, B. G., Lustbader, E. D., Payet, M., and Blumberg, B. S. (1976). Lancet ii, 534.
29. Schweitzer, I. L. (1975). Prog. in Med. Virol. 20, 27.
30. Stevens, C. E., Beasley, R. P., Tsui, J., and Lee, W. C. (1975). New Engl. J. Med. 292, 771.
31. Beasley, R. P., Trepo, C., Stevens, C. E., and Szmuness, W. (1977). Am. J. Epidemiol. 105, 94.
32. Peters, R. L., Afroudakis, A. P., and Tatter, D. (1976). Am. J. Clin. Pathol. 66, 462.
33. Popper, H. (1977). Personal communication.
34. MacNab, G. M., Alexander, J. J., Lecatsas, G., Bey, E. M., and Urbanowicz, J. M. (1976). Brit. J. Cancer 34, 509.
35. Robinson, W. S., and Lutwick, L. I. (1976). New Engl. J. Med. 295, 1168 and 1232.

DETECTION OF A NOVEL ANTIGEN IN TWO CASES OF NON-A NON-B HEPATITIS, A RECENTLY RECOGNIZED PERSISTENT INFECTION OF PROBABLE VIRAL ORIGIN

Robert H. Purcell[1], Stephen M. Feinstone[1], Harvey J. Alter[2], and Doris C. Wong[1]

National Institutes of Health, National Institute of Allergy and Infectious Diseases, Laboratory of Infectious Diseases, Bethesda, Maryland 20014

ABSTRACT Non-A non-B hepatitis has emerged as an important cause of hepatitis in the United States. Diagnosis is by exclusion, since an antigen-antibody system specific for this disease has not been discovered. The disease may have multiple etiologies. We report the detection and partial characterization of an antigen (or antigens) detected in two cases of non-A non-B hepatitis. At least one of the antigens has characteristics consistent with its being a virus or virus-like entity.

INTRODUCTION

Until recently only two types of human viral hepatitis were recognized: Type A (infectious) hepatitis and type B (serum) hepatitis. Although so designated in the 1940's on the basis of results of epidemiologic and volunteer studies, it was not until the development of sensitive assays for antigens and antibodies associated with these two types of hepatitis that their true nature was delineated. The diagnosis of hepatitis B virus (HBV) infection can now be made with considerable certainty by testing for the appropriate viral antigens or antibodies. Among the markers available for detection of HBV infection are hepatitis B surface antigen (HBsAg) and antibody (anti-HBs), hepatitis B core antigen (HBcAg) and antibody, (anti-HBc), hepatitis B e antigen (HBeAg) and antibody (anti-HBe) and in addition, a virus-specific DNA-dependent DNA polymerase activity (1, 2).

[1]Present address: National Institutes of Health, Building 7, Room 202, Bethesda, Maryland 20014.

[2]Present address: National Institutes of Health, Clinical Center Blood Bank, Bethesda, Maryland 20014.

ISBN 0-12-668350-6

Hepatitis A virus (HAV) infection can be diagnosed by detecting hepatitis A antigen (HAAg) or antibody, anti-HA (3). Radioimmunoassays (RIA) and comparably sensitive techniques have been developed for all of these markers of hepatitis virus infection.

When these sensitive diagnostic tools were applied to the analysis of posttransfusion hepatitis, it was found that over half of the cases could not be ascribed to HAV or HBV (4,5). Nor could they be diagnosed as being due to cytomegalovirus (CMV) infection or infection with the Epstein-Barr virus (EBV). Thus, a "new" type of viral hepatitis was discovered. Subsequent epidemiologic studies suggest that more than one agent may be etiologically responsible for this "non-A non-B" hepatitis (6). At present, 80-90% of transfusion-associated hepatitis appears to be caused by this agent or agents. Limited epidemiologic studies also indicate that perhaps as much as 25% of acute clinical hepatitis in the United States may be type non-A non-B (7). Because the diagnosis at present is one of exclusion and because more than one agent may exist, the term non-A non-B has been applied to this disease complex.

Epidemiologically, non-A non-B hepatitis differs in several important respects from type B hepatitis which it most closely resembles (5). Its mean incubation period is somewhat shorter than that of type B disease (7-8 weeks vs. 14-16 weeks) following exposure via blood transfusion. In addition, clinical illness associated with non-A non-B hepatitis tends to be milder, with lower mean maximum liver enzyme abnormalities and a smaller proportion of patients with jaundice than is seen in type B hepatitis. However, the long-term prognosis is probably worse than that for type B hepatitis: whereas approximately 5-10% of patients with clinical type B hepatitis progress to a chronic infection, 30-50% of patients with non-A non-B hepatitis progress to chronicity. This may be an underestimate, since only patients with associated chronic hepatitis can be identified at present, and patients with an inapparent carrier state, analogous to that seen in a proportion of patients with chronic hepatitis B virus infection, may exist for type non-A non-B hepatitis. Thus, non-A non-B hepatitis viruses account for a significant proportion of chronic hepatitis in the United States.

Considerable effort is being expended in a search for markers specific for non-A non-B hepatitis viruses. Approaches include those that have been most useful in the study of viral hepatitis types A and B: immune electronmicroscopy (IEM), RIA and transmission to non-human primate species. Until recently, progress with these techniques has been modest. However, non-A non-B hepatitis has recently been successfully transmitted to chimpanzees (8, 9). This animal model system should prove invaluable for detailed characterization of the syndrome and attempts to identify viral antigens or antibodies.

Radioimmunoassay procedures are being applied to a search for antigens in the plasma of patients with non-A non-B hepatitis. Here, too, progress has been modest, but an antigen (or antigens) has been detected in the serum of two patients who developed non-A non-B hepatitis following blood transfusion. We report herein a preliminary characterization of the antigen and its antibody.

METHODS

Patients. Patients over 18 years of age undergoing open-heart surgery at the Clinical Center, National Institutes of Health, Bethesda, Maryland were followed prospectively for evidence of hepatitis. Detailed descriptions of this prospective study, which began in 1964, have been published previously (10-14). Briefly, serum samples were obtained prior to transfusion and at weekly to monthly intervals for six months or longer and in some cases at yearly intervals thereafter from patients entered into the study. These serum samples were tested for biochemical evidence of hepatitis (alanine aminotransferase: ALT; aspartate amino transferase: AST), HBsAg by RIA (15), anti-HBs by passive hemagglutination (16) and/or RIA (Ausab, Abbott Laboratories), anti-HBc by immune adherence hemagglutination (IAHA, 17) and/or RIA (18), anti-HA by IAHA (19) and/or RIA (20), antibody to cytomegalovirus (CMV) by complement fixation (21) and antibody to Epstein-Barr virus (EBV) by immunofluorescence (22).

Liver Tissue. Normal liver tissue was obtained at autopsy from a patient who died from causes unrelated to liver disease. Liver tissue from a fatal case of fulminant hepatitis of presumed non-A non-B etiology was obtained from Dr. Ruth M. Dalton. Both samples of liver were stored frozen at -70°C.

Radioimmunoassays. Micro-solid phase radioimmunoassays were performed essentially as described previously for other antigen/antibody systems (18, 20, 23). Briefly, IgG was purified from convalescent sera of non-A non-B hepatitis patients (16 patients) or from sera obtained from patients who had been transfused many times for various conditions including hemophilia, thalassemia and malignancy (eight patients). The purified IgG was labeled with ^{125}iodine by the chloramine T procedure (24). Wells of polyvinyl microtiter plates were coated with an appropriate dilution of convalescent serum or serum from a multiply transfused patient, and excess binding sites on the plastic were quenched with 1% bovine serum albumin (BSA). After washing, dilutions of acute phase serum samples from patients with non-A non-B hepatitis were added to the wells. After incubation, the wells were washed, and convalescent IgG labeled with ^{125}I was added. After additional incubation and washing, the wells of the microtiter plates were cut apart and individually counted in a gamma spectrometer. The ratio of counts bound by the test sample (positive, P) to the counts bound by the patient's own pre-transfusion sera (negative, N) was calculated. Sera with a P/N value greater than 2.1 were considered positive for a putative antigen, based upon the convention for positivity applied to previously described solid-phase radioimmunoassays performed in a similar manner. Antibody was detected by its ability to competitively inhibit ^{125}I-labeled IgG in a test containing a known positive antigen (18, 20).

Biophysical Studies. Antigen-containing serum was subjected to ultracentrifugal sedimentation, isopycnic banding in cesium chloride and rate-zonal separation as described previously (25) in a Beckman L5-50 ultracentrifuge. Fractions were collected from the bottom of the tube and analyzed for density by refractometry and for antigen by RIA. One gram of liver was thawed and homogenized with a mortar, pestle and sea sand in 5 ml 0.2M Tris NaCl, pH 7.2 and 0.5% BSA. This was clarified by centrifuging at 10,000 rpm for 60 min. in a 40.2 rotor. Four ml of clarified liver homogenate was layered on a 1.1-1.6 gm/cm^3 discontinuous cesium chloride gradient and centrifuged in a SW 40 rotor for 22 hours at 37,000 rpm. Fractions were collected from the bottom of the tube and assayed for antigen by RIA.

RESULTS

Search for Non-A Non-B Antigens. Serial acute phase serum samples from 37 cases of non-A non-B hepatitis were tested, in various combinations, against ^{125}I-labeled IgG from 16 non-A non-B patients and eight multiply-transfused patients. However, all acute phase sera were not tested against all convalescent IgG's.

Acute phase sera from only two non-A non-B patients were reproducibly positive when tested against the ^{125}I-labeled IgG probes. Two of the probes prepared from convalescent non-A non-B hepatitis sera [patient B, who also developed type B hepatitis approximately 12 weeks after onset of non-A non-B hepatitis (14), and patient C, who had only non-A non-B disease] reacted well with the acute phase sera, and one probe from a multiply-transfused patient reacted weakly. No other antigen-antibody combination yielded a reproducibly positive result. All serial sera from the two antigen-positive patients (patients En and St) were tested by RIA against one of the reactive ^{125}I-labeled IgG preparations (from patient B).

As seen in Figure 1, both patients developed transfusion-associated hepatitis in 1969. The incubation period to first transferase elevation in patient En was approximately 12 weeks. The disease was anicteric but the patient had chronically elevated ALT levels for well over a year. The ALT activity demonstrated a cyclic pattern characteristic of some cases of non-A non-B hepatitis. By 1974 it had returned to normal. This patient lacked all markers of HBV infection and remained seronegative for anti-HA. She had antibody to

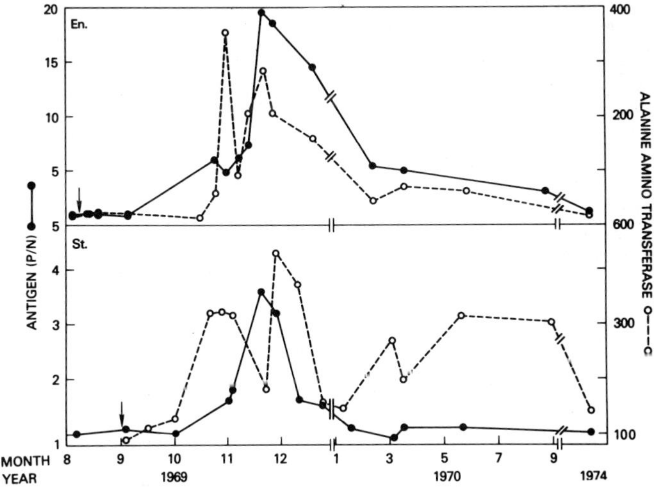

FIGURE 1. Pattern of alanine aminotransferase levels and new antigen(s) in two cases of transfusion-associated non-A non-B hepatitis. The arrows indicate time of transfusion.

CMV and EBV prior to transfusion. Radioimmunoassays for a putative non-A non-B hepatitis antigen were negative prior to transfusion and during the incubation period but became positive at approximately the same time that ALT levels became elevated. The concentration of antigen rose to relatively high levels (P/N of approximately 20) and roughly paralleled the levels of ALT activity. Antigen was still detectable at low levels one year after surgery but had disappeared when retested four years later.

Patient St, transfused within one month of patient En, had a similar clinical course. The incubation period in this patient was approximately seven weeks to first ALT elevation, and the patient's course was marked by cyclic elevations of ALT and progression to chronicity as evidenced by ALT elevations a year after transfusion. However, as with patient En, ALT had returned to normal four years later. An antigen was first detected in the serum of this patient after ALT had become elevated. The antigen remained detectable, although at relatively low levels (maximum P/N = 3.6) during the time of maximum ALT elevation but became undetectable within four months of surgery and remained so despite continued elevation of ALT activity.

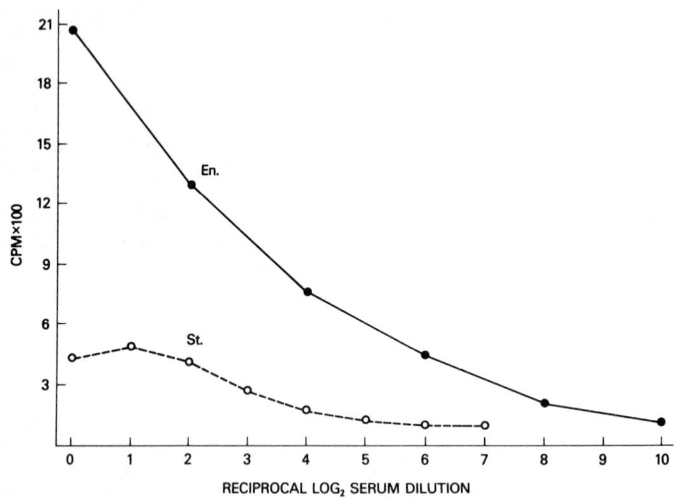

FIGURE 2. RIA titration curves obtained with antigens from patients En and St depicted in Figure 1.

<u>Titration of Antigens From Patients En and St.</u> Serial two-fold or four-fold dilutions of acute phase serum samples containing peak antigen activity from patients En and St were prepared and tested by RIA against the convalescent IgG of patient B. As seen in Figure 2, the results described typical RIA titration curves. However, a distinctive difference in the shapes of the two curves was noted. Antigen in undiluted serum from patient En was present in insufficient concentration to saturate the radiolabeled antibody, and the P/N value so obtained was in the linear portion of the RIA curve. In contrast, the antigen from patient St appeared to saturate the available labeled antibody despite the fact that this antigen was present at lower titer. These results suggest that the two antigens are not identical and, indeed, may not be antigenically related. However, analysis of antibody patterns (see below) suggests a relationship. Because of the extremely small quantities of antigen-positive serum available and because of the low titer of the antigen from patient St further characterization was carried out with samples of antigen from patient En only.

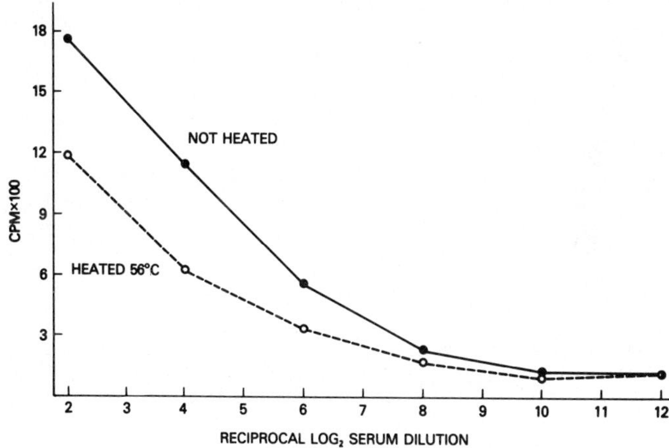

FIGURE 3. RIA titration curve of antigen from patient EN before and after heating at 56°C for 30 minutes.

Two-Tenths ml of antigen-positive serum (diluted 1:4 in saline) from patient En was heated at 56°C for 30 minutes in a water bath. Serial four-fold dilutions of the heated serum and an unheated sample of the same serum dilution were prepared and tested by RIA against radiolabeled IgG from patient C. As seen in Figure 3, approximately 75% of the antigen activity was destroyed by heating. The RIA curve of the heated antigen paralleled the curve of the unheated sample but was simply displaced (compare with antigen from patient Sn, Figure 2).

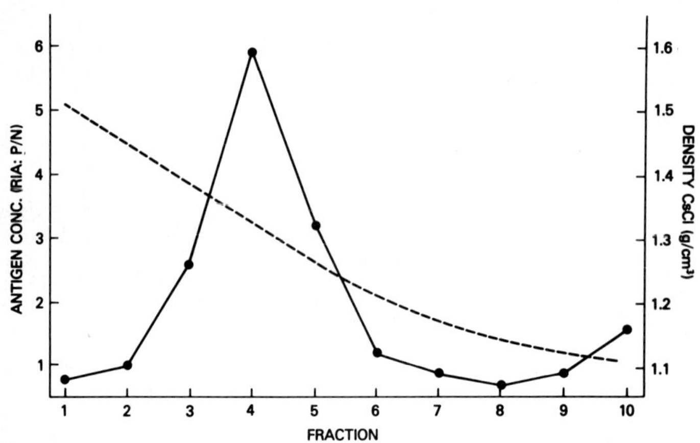

FIGURE 4. Isopynic banding of En antigen in a cesium chloride gradient. Antigen-positive serum from patient En was clarified by low speed centrifugation, diluted in 0.5 M Tris buffer, pH 7.4 (TB) and centrifuged through 25% (w/w) sucrose in TB at 47,000 rpm for 14 hours at 5°C in a 50.1 Beckman rotor. The pellet, which contained most of the RIA reactivity, was resuspended in 1/2 the original serum volume and layered on a preformed CsCl stepwise gradient (1.1 to 1.5 gm/cm^3). After centrifugation at 47,000 rpm for 20 hours at 5°C in a 50.1 Beckman rotor, fractions were collected from the bottom of the tube and assayed by RIA.

Biophysical Characterization. Antigen from patient En, either in serum or pelleted from serum as described below, banded in CsCl at a buoyant density of 1.33 to 1.35 gm/cm^3 (Figure 4). En antigen which had been banded in CsCl was layered over four ml of 25% sucrose and centrifuged in a SW 40 rotor at 39,000 rpm for five hours. The tube was unloaded from the top and fractions were tested by RIA for En antigen.

Virtually all of the antigen detectable by RIA was present in the bottom two fractions of the tube, most of it associated with the pellet. Rate-zonal sedimentation of En antigen (following pelleting or directly from serum) resulted in a broad band of reactivity suggesting heterogeneity of size, possibly due to aggregation. Thus, En antigen had biophysical characteristics of a particulate substance, similar to certain viruses.

Lack of Relationship of En and St Antigens to Liver Components. The CsCl gradient of homogenate of normal liver was completely non-reactive when tested in RIA against radiolabeled IgG from patient B. A low level of reactivity (P/N values of 1.7 and 1.9) was detected in two adjacent fractions from the CsCl gradient of the homogenate of liver from a patient with fulminant hepatitis. However, the densities of the two immunoreactive fractions were 1.125 and 1.151 gm/cm^3, markedly different then the buoyant density of En antigen in serum. The significance, if any, of the immunoreactivity of low density materials from this liver remains to be determined. Nevertheless, En antigen did not appear to be a normal or abnormal liver component.

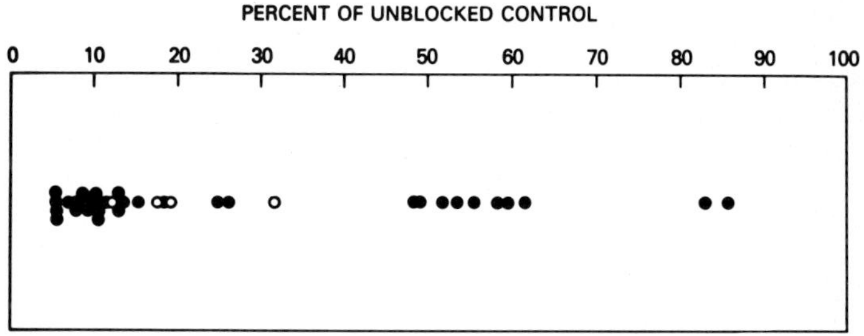

FIGURE 5. Percent radioactive counts bound after blocking with serum samples from patients with non-A non-B hepatitis (closed circles) and from chimpanzees (open circles), compared to a saline control, in an RIA blocking test for antibody to En antigen.

Antibody to "New" Antigen(s). Convalescent sera from 32 patients with non-A non-B hepatitis were tested at a 1:10 dilution for RIA-blocking antibody as measured by percent of RIA counts bound compared to "blocking" of binding by a saline control. As seen in Figure 5, the results obtained with the sera clustered into three groups. The majority (22/32) of the sera markedly inhibited binding of the radiolabeled IgG to the standard antigen. Two sera showed little of no inhibition and eight sera fell into an intermediate zone. Since one each of the pre-exposure sera from the antigen-positive patients En and St fell into the non-inhibitory and the intermediate-inhibitory clusters, we designated sera inhibiting RIA binding by 60% of more (i.e., yielding RIA counts of 40% or less of the saline control) as positive for antibody and those demonstrating less inhibition as negative. Thus, 10 of the 32 pre-exposure sera from non-A non-B hepatitis patients lacked significant antibody to the antigen and 22 were positive for antibody. Serum samples from four chimpanzees were tested in a similar manner (open-circles in Figure 5) and all were positive for antibody.

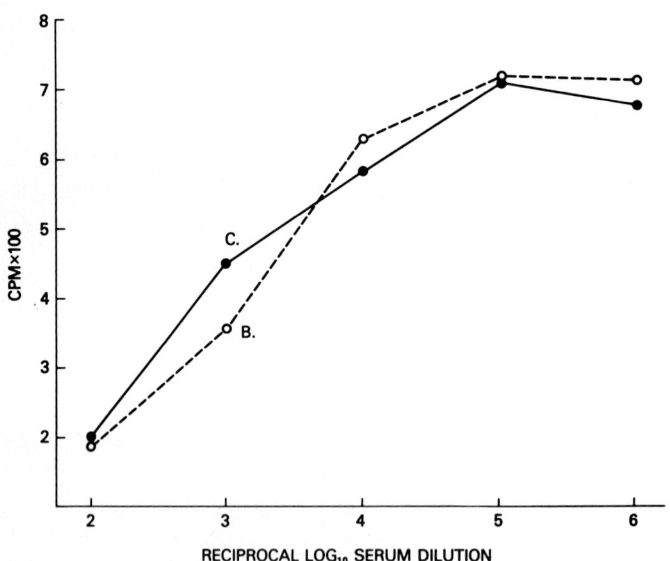

FIGURE 6. Titration of antibody to En antigen in serum from patients B and C, as measured by RIA blocking.

Using the RIA blocking assay, with antigen from patient En and ^{125}I-labeled IgG from patient B, we titered convalescent serum samples from patients B and C in 10-fold dilutions for RIA-blocking antibody to the antigen. As seen in Figure 6, the two sera described similar antibody titration curves. Using the criteria for positivity described above, both sera had RIA blocking titers of approximately 1:1000.

To determine if the antigens from patients En and St could be distinguished serologically, sera from 11 patients with various types of hepatitis were tested for antibody against both antigen from patient En and antigen from patient St. With one exception, the patterns of antibodies to the two antigens were identical (Table I). The one exception was a serum weakly positive for antibody to antigen En but negative for antibody to antigen St. However, this difference may be artifactual, since both sera showed some degree of inhibition. Thus, the two antigens could not be readily distinguished on the basis of patterns of antibody to them in a small sampling of hepatitis patients.

TABLE I

SEROLOGIC COMPARISON OF EN AND ST ANTIGENS: NUMBER OF SERA WITH INDICATED ANTIBODY PATTERN

Anti-En+ Anti-St+	Anti-En- Anti-St-	Anti-En+ Anti-St-	Anti-En- Anti-St+
7	3	1[a]	0

[a]Weakly positive for anti-En.

Additional evidence for the uniqueness of the En antigen was sought by comparing the pre-infection/convalescent status of antibody to selected hepatitis antigens in patients with type A, type B or non-A non-B hepatitis. As seen in Table II, there was no evidence of a serologic relationship between the En antigen and hepatitis A antigen, hepatitis B surface antigen, CMV antigen or EBV antigen. Of particular note is patient K, a hepatitis A patient, who seroconverted for anti-HA but lost antibody to the En antigen during the relatively long interval between collection of the preinfection and convalescent serum samples. Surprisingly, neither of the antigen-positive patients, patients En and St, developed antibody to the En antigen, possibly because both developed prolonged liver disease. Even more surprising was the finding that antibody-positive patients B and C both had high levels of antibody prior to transfusion and thus prior to their non-A non-B hepatitis. Thus, in both cases, their antibody was the result of a prior exposure to the antigen, and their prospectively followed non-A non-B hepatitis must have been unrelated to the En or St antigens. Indeed, patient B had serologic evidence of recent infection with CMV, making the etiology of his non-A non-B illness uncertain.

TABLE II

PATTERNS OF ANTIBODY TO SELECTED HEPATITIS-ASSOCIATED ANTIGENS IN HEPATITIS PATIENTS: EVIDENCE OF LACK OF SEROLOGICAL RELATEDNESS

Patient	Type of hepatitis	Preinfection/convalescent status of antibody to indicated antigen				
		HAAg	HBsAg	EnAg	CMV	EBV
K.	A	-/+	-/-	+/-	N.T.	N.T.
W.	B	-/-	-/+	+/+	N.T.	N.T.
B.[a]	non A or B, B	+/+	-/+	+/+	-/+	+/+
C.[a]	non-A non-B	-/-	-/-	+/+	+/+	+/+
J.	" "	-/-	-/-	+/+	-/-	+/+
K.	" "	-/-	+/+	-/-	+/+	+/+
En.[b]	" "	-/-	-/-	-/-	+/+	+/+
St.[b]	" "	+/+	+/+	-/-	-/-	+/+

[a]Sources of antibody to "new" antigen(s).
[b]Sources of "new" antigen(s).

experimental animals as well as direct invasion of meninges from infection of the submucosal nasal tissues (4).

In most spontaneous and experimental infections virus invasion of the nervous system occurs from the blood. Unlike other endothelial cells, the cerebrovascular endothelial cells form tight junctions and lack pores. In addition, these cells are surrounded by a dense basement membrane, and this membrane is tightly apposed to the endfeet of astroglial cells. Within the brain there is little extracellular space. The usual gap between plasma membranes measures only 200A^{o}, which would appear to be inadequate for free movement of most viral particles. Despite this unique structure of vasculature and tightly-packed cellular components, viruses can penetrate into the CNS from the blood with or without infection of vascular endothelial cells. Alternatively, some viruses can pass through or replicate in the choroid plexus seeding virus into the cerebrospinal fluid (5).

Once within the CNS the environment is optimal for the sequestration of viruses. The brain contains no lymphatic system. Thus, macrophages and other leukocytes do not normally monitor the tissue spaces, and the clearance of virus particles or recognition of foreign antigens is impeded. Furthermore, neural cells are the most specialized in the body with plasma membranes and biosynthetic mechanisms adapted for the transmission and receipt of specific neural signals. Neurons also have complex, functionally integrated cell-to-cell connections which extend over great distances. This intense specialization of cells may explain the varied susceptibility of different subpopulations to infection, and their conductivity may provide a transport system for the movement of viruses between distant but functionally related areas of the brain.

Metabolic factors may also affect the CNS susceptibility to disease or cause the selective expression of disease within the nervous system. Although the brain represents only 2% of the total body weight, it accounts for 20% of the resting total body oxygen consumption; and because of a paucity of oxygen and glucose stores, 15% of the basal cardiac output goes to the brain. Furthermore, the brain has a relative lack of regenerative capacity. Proliferation of astroglia occurs in response to injury, and generation of new oligodendroglia may follow demyelination, but these glial cells do not appear to divide during normal homeostasis (6). On the other hand, neurons totally lack the capacity to regenerate. In view of these metabolic requirements and the lower regenerative capacity, it seems reasonable that a virus causing only minor attenuation of cell function or moderately shortening the time of cell viability would manifest disease first, or solely, in the CNS. These factors may explain why the unconventional viruses associated with subacute spongiform encephalopathies cause pathological changes and clinical symptomatology limited to the CNS, even though infectivity is widespread in other tissues.

ACUTE INFECTION AND CHRONIC DISEASE

Acute infection is usually associated with acute disease resolved by recovery or host death. Often, however, severe acute infections of the CNS leave sequelae; that is, a chronic static residua such as the paralysis which follows acute infection of motor neurons with polioviruses. However, following acute infection of the fetal or neonatal CNS, static sequelae may give the appearance of progressive disease. For example, parvoviruses, which selectively infect cells in mitosis, cause gastrointestinal or hemotopoetic cell destruction in some adult animals, but do not cause CNS disease. In contrast, fetal infection can result in destruction of populations of neurons undergoing mitotic division during development. Thus, a newborn or late-gestational kitten infected with feline panleukopenia virus can have destruction not only of epithelial cells in the gut and hematopoetic cells of marrow, which are rapidly replenished, but of the developing granular neurons of the cerebellum, a non-regenerating cell population (7). These kittens develop post-natally without the sole excitatory cell population of the cerebellum; and, therefore, lack the potential to develop normal synaptic connections to provide smooth, coordinated motor activity (8). Kittens at birth are normally uncoordinated. However, as they develop the failure to acquire coordination gives the appearance of a progressive degenerative neurological disease.

Similarly in man, arbovirus encephalitis in infancy has been noted to cause apparent progressive disease. Studies of these children suggest that the acute destruction of neural cells may simply leave the child unable to achieve subsequent developmental landmarks. Therefore, neurologic deficits appear to intensify over time, even though infection is not persistent, and further cell destruction does not occur (9).

Chronic disease with progressive tissue destruction can follow acute viral infections. For example, progressive obstructive hydrocephalus has been found to develop following a variety of viral infections which selectively involve the ependymal cells lining the ventricles of the brain (10).

This was first demonstrated with experimental mumps virus infections of neonatal hamsters (11). The young animals show no clinical signs of disease during the acute infection. The ependymal cells are destroyed, the virus is cleared, the attendant inflammatory response resolves, and immunity develops. Subsequent to these events, the aqueduct of Sylvius, a narrow ependymal cell-lined channel connecting the third and fourth ventricles of the brain, is reduced in size and eventually becomes markedly stenotic or totally occluded. This leads to increased pressure within the lateral ventricles of the brain, ventricular dilatation, compression of cerebral cortex, distortion of the skull and paralysis. This progressive hydrocephalus usually causes death within 3 months after the infection. However, at the time of even early clinical disease recoverable virus, viral antigen, and inflammatory response are absent, so no clues remain to associate this progressive disease with prior infection. Subsequent to the experimental studies a number of children have been reported with hydrocephalus following mumps virus meningitis or encephalitis (10), and electron microscopic studies of cerebrospinal fluid cells during mumps infections indicate that virus selectively infects and causes loss of ependymal cells in man (12).

A progressive sequela with even greater latency may occur in patients who develop progressive weakness many years after paralytic poliomyelitis. The loss of anterior horn cells during poliovirus infections probably results in sprouting and an increase in the peripheral field of innervation of the remaining motor neurons. This overloading of residual motor neurons combined with the natural attrition of these cells with age apparently can lead to an increased paralysis in late life (13). Thus, chronic disease need not depend on the persistence of virus but can result from the anatomic residua of a remote acute infection (Figure 1).

PERSISTENT INFECTIONS AND CHRONIC DISEASE

Persistent infections can cause acute monophasic disease followed by a chronic asymptomatic carrier state with a high or low rate of virus replication. Acute respiratory infections with adenoviruses and acute hepatitis can follow such patterns.

Chronic carrier states can also lead to chronic disease as in human fetal infection with rubella virus. In the early gestational fetus rubella virus presumably undergoes replication in some fetal cells, but subsequent infection of further cells is limited by fetal or maternal antibodies.

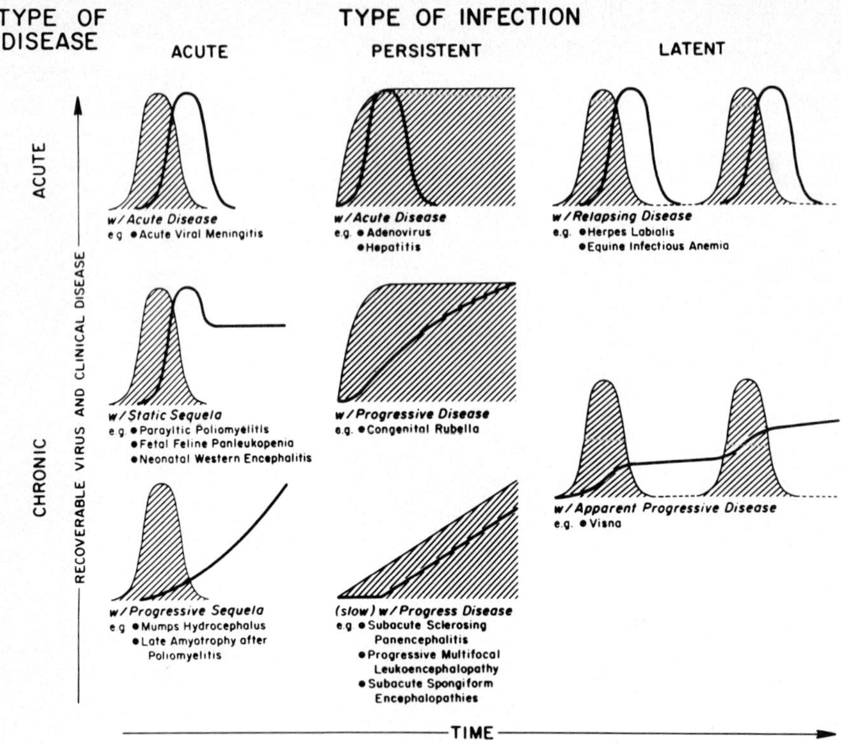

FIGURE 1. Schematic representation of patterns of infection (shaded) and disease (solid line).

In general, rubella virus fails to lyse cells but slows mitotic activity. Therefore, in the immune developing fetus only daughter cells of infected progenitors presumably perpetuate the infection giving rise to focal clones of cells with decreased mitotic activity. The observations that children born with rubella syndrome are chronically infected (14), are small for gestational age with an actual decrease in cell numbers (15), have focal collections of viral antigen-containing cells in histologically normal tissues (16), and have focal asymmetrical malformations are all consistent with this presumed pathogenesis. Furthermore, cloning of cells *in vitro* from infants with congenital rubella show that only a small number of cells are infected, but these cells give rise to cultures of uniformly infected cells with slower growth rates (17). The clinical manifestations of congenital rubella may be static or progressive, but usually virus shedding slowly resolves presumably because cells with a lower doubling time eventually disappear from most organs (18). However, neural cells with their lack of turnover again appear to be the locus for the longest persistence of rubella virus. Reports of a progressive rubella panencephalitis in adolescents bearing stigmata of congenital rubella supports selective neural persistence as well as suggesting that, for unexplained reasons, a delayed progressive disease process can be rekindled from this persistent infection (19, 20).

In slow infections a gradual development of infectivity has been demonstrated or is assumed to occur within the CNS. This is followed by a parallel slow progression of histological changes and neurological deficits (Figure 1). This phenomenon is best demonstrated with the subacute spongiform encephalopathy of scrapie in mice (21), although the mode of replication within individual cells remains an enigma. With the slow infections due to conventional viruses, i.e., subacute sclerosing panencephalitis and progressive multifocal leukoencephalopathy, large quantities of virus are found morphologically in limited numbers of CNS cells. In subacute sclerosing panencephalitis virions are not seen, but nucleocapsid structures often fill and distort nuclei of neurons, astrocytes or oligodendrocytes (22). Presumably there is slow cell-to-cell spread of this virus because of a defect in assembly. In progressive multifocal leukoencephalopathy, however, the viral particles, limited largely to the nuclei of oligodendrocytes, appear to be morphologically intact virions (23). Slow elaboration of papovaviruses within individual cells is difficult to imagine, but it is reasonable to speculate that slowness of disease might result from delayed release of virions into a tightly compacted neuropil. If depression of cellular DNA and protein synthesis are major factors leading to cell destruction in papovavirus infected cells, this depression might have less impact on oligodendrocytes. These cells do not divide and have a slow turnover rate of highly stable membranes. Thus, a long period of virus-induced depressed metabolic activity might be tolerated by these cells before physical disruption of cell membranes occurs releasing virions into a confined space where only neighboring oligodendrocytes would then become infected.

Since progressive multifocal leukoencephalopathy is the one human slow infection in which large nubmers of virions are seen, it should be the prime candidate for relating a human slow infection to defective interfering particles. However, there is some evidence that this is not the case. JC virus extracted directly from the brain of a patient with progressive multifocal leukoencephalopathy and centrifuged to equilibrium in CsCl showed a single band of virus at a density of 1.345 g/ml, a slightly greater density than SV40. Surprisingly a second peak or shoulder of less dense virus, as routinely found in SV40 and BK virus grown in cell culture, was not present. Thus, there was no evidence for two populations of virions corresponding to defective interfering particles and standard particles (24).

Immune mediated pathological changes do not appear to be a factor in these slow infections in man. No immune response can be demonstrated in the spongiform encephalopathies. The immune response in subacute sclerosing panencephalitis appears to be normal, but disease appears to be related to the failure of assembly of the mutant virus. In progressive multifocal leukoencephalopathy, where the virus may be fully virulent, disease seems related with immunological impairment of the host.

LATENT INFECTIONS AND CHRONIC DISEASE

Latent infections, where virus is harbored without continuous production of infectious virus, are expected to produce disease only with reactivation of latent genomes. This usually leads to acute exacerbations of disease such as exacerbations of cold-sores caused by herpes simplex virus, shingles caused by varicella-zoster virus, or the early acute hemolytic crises caused by equine infectious anemia virus.

Recent findings in studies of visna, a classic slow infection of sheep, have suggested novel mechanisms relating virus latency to chronic disease which may have relevance for human disease. In sheep infected with visna virus sequestration of proviral DNA in cells provides a logical explanation for viral persistence (25), but fails to explain the chronic disease. Studies of virus replication in cell cultures suggest that the amount of virus produced by a cell is dependent on the amount of proviral DNA associated with cellular DNA (Clements, unpubl.). The data further suggest that the high levels of viral DNA in lytically infected cells may be dependent on reinfection of cells with progeny virus from the supernatant fluid (26). In the infected animal, in which only relatively few cells are infected and dilution factors are immense; such reinfection would occur only rarely. Therefore, replication of infectious virus in sheep might be severely limited explaining the paucity of cell-free virus recoverable from tissues (27, 28) and the failure to find virions by electron microscopic examination of cerebral lesions (29, 30).

The demonstration of antigenic mutation of visna virus in persisently infected sheep (31) and the similar observations in equine infectious anemia (32) provide a further mechanism for chronic disease. Antigenic drift within the individual host with chronic protozoal disease such as malaria, trypansomiasis and leishmaniasis is well established (33). However, in virology antigenic drift has been considered primarily in terms of acute infections of whole populations as seen in epidemics of influenza. The continuous evolution of virulent mutants within the same individual might result in exacerbations of acute disease or cause clinically progressive disease resulting from cumulative lesions (Figure 1). Analogous mechanisms need consideration in studies of a variety of chronic or remitting and relapsing diseases of suspected viral etiology in man.

CONCLUSION

The complexity of the virus-host relationships and pathogenesis of disease in animals are not adequately reflected in cell culture systems. Not only does the animal host provide a mosaic of different cells with varying susceptibility and varying response to infection but a myriad of host responses and defense mechanisms. The pattern of virus replication may bear little apparent relationship to the pattern of disease. However, some of this apparent contrariety can be resolved by considering the morphological and functional features of the infected organ. In discussing viral diseases of the CNS over 40 years ago Weston Hurst (34) stressed the need to take "Into account both the viruses and the terrain on which they manifest their activity."

REFERENCES

1. Johnson, R. T., and ter Meulen, V. (1978). Adv. Intern. Med. 23, 353.
2. Stevens, J. F. (1978). Adv. Cancer Res. 26, 227.
3. Murphy, F. A. (1977). Arch. Virol. 54, 297.
4. Johnson, R. T. (1964). J. Exp. Med. 119, 343.
5. Johnson, R. T., and Griffin, D. E. (1978). In "Handbook of Clinical Neurology" (P. J. Vinken and G. W. Bruyn, eds.), Vol. 33, North Holland Publ. Co., Amsterdam.

6. Herndon, R. M., Price, D. L., and Weiner, L. P. (1977). Science 195, 693.
7. Kilham, L., and Margolis, G. (1966). Amer. J. Path. 48, 991.
8. Herndon, R. M., Margolis, G., and Kilham, L. (1971). J. Neuropath. Exp. Neurol. 30, 196.
9. Finley, F. H., Fitzgerald, L. H., Richter, R. W., Riggs, N., and Shelton, J. T. (1967). Arch. Neurol. 16, 140.
10. Johnson, R. T. (1975). Develop. Med. Child Neurol. 17, 807.
11. Johnson, R. T., Johnson, K. P., and Edmonds, C. J. (1967). Science 157, 1066.
12. Herndon, R. M., Johnson, R. T., Davis, L. E., and Descalzi, L. R. (1974). Arch. Neurol. 30, 475.
13. Mulder, D. W., Rosenbaum, R. A., and Layton, D. D. (1972). Mayo Clinic Proc., 47, 756.
14. Desmond, M. M., Wilson, G. S., Melnick, J. L., Singer, D. B., Zion, T. E., Rudolph, A. J., Pineda, R. G., Ziai, M. H., and Blattner, R. J. (1967). J. Pediatr. 71, 311.
15. Naeye, R. L., and Blanc, W. (1965). J. Amer. Med. Assoc. 194, 1277.
16. Woods, W. A., Johnson, R. T., Hostettler, D. D., Lepow, M. L., and Robbins, F. C. (1966). J. Immunol. 96, 253.
17. Rawls, W. E., and Melnick, J. L. (1966). J. Exp. Med, 123, 795.
18. Rawls, W. E., (1974). Progr. Med. Virol. 18, 273.
19. Townsend, J. J., Baringer, J. R., Wolinsky, J. S., Malamud, N., Mednick, J. P., Panitch, H. S., Scott, R. A. T., Oshiro, L. S., and Cremer, N. E. (1975), New Engl. J. Med. 292, 990.
20. Weil, M. L., Itabashi, H. H., Cremer, N. E., Oshiro, L. S., Lenette, E. H., and Carnay, L. (1975). New Engl. J. Med. 292, 994.
21. Eklund, C. M., Hadlow, W. J., and Kennedy, R. C. (1963). Proc. Soc. Exp. Biol., N. Y. 112, 974.
22. ter Meulen, B., Katz, M., and Muller, D. (1972). Curr. Top, Microbiol. Immunol. 57, 1.
23. ZuRhein, G. M. (1969). Progr. Med. Virol. 11, 185.
24. Dörries, K., and Johnson, R. T. (1977). Proc. XVIth Symp. European Assoc. Virus Diseases. Amsterdam.
25. Haase, A. T., Stowring, L., Narayan, O., Griffin, D., and Price, D. (1977). Science 195, 175.
26. Clements, J., Narayan, O., and Griffin, D. E. (this volume).

27. Petursson, G., Nathanson, N., Georgsson, G., Panitch, H., and Palsson, P. A. (1976). Lab. Invest. 35, 402.
28. Narayan, O., Griffin, D. E., and Silverstein, A. M. (1977). J. Infec. Dis. 135, 800.
29. Narayan, O., Silverstein, A. M., Price, D., and Johnson, R. T. (1974). Science 183, 1202.
30. Georgsson, G., Palsson, P. A., Panitch, H., Nathanson, N., and Petursson, G. (1977). Acta Neuropath. 37, 127.
31. Narayan, O., Griffin, D. E., and Chase, J. (1977). Science 197, 376.
32. Kuno, Y., Kobayashi, K., and Fukunga, Y. (1973). Arch. Ges. Virusforsch. 41, 1.
33. Brown, K. N., and Hills, L. A. (1968). Immunology 14, 127.
34. Hurst, E. W. (1936). Brain 59, 1.

VIROLOGICAL STUDIES ON CENTRAL NERVOUS SYSTEM DISEASES OF UNKNOWN ETIOLOGY[1]

R. I. Carp,[2] R. J. Kascsak,[2] H. Donnenfeld[3] and H. Bartfeld[3]

N. Y. S. Institute for Basic Research in Mental Retardation
Staten Island, New York 10314[2]
and
St. Vincent's Hospital and Medical Center, New York, N.Y.[3] 10011

Serological studies were performed on the sera of patients with amyotrophic lateral sclerosis (ALS), their contacts, neurological controls and age-matched normal controls. Antibody response to polio types 1, 2 and 3, herpes simplex, adeno and measles viruses were similar for the various groups. The spectrum of viral antigens was expanded to include agents never before examined serologically in this disease. Complement fixation (CF) and hemagglutination inhibition antibody responses to a wide range of arboviruses and a number of unclassified viruses were studied. Again no difference in responses was noted among the different groups. CF antibodies to adeno-associated viruses (AAV) were also analyzed. The proportion of positive individuals for the ALS group was significantly higher than the values obtained for each of the control groups. Also, the proportion of positives in the ALS contact group was higher than that seen for the control groups. Eluates prepared from kidneys of ALS patients were examined for IgG content. Three of 8 cases showed a significantly elevated IgG content, indicative of the presence of immune complexes, but there was no antibody activity against AAV or polio antigens. Explant cultures derived at autopsy from the CNS tissues of 12 ALS cases were established and examined for: CPE, changes after fusion with various indicator cells and standard and intrinsic interference. These tests were negative. With the serologic data serving as the basis, explant cultures were examined for the presence of AAV. Adeno-associated virus was isolated, after the addition of helper adeno virus, from the cervical cord and muscle explant cultures of one ALS case and from the lumbar cord culture of another case. All other ALS cases (4) and all

[1]This work was supported by the National Institute of Health, Grant No. 1 P01 NS 11605-2 and grants from the Muscular Dystrophy Association.

ISBN 0-12-668350-6

control cases (4), tested similarly, have been negative. A series of experiments were initiated to study molecular alterations occurring either in explant cultures or cultures treated with CNS material from ALS or multiple sclerosis (MS) patients. A radioactive peak (TCA insoluble) was observed in the cytoplasmic extracts of a high proportion of MS-treated cultures but only rarely seen in cultures infected with control material. The peak was seen employing DNA and protein but not RNA precursors, in rate zonal but not isopycnic density gradients, in cytoplasmic but not nuclear extracts. In several experiments, brain explant cultures were established from suckling mice inoculated with MS and normal human brain homogenates. Three of 4 MS explant cultures demonstrated a rate zonal peak of similar sedimentation properties to that seen in *in vitro* treated cultures. None of 4 control cultures exhibited a similar peak. The presence of actinomycin-D resistant RNA was monitored in ALS and MS CNS explant cultures. A small increase in the ratio of the RNA was detected in 1 of 10 ALS cultures. Current work is aimed at investigating the presence of the rate zonal peak or other molecular changes in cell cultures established from various tissues of MS, ALS and control individuals.

INTRODUCTION

Amyotrophic lateral sclerosis (ALS) and multiple sclerosis (MS) are extremely serious human diseases of unknown etiology. There is circumstantial evidence which suggests that these diseases are caused by viruses which initiate slow infection processes, the type of infections originally described by Sigurdsson[1]. ALS usually affects individuals during the 5th or 6th decade. The course of the disease is usually 2-3 years, although it can be longer, and invariably ends in death. There is severe muscle atrophy and the CNS histopathology is very similar to that described for poliomyelitis. For ALS there have been comparatively few attempts to search for a virus and thus far these attempts have failed to reveal a candidate virus. The limited attempts at virus isolation have prompted us to stress the approaches that have been used in the study of a number of other presumptive human slow infections. Our work on ALS is part of a multidisciplinary effort which is headed by Dr. Harry Bartfeld of St. Vincent's Hospital in New York.

MS usually strikes individuals 20-40 years old. The course varies considerably and is often characterized by a series of exacerbations and remissions of variable length.

The histopathology is characterized by plaques in the white matter, usually perivascular or periventricular. Early or young plaques show extensive active demyelination with perivascular cuffing and infiltration of lymphocytes and macrophages. Old plaques are made up of demyelinated nerve fibers with an increase in fibrous astrocytes. Many laboratories have expended much time and effort in the search for the etiologic agent in this disease. These studies have primarily involved those standard approaches used in slow infection research. Thus far, there has been no consistent reproducible finding. Our current efforts have stressed the search for unconventional agents or indirect evidence of virus presence.

MATERIALS AND METHODS

Serology. All sera were obtained from St. Vincent's Hospital, Manhattan as part of an ongoing NIH funded research project under the direction of Dr. H. Bartfeld. The microtiter complement-fixation test was conducted as described by Hsuing[2] employing 4 units of antigen or 8 units of antibody. Viral antibodies to polio, types 1, 2 or 3, adenovirus, measles and herpes simplex were commercially obtained from either Flow Laboratories or Microbiological Associates, Rockville, Maryland. Antigens and antibodies to adeno-associated viruses were kindly supplied by Dr. M. Hoggan, NIH.

Elution of Antibody from Immune-Complexes. Antibody was eluted using the procedure of Hollis *et al*[3]. A 0.02 M glycine-HCl (pH 2.8) elution buffer containing 50 ng/ml of pepstatin was employed. Following elution, antibody was dialyzed and concentrated by lyophilization.

Fluorescent Focus Assay. Hep-2 cells, in Leighton tubes, were infected with adenovirus, type 5 (0.1 PFU/cell) and AAV, type 2 (a 1-100 dilution of a virus stock with a 1-64 CF titer). When CPE reached the 2+ level, cultures were fixed in 95% ethanol. Antibody eluted from immune-complexes was diluted 1:4 and absorbed with adeno-infected Hep-2 cells. This antibody was then utilized in an indirect fluorescent antibody test. Fluorescent conjugated anti-human gammaglobulin was obtained from Behring Diagnostics, Somerville, N. J. Cultures were examined with a dark field Zeiss fluorescent microscope using a BG 12 excitation filter and a K530 barrier filter.

Neutralization Test. Polio virus, types 1, 2 and 3 (10^4 PFU) was reacted with an equal volume of eluted antibody for 1 hr at room temperature. Virus neutralization was determined by plaque assay on Vero cells as described by Kascsak and Lyons[4].

Establishment of Explants. Autopsies were performed at St. Vincent's Hospital, Manhattan under the direction of Dr. H. Donnenfeld. The autopsies were performed as soon as possible after death, generally 2-6 hrs. Sections of motor cortex, medulla, spinal cord, kidney and muscle were removed sterily, washed in growth media, minced into 1-2 mm^2 pieces and placed in 25 cm^2 tissue culture flasks. These flasks were transported back to the laboratory where the media was removed, the tissue pieces air dried onto the flasks and fresh growth media added. Growth media consisted of Eagle's minimal essential media with Hanks' salts supplemented with 20% heat-inactivated fetal calf serum, 0.6% glucose, 0.1% chicken embryo extract, 2X glutamine, penicillin (100 units/ml) and streptomycin (100 μg/ml). Growth media was removed and fresh media added once a week. Cell outgrowth generally occurs 2-3 weeks after planting.

Isolation of AAV from Explant Cultures. Primary cultures were infected with adenovirus, type 5, at a multiplicity of infection (MOI) of 0.1 PFU/cell. This adeno stock was shown to be free of AAV by performing 5 blind passages in WI-38 cells followed by electron microscopic and CF analysis. When cultures exhibited 4+ CPE, they were frozen and thawed three times, the lysate concentrated by ultracentrifugation (140,000 g for 3 hrs), and blind passaged twice in WI-38 cells. Final lysates were again concentrated by ultracentrifugation and the pellets examined for CF activity against AAV antisera.

Exposure of Cells to MS and Non-MS Material. The technique used for exposure of cells to MS and non-MS material and their maintenance thereafter have been described[5].

Animal Inoculation. Three day old C57/BL mice were injected intracerebrally with 0.02 ml of a 1-4 dilution of 10% MS or non-MS brain homogenates. The dilution fluid and the diluent for the brain homogenates was Eagle's medium in Earle's balanced salt solution (MEM). Explant cultures were established 16 days post injection as described previously.

Radiolabeling and Cell Extraction. Cultures were labeled with either 5 μCi/ml of ^{3}H-thymidine, 5 μCi/ml ^{3}H-leucine or 1 μCi/ml of ^{3}H-protein hydrolysate. Cultures were then maintained for five days in MEM with 5% fetal calf serum. Cells were then trypsinized (.05% trypsin in EDTA) and cytoplasmic extracts prepared by suspending whole cells in 0.01 M

NaCl, 0.003 M $MgCl_2$ and 0.01 M Tris, pH 7.4. The cells were allowed to swell for 20 min in the hypotonic buffer and then homogenized with a tight fitting ball homogenizer. The homogenate was then centrifuged at 2,000 rpm for 10 min and the supernatant used as the cytoplasmic extract.

Rate Zonal Centrifugation. A 10 ml linear 5-20% w/w sucrose gradient in TNE (0.05 M Tris, 0.1 M NaCl, 0.001 M EDTA, pH 7.4) was used with a 1 ml cushion of 65% w/v sucrose in TNE. Centrifugation was at 120,000 g for 1.5 hrs at 4°C. The sample volume was 0.3 ml. Gradients were fractionated from the top, using a Buchler Densiflow apparatus.

TCA Precipitation and Radioactive Counting. TCA precipitable activity was determined by spotting 50 µl aliquots of each fraction on Whatman scintillation pads. These pads were then processed as described by Mans and Novelli[6]. Density determinations were made using an Abbe refractometer.

Actinomycin-D Resistant RNA. Actinomycin-D was added to the explant cultures at 1.0 µg/ml 1 hr prior to the addition of 2.5 µC/ml ^{3}H-uridine. After a 24 hr labeling period, cells were solubilized in 2% SDS and 1% 2-mercaptoethanol. An equal volume of 10% ice cold TCA was added and RNA allowed to precipitate overnight. TCA precipitable activity was determined by collection of precipitate in 0.45 nm millipore filters and addition of PPO-POPOP scintillation cocktail. The % resistant counts were determined by the ratio of activity in cultures treated with actinomycin-D to activity in untreated cultures.

Actinomycin-D resistant RNA was further analyzed on a 15-30% w/v sucrose gradient containing 0.2% SDS. Centrifugation conditions were 66,000 g for 18 hrs and Hep-2 ribosomal RNA served as sedimentation value markers.

RESULTS

Serologic Studies on Sera of ALS Patients. Humoral antibody activity to various viral antigens as measured by complement fixation was examined in 4 groups: ALS patients, contacts of ALS patients, neurological controls and age-matched normal individuals. Antibody responses to polio types 1, 2 and 3, adeno, herpes simplex, measles, LCM, reovirus type 3, rabies and NDV were similar in all 4 groups (results not shown). In a study conducted at the Yale Arbovirus Research Center, antibody titer to 'arboviruses' were monitored in these 4 groups and also for multiple sclerosis (MS) patients.

Tests were performed using 78 CF antigens and 15 HI antigens. Togaviruses, members of Bunyaviridae and several ungrouped and unclassified agents were examined. No differences were found among the various groups. These results have been accepted for publication in Archives of Neurology.

Response to the type 2-3 complex of adeno-associated viruses (AAV), shown in Table 1, was increased in the ALS group compared to the neurological controls and normal individuals. ALS patients exhibited statistically higher serum CF antibody titers against AAV (2-3 complex) than did age-matched controls ($p < 0.05$). The incidence of antibody response for the normal controls fell into the range of 5-10% which is expected for this age group[7]. The neurological controls exhibited a slightly higher ratio of positive individuals than did the control group (16% vs. 9%), not statistically significant, but a significantly lower ratio than the ALS group (35% vs. 16%, $p < 0.05$). The ALS contact population also possessed a higher incidence of positiveness to AAV than their age-matched controls (35% vs. 9%, $p < 0.10$). These contacts included one blood relative in addition to husbands, wives, doctors and nurses. Several ALS patients possessed chronically high CF titers to these viruses. One patient examined over 2.5 years showed titers of 1:16 or

TABLE 1
COMPARISON OF SERUM CF ANTIBODY TITERS TO AAV BETWEEN ALS PATIENTS AND CONTROL GROUPS

	AAV 1	AAV 2**	AAV 3**
ALS	4/28 (14%)*	13/40 (33%)	8/28 (29%)
ALS Contacts	1/12 (8%)	6/17 (35%)	5/15 (33%)
Neuro. Controls	1/41 (3%)	10/62 (16%)	7/46 (15%)
Normal Controls	0/20 (0%)	0/22 (0%)	2/22 (9%)

*percentage possessing CF titer of 1:2 or greater
**p values based on 2-3 complex of AAV
$p < 0.05$ between ALS and Normal Controls
$p < 0.05$ between ALS and Neurological Controls
$p < 0.10$ between ALS contact group and Normal Controls

greater for all five sera tested. No statistically significant correlation was demonstrated between seroconversion to AAV and clinical status.

Cerebrospinal fluids from 17 ALS cases have failed to exhibit any activity against AAV. Spinal fluids were tested both unconcentrated and after concentration by Amicon filtration. We have been unable to detect any viral-specific antibody in the CSF of ALS patients.

Elution of Antibody from Immune-Complexes in Kidney. The presence of immune-complexes in the kidneys of ALS patients has been demonstrated by Oldstone *et al*[8] and by our group. We have eluted the antibody from these complexes

TABLE 2
ELUTION OF ANTIBODY FROM IMMUNE-COMPLEXES IN KIDNEY

Case	Diag.	mg% IgG/protein	Fluorescent focus assay	Neutralization		
			Adeno-associated virus	P1	P2	P3
1	ALS	2.7*	- **	+0.2+	-0.4	+0.2
2	ALS	1.0	-	+0.1	-0.3	-0.2
21	ALS	0.2	ND	-0.3	0.0	-0.1
22	ALS	11.5	-	+0.1	-0.7	-0.2
25	Alzheimers	1.4	ND	0.0	-0.3	0.0
29	ALS	0.0	ND	-0.1	-0.1	+0.2
30	ALS	2.8	-	-0.1	-0.3	-0.2
31	ALS	0.0	ND	ND	ND	ND
32	Multiple Sclerosis	3.2	-	ND	ND	ND
34	ALS	0.0	ND	ND	ND	ND

*value determined by subtraction of % IgG/protein in final wash from % IgG/protein in kidney eluate
**dilution of 1:4 of eluate antibody
+log change in plaque titer
ND = not done, P1, polio type 1; P2, polio type 2; P3, polio type 3

and examined its specificity to AAV and polio antigens. Antibody was eluted as described in Materials and Methods. Protein concentrations were monitored by the Lowry test and gamma-globulin concentrations by radial immune diffusion. Eluates from 8 ALS and 2 neurological control kidneys have been examined as shown in Table 2. A significant amount of bound IgG, as indicated by mg% IgG/protein, was seen in 3 of 8 ALS cases (Nos. 1, 22 and 30). One neurological control (No. 32, MS) also exhibited a high concentration of bound IgG in the eluate, indicative of the presence of immune complexes.

The eluated antibody was initially tested for viral specificity in a CF test. No response to any of the following antigens was evident: polio, adeno, herpes, measles, AAV. The F_c portion of the antibody may be damaged in our elution procedure[9] and the antibody no longer functional in a CF test. Specificity to AAV was then investigated using a fluorescent focus assay, and specificity to polio types 1, 2 and 3 by a plaque neutralization test. Results, shown in Table 2, indicate lack of AAV or polio specificity for eluted antibody from any of these cases.

Establishment of Explant Cultures. The establishment of a large number of explant cultures from autopsy specimens was one of our major objectives in the search for a viral etiology for ALS. Tissues from all sites where the primary affect of a causative agent might be present were explanted: spinal cord (cervical and lumbar), motor cortex, medulla, muscle and kidney. Cultures have been established from 12 ALS, 2 neurological control (Alzheimer and MS) and 3 control autopsy cases. Over 190 cultures established from tissues of ALS patients (cervical cord (53), lumbar cord (48), motor cortex (18), medulla (25), kidney (22), muscle (27)) have been examined, using a variety of parameters, for the presence of virus. Techniques such as cell fusion, standard and intrinsic interference, analogue induction, assay for interferon hemadsorption and fluorescent antibody staining have failed to indicate viral involvement. A more detailed account of these procedures and their results will be published elsewhere.

Isolation of Adeno-Associated Virus from Explant Cultures. Serologic data suggested an involvement between AAVs and ALS. Since these viruses are conditionally defective and require a helper function for replication, their presence in explant cultures would not be detectable by procedures outlined in the previous section. Adenovirus is capable of inducing the helper function required by these viruses. Primary explant cultures were exposed to adenovirus, type 5, and processed as described in Materials and Methods. We have

to establish the presence of coincidence of the thymidine and amino acid peaks in that this result would indicate some type of nucleic acid-protein complex, possibly a virus-like particle. This should be possible with double labeling experiments.

The demonstration of the peak in cytoplasmic extracts of brain explants derived from mice injected with MS material suggested that agent can replicate or, at least, induce an effect _in vivo_ as well as _in vitro_. Previous results using biological parameters have shown MS induced effects in mice injected with MS material[15, 16] and in mouse tissue culture cells exposed to MS material[5]. In each instance the findings have involved the measurement of subtle changes in physiological parameters. These results have been difficult to confirm and, for reasons that are not apparent at present, early success has been followed by inconsistent findings[17, 18]. The presence of a biochemical change in mouse cells exposed to MS material _in vitro_ as well as in brain explant cells derived several weeks after injection of mice with MS material would suggest that there are MS induced effects in these systems.

One discouraging result has been the absence of the ^{3}H-thymidine peak in cytoplasmic extracts of explant cultures obtained from the brain and cord of one MS autopsy. It is quite possible that the cells in the human which support the growth or persistence of the presumptive agent did not survive the explant culturing procedures. We hope to do more work with explants derived from MS autopsy material.

We can not explain the failure to consistently demonstrate the ^{3}H-thymidine peak, even in repeat experiments in which the same MS brain homogenate and the same cell type were used. Certainly, the peak obtained with ^{3}H-thymidine is not large. It is clear that if the cause of the peak is a replicating agent we know nothing about the kinetics of its replication or the effects of the physiological state of the cells upon virus synthesis. For this reason the labeling conditions which we have used may be far from ideal for this particular virus-cell interaction. We are currently experimenting with a number of parameters in an attempt to enhance the difference in labeling in the peak area between MS and non-MS treated cultures. The parameters being checked include: multiple exposure to brain homogenates, the lengths of the labeling period, infection and/or labeling of synchronized cells at various stages of cell growth and preparation of cytoplasmic extracts by various means, such as mild detergent treatment. Another approach to the problem being explored is the use of more sensitive means of analyzing the nucleic acid and protein contents of the extracts, such as polyacrylamide gel electrophoresis.

REFERENCES

1. Sigurdsson, B. (1954). Br. Vet. J. 110, 341.
2. Hsiung, G. D. (1973). "Diagnostic Virology", Yale University Press, New Haven, Conn.
3. Hollis, V. W., Aoki, T., Barrera, O., Oldstone, M. B. A. and Dixon, F. J. (1974). J. of Virol. 13, 448.
4. Kascsak, R. J. and Lyons, M. J. (1977). Virology 82, 37.
5. Carp, R. I., Merz, G. S. and Licursi, P. C. (1974). Infect. Imm. 9, 1011.
6. Mans, R. J. and Novelli, C. D. (1961). Arch. Biochem. Biophys. 94, 48.
7. Parks, W. P., Boucher, D. W., Melnick, J. L., Taber, L. H. and Yow, M. D. (1970). Infect. Imm. 2, 716.
8. Oldstone, M. B. A., Perrin, L. H., Wilson, C. B., and Norris, F. H. (1976). Lancet ii, 167.
9. Woodroffe, A. J. and Wilson, C. B. (1977). J. of Immunol. 118, 1788.
10. Cremer, N. E., Oshero, L. S., Norris, F. H. and Lennette, E. H. (1973). Arch. Neurol. 29, 331.
11. Berns, K. I., Pinkerton, T. C., Thomas, G. F. and Hoggan, M. D. (1975). Virology 68, 556.
12. Mayor, H. D. (1973). Meth. in Can. Res. 8, 203.
13. Oldstone, M. B. A., Holmstoen, J. and Welsh, R. M. (1977). J. of Cell. Phys. 91, 459.
14. Joseph, B. S. and Oldstone, M. B. A. (1975). J. of Exp. Med. 142, 364.
15. Carp, R. I., Licursi, P. A., Merz, P. A. and Merz, G. S. (1972). J. Exp. Med. 136, 618.
16. Koldovsky, U., Koldovsky, P., Henle, G., Henle, W., Ackerman, R. and Haase, G. (1975). Infect. Imm. 12, 1355.
17. Carp, R. I., Licursi, P. A., Merz, P. A., Merz, G. S., Koldovsky, U., Koldovsky, P., Henle, G. and Henle, W. (1977). Lancet ii, 814.
18. Madden, D. L., Krezlewicz, A. G., Gravell, M., Sever, J. L. and Tourtellotte, W. W. (1977). Lancet ii, 976.

SCRAPIE: VIRUS, VIROID, OR VOODOO?[1]

R.F. Marsh, Terry G. Malone*, Wayne D. Lancaster[2], R.P. Hanson and J.S. Semancik*

Department of Veterinary Science, University of Wisconsin, Madison, Wisconsin 53706; and *Department of Plant Pathology, University of California, Riverside, California 92502

ABSTRACT Membrane-free preparations from scrapie-infected hamster brain were partially purified by ultracentrifugation or hydroxyapatite chromatography and found to be more sensitive to heat or UV irradiation inactivation. These results indicate that as the minimal scrapie infectious unit is separated from nonessential host proteins and macromolecules it becomes more radiosensitive suggesting a nucleic acid component.

INTRODUCTION

The biophysical nature of scrapie agent has confounded investigators for at least two decades. The unusual resistance of crude brain extracts to chemical and physical inactivation, the inability to visualize virus particles in infected tissues, and the unorthodox biological properties of this pathogen have perpetuated an almost "supernatural" aire to scrapie and its related diseases.

Numerous hypotheses have been put forth on the nature of scrapie agent including speculation as to it being a replicating membrane (1), molecular complex devoid of nucleic acid (2), protein (3), polysaccharide (4), or viroid (5). Diener's discovery of plant viroids demonstrated for the first time that small nucleic acids could produce disease by themselves, and this finding stimulated a reappraisal of scrapie infectivity. However, subsequent experiments failed to show that scrapie disease could be produced by free nucleic acids (6,7). This proved to be only a minor setback for it was simultaneously speculated that viroids of animals may require membrane components for protection or recognition of suscep-

[1]This work was supported in part by NIH grant AI 11250 and NSF grant GB 39605.

[2]Present address: Division of Otolaryngology, Case Western Reserve University, Cleveland, Ohio 44106

ISBN 0-12-668350-6

tible cells (6,8).

This hypothesis can now be examined using a new hamster scrapie model having incubation periods only one-half the length of those in the mouse and concentrations of scrapie agent in brain which are 10 to 100 fold higher than any other animal model (9). Previous studies of subcellular fractions of scrapie-infected hamster brain have confirmed the membrane association of scrapie agent, but have been unsuccessful in ascribing scrapie activity to any particular type of membrane (10). The ubiquitous distribution of scrapie infectivity in plasma membrane, and in smooth and rough endoplasmic reticulum, suggests a nonessential association with membrane which we have now confirmed (11). This report describes the partial purification and characterization of the minimal scrapie infectious unit present in hamster brain preparations containing no intact membranes.

METHODS

Scrapie Agent. These experiments were performed using 16th to 18th passage hamster-adapted scrapie, the history of which has been previously described (9,12).

Bioassay. Scrapie infectivity was determined by intracerebral inoculation of weanling, outbred (LVG/LAK) hamsters purchased from Lakeview Hamster Colony, Newfield, New Jersey. Ten-fold dilutions of samples to be tested were injected (0.05 ml) into three or four animals each and endpoints calculated after 16 weeks by the Spearman-Kärber method (13).

Subcellular Fractionation of Brain. Detailed procedures have been published (10). Briefly, fresh brain tissue was homogenized to 5% in either a Virtis or Polytron blender and separated by serial ultracentrifugation. The buffer used for all experiments, with the exception of the linear sucrose gradient, was composed of .01M Tris, .15M NaCl, .012 sodium phosphate, pH 7.2. The principal membrane-free fractions are high speed supernatants (HSS) which are centrifuged (Beckman 42.1) at 100,000 x g for either 2 or 18 hours and will be referred to hereafter as HSS-2hr or HSS-18hr. Protein determinations were made according to Lowry et al. (14) using bovine serum albumin as a standard.

Sucrose Gradient Sedimentation. Gradients were prepared using the method of Brakke and Van Pelt (15). The diluent and homogenization buffer consisted of 5mM Tris, pH 7.5, 2mM EDTA, and 1mM dithiothreitol. 750 μl of HSS-2hr containing

$10^{7.5}LD_{50}$ was layered on top (total volume 5.4 ml) and the gradients centrifuged (Beckman SW 50.1) at 45,000 rpm and 10C for 18, 24 or 39.5 hours. 900 µl fractions were collected from the top of the tube using an Isco Fractionator. Density of consecutive samples was determined with an Abbe refractometer.

Hydroxyapatite Chromatography. A suspension of hydroxyapatite in .012M sodium phosphate was poured into a 1.0 by 40.0 cm column attaining a final packed volume of 4.5 cm. Nine 4 ml aliquots of HSS-18hr containing 10^6LD_{50}/ml were applied to the column at room temperature, and each 4 ml exclusion volume collected for bioassay. The column was then washed with five 4 ml aliquots of .012M sodium phosphate, seventeen 4 ml aliquots of .12M sodium phosphate (pH 6.8), and five 4 ml aliquots of .48M sodium phosphate (pH 6.8). The 2nd, 3rd, 4th and 5th four ml effluents from each series of washes were collected for bioassay. All samples were dialyzed against distilled water overnight at 4C before animal inoculation.

Heat Inactivation. One ml samples of HSS-2hr, HSS-18hr, or infectious material eluted and dialyzed from the 2nd .48M sodium phosphate wash of the hydroxyapatite column (HA-S) were placed in ampoules, sealed, and heated in a water bath at 60, 80, or 100 C. At the end of 15 minutes, the samples were removed and immediately cooled in an ice bath.

UV Inactivation. Two ml samples of a 5% suspension of membranes (40,000 x g pellet) from scrapie brain (see ref. 10), HSS-18hr, or HA-S were placed in open 60 by 15 mm plastic dishes and irradiated with agitation using a 115 volt, 6 amp Mineralight, model R-51 (Ultra-Violet Products, Inc., San Gabriel, California). The dose rate was 1000 ergs/sec/mm^2 as measured by a General Electric germicidal UV intensity meter. Samples were removed for bioassay after 10, 100, or 1000 seconds exposure.

Enzyme Treatment. HSS-18hr and HA-S were treated with DNase I (Worthington), RNase A (Worthington), or proteinase K (Beckman) at concentrations of 100 µg/ml, 50 µg/ml, and 62 µg/ml, respectively. The buffer contained .096M sodium phosphate; .02M $MgCl_2$ was added to the DNase I reaction. Tubes were incubated at 37C for 60 minutes, then at 57C for 30 min to inactivate the enzymes. Enzymes were tested for effect by digestion of appropriate substrates under identical conditions in the presence of HSS-18hr from normal hamster brain. Controls were composed of tubes containing no enzyme treatment but incubated as described before bioassay.

RESULTS

Sucrose Gradient Sedimentation. The results of bioassay of fractions collected from linear-log sucrose gradients of HSS-2hr show almost equal concentrations of infectivity in all samples and no visable pelleting thorough 30% (1.13) sucrose (Fig. 1). There were no differences in the infectivity distributions observed relative to the period of centrifugation.

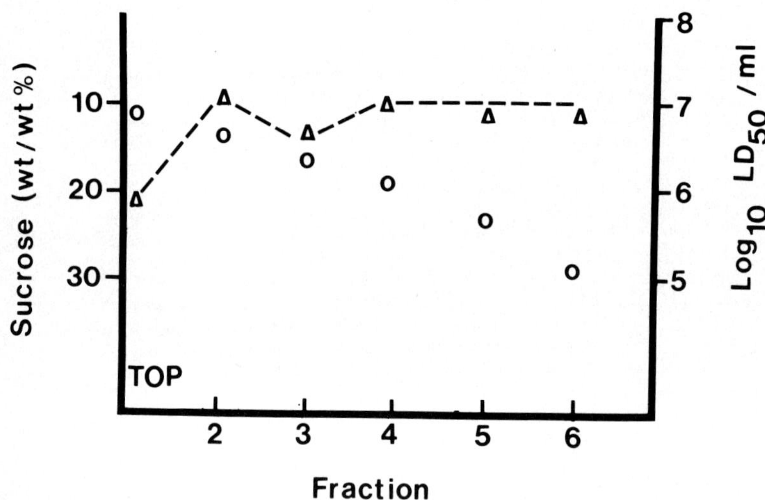

FIGURE 1. Infectivity (▲) distribution in 39.5 hr linear sucrose gradient.

Hydroxyapatite Chromatography. The results of separation of HSS-18hr on hydroxyapatite are summarized in Figure 2. Only the .48M sodium phosphate washes were infectious, and most of the infectivity was contained in the second 4 ml fraction which had a protein concentration of 84 μg/ml (see Table III) and 8%/16% of the input *260/280* OD.

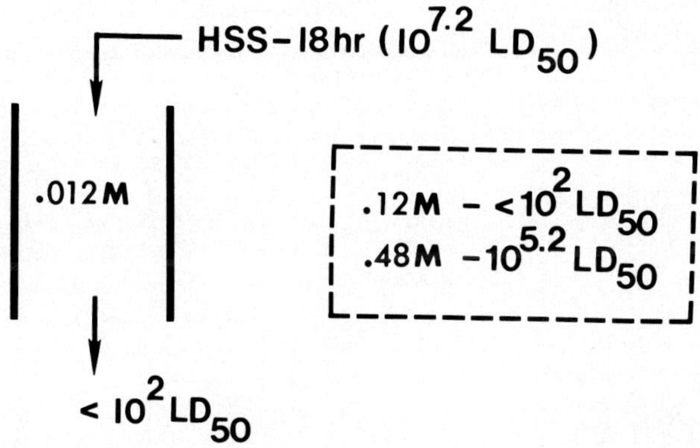

FIGURE 2. Recovery of scrapie infectivity in HSS-18hr after adsorption to hydroxyapatite.

Heat Inactivation. Ten per cent of the infectivity remained in the HSS-2hr after boiling for 15 minutes (Table I). HSS-18hr was inactivated below 16 week bioassay detection levels by boiling as was HA-S, which in addition showed some loss of infectivity at 80 C.

TABLE I
HEAT INACTIVATION

Temp.(15min.)	HSS-2hr	HSS-18hr	HA-S
60C	7[a]	6	4
80 C	7.3	5.5	3
100 C	6	<2	<2

[a] Log_{10} LD_{50} /ml

<u>UV Inactivation</u>. The 5% 40,000 x g pellet was relatively resistant to UV inactivation at levels as high as 10^6 ergs/mm^2, whereas, the HSS-18hr and HA-S were inactivated to less than $10^2 LD_{50}$/ml by the same amount of irradiation (Fig. 3). The HA-S was most sensitive; 10^5 ergs/mm^2 resulting in inactivation to 1%.

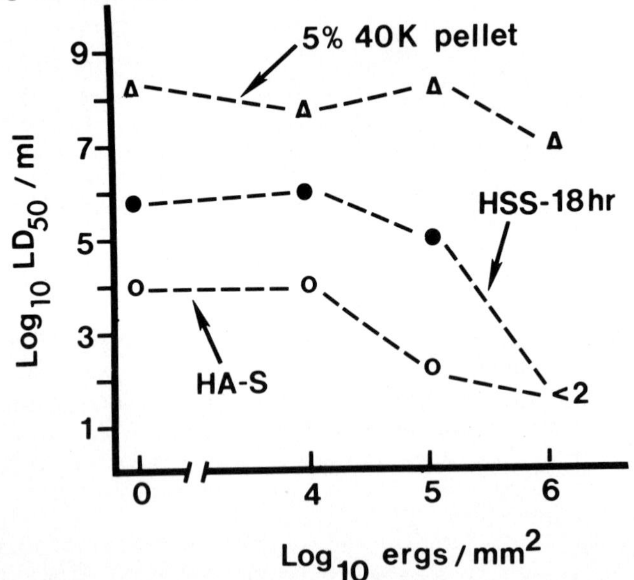

FIGURE 3. UV inactivation at 254 nm of 5% 40,000 x g pellet (**Δ**), HSS-18hr (●), and HA-S (**O**).

Enzyme Treatment. DNase I, RNase A, or proteinase K did not significantly reduce the infectivity of HSS-18hr under the conditions described (Table II). Bioassay data on the HA-S are not available at this time (March 3, 1978).

TABLE II
ENZYME TREATMENT

Rx (37C/60min)	HSS-18hr	HA-S
None	6[a]	NA[b]
DNase (100μg/ml)	6.2	NA
RNase (50μg/ml)	6	NA
proteinase K (62μg/ml)	5.5	NA

[a] Log_{10} LD_{50} / ml

[b] Bioassay data not yet available

DISCUSSION

Figure 4 stylistically illustrates the fragmentation of scrapie-infected membrane with reduction to smaller and smaller complexes. Our previous studies have shown a rapid increase in HSS titer midway into the incubation period (11). This 1000 fold increase in infectivity coincides with the onset of membrane dissolution in brain which produces the spongiform degeneration so characteristic of these diseases. Other studies have shown that membrane destruction is a secondary effect of the disease process, possibly caused by lysosomal enzymes, rather than due to the direct result of agent replication (16). Therefore, there is no reason to believe that the infectivity in the HSS represents a qualitatively different or more "mature" complex than the membrane-associated agent.

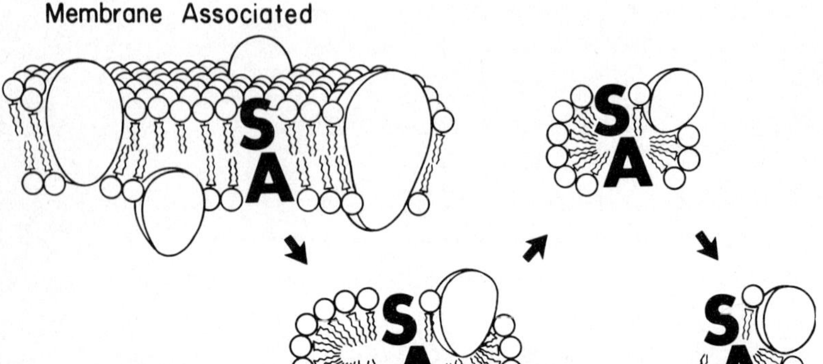

FIGURE 4. Schematic representation of membrane-associated scrapie agent (SA) and its reduction to smaller and smaller complexes in the HSS.

Results of the sucrose density gradient separation show that the infectious material in the HSS-2hr is heterogeneous, but it should now be possible to partition and concentrate the minimal scrapie infectious unit using classical methods of virus purification. This paper examines one procedure, hydroxyapatite chromatography. This method of affinity binding does not appear to offer any substantial selective advantage in separating scrapie infectivity from noninfectious macromolecules in the HSS-18hr. Although some degree of purification was attained, only 1% of the input infectivity was recovered from the column (see Fig. 2 and Table III). More efficient means of separation are going to be needed to obtain the quantities necessary for biochemical analysis. However, it should also be remembered that we do not yet know which subpopulation(s) of the HSS-18hr the HA-S represents, nor how homogeneous. Since the HSS-18hr appears to be substantially resistant to proteolytic digestion under the conditions of this experiment, it will be important to determine if hydroxyapatite binding is more or less efficient at removing infectivity from predigested preparations.

TABLE III

Preparation	Protein Conc.	$Log_{10} LD_{50}$/ml
5% Brain	10mg/ml	9
HSS-2hr	650µg/ml	7.3
HSS-18hr	220µg/ml	6
HA-S	84µg/ml	4.2

The most significant finding in these studies is that scrapie infectivity becomes more radiosensitive as protective host macromolecules are removed. This would be expected if the agent contained an essential nucleic acid component. Conclusive proof, however, must eventually come in the form of inactivation of *scrapie infectivity* with nuclease. Any other experiment is equivocal.

The title of this paper poses a question which we can not answer. It is still possible that scrapie could be caused by a virus. Resistance to enzymatic digestion could be explained by existence of infectivity in the form of an intact viral particle. We also do not know if scrapie has a viroid-like informational moiety. This remains the most likely possibility, albeit the evidence to date has been more intellectually stimulating than scientifically convincing. But while we have yet to succeed in our ultimate goal of describing the molecular nature of scrapie agent, we have certainly made some progress in removing a little of the mystery from this pathogen.

REFERENCES

1. Gibbons, R. A., and Hunter, G. D. (1967). Nature (London) 215, 1041.
2. Alper, T., Cramp, W. A., Haig, D. A., and Clarke, M. C. (1967). Nature (London) 214, 764.
3. Griffith, J. S. (1967). Nature (London) 215, 1043.
4. Field, E. J. (1966). Brit. Med. J. 2, 564.
5. Diener, T. O. (1972). Nature (London) 235, 218.
6. Marsh, R. F., Semancik, J. S., Medappa, K. C., Hanson, R. P., and Rueckert, R. R. (1974). J. Virol. 13, 993.
7. Ward, R. L., Porter, D. D., and Stevens, J. G. (1974). J. Virol. 14, 1099.
8. Marsh, R. F. (1974). In "Advances in Veterinary Science and Comparative Medicine," Vol. 18 (C. A. Brandly and C. E. Cornelius, eds.), pp. 155-178. Academic Press, New York.
9. Marsh, R. F., and Hanson, R. P. (1978). Fed. Proc. in press.
10. Semancik, J. S., Marsh, R. F., Geelen, J. L. M. C., and Hanson, R. P. (1976). J. Virol. 18, 693.
11. Malone, T. G., Marsh, R. F., Hanson, R. P., and Semancik, J. S. (1978). J. Virol. 25, in press (March).
12. Kimberlin, R. H., and Marsh, R. F. (1975). J. Infect. Dis. 131, 97.
13. Dougherty, R. M. (1964). In "Techniques in Experimental Virology" (R. J. C. Harris, ed.), p. 183. Academic Press, New York.
14. Lowry, O. H., Rosebrough, N. J., Farr, A. L., and Randall, R. J. (1951). J. Biol. Chem. 193, 265.
15. Brakke, M. K., and Van Pelt, N. (1970). Anal. Biochem. 38, 56.
16. Marsh, R. F., Sipe, J. C., Morse, S. S., and Hanson, R. P. (1976). Lab. Invest. 34, 492.

EVIDENCE FOR MULTIPLE MOLECULAR FORMS OF THE SCRAPIE AGENT[1]

Stanley B. Prusiner[2], David E. Garfin, J. Richard Baringer*, S. Patricia Cochran

Departments of Neurology and Biochemistry and Biophysics, University of California and *Ft. Miley V.A. Hospital, San Francisco, California 94143

William J. Hadlow, Richard E. Race, and Carl M. Eklund[3]

Rocky Mountain Laboratory, National Institute of Allergy and Infectious Disease, National Institutes of Health, Hamilton, Montana 59840

ABSTRACT The scrapie agent has eluded isolation and identification to date, and appears to be a novel infectious entity differing from conventional viruses in many respects. Preparatory to developing a purification protocol, the sedimentation behavior of the agent in extracts of murine spleen was studied in fixed-angle rotors. Sedimentation profiles for the scrapie agent suggested that the agent is a particle or a series of particles with sedimentation coefficients ($S_{20,w}$) between 50 and 450S. These profiles were not substantially changed by sonication or detergent treatment. Similar sedimentation profiles for the agent in extracts of murine brain were observed, indicating that the size of the agent is independent of the tissue in which it replicates. From the profiles a partial purification scheme was developed which yielded a fraction designated "P_5" enriched for scrapie infectivity ∿20-fold with respect to cellular protein. The P_5 fraction was devoid of cellular membranes but heavily contaminated with ribosomes. In an attempt to separate the agent from ribosomes, P_5 was subjected to sucrose gradient centrifugation after storage at -70° under conditions where infec-

[1]This work was supported in part by research grants from the National Institutes of Health NS 11917 and NS 14069 and from the National Science Foundation PCM 75-22806 as well as by a basic research fellowship from the Alfred J. Sloan foundation to S.B.P.

[2]Investigator, Howard Hughes Medical Institute.

[3]Deceased, November 25, 1977.

ISBN 0-12-668350-6

tivity is preserved. In a near equilibrium gradient, the scrapie infectivity was found distributed over a range of densities from 1.08 to 1.30 gm/cm^3. A rate-zonal gradient showed the infectivity was distributed over a range of particle sizes with $S_{20,w}$ values from $\sim$40S to >500S. Incubation of P_5 at 37° or 80°, under conditions which disrupt ribosomes, was found to alter dramatically the rate-zonal gradient profile of the agent. Under these conditions the agent sedimented as a particle with an $S_{20,w}$ >500S. The apparent heterogeneity of the scrapie agent with respect to both size and density and its ability to shift from one form to another suggest that the agent may contain non-polar regions or hydrophobic patches on its surface. A variety of hydrophobic proteins are known to undergo aggregation and dissociation as well as bind lipids.

INTRODUCTION

Scrapie infection is attended by a slow, progressive spongiform degeneration of the central nervous system. To date, the scrapie agent has eluded isolation and identification; and it appears to be a novel infectious entity differing from conventional viruses in many respects (1-4). In large part, attempts at purification of the agent have been hampered by the slow mouse titration assay which requires incubation periods up to twelve months prior to scoring.

The unusual resistance of the scrapie agent to inactivation by heat, formalin and ultraviolet radiation as well as its small ionizing radiation target size have prompted several investigators to suggest that the scrapie agent may not contain a nucleic acid genome (1,3,4). However, any confirmation of such a suggestion must await purification of the agent to homogeneity.

The elusiveness of the scrapie agent to isolation and identification for nearly two decades and the recognition of analogous agents causing the diseases Kuru and Creutzfeldt-Jakob disease in humans and transmissible encephalopathy in mink have intensified the interest in this agent (1,2) It has been suggested that similar agents may play a role in the causation of a wide variety of human diseases for which etiologies remain obscure. Such diseases include multiple sclerosis, amyotrophic lateral sclerosis, rheumatoid arthritis, systemic lupus erythematosus, juvenile diabetes, many forms of human cancer, Parkinson's disease, progressive supranuclear palsy and the senile and presenile dementias of Alzheimer's disease (1). Whether or not agents similar to that which causes scrapie will be found to play a causative role in these diseases remains to be established. By analogy, the elusiveness of the scrapie agent to identification is not unlike the putative agents whose existence has been inferred from a variety of epidemiological studies for the diseases noted above. Indeed, elucidation of the molecular structure of the scrapie agent and the development of rapid and sensitive assay techniques for detecting similar agents may have enormous implications for studies on a wide variety of so-called "degenerative diseases."

In order to develop a preparatory procedure for the purification of the scrapie agent, we initially studied the sedimentation characteristics of the agent in fixed-angle rotors using the technique of analytical differential centrifugation to generate a series of sedimentation profiles (5-7). The sedimentation profiles were observed to be similar for the agent in homogenates of both murine spleen and brain, and were not substantially altered by the addition of detergent. From these sedimentation profile studies a purification procedure was developed which yielded a fraction enriched for the scrapie agent approximately 20-fold with respect to cellular proteins. As predicted from the sedimentation profile data, the partially purified preparation of the agent contained many ribosomal structures. To separate the agent from the contaminating ribosomes sucrose gradient centrifugation was employed. These studies indicate that after a partial purification of the agent from spleen and storage at -70° the agent exhibits heterogeneity with respect to both size and density. When preparations of the agent were heated to 37° and 80° under conditions which disrupt ribosomal structures, the agent appeared to sediment in an aggregated form. The data presented suggest that the scrapie agent may exist in multiple molecular forms.

METHODS

Materials. All materials were of the highest grade commercially available. [^{35}S]-Simian Virus 40 (SV40) was a generous gift from Drs. J. Reiser and G. Stark.

Source of the Scrapie Agent. The inoculum used in studies reported here represented the 4th passage of a "Chandler" strain in mice (8). Brain tissue from 50 Swiss mice with clinical signs of scrapie was homogenized in 9 vol of 0.32 M sucrose and centrifuged at 121 g for 10 min at 4°. Inocula were prepared from this supernatant fluid. Control mice were similarly inoculated with analogous suspensions prepared from the brains of healthy Swiss mice. The controls neither developed signs of neurological dysfunction nor histopathological changes of the brain as judged by light and electron microscopy.

Preparation of Homogenates. Female Swiss mice, 1 month old, were inoculated intracerebrally with 0.03 ml of mouse-adapted scrapie agent (10^6 ID_{50} units). For studies on the agent in spleen, mice were killed 40 days after inoculation; their spleens were removed immediately and washed in 20 mM Tris-HCl, pH 7.4, containing ice-cold 250 mM sucrose. A Potter-Elvehjem glass homogenizer equipped with a motor-driven Teflon pestle was used to prepare a 20% homogenate of the tissue (w/v). The homogenizing media for spleen consisted of 20 mM Tris-HCl, pH 7.2/250 mM sucrose and unless otherwise noted, all procedures were performed at 4°. The homogenates were centrifuged for 10 min at 121 g in a Sorvall RC2B centrifuge equipped with an SS34 rotor, and the supernatant fluids removed. The pellets were rehomogenized in additional buffer and centrifuged again. After the pellets from the second centrifugation were washed, the three portions of supernatant fluids were combined; the final suspensions were 10% (w/v). The 121 g combined supernatant fluids designated "S_1" were used as the starting material for all of the analytical differential centrifugation studies reported here.

Sedimentation Analysis in Fixed-Angle Rotors. All analytical differential centrifugation studies were performed in a Spinco L2-65B ultracentrifuge equipped with a 50 Ti fixed-angle rotor. Two-ml aliquots of the 121 g combined supernatant fluid were centrifuged at specified speeds for given time periods in cellulose nitrate tubes (7.9 mm diameter X 49 mm length). The total time of centrifugation included 33% of the time required for acceleration and deceleration of the rotor. After centrifugation was completed, a long-tip Pasteur pipet was introduced along the wall of the tube that had been positioned toward the interior of the rotor during centrifugation to minimize contamination of the supernatant fraction by pelleted material. About 1.7 ml of supernatant fluid were slowly aspirated; fluid overlying the pellet was left undisturbed (5,6).

Sucrose Gradient Centrifugation. Partially purified suspensions (P_5) containing the scrapie agent were stored at -70°. Before use they were rapidly thawed in a 37° bath and held at 4° prior to sonication for 15 sec with a Bronwill sonicator. For near equilibrium centrifugation a Buchler conical gradient maker was used to construct the gradients; each gradient had a total volume of 13.2 ml. A sample (0.6 ml) was layered on the 20-65% (w/w) linear sucrose gradient buffered with 20 mM Tris-HCl, pH 7.4. The gradients were centrifuged in an L2-65B Spinco ultracentrifuge, using an SW41 rotor at 41,000 rpm for 19 h.

For the rate-zonal gradient centrifugation studies, gradients were constructed by layering in order 3.6 ml each of 65%, 30%, 25%, 20% and 15% sucrose in 0.02 mM Tris-HCl, pH 7.4. They were left standing overnight at 4° before use. A sample (0.45 ml) was floated on the top of the 15-30% (w/w) linear sucrose gradient with a 65% (w/w) sucrose cushion. Centrifugation in an SW 27.1 rotor at 27,000 rpm was for 3.5 h. All gradients were separated into 1-ml fractions from the top, using a Buchler Auto Densi-Flow apparatus.

Estimation of $S_{20,w}$ values for particles sedimenting in rate-zonal sucrose gradients was performed as described by McEwen (9).

Assay of Biochemical Markers. RNA was measured by the procedure of Schneider using orcinol reagent after repeated ethanol precipitation of the RNA (10). Commercial preparations of yeast RNA were extracted with phenol and used as standards assuming an $E_{260}^{1\%} = 250$ for RNA. Protein was determined by the method of Lowry *et al.* (11); bovine serum albumin was used as a standard. Succinic dehydrogenase (E.C.1.3.99.1) and lactic dehydrogenase (E.C.1.1.1.27) were measured spectrophotometrically as previously described (7). Absorption at 260 nm was measured in a Gilford spectrophotometer 252, equipped with a rapid sampling device having a 1 cm pathlength. Sucrose densities were determined by refractometry, using a Bausch and Lomb Abbé refractometer.

Assay of Scrapie Agent Infectivity. Female Swiss mice of Rocky Mountain Laboratory stock were used to titrate the amount of scrapie agent in the fractions of supernatant fluid. Serial 10-fold dilutions of the fractions were prepared with phosphate buffered saline that contained 10% fetal calf serum, 0.5 unit/ml penicillin, 0.5 μg/ml streptomycin and 2.5 μg/ml amphotericin. For each dilution, 6 mice were inoculated intracerebrally with 0.03 ml of the diluted suspension. The animals were examined weekly during the next 12 months for clinical signs of scrapie, i.e., bradykinesia, plasticity and rigidity of the tail, waddling gait, and a coarse, ruffled appearance of the coat (Fig. 1). Histologic examination of the brain and spinal cord was done occasionally to confirm the clinical diagnosis. Titers were calculated by the method of Spearman and Kärber (12). The standard errors of these titrations varied between ±0.2 and ±0.4 log units.

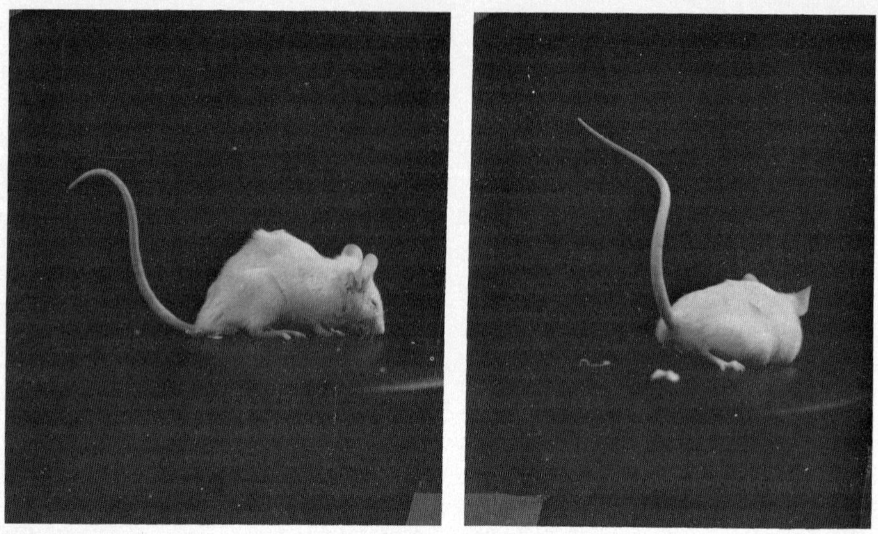

FIGURE 1. A Swiss mouse with clinical signs of scrapie. Note the marked rigidity and posturing of the tail.

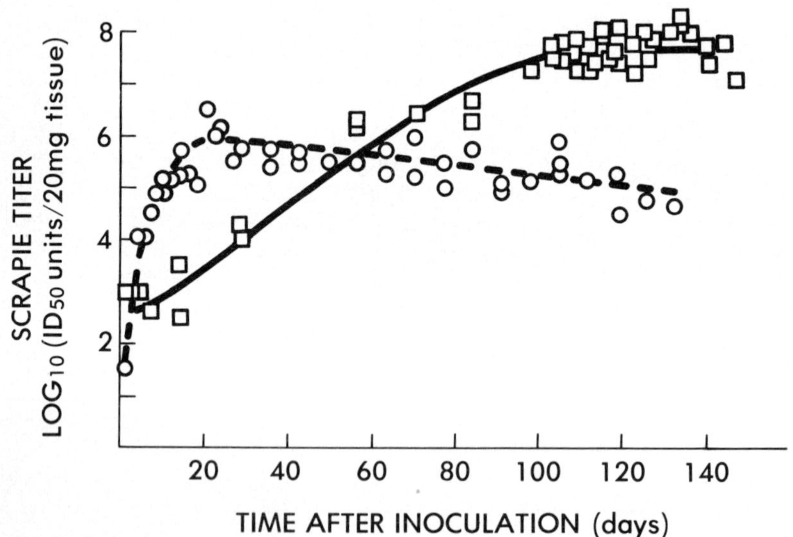

FIGURE 2. Time course for replication of the scrapie agent in murine spleen (--0--) and brain (—□—). Mice were inoculated intracerebrally at time zero (adapted from Kimberlin, Ref. 14).

RESULTS

Almost a decade ago studies on the pathogenesis of scrapie infection in mice by Eklund, Hadlow and Kennedy showed that the agent was found in high titers in the spleen early after intracerebral inoculation (13). Those studies were extended by others as shown in Fig.2 where numerous determinations of scrapie infectivity in spleen and brain during the pathogenesis of the disease have been made (14). The titer of scrapie in the spleen reaches a maximum of $\sim 10^6$ ID_{50} units per 20 mg of tissue within 20 days after intracerebral inoculation and remains almost constant throughout the remaining course of the disease. Coordinate with this early rise in scrapie titer is the development of a modest yet significant degree of splenomegaly as well as a depression of the polyclonal mitogenic response of splenocytes from C3H/HeJ and Balb/c mice to stimulation by lipopolysaccharide (15,16). In contrast, more than 100 days are required for scrapie titer to reach a maximum in brain. At this time the mice begin to exhibit the clinical features of scrapie described above. As shown in Fig.2 the final titer of scrapie in brain tissue is significantly higher than that found in spleen.

The Pathology of Scrapie Infection. The pathology of scrapie infection in the mouse is characterized by spongiform changes consisting of vacuolation and degeneration of neurons, marked proliferation of astroglial cells, and the absence of an inflammatory reaction (17,18). It is noteworthy that the histological changes seen during scrapie infection at the light microscopic level as shown in Fig.3 can all be seen in other diseases. However, it is the distribution of these lesions, their time of appearance, and their correlation with a specific clinical syndrome that make the diagnosis of scrapie inescapable. By electron microscopy, collections of particles found within evaginated post-synaptic processes do appear to be characteristic of scrapie infection in the mouse (18). The particles shown in Fig. 4 were first reported in association with scrapie infection by David-Ferriera and colleagues, but their consistent presence with scrapie may not have been appreciated (18,19). These particles are at times found in crystalline arrays. The core of the particles measures 23 nm in diameter, while the center-to-center distance of the arrays is 37 nm. Whether or not these particles are infectious and indeed represent the scrapie agent remains to be established. As discussed below, their size is too large for each of the particles to represent a single infectious unit of the agent. They may indeed represent aggregated forms of the agent with a regularized structure.

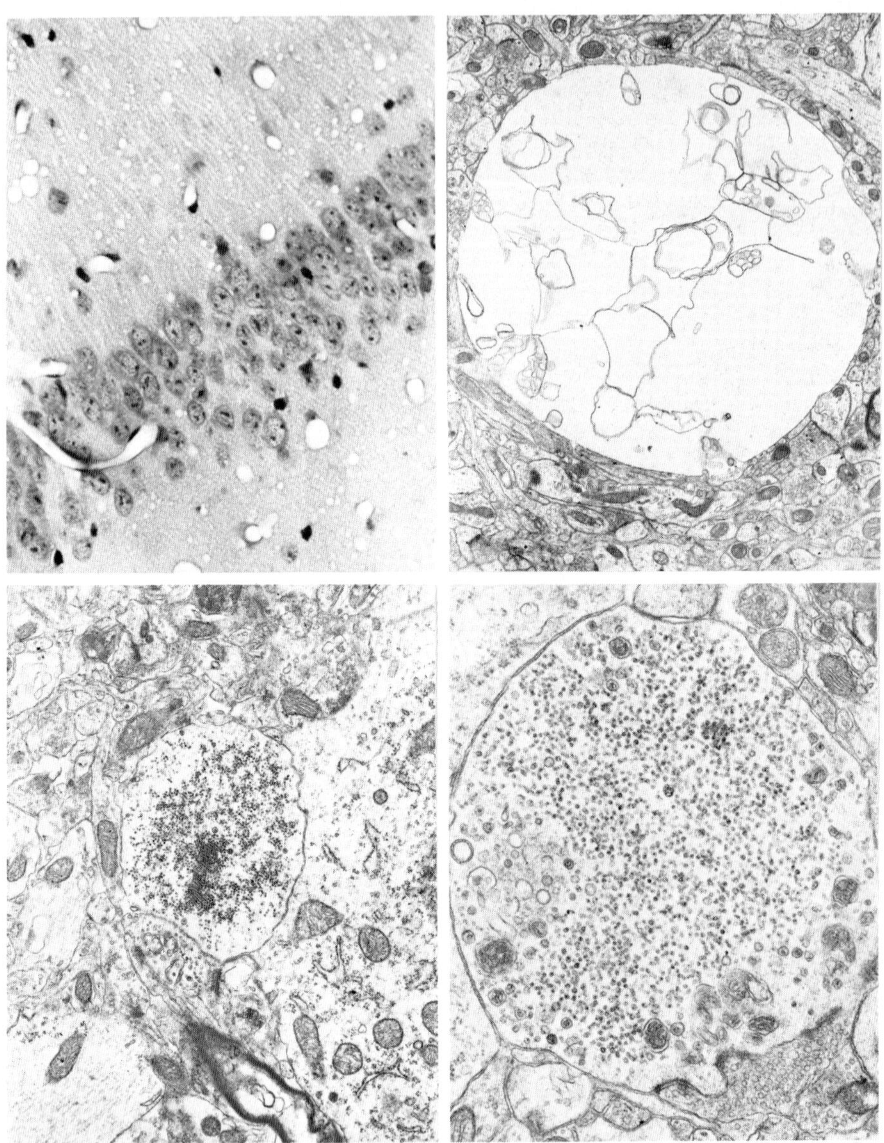

FIGURE 3. Light micrograph showing extensive vacuolation in neuropil of mouse hippocampus at 5 months after scrapie infection. Hematoxylin and eosin stain X500 (upper left).

FIGURE 4. Electron micrographs of mouse cortex at 5 months after scrapie infection. Appearance of a vacuole in cortex X7800 (upper right). Post-synaptic process containing dense particles X11,600 (lower left) and another process containing particles X22,300 (lower right).

Strategy for Developing a Purification Procedure for the Scrapie Agent. Multiple attempts to isolate the scrapie agent have been unsuccessful, but these have used murine brain as a source of the agent because of the high titers found in this tissue late in the pathogenesis of the disease (1,3,4). We chose to concentrate our initial efforts on the spleen because of the relatively rapid replication of the agent in the spleen and the ease of fractionation of this tissue compared to brain. Most biological macromolecules can be purified at least in part by a series of differential centrifugations; but because of the enormously long incubation periods required to assay the agent, the usual empirical approach to find optimal conditions for purification by differential centrifugation seemed inadequate. In order to optimize systematically our choice of centrifugation conditions, we chose to explore the sedimentation behavior of the scrapie agent in fixed-angle rotors using the technique of analytical differential centrifugation (5,6). In these studies homogenates of both spleen and brain were subjected to low speed centrifugation to remove nuclei and unbroken cellular debris. The supernatant from this low speed centrifugation was designated S_1. Aliquots of the fraction S_1 were then subjected to centrifugation in a fixed-angle rotor for increasing speeds and times. In these studies the activities found in the supernatant fraction after centrifugation were plotted as a function of $\omega^2 t$ where ω is the angular velocity of the rotor in rad/sec and t is the time of centrifugation in sec.

Sedimentation Analysis in Fixed-Angle Rotors. The sedimentation profiles for three subcellular fraction markers are shown for the S_1 fraction from murine spleen in Figure 5. At low $\omega^2 t$ values succinic dehydrogenase, a mitochondrial marker, is readily removed from the supernatant fraction while higher values of $\omega^2 t$ are required to remove RNA which is a marker for ribosomes in the main. Even greater values of $\omega^2 t$ are required to remove from the supernatant fluid lactic dehydrogenase which is a soluble enzyme marker. The addition of 0.5% (w/v) sodium deoxycholate (DOC) caused a dramatic shift in the profile of succinic dehydrogenase. Its profile became virtually indistinguishable from that of lactic dehydrogenase, illustrating the release of succinic dehydrogenase from mitochondria upon detergent solubilization of mitochondrial membranes. In contrast, the RNA marker for ribosomes showed only a minor shift to the right indicating that relatively few of the ribosomes found in the S_1 fraction were membrane bound.

In Figure 6, the sedimentation profiles for the scrapie agent in the S_1 fraction of murine spleen are depicted in the absence and presence of 0.5% deoxycholate. As shown, the scrapie agent remains in the supernatant fraction at $\omega^2 t$ values less than 10^9 rad^2/sec while it is sedimented at ω^2t values at 10^{11} rad^2/sec or greater. The addition of DOC did not result in a substantial shift of the sedimentation profile for the agent. Similar sedimentation profiles have been observed for the scrapie agent from murine brain indicating that the sedimentation behavior of the agent is independent of the tissue in which it is found.

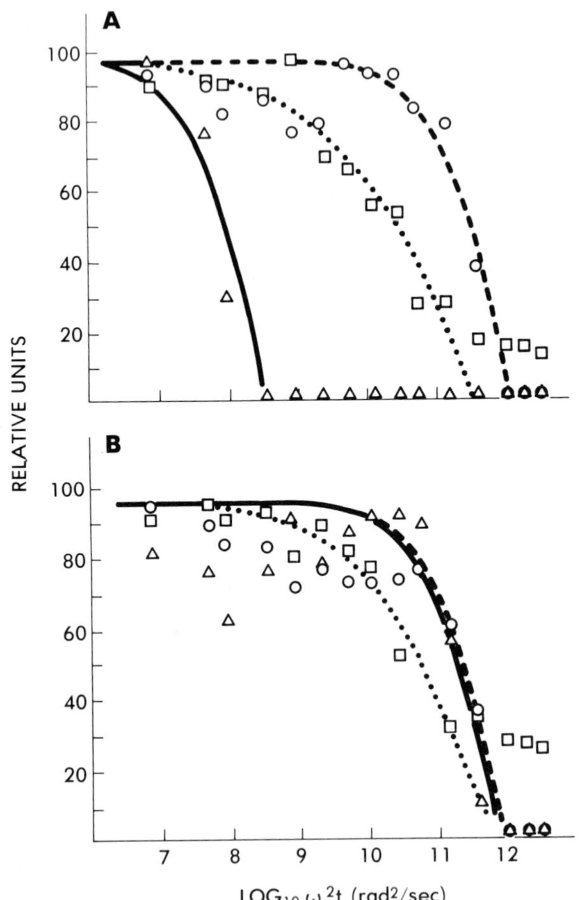

FIGURE 5. Sedimentation profiles for succinic dehydrogenase (—Δ—), RNA (···□···) and lactic dehydrogenase (--O--) in the S_1 fraction from scrapie-infected murine spleen. A) Samples were untreated and B) treated with 0.5% (w/v) sodium deoxycholate.

Because the sedimentation of particles in fixed-angle rotors differs from that in sectored cells, the sedimentation behavior of a spherical non-enveloped virus, SV40, was stud-

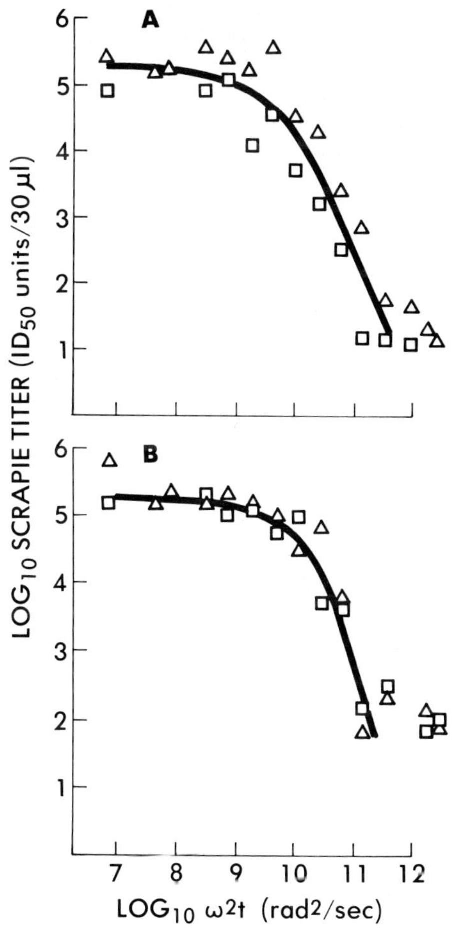

FIGURE 6. Sedimentation profiles of the scrapie agent in the S_1 fraction from murine spleen. A) Samples were untreated and B) treated with 0.5% (w/v) sodium deoxycholate. Data from two separate experiments denoted by (-Δ-) and (-□-) symbols.

ied. The SV40 was propagated in cultured cells in the presence of [^{35}S]-methionine and then purified by CsCl density gradient and sucrose rate-zonal gradient centrifugation. The purified radioisotopically-labelled virus was then added to S1 fractions of murine spleen from uninfected mice. As shown in Figure 7, the curve is quite similar to that observed for the scrapie agent though the scatter of the data is considerably less. Calculation of the sedimentation coefficient

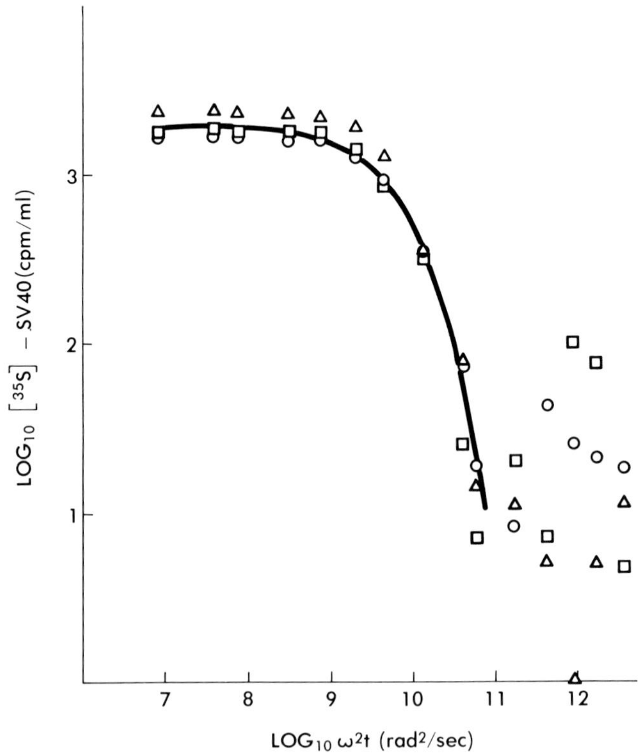

FIGURE 7. Sedimentation profiles of SV40. The [^{35}S] labelled virus was mixed with the Tris-HCl sucrose buffer containing DOC (-Δ-), the S_1 fraction from control mice (-□-) or the S_1 fraction treated with DOC (-O-).

PATHOGENETIC ASPECTS OF SUBACUTE SCLEROSING PANENCEPHALITIS[1]

V. ter Meulen, W.W. Hall,[2] and H.W. Kreth

Institut für Virologie und Immunbiologie
der Universität Würzburg
D-8700 Würzburg, W. Germany

Subacute sclerosing panencephalitis (SSPE) is an uncommon slowly progressing disorder of the central nervous system (CNS) in children and young adults which has been associated with a measles virus infection (1, 2). This disease has certain clinical and laboratory features which are pathognomonic for this CNS infection. The clinical course is usually stereotype beginning with insidious behavioural changes which, weeks or months later, are followed by characteristic neurological symptoms accompanied by a typical electroencephologram pattern. Laboratory investigations have revealed high measles antibody titers in serum and cerebral spinal fluid (CSF) specimens with a reduced serum/CSF ratio indicating intracerebral production of IgG antibodies. In addition, the antibodies in these CSF represents oligoclonal IgG corresponding in large part to antibodies against measles virus. Virological studies have demonstrated measles antigen in SSPE brain cells and led to the isolation of a measles-like virus (referred to as SSPE virus) from brain and lymphnode material.

These findings incriminate measles virus as the etiological agent in this disease, however, they do not explain the pathogenesis of SSPE. If measles virus is involved, additional factors either host or virus derived, must play a pathogenetic role since rarity and rural prevalence of SSPE cannot be correlated to the ubiquitous measles infection. Moreover, the mechanisms have to be explained by which measles virus infection persists in the CNS and is activated years after onset of acute measles. The understanding of these disease processes will depend to a great extent not

[1]This work was supported in part by the Deutsche Forschungsgemeinschaft.

[2]Present address: The Rockefeller University, New York, N.Y.,U.S.A.

ISBN 0-12-668350-6

only on the role of the immune response to this CNS infection but also on our knowledge about biological and biochemical properties of SSPE and measles virus.

IMMUNE REACTIONS IN SSPE

Measles virus infection of a sero-negative immunocompetent host results in an acute disease with a life-long immunity. During this acute illness, the virus infection has a direct effect on the immune system which indicates a high tropism of measles virus to the lymphoid organs of the host. It has been shown that giant cell formations are present in lymphnodes as well as thymus resembling cytopathic changes in measles infected tissue cultures (3). Measles virus can easily be isolated from washed leukocytes of patients with acute measles which also reveal chromosome breakage in lymphocytes (4). Moreover, during this infection a partial or total tuberculin skin test anergy develops in tuberculous children which is sometimes accompanied by an activation and spread of this bacterial infection (5). Moreover, in malmourished children, who have been shown to be T-lymphocyte deficient, a measles infection often leads to the development of kwashiorkor (6). Similar severe complications are observed in individuals suffering from congenital, acquired or iatrogenic T-cell deficiency (7). These patients often develop giant cell pneumonia, abnormal rash or a subacute CNS infection. Similar complications have been found in children with leukemia who sometimes develop fatal lung or CNS complications associated with a persistent measles virus infection months after the initial measles infection (8). These observations suggest that an intact cell mediated immune system is vital for the host to overcome and control an infection with measles virus.

Therefore, the demonstration of measles virus in SSPE brain has led to the interpretation that this CNS infection might be a late complication of measles virus infection in an immuno-incompetent host. Accordingly, many investigations have been carried out to analyze the humoral and cell-mediated immune reactions in SSPE.

Humoral Immunity. So far, no immunologic deficiencies in the humoral immunity have been detected in SSPE (2). All standard tests employed to assess humoral immune responses are within the normal limits. Normal counts of peripheral B and T lymphocytes and normal levels of immunoglobulins are found in most patients. Active immunization with different viral or bacterial antigens led to normal antibody responses. However, each SSPE patient tested showed

Antigenic studies on isolated M proteins. In order to provide further and more definite evidence that there are real antigenic differences between measles and SSPE viruses, we have recently carried out studies on M proteins isolated from purified virus preparations (26). Such studies have proven to be particularly difficult since large amounts of purified virus are difficult to prepare and methods for the isolation and purification of M proteins are relatively inefficient. M protein was isolated from purified Edmonston and LEC viruses using a modification of the method of Scheid et al., (27). Virus envelope fragments were prepared by disrupting the virus with 2% triton-X 100 in the presence of 1 M KCl. After removal of the salt by dialysis, M protein could be collected by centrifugation through a sucrose cushion. M protein was purified from this "partially pure" preparation by repeated cycles of solubilization, re-precipitation and centrifugation through sucrose cushions. Usually after 5 to 7 of such cycles, pure M protein could be obtained which was free of viral NP-protein. The purified preparations used contained a single polypeptide of molecular weight of approximately 38.000 and had an amino acid composition which would be expected for a hydrophobic protein (28). Electron microscopic examination revealed no definable structural organization but rather a heterogenous mixture of subunit structures.

Antisera against the purified protein were prepared in rabbits using purified protein in a H_2O - deoxycholate (o,2%) suspension. Specific antibodies were detected in a solid phase indirect radioimmuno assay and antibody titers up to 2×10^{-15} were determined. Inhibition tests which included HI, HLI and neutralization assays demonstrated that the antisera did not contain any of these activities. The antigenic relatedness of the M proteins of SSPE and measles viruses were then assayed by immuno-diffusion and immuno-electrophoresis according to Laurell (29). It could be shown that the M proteins of these viruses were immunologically distinct. Fig. 1 reveals an antigen-antibody reaction resulting in a "rocket" formation only in the homologous system whereas the heterologous antigens were not immuno-precipitated. These results provide a firm basis for the antigenic differentiation of measles and SSPE viruses used in this study.

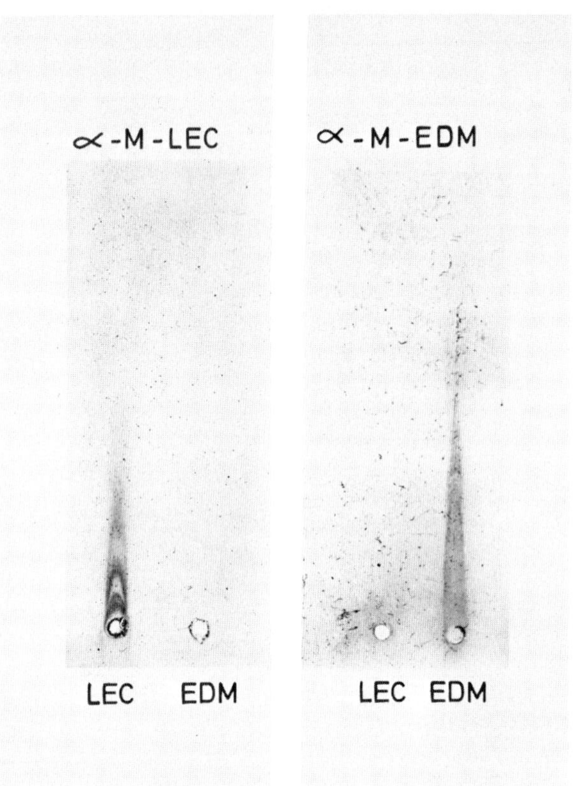

Fig. 1. Rocket immuno-electrophoresis according to the method of Laurell (29). A precipitation occurs only in the homologous antigen-antibody system.

LEC = LEC M protein
EDM = Edmonston M protein
∝M-LEC = rabbit antiserum against LEC M protein
∝M-EDM = rabbit antiserum against LEC M protein.

Each antigen well obtained 1000 antigen units as determined by a solid phase indirect radioimmune assay. Reprinted with permission from Hall et al., (26).

Biochemical Characterization. On the basis of its morphological, biological and biochemical properties, measles virus has been classified as a member of the morbilli virus group of paramyxoviruses. The general characteristics of paramyxovirus replication have been established over the last decade and, recently, it was shown that measles virus replication in a permissive infection follows a closely similar pattern (30-32, 25). The genetic information of the Virus is contained in a single-stranded RNA which sediments at 50 S and has a molecular weight of approximately 5 x 10^6 daltons. In a permissive infection, this genome RNA is copied into complementary messenger (m) RNA molecules sedimenting heterogeneously at 12 to 36 S in aqueous sucrose gradient. They serve as templates for virus-specific protein synthesis. The genomic RNA is self-replicated through a complementary "positive-strand" 50 S RNA (anti-genome) which in turn acts as a template for new 50 S negative strands. These events are continuously repeated until at some, as yet unknown, point the macro-molecules are assembled and new virus particles are released from the cell membrane. These well defined steps of measles virus replication offer a chance to compare them to the corresponding steps of SSPE virus replication.

Genetic relatedness of measles and SSPE viruses. Studies have been carried out to compare several SSPE isolates with measles virus in terms of genome homology (33). In these investigations, the ability of unlabeled 12 S to 36 S mRNAs from different virus strains to compete with the annealing of tritium-labeled mRNA to its corresponding genome were assayed. The results showed that, although the viruses were highly related, minor differences in genome homology did exist. Moreover, the results suggested that the SSPE genome may contain some additional information over and above that contained in the measles virus genome.

Comparison of mRNA synthesis in measles and SSPE virus-infected cells. Fig. 2 shows sucrose gradient sedimentation profiles of RNA synthesized in Vero cells infected with measles (Edmonston) and SSPE (Lec) viruses. The majority of the RNA sedimented at 12 to 36 S with a major peak at 18 S. These RNAs are considered to be the virus messengers since they contain poly (A) sequences and are associated with polyribosomes (30, 34). The mRNAs were further purified by chromotography on oligo d(T) cellulose, heat denaturation and resedimentation (Fig. 1). In order to detect differences between the viruses, the purified mRNAs were subjected to electrophoresis on polyacrylamide gels in the presence of formamide and urea, the idea being that if a difference in their genetic and antigenic composition exists then it should be reflected in at least one of the complementary messenger species. Fig. 3 shows a densitometer scan of a prepared fluorogram of such an experiment. It can be seen that six major mRNAs can be resolved, and these had molecular weights of approximately (1) 0.39 x 10^6; (2) 0.49 x 10^6; (3) 0.58 x 10^6; (4) 0.62 x 10^6; (5) 0.72 x 10^6; (6) 0.79 x 10^6 relative to 16 S, 23 S and 18 S ribosomal RNAs. The molecular

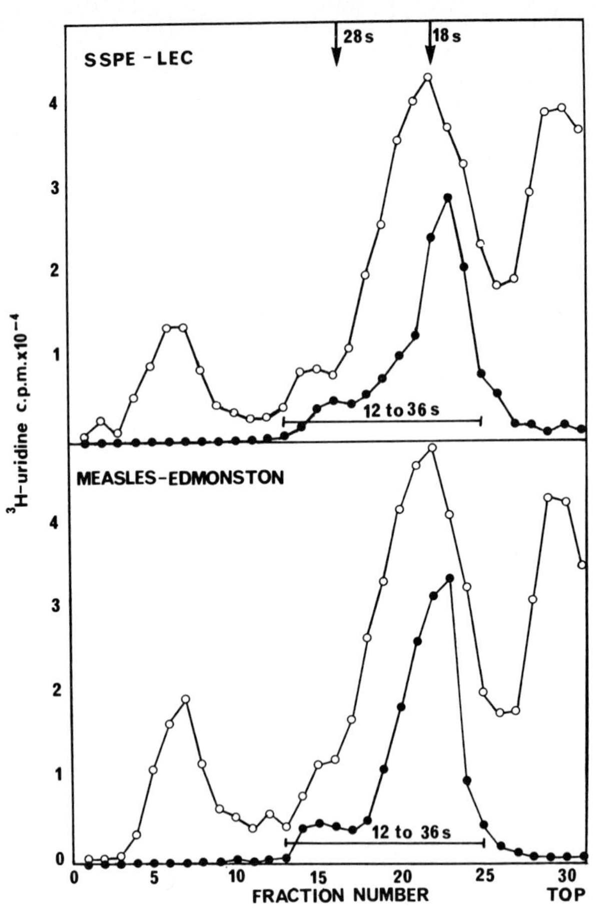

Fig. 2. Sucrose gradient sedimentation of RNAs synthesized in measles (Edmonston) and SSPE (Lec) virus-infected Vero cells. RNA was isolated as described by Hall & ter Meulen (30). 12 to 36 S RNAs were pooled, precipitated in ethanol, chromatographed on oligo d(T) cellulose, heat denatured and re-sedimented on sucrose gradients.
o—o, ^{3}H-uridine ct/min of total RNA; ●—●, ^{3}H-uridine ct/min of purified poly (A) rich mRNAs.

weights have been corrected to account for 100 nucleotides in their poly (A) sequences. In addition, one or more minor high molecular weight species (designated L_1 L_2 and L_3) were usually present in the gel. It is thought that at least one of these represents a large messenger molecule since a large protein has been detected both in purified virus particles and in virus-infected cells (32, 31). The function of the remaining L RNAs remains unknown. The similarity of the migration of the RNAs was very striking in spite of the fact that RNAs (3) and (4) were not always well resolved. The one minor difference was associated with mRNA (1), and was consistently observed in SSPE virus-infected cells where it revealed a slightly slower migration (26).

Correlation of mRNAs with virus polypeptides. In order to correlate the various mRNAs with the proteins present in the infectious virus particles, the polypeptide composition of the Edmonston and Lec viruses were examined by electrophoresis on 7,5% polyacrylamide gels. Fig. 4 shows the comparative profiles obtained. In the following, the designations proposed by Graves et al., (31) for the nomenclature of measles virus polypeptides have been adopted. The polypeptide profiles are similar to those recently described by other workers (35, 32, 31). The material at the top of the gel probably represents a mixture of aggregation products and/or a large polypeptide (L). The latter is presumably coded for by one of the large mRNAs mentioned above. Other polypeptides clearly resolved had the following molecular weights: H, a glycoprotein (78,000); P, a phosphoprotein (69,000), (36, 32), the nucleocapsid subunit protein NP (60,000); and the membrane protein M (38,000). The origin and function of P53 (our designation) is unknown; whether this is unique or represents a degradation product has not been established. The other polypeptide peak migrating in the 40-44,000 dalton range apparently represents a mixture of host-cell actin (A) and F_1 (31). In recent studies, workers at the Rockefeller University in New York (31, 37) have demonstrated in measles virus-infected cells the existence of a glycosylated protein not found in purified virus. This is thought to be F_o, the precursor to F_1 and F_2 which appear in pulse-chase experiments. Further studies by these workers have suggested that F_1 and F_2 are produced as a result of proteolytic cleavage of the F_o precursor (molecular weight 60,000) and that the former are also linked by disulphide bonds. This processing is similar to that of the classical paramyxoviruses except that in these viruses, the F_1 protein possesses carbohydrate side chains, whereas in measles virus this seems not to be the case. On the basis of only molecular weights, the following

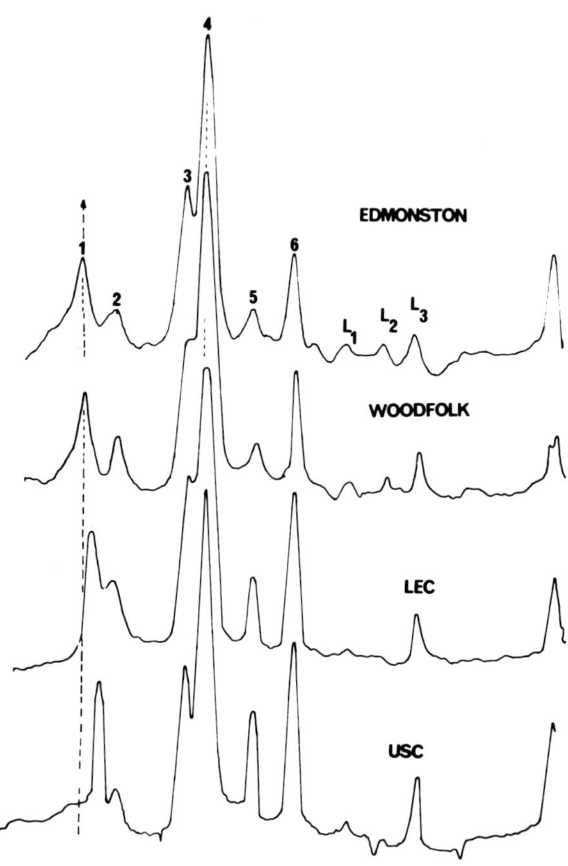

Fig. 3. Gel electrophoresis of the purified 12 to 36 S mRNAs from two measles (Edmonston, Woodfolk) and two SSPE (LEC, USC) viruses on 2.5% (w/v) polyacrylamide slab gels in the presence of 6M urea and 97% formamide. Conditions of electrophoresis were described by Villareal et al. (51) and migration is from right to left. Gels were prepared for fluorography and prepared fluorograms were scanned using a Zeiss spectrophotometer. Mol. wts of individual RNAs were calculated relative to 16 S, 18 S and 23 S rRNAs. In the case of 28 S rRNA the linear relationship between migration and log mol. wt no longer existed. (Reprinted with permission from Hall et al., (26) and Nature)

correlations between mRNAs and virus polypeptides are proposed: 1 = M; 2 = not yet known; 3 = F_O; 4 = NP; 5 = P; 6 = H, and $L_{1/2/3}$ = L. It must be stated that, at the present time, these correlations are tentative and will only be definitely established when in vitro translation studies of the individual RNAs are carried out. However, on the basis of such a correlation the difference between the measles and SSPE viruses would seem to be associated with the membrane (M) protein. Evidence that a difference may exist in the M proteins was first indicated by Schluederberg et al., (38). These workers reported that on polyacrylamide slab gels, the M protein of one SSPE strain migrated more slowly than that of a wild-type measles virus. Confirmation and extension of this finding has been reported by Wechsler and Fields (39) who compared polypeptide synthesis in CV1 cells infected with measles and SSPE viruses all synthesized M proteins which migrated more slowly than the M protein of wild-type measles. These workers also demonstrated similar migrational differences in purified virus preparations and found some indication of differences in migration of the P protein. The significance of this remains uncertain in that this protein is phosphorylated and different degrees of phosphorylation may produce migration differences. The comparison of the mRNA synthesized in infected cells indicate that the mRNA for the SSPE M protein is larger than that of the measles mRNA. This may well account for the differences in genome homology, as well as for the seemingly larger molecular weight of SSPE M protein. Moreover, the M proteins of these two viruses studied are antigenically different demonstrating that SSPE virus is, in fact, not identical with measles virus.

DISCUSSION

The epidemiological observation that SSPE is preceded in average 7 to 8 years by an acute measles virus infection or measles vaccination has led to many hypotheses in the attempt to explain the etiology and pathogenesis of this disorder. Under normal conditions, on infectious virus entering a sero-negative host induces an acute infection which can be overcome by the defense mechanisms of the host, whereby the immune system plays a major role. In SSPE, however, provided the acute measles infection is etiologically related to the disease process, the agent escapes these defense mechanisms and induces a chronic CNS disorder, which cannot be controlled by the organism. The question therefore arises whether this escape is due to a functional impairment of the host immune system or is related to unique

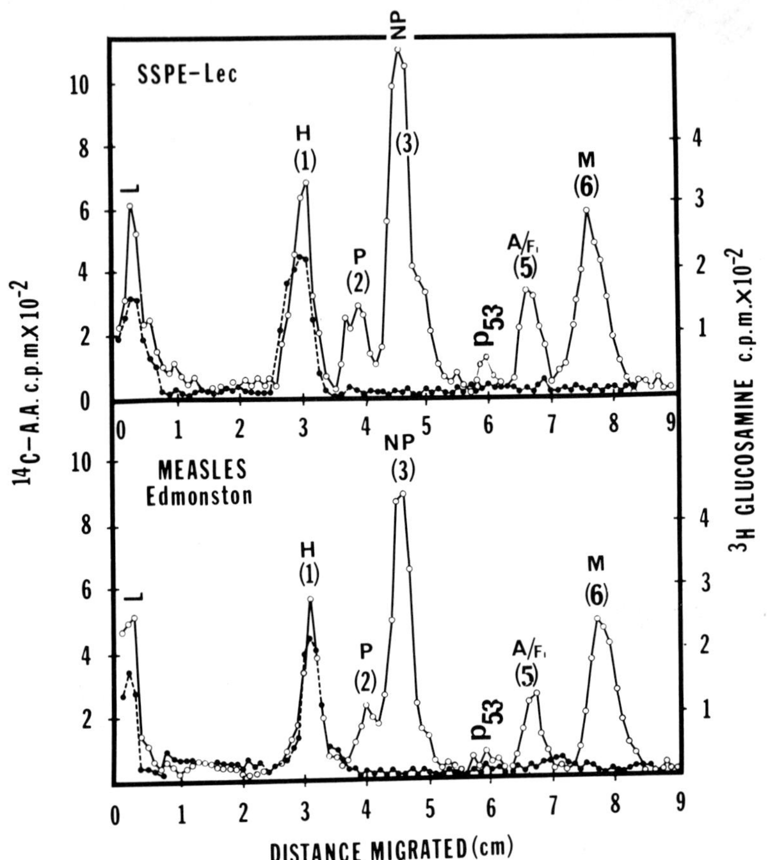

Fig. 4. Gel electrophoresis of virus polypeptides on 7.5% polyacrylamide gels (52). Viruses were grown in Vero cells in MEM containing ^{14}C-protein hydrolysate (10 µCi/ml) and ^{3}H-glucosamine (5 µCi/ml) for 60 hr. Virus was purified from tissue culture fluids as follows. Cell debris was removed by centrifugation at 3000 rpm for 20 min and virus was pelleted from the supernatant through a 25% (w/w) sucrose cushion for 2 hr at 25,000 rpm in a SW27 rotor. Pellets were resuspended in NTE and banded on 15 to 40% (w/w) potassium tartrate gradients by centrifugation at 25,000 rpm for 3 hr. Virus bands were collected, diluted, pelleted and directly electrophoresed. The nomenclature used for the polypeptides was after Graves et al. (31) and is detailed in the text.

properties of the agent involved.

SSPE patients display on oligoclonal hyperimmune response against measles. This type of immunological response is very unusual for a viral infection since normally a heterogeneous antibody response is found which, for most viruses, is T-cell dependent. Since no immunological data are available during the period between acute measles and the onset of SSPE, one has no information whether SSPE patients may be genetically high responders to measles antigens. A genetically-determined high susceptibility has been discussed but no definite HLA pattern has been observed in SSPE (2). Moreover, there are three reports of the occurrence of SSPE in one of identical twins arguing against a genetic basis for this disease (2). The observation that measles virus infects lymphoid cells and, in acute measles, causes a marked impairment of CMI suggests that the specific unresponsiveness to measles virus found in skin tests with SSPE patients could result from a selective depression of the respective T cell clones. If suppressor cells were affected, normally controlling antibody synthesis to measles virus, then a hyperimmune reaction would occur as it is indeed found in cases of SSPE. An explanation for the oligoclonal hyperimmune reaction could be the regulatory involvement of anti-idiotypic antibodies. Based on preliminary experimental evidence in small laboratory animals, anti-idiotypic antibodies may lead to either an inhibition of the corresponding idiotype production or to a marked stimulation of its synthesis, depending on the IgG sub-class of the anti-idiotype antibodies injected (40). So far, no evidence for such immunological regulatory mechanism has been presented in SSPE.

It is also conceivable that an oligoclonal antiviral hyperimmune response could favor the occurrence of virus mutants, since antibodies only directed against certain determinants of the parent virus would increase the probability of a mutant to escape neutralization. In the presence of antiviral antibodies, it could be shown that influenza viruses undergo antigenic changes of their surface proteins. Under suitable experimental conditions, influenza mutants have been selected *in vivo* and *in vitro* (41). However, in the same host, a recurrent influenza infection has never been linked to a mutant virus. In contrast to these observations in Visna, a slow virus disease of sheep, a mutant derives in the infected host which is responsible for this disease. Gudnadottier and co-workers (42) and Narayn and co-workers (43) have shown that an antigenic shift occurs in Visna which allows the virus to escape neutralization. Isolated Visna virus strains from ex-

perimentally infected animals differ in their antigenicity to the virus inoculum. Moreover, propagation of Visna virus in tissue cultures in the presence of antibodies yielded an infectious virus which was antigenically different from the parent strain. The occurrence of an antigenic shift in Visna during virus replication permits the persistence of extra-cellular infectious virus, a phenomenon known only in chronic protozoal infections and in infections with equine infectious anemia virus (44). The latter disease is characterized by hemolysis and fever which occurs during viremia. The development of neutralizing antibodies leads to remission of the disease which relapses when antigenically altered virus appears in the blood stream. In SSPE such phenomena have not been observed but it could be shown that SSPE viruses revealed a significant lower reactivity in neutralization assays against anti-measles and SSPE sera, which has been interpreted as antigenic variation and possible cause of SSPE (24).

Obviously, the virus-host relationship in SSPE plays a major role in this disease. From a virological point of view, the long incubation period, the impairment of virus replication in SSPE brain cells as well as the biochemical and immunological differences between SSPE and measles viruses have to be explained and interpreted. It has been suggested that measles virus can persist by means of transferring the genetic information into a DNA copy, as it has been described for Visna virus (45). Visna virus can persist in the CNS in form of a DNA pro-virus without being recognized by the immune system. However, since measles virus does not contain on RNA-dependent DNA polymerase as does Visna virus, the enzyme would have to be provided by the cell, e.g., by an endogenous virus. No laboratory evidence does support this hypothesis and the reported finding in persistently infected tissue culture cells with measles virus are unconfirmed.

Biological and biochemical characterizations of tissue cultures latently infected with SSPE virus have revealed a state of viral defectiveness that may represent a mechanism similar to that responsible for the virus persistency in the brain of SSPE patients. All attempts to initiate a reactivation of virus synthesis in those cultures were unsuccessful. These cultures contained nucleocapsid antigen and revealed salt-dependent hemagglutinin antigen on their surface. Moreover, biochemical analysis demonstrated that the majority of viral RNA produced was defective or subgenomic and 50 S RNA normally associated with the infectious virus was present in only small amounts (46). These observations suggest an impaired state of replication which could be related to the presence of defective interfering (DI) particles. Studies on VSV in vitro and in vivo have shown that DI particles interfere with replication of infectious virus (47). In infected animals, VSV-DI particles are capable of influencing the disease process in prolonging the incubation period. In measles, DI particles cannot be separated biochemically from complete virus and, therefore, their presence cannot be directly demonstrated in infected tissue.

The observations of biochemical and immunological differences of SSPE and measles virus M proteins raises an interesting point concerning the etiology of this disease. If SSPE viruses were derived from measles virus, it is conceivable that a mutation or modification of the gene region coding for the M protein has occurred during latency. During this modification process it is possible that a non-functional M protein becomes produced incompatible with normal virus assembly. This would explain both the genesis and maintenance of a persistent virus infection in that no progeny virus could be assembled and released. However, one limitation of the M protein analysis has been the use of only two viruses in the antigenic comparison. The possibility exists that the differences found are related to successive mutations of the one SSPE virus tested which are not typical for other SSPE strains. Alternatively, if SSPE viruses primarily represent independent strains of the measles virus group infecting SSPE patients, one would expect a clustering of this disease which has not been observed.

The characteristic course of SSPE suggests, as in Visna, an immunopathological component in the disease process. However, the available immunological data are insufficient to prove this notion (2). Humoral antibody against brain cell membranes or brain antigens could not be detected as one would expect in a situation where an autoimmune phenomenon had been triggered by the virus infection. The blocking factor hypothesis could not be proven effective in vivo when patients were treated with immunosuppressive drugs or by drainage of cerebro spinal fluid. Moreover, the interpretation of the possible pathogenetic role of antibodies or antigen antibody complexes is complicated by reports that SSPE has been observed in patients with hypogammaglobulinaemia and a combined immunodeficiency syndrome (48, 49). On the other hand, SSPE brain tissue reveals a high content of measles antibodies which would be expected to eliminate measles virus antigens from infected brain cells by the process of antigen capping. This phenomenon would prevent the immune system from destroying the infected cells and arrest the disease process. Other reports stress the importance of measles infections in the presence of protective maternal antibodies which could lead to atypical measles infection providing a ground for the development of chronic CNS infection (2).

In spite of the intensive research carried out on this rare slow virus disease which is associated with a conventional virus, the underlying pathogenetic mechanisms are still not unravelled. An explanation of the pathogenesis of SSPE must account for its development after a presumably normal measles infection in a presumably normal host by a process that changes both the virus and the host response.

REFERENCES

1. ter Meulen, V., Katz, M., and Müller, D. (1972). Current Topics in Microbiology 57, 1.
2. Agnarsdottir, G. (1977). In "Recent Advances in Clinical Virology" Churchill Livingstom, London.
3. White, R. G., and Boyd, J. E. (1973). Clin. exp. Immunol. 13. 343.
4. Nichols, W. W., Levan, A., Hall, B., and Östergren, G. (1962). Hereditas 48, 367.
5. Bech, V. (1962). Am. J. Dis. Child. 103, 252.
6. Dossetor, J. F. B., and Whittle, H. C. (1975). Br. Med. J. 2, 592.
7. Gatti, J. M., and Good, R. A. (1970). Med. Clin. North Am. 54, 281.
8. Breitfeld, M. D., Hashida, Y., Sherman, F. E., Odagiri, K., and Yunis, E. J. (1973). Lab. Invest. 28, 279.
9. Mehta, P. D., Kane, A., and Thormar, H. (1977). J. Immunol. 118, 2254.
10. Vandvik, B., and Norrby, E. 1973. Proc. Nat. Acad. Sci. 70, 1060.
11. Strosberg, A. D., Karcher, D., and Lowenthal, A. (1975). J. Immunol. 115, 157.
12. Kiessling, W. R., Hall, W. W., Yung, L. L., and ter Meuleń, V. (1977). Lancet. 324.
13. Gerson, K. L., and Haslam, R. H. A. (1971). N. Engl. J. Med. 285, 78.
14. Kreth, H. W., Käckell, Y. M., and ter Meulen, V. (1974). Med. Microbiol. and Immunol. 160, 191.
15. Sell, K. W., Thurman, G. B., Ahmed, A., and Strong, D. M. (1973). N. Engl. J. Med. 288, 215.
16. Ahmed, A., Strong, D. M., Sell, K. W., Thurman, G. B., Knudsen, R. C., Wistar, R., Jr and Grace, W. R. (1974). J. exp. Med. 139, 902-924.

17. Allen, J., Oppenheim, J., Brody, J. A., and Miller, J. (1973). Infection and Immunity 8, 80.
18. Valdimarsson, H., Agnarsdóttir, G., and Lachmann, P. J. (1974). Proc. R. Soc. Med. 67, 1125.
19. Kreth, H. W., and ter Meulen, V. (1977). J. Immunol. 118, 129.
20. Doherty, P. C., Blanden, R. V., and Zinkernagel, R. M. 1976. Transpl. Rev. 29, 89.
21. Perrin, L. H., Tishon, A., and Oldstone, M. B. A. (1977). J. Immunol. 118, 282.
22. Katz, M., and Koprowski, H. (1973). Arch. Ges. Virusforsch. 41, 390.
23. Dubois-Dalcq, D. M., Barbosa, L. H., Hamilton, R., and Sever, J. L. (1974). Lab. Invest. 30, 241.
24. Payne, F. E., and Baublis, J. V. (1973). J. Infect. Dis. 127, 505.
25. Hall, W. W., Kiessling, W. R., and ter Meulen, V. (1978). In "Negative Strand Viruses and the Host Cell." Academic Press, in press.
26. Hall, W. W., Kiessling, R. R., and ter Meulen, V. (1978). Nature, in press.
27. Scheid, A., Caligun, L. A., Compans, R. W., and Choppin, P. W. (1972). Virology 50, 640.
28. Hall, W. W., Kiessling, W. R., Nagashima, K., and ter Meulen, V. (1978). In "Negative Strand Viruses and the Host Cell." Academic Press, in press.
29. Laurell, C. B. (1967). Prot. Biol. Fluids 14., 499.
30. Hall, W. W., and ter Meulen, V. (1977). J. gen. Virol. 35, 497.
31. Graves, M. C., Silver, S. M., Choppin, P. W. (1978). Virology, in press.
32. Wechsler, S., and Fields, B. N. (1978). J. Virol. 25, 285.
33. Hall, W. W., and ter Meulen, V. (1976). Nature 264, 474.
34. Hall, W. W., Genius, D., and ter Meulen, V. (1977). J. gen. Virol. 35, 579.
35. Mountcastle, W. E., and Choppin, P. W. (1977). Virology 78, 463.
36. Bussell, R. H., Wataro, D. J., Seals, M. K., Robinson, W. S. (1974). Med. Microbiol. Immunol. 160, 105.
37. Scheid, A., Graves, M. C., Silvers, S. M., Choppin, P. W. (1978). In "Negative Strand Viruses and the Host Cell." Academic Press, in press.
38. Schluederberg, A., Chavanich, S., Lipman, M. B., Corter, C. (1974). Biochem. Biophy. Res. Commun. 58, 647.
39. Wechsler, S., and Fields, B. N. (1978). Nature, in press.
40. Eichmann, K., and Rajewsky, K. (1975). Eur. J. Immunol. 5, 661.

41. Laver, W. G., and Webster, R. C. (1968). Virology 34, 193.
42. Gudnadottir, M. (1974). Progr. med. Virol. 18, 336.
43. Narayan, O., Griffin, D. E., and Silverstain, A. M. (1977). J. Infect. Dis. 135, 800.
44. Kono, Y., Kabayashi, K., Fukunaga, Y. (1973). Arch. Ges. Virusforsch. 41, 1.
45. Zhdanov, V. M., and Parfonovich, M. I. (1974). Arch. Ges. Virusforsch. 45, 225.
46. Kratzsch, V., Hall, W. W., Nagashima, K., and ter Meulen, V. (1977). J. Med. Virol. 1, 139.
47. Huang, A. S. (1977). Bacteriol. Rev. 41, 811.
48. Hanissian, A. S., Jabbour, J. T., de Lamerens, S., Garcia, J. H., and Horta- Barbos, L. (1972). Am. J. Dis. Child. 123, 151.
49. Allison, A. C. (1972). J. Clin. Pathol. 25, 121.
50. Joseph, B. S., and Oldstone, M. B. A. (1975). J. exp. Med. 142, 864.
51. Villareal, L. P., Breindl, M., and Holland, J. J. (1976). Biochemistry 15, 1663.
52. Laemmli, U. K. 1970. Nature 227, 680.

SUBACUTE SCLEROSING PANENCEPHALITIS: AN ABERRANT INFECTION BY A MEASLES-LIKE VIRUS

J. C. Ramsey, L. Eron, J. Sprague, P. Bensky, D. Jones, R. Dunlap, and P. Albrecht

Food and Drug Administration, Bureau of Biologics, Division of Virology, Bethesda, MD 20014

ABSTRACT Comparison of properties of measles virus with a measles-like virus, IP-3, isolated from the brain of a patient with subacute sclerosing panencephalitis revealed substantial differences between the virions. IP-3 infectious particles possess a lighter buoyant density than strain EL measles virus (1.21 vs. 1.24 g/cc). The IP-3 progeny contains a nucleocapsid and glycoprotein similar in size to EL measles virus, but apparently is deficient in the measles membrane and phosphoproteins. While their nucleocapsids cross-react antigenically, their primary structures are slightly different as shown by polyacrylamide gel electrophoresis. The IP-3 genome is 50S and its nucleocapsid is 200S, the same as nondefective measles particles and different from defective particles (18S and 130S, respectively). These data are consistent with an aberrant infection by IP-3, and not with infection by defective particles, as the cause of subacute sclerosing panencephalitis.

INTRODUCTION

Subacute sclerosing panencephalitis (SSPE) is a rare disease of children that emerges on the average 7 years after an uneventful measles infection. Patients with SSPE have high titers of antibody to measles in their cerebrospinal fluid (1,2), have measles antigen detectable by immunofluorescence in brain nerve cells (1), and have had measles-like viruses isolated from their brain tissue (3,4). The rarity of the disease suggests that some unusual factors, host or virus dependent or both, play a role in its occurrence; however, the mechanism of pathogenesis remains unknown. This report describes properties of the SSPE isolate IP-3 which produces a chronic encephalitis in monkeys that appears histologically and clinically similar to SSPE in man (5).

ISBN 0-12-668350-6

MATERIALS AND METHODS

Vero cells, maintained on Eagle's MEM with 10% fetal calf serum were infected with IP-3 or one of two attenuated vaccine strains of measles virus: strain MSD (Merck, Sharpe, & Dohme, West Point, PA) or strain EL (Eli Lilly & Co., Indianapolis, IN). Infectivity titers were assayed on Vero cell monolayers (6). Viral polypeptides were labeled with ^{35}S-methionine (5 μC/ml) added 24 hr before harvest. For labeling of viral nucleic acids, cells were treated with actinomycin D (10 μg/ml) immediately prior to the addition of ^{3}H-uridine (10 μC/ml) 24 hr before harvest. Labeled cytoplasmic RNA was extracted and analyzed on sucrose gradients (7). Virus particles released into the culture medium were concentrated by precipitation with polyethylene glycol (8). Particles remaining cell-associated were harvested by centrifugation of post-nuclear supernatant (7) on 20-65% discontinuous sucrose gradients; virus material banded at the 20-65% sucrose interface was density equilibrium banded in linear sucrose gradients. Sedimentation coefficients of nucleocapsids derived from purified virus were estimated using ribosome monomers as markers (9). Virus particles and nucleocapsids, negatively stained on formvar-carbon coated grids with phosphotungstic acid, were examined in an RCA EMU-3G electron microscope. Viral polypeptides as well as polypeptides immunoprecipitated (10) with whole measles or measles nucleocapsid antiserum were subjected to sodium dodecyl sulfate-polyacrylamide gel electrophoresis (SDS-PAGE) on 10% gels (11) and autoradiographed. Molecular weights of measles polypeptides were estimated by comparison to known adenovirus polypeptide standards.

RESULTS

Vero cells infected either with the attenuated strains MSD or EL of measles virus or with the IP-3 SSPE isolate (12), were harvested when 90% of the cells showed a cytopathic effect. Cell homogenates (cell-associated virus) and culture media (released virus) were assayed for infectious particles by plaque titration. While EL and MSD produced 100 infectious particles per cell, more than 50% of which remained cell-assicated, IP-3 formed 4 logs less of infectious virus, essentially all of which remained cell-associated (results not shown).

Electron microscopy of purified virions revealed typical paramyxovirus projections on the surface of EL particles, thought to be hemagglutinin (Fig. 1A); such projections were only rarely observed on IP-3 virions (Fig. 1B,C). In addition,

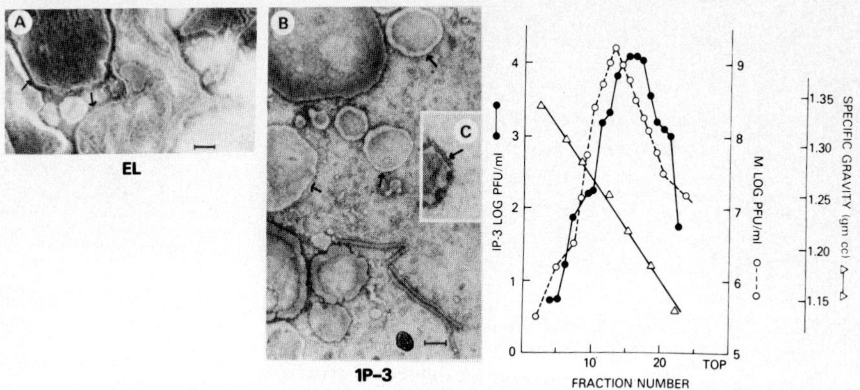

FIGURE 1. Electron micrographs of negatively stained virion particles. A. Purified EL particles. Nucleocapsid strands appear inside the virion. Hemagglutinin spikes are visible on the surface (arrows). B. Purified IP-3 particles. Virions lack hemagglutinin spikes (arrows). C. Rare IP-3 particle with surface projections (arrow). Bars = 100 nm.

FIGURE 2. Equilibrium sedimentation of IP-3 and EL virus suspensions. Density and titers of infectious virus were determined by weighing 100 µl aliquots and by plaque titration of gradient fractions, respectively.

density equilibrium sedimentation in sucrose gradients (Fig. 2) demonstrated that IP-3 virions were less dense than EL virions (1.21 vs. 1.24 g/cc, respectively) further suggesting differences between IP-3 and measles virus structure.

To determine if the above differences were associated with the formation of defective viral particles, an examination of RNA and nucleocapsids in EL and IP-3 infected cells was undertaken. An analysis of cytoplasmic RNA extracted from EL infected (Fig. 3) and IP-3 infected cells (Fig. 4) by sedimentation through linear sucrose gradients showed that 50S RNA was synthesized in both infected cell populations, suggesting that the complete IP-3 genome was represented. Furthermore, rate zonal sedimentation of EL and IP-3 nucleocapsids (Fig. 5) demonstrated that both migrate as 200S particles. The attenuated measles virus, but not IP-3 nucleocapsid preparation contained a second smaller peak sedimenting as 140S. Electron microscopic measurements of negatively stained nucleocapsids confirmed the size similarity. The majority of both IP-3 and EL nucleocapsids

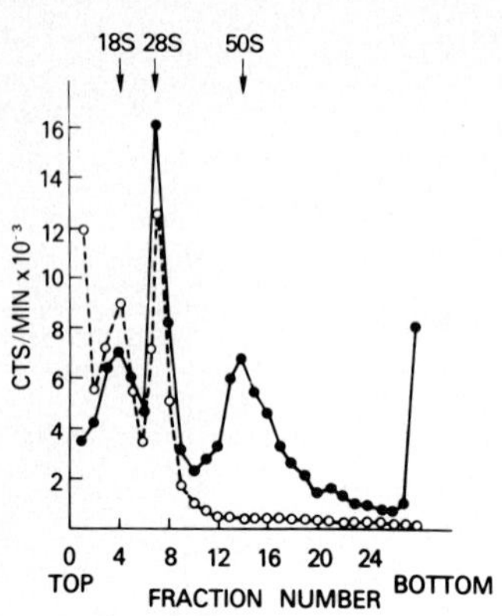

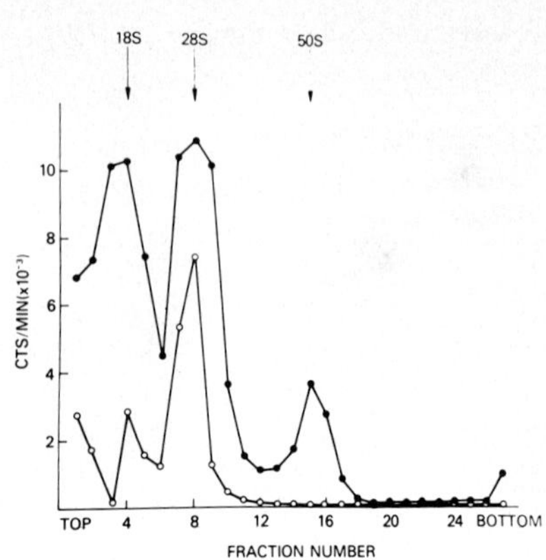

FIGURE 3. Sucrose gradient analysis of cytoplasmic RNA of EL infected cells. Cells were labeled with ^{3}H-uridine as described in Materials and Methods. Under these conditions of labeling, some cellular 18S and 28S ribosomal RNA becomes labeled. RNA of infected (o-o) and uninfected (o-o) cells.

FIGURE 4. Sucrose gradient analysis of cytoplasmic RNA of IP-3 infected cells. Conditions were the same as described in Fig. 3. RNA of infected (o-o) and uninfected (o-o) cells.

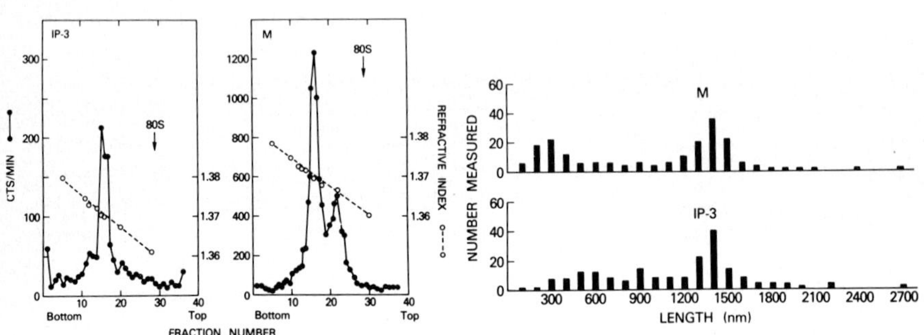

FIGURE 5. Rate-zonal sedimentation of EL and IP-3 nucleocapsids. Both attenuated measles and IP-3 nucleocapsids migrate as 200S particles. The EL preparation contains a second peak migrating as 140S. The arrow represents the position of 80S ribosomal monomer in the gradient.

FIGURE 6. Histogram of length measurement of EL and IP-3 nucleocapsids. The majority of both EL and IP-3 nucleocapsids (185 strands per sample) measured 1.4 μ.

measured 1.4 μ, despite a small degree of fragmentation (Fig. 6). In addition, there appeared to be a much smaller peak of EL nucleocapsid, measuring 0.2 to 0.3 μ.

An analysis of viral structural polypeptides by SDS-PAGE of EL and IP-3 virions revealed substantial differences between the two virus strains (Fig. 7). Polypeptides observed in EL preparations included the measles glycoprotein (G), phosphoprotein (P), nucleocapsid protein (NC), and membrane (M) components of molecular weight 80, 69, 60, and 37 X 10^3 daltons, respectively. In addition, a major polypeptide of 46,000 daltons molecular weight (NC_1) was detected. Tryptic digest fingerprints of ^{35}S-methionine labeled NC_1 and NC revealed regions of identical primary structure (Eron, unpublished) suggesting that NC_1 may be structurally and/or functionally related to NC and in fact may represent a proteolytic cleavage product of NC. In contrast to the EL strain, IP-3 viral polypeptides consist primarily of NC_1' which appeared to migrate more rapidly than EL NC_1. A second polypeptide (G') migrating more rapidly than EL glycoprotein was also visible.

Since the NC_1 polypeptides of measles and IP-3 migrated at different rates in SDS-PAGE and since EL NC_1 appeared to

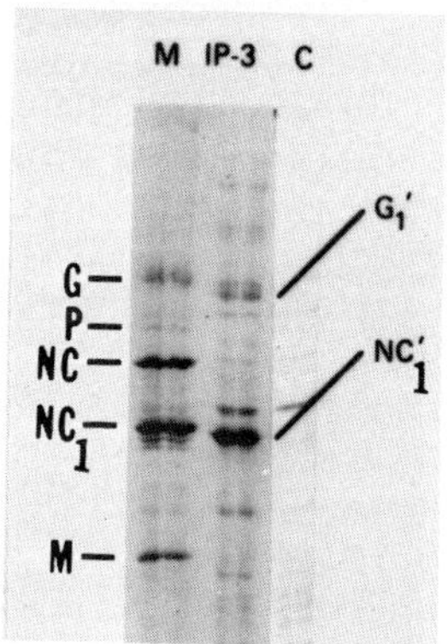

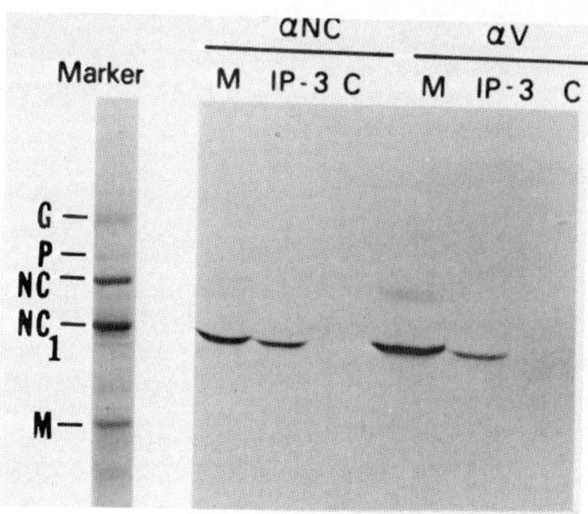

FIGURE 7. SDS-PAGE of viral polypeptides. Polypeptides of purified virus were electrophoresed and autoradiographed as in Materials and Methods. M, attenuated measles virus (strain EL); IP-3, SSPE isolate; C, uninfected control.

FIGURE 8. SDS-PAGE of immunoprecipitated viral nucleocapsids. M, measles virus (strain EL); IP-3, SSPE isolate; αV, antiserum to whole natural measles virus; αNC, antiserum to nucleocapsid of natural measles virus. Measles marker (left lane) and uninfected control (C) were prepared as in Materials and Methods.

be related to NC, the antigenic relatedness of measles and IP-3 nucleocapsids was analyzed. Purified measles and IP-3 native nucleocapsids were mixed with antisera prepared against measles nucleocapsid or virion particles and the resulting immunoprecipitates analyzed by SDS-PAGE (Fig. 8). Both IP-3 and measles nucleocapsids were precipitated by measles antisera.

DISCUSSION

The facts that both the IP-3 and non-neurotropic measles virus genomes are 50S and that their 200S nucleocapsids can be precipitated with measles whole virion or nucleocapsid antiserum suggest a high degree of relatedness between these two viruses. However, compared to the EL strain of measles virus, IP-3 particles have a different buoyant density, have a morphologically absent or altered hemagglutinin as determined by electron microscopy, produce mainly nucleocapsid and glycoprotein polypeptides in infected cells, and apparently lack the measles membrane and phosphoproteins. On the other hand, several observations suggest that limited quantities of envelope proteins are synthesized in IP-3 infected cell cultures. IP-3 spreads by the process of cell fusion, suggesting the presence of a fusion factor (12). In addition, rare IP-3 virion particles possess surface projections suggestive of hemagglutinin and in infected cell cultures large areas of syncytia show hemadsorption of monkey red blood cells which can be blocked with measles-positive serum (Albrecht, unpublished). It is possible that one or several of these activities are represented by G'.

The imbalance or deficiency of IP-3 viral structural proteins may interfere with the nucleocapsid's association with the cell surface required for normal maturation (13) and may explain its different *in vitro* growth characteristics (12) and pathogenesis (5). This mode of replication seems to be characteristic for the slowly progressing infection in the central nervous system of SSPE patients and of experimentally infected rhesus monkeys (5). Certainly, IP-3 particles do not satisfy the criteria established for defective particles which possess the same structural polypeptides as wild-type particles but only a fragment of the wild-type genome. However, the possibility that the IP-3 genome has had a portion of its measles information replaced with non-measles information cannot be excluded.

ACKNOWLEDMENTS

We thank Dr. David J. Waters for his gift of antiserum directed against measles nucleocapsid and Heiner Westphal for ^{35}S-methionine labeled adenovirus polypeptides.

REFERENCES

1. Connolly, J. H., Allen, I. V., Hurwitz, L. J., and Millar, J. H. (1967). *Lancet 1,* 542.
2. Sever, J. L., Krebs, H., Ley, A., Barbosa, L. H., and Rubenstein, D. (1974). *J. Am. Med. Assoc. 228,* 604.
3. Horta-Barbosa, L., Fuccillo, D. A., Sever, J. L., and Zeman, W. (1969). *Nature (London) 221,* 974.
4. Payne, F. E., Baublis, J. V., and Itabashi, H. H. (1969). *N. Eng. J. Med. 281,* 585.
5. Albrecht, P., Burnstein, T., Klutch, M. J., Hicks, J. T., and Ennis, F. A. (1977). *Science 195,* 64.
6. Schumacher, H. P., Albrecht, P., and Tauraso, N. M. (1972). *Arch. Gesamte Virusforsch. 36,* 296.
7. Penman, S., Greenberg, H., and Willems, M. (1969). In "Fundamental Techniques of Virology" (K. Habel and N. Salzman, eds.). pp. 49-58. Academic Press, New York.
8. McSharry, J. J. and Benzinger, R. (1970). *Virology 40,* 745.
9. Kiley, M. P., Gray, R. H., and Payne, F. E. (1974). *J. Virol. 13,* 721.
10. Eron, L., Callahan, R., and Westphal, H. (1974). *J. Biol. Chem. 249,* 6331.
11. Maizel, J. V., Jr. (1971). In "Methods in Virology" (K. Maramorosch and J. Koprowski, eds.). Vol. V, pp. 180-246. Academic Press, New York.
12. Burnstein, T., Jacobsen, L. B., Zeman, W., and Chen, T. T. (1974). *Infect. Immun. 10,* 1378.
13. Wagner, R. R., Emerson, S. U., Imblum, R. L., and Kelley, J. M. (1975). In "Negative Strand Viruses" (B. W. J. Mahy and R. D. Barry, eds.). Vol. 1, pp. 1-19. Academic Press, New York.

VISNA: AN ANIMAL MODEL FOR STUDIES OF VIRUS PERSISTENCE[1]

A.T. Haase, M. Brahic, D. Carroll, J. Scott,
L. Stowring, B. Traynor, P. Ventura

Infectious Disease Section, Veterans Administration Hospital
Departments of Medicine and Microbiology, University of
California, San Francisco 94121

O. Narayan

Department of Neurology and Animal Medicine, Johns Hopkins
University, School of Medicine, Baltimore, Maryland 21205

ABSTRACT Certain aspects of the structure and replication of visna virus are particularly well suited to the perpetuation of infections *in vivo*. During the initial phase of experimental infection of sheep, the virus genome is introduced into a number of cells, but virus genetic expression is sufficiently restricted to allow survival of the host. The lysogenic relationship of virus and cell *in vivo* provide a protected intracellular site for virus to escape host defense mechanisms. Recent work indicates that neither host immunity nor interferon mediate the limitation in gene expression; and that the restriction is at a transcriptional level. The distant spread of virus despite neutralizing antibody can be attributed in part to the emergence of antigenic variants in the course of infection. The genetic mechanism generating these variants is not understood at present, but may involve only the portion of the viral genome coding for the viral glycoprotein.

INTRODUCTION

Visna is an inflammatory disease of the central nervous system (CNS) of sheep caused by a lentivirus in the family of retraviruses (1). Both the long incubation period and protracted course of disease are characteristic of slow infections, and the presence of virus in its host for periods often of years make this agent a prototype of persistent viruses. Here we describe the current status of work on the

[1]This work was supported by grants from the American Cancer Society (VC120C and D-281) and the PHS (NS11782, NS12127-03, and NS10920-05). This is project MRIS 3367 within the Veterans Administration.

ISBN 0-12-668350-6

structure and replication of visna virus in its animal host relevant to understanding persistence of this virus.

Background: Disease, Virus, and Experimental Models. Visna is a neurological affliction of sheep in which infiltration of the CNS by inflammatory cells is associated with destruction of cells and dysfunction, generally paralysis. The causative agent was isolated by Sigurdsson in cultures of cells derived from sheep choroid plexus (SCP) (2). Growth of virus in SCP is attended by two kinds of distinctive cytopathic effects (CPE): formation of polykaryocytes and of stellate cells. In contrast to the situation in the animal, the replication of visna virus *in vitro* is not at all slow. In the lytic cycle of growth, progeny virus appears at about 24 hours and the cycle is complete by 72 hours when the cells have degenerated and produced 50-100 PFU/cell (3).

Visna virus is a member of the retravirus family: virus matures by budding from the membrane of the infected cell (4), the virus genome is polyploid (5,6), comprised of two to four RNA subunits (7,8) enclosed in a core whose major structural polypeptide P30 (9) bears antigenic determinants that define the group of lentiviruses, visna, Maedi, and progressive pneumonia virus (10); and during replication the virion associated DNA polymerase transfers information from viral RNA to a DNA intermediate in the cell (11, 12).

Investigations of sheep infected under controlled experimental conditions in Iceland (13) and in this country (4) have provided us with a clear picture of the pathogenesis of visna, and in turn have raised two fundamental questions, how the virus persists in its host, and why the evolution of disease is so slow.

When visna virus is inoculated intracranially (i.c.), it is introduced into the choroid plexus and ependymal cells lining the ventricular system, and cells along the tract followed by the needle during inoculation of virus. In response to infection, host immune and inflammatory cells infiltrate the infected areas, and spill into the CSF. Animals do not die of acute meningoencephalitis, because as we shall see virus replication is limited, and because host defense mechanisms are brought into play; but virus is not wholly eliminated either. The viremia that occurs in the initial inoculation, and subsequently from local production of virus leads to infection of cells in many organ systems, especially those in the macrophagocytic system, and in the lungs. In time there is sufficient viral antigen to elicit a neutralizing antibody response; this sharpley curtails the spread of virus extracellularly. An exceptional animal may recover; but usually there is persistent infection despite the host immune and inflammatory response.

Persistence and the Lysogeny Hypothesis. How does visna virus survive in the face of host defense mechanisms? One of us suggested several years ago (1) that the explanation for persistence is analogous to gene regulation and lysogeny in prokaryotes; once the viral genome is introduced into cells and stably encoded in DNA, repression of gene expression would lead to a cohort of latently infected cells that would go undetected by the immune surveillance system of the host. These cells would serve as refuge and reservoir for viral genetic information. Infectious virus could emerge from time to time to perpetuate infection in the occasional cell where the restriction in viral genetic expression is overcome.

RESULTS AND DISCUSSION

Experimental Predictions and Tests: 1. Recovery of Infectious Virus and Virus Particles. This hypothesis makes a number of predictions that we have undertaken to test experimentally. The first prediction is a relatively simple one, that virus is associated with many more cells in the animal than produce infectious virions at any particular time.

There is now ample evidence to satisfy this prediction (15). Visna virus is present in low titer in most of the tissues and fluids of an infected animal. Since virus is most frequently isolated from SCP (13), we chose to analyze this tissue in detail, comparing first the amount of infectious virus contained in the tissue with the amount that we hoped to obtain from the postulated latently infected cells. As expected (FIGURE 1) little if any infectious virus was recovered directly from homogenates of SCP, but virus could be recovered by explanting fragments of tissue. Within the first few days of cultivation, fibroblasts migrate from the explant. These outgrowths show evidence of virus CPE, and infectious virus in high titer is found in the tissue culture fluid. This observation suggests that visna virus *in vivo* is associated with cells without production of virus, an interpretation supported by the failure to find virus particles in the tissue in exhaustive surveys by electron microscopy.

The quantitative aspects of the argument can be examined as follows: In a recent representative experiment, we injected an animal i.c. with 10^9 PFU of visna virus. Two

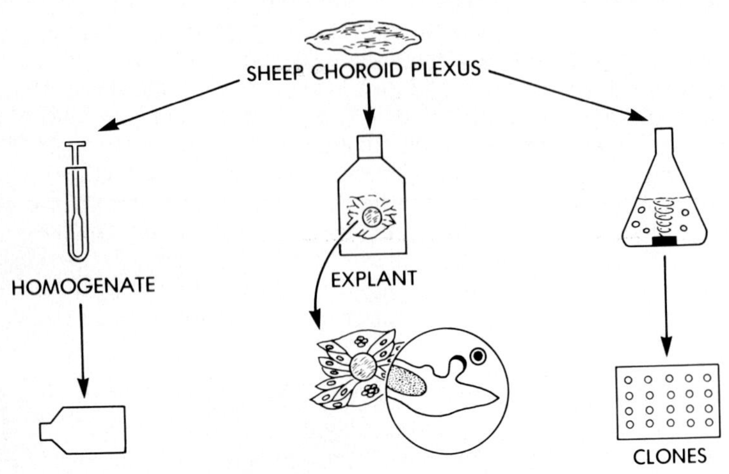

FIGURE 1. Scheme for isolation of visna virus from infected sheep choroid plexus. See text for explanation.

weeks later, the animal was sacrificed. A portion of SCP was trypsinized, and a known number of cells were placed in the wells of microtiter trays. The cells that formed monolayers were scrutinized for viral CPE. Although this procedure does not identify strictly speaking "clones", we found that all of the wells with colonies derived from one to a few cells gave rise to virus. Thus most cells in SCP were infected, but did not produce virus until induced to do so in some way by the conditions of cultivation _in vitro_.

Experimental Predictions and Tests: 2. Viral DNA and Antigens. The second prediction is that we expect to find the viral genome as a DNA intermediate resident in at least as high a proportion of cells as can be induced to give rise to virus _in vitro_, but because of the postulated restriction in viral gene expression in the animal, only a small proportion of the cells with vDNA will also contain the viral gene products. Accordingly we designed experiments to simultaneously assess vDNA and viral proteins in individual cells in tissues.

Viral DNA can be detected in single cells by *in situ* hybridization of a virus specific probe labelled to high specific radioactivity to tissue sections. In reported studies of this kind (15), we found that about the same proportion of cells in the SCP of an animal infected with 10^7 PFU of virus contained vDNA as scored as infectious foci in cloning experiments. Only a small fraction (0.1-0.5%) of cells containing vDNA also had P30 as measured by immunofluorescence with a monospecific antiserum to P30; this contrasts dramatically with the lytic and permissive infection *in vitro* where this same reagent readily demonstrates P30 in nearly every cell.

In recent extensions of these studies, we find that the virion glycoprotein gP135 is also not synthesized at detectible levels by immunofluorescence assay employing antisera directed against this gene product. Moreover, the restriction is stringent; in the animal inoculated with 10^9 PFU of virus, in which nearly all of the SCP cells probably contain a resistant viral genome, less than 0.1% of the cells contain viral antigens.

Experimental Predictions and Tests: 3. The Level of Restriction: Quantitative Studies of Transcription In Vivo. We have implied in the lysogenic or gene repression model of virus persistence *in vivo* that the block in gene expression is effected at the level of transcription. Experiments carried out in the past year have been directed at developing an assay to establish that this is so.

As in prior investigations of host restriction we wish to know what proportion of cells carry vDNA and vRNA, and how much vRNA *vis a vis* levels transcribed under permissive conditions *in vitro*. As a first step to satisfy these conditions, we have developed a quantitative *in situ* hybridization assay for vRNA in single cells using permissively infected cells (FIGURE 2). Cells are collected by trypsinization at increasingly later times after infection, and either deposited on slides by cytocentrifugation, or lysed with detergent. RNA is isolated from the latter aliquot, and the average number of copies of vRNA/cell are determined by hybridization in solution to cDNA. In the course of lytic infection, viral RNA increases from 10-15 copies/cell, introduced during adsorption of 3 PFU/cell, to several thousand copies/cell at the end of the growth cycle (16). The cells on the slides are fixed, treated to increase the diffusion of probe, and then hybridized *in situ* in formamide to a cDNA probe labelled to $3x10^8$ dpm/µg. Following hybridization, the slides are washed extensively, coated with emulsion, and stained after appropriate radioautographic

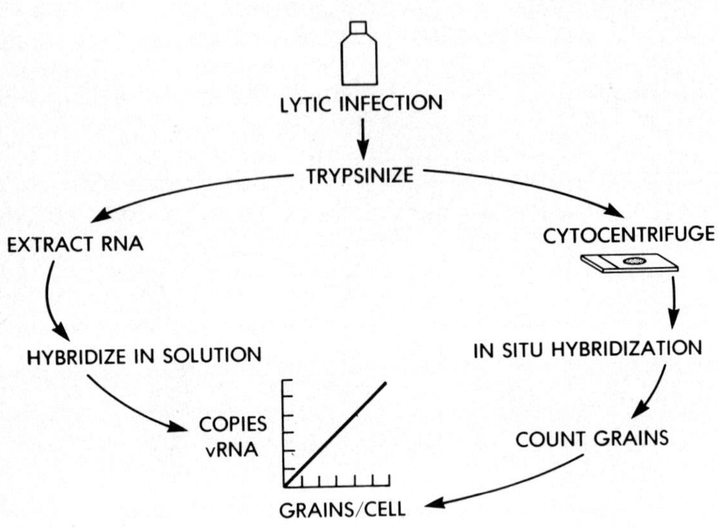

FIGURE 2. Experimental protocol for quantitative assay of viral RNA in tissues by *in situ* hybridization.

exposures. The number of copies of vDNA/cell can be computed, and compared with the experimentally determined copy number from liquid hybridization to validate the method as a quantitative assay of transcription in tissues.

By systematically improving the efficiency of *in situ* hybridization, we can now demonstrate 1-10 copies of vRNA/cell with radioautographic exposures of a few weeks. In permissively infected cells with 1000-2000 copies/cell a few hours of exposure reveals extensive labelling of both nucleus and cytoplasm (FIGURE 3).

Although preliminary, studies conducted in one animal suggest that the block in virus gene expression is at a transcriptional level. As FIGURE 4 illustrates, there are only a few cells in the section of SCP that contain vRNA with this sensitive assay, but a high proportion of SCP cells give rise to virus on cultivation *in vitro*. Estimates of copy numbers in the positive cells suggest that the cells with vRNA have levels comparable to those found *in vitro* under permissive conditions. These cells therefore may be the source of the small amount of free virus found in tissues.

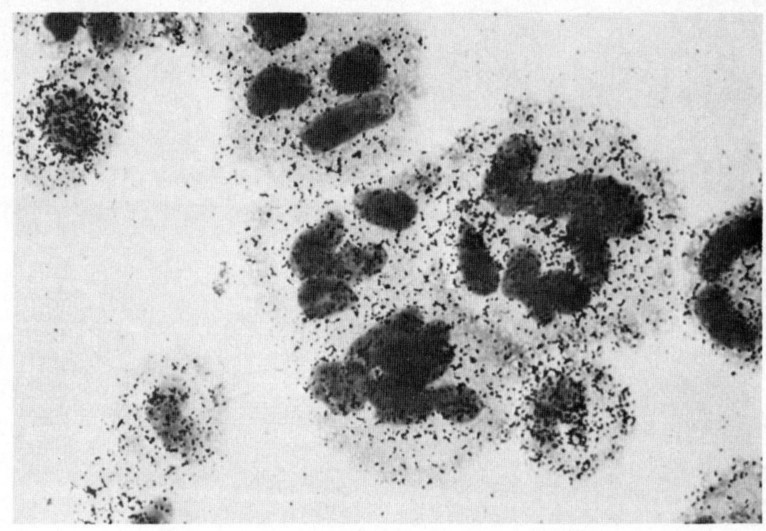

FIGURE 3. Demonstration of viral RNA in cells infected with visna virus under permissive conditions. Cells were collected 72 hours after infection and processed as illustrated in FIGURE 2 (details in text). After hybridization *in situ* to visna cDNA (calculated specific activity $3x10^8$ dpm/μg), the slides were washed, coated with emulsion, and exposed for 4 hours. Giemsa stain, original magnification x 400.

Experimental Predictions and Tests. 4. Host Immunity and Interferon. It seems quite unlikely that restriction is mediated by interaction with the host immune system or by interferon. Immunity is unlikely as the same restriction in virus production obtains in animals that have been thymectomized (17), or immunosuppressed (18). Interferon is unlikely as a mechanism of restriction, as visna virus is remarkably resistant to interferon.

In these studies we prepared interferon in high

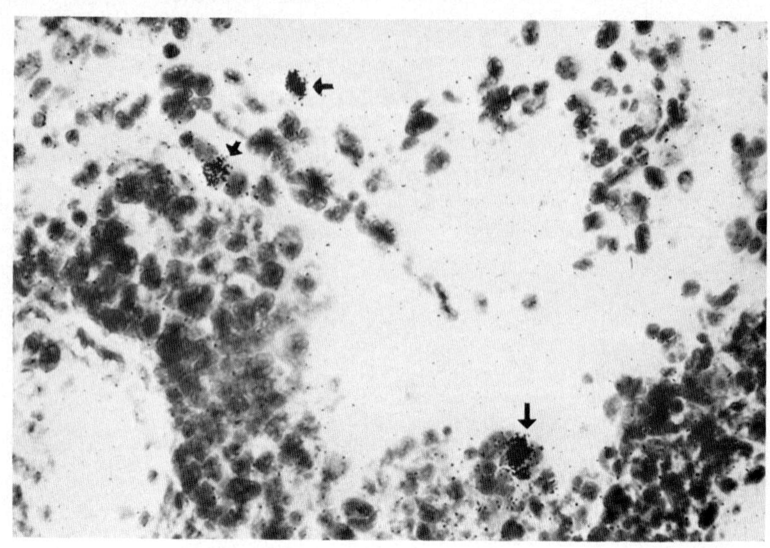

FIGURE 4. Visna virus RNA in cells in tissues. Frozen sections were cut from SCP tissue obtained two weeks after inoculation i.c. of 10^9 PFU of virus. Hybridized in situ as in FIGURE 3. Radioautographic exposure 4 days; x 250. Arrows point to cells containing vRNA.

concentrations by injecting fetal sheep in utero with poly I:C (19). Serum obtained 2 hours later had a titer of several thousand units/ml of sheep interferon which inhibited plaque formation by relatively sensitive viruses such as VSV and sindbis virus, and relatively resistant DNA viruses such as Herpes simplex virus, albeit at lower titers. Visna virus plque formation was not inhibited by interferon at the highest levels obtainable, orders of magnitude greater than that required to inhibit the growth of the other viruses. Interferon also failed to inhibit visna virus in a single growth cycle of the virus, and when interferon was renewed at later times in the virus life cycle, the point at which interferon acts in inhibiting production of other retroviruses (20). These experiments establish an unusual degree of resistance to interferon, and suggest that other mechanisms must be sought to explain restriction. They also reveal another aspect of visna virus that adapts the virus to survival in the host.

Recapitulation, Interpretation, and the Enigma of Virus Spread. Thus far we have shown that virus gene expression is altered in the animal with some striking parallels to lysogeny. This also provides us with a plausible basis for persistence: in the initial phase of infection, virus is introduced into many cells but little virus or virus antigens are produced. The host survives, as the restriction damps down the acute infectious process; the virus survives, as the latently infected cells go undetected by the host. From time to time an occasional cell spontaneously produces virus to perpetuate infection.

At this point our construct of pathogenesis encounters a major difficulty: in time the host produces neutralizing antibody; this should and does limit spread of virus and should convert infection to a covert state, or one in which virus spreads only focally from cell to cell. This is probably not the case, as virus is widely dispersed in many organ systems long after the appearance of neutralizing antibody.

This paradoxical spread of virus in the presence of antibody might be explained in two ways; the first is vectorial transmission in which the virus takes up temporary residence inside one of the cells comprising the formed elements of the blood, and in this fashion is carried to some distant site. The second mechanism is some modification in antigens of the virus that renders neutralizing antibody ineffective. In the concluding section, we take up what is known about the second mechanism.

Antigenic Shift. Narayan and his coworkers recently discovered that when an animal is inoculated with a single genotype of visna virus, the animal responds in time with production of antibody which neutralizes the inoculum strain of virus but not other strains of virus (21). In time antigenic variants appear, isolated from buffy coat cells, that are not neutralized by extant antibody to the inoculum strain of virus. These variants likely arise *de novo* rather than from immunological selection of variants contaminating the inoculum strain. Evidence in support of this contention comes from examination of a large number of plque isolates of the inoculum strain; antibody to the inoculum strain which does not neutralize the variant strains, neutralizes each of these isolates.

Antigenic variants of visna provide the virus with an opportunity periodically to spread until the host can re-exert immunological controls. As schematically depicted in FIGURE 5, the situation is reminiscent of influenza epidemics in populations, or borrelia or malaria infections in individual animals; host immunity limits spread of virus, control is temporarily lost as a variant arises and advances the infectious process, the host then regains control again with production of antibody to the variant.

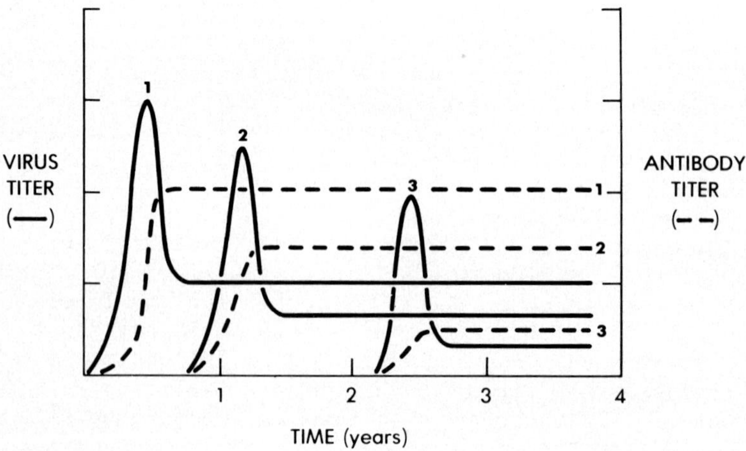

FIGURE 5. Theoretical scenario of the role of antigenic drift in the spread of visna virus. Curve 1 (-) represents the increase in titer of the inoculum strain of virus, and fall in titer with the appearance of neutralizing antibody to virus 1 (----1); spread of virus 1 is limited until antigenic variants 2 and 3 arise which reproduce and spread until restricted by antibodies in the manner described for virus 1.

Preliminary investigations of these variants by peptide mapping the structural polypeptides indicate that variation occurs in the virion glycoprotein, gP135, predictably so, as we have shown that this is the antigen that elicits a neutralizing antibody response. Although limited thus far, analysis of other structural polypeptides such as P30 suggests that the variants might be derived from a change in only that portion of the genome coding for the glycoprotein. We plan to extend this work with the objective of understanding the genetic mechanism that generates these variants.

ACKNOWLEDGMENTS

We wish to thank H. Lukes for the typing of this manuscript.

REFERENCES

1. Haase, A. T. (1975). Cur. Topics in Microbiol. and Immun. 72, 102.
2. Sigurdsson, B., Thormar, H., and Palsson, P. A. (1960). Arch. ges. Virusforsch 10, 368.
3. Thormar, H. (1963). Virol. 19, 273.
4. Thormar, H. (1963). Virol. 14, 463.
5. Beemon, K. L., Faras, A. J., Haase, A. T., Duesberg, P. H., and Maisel, J. E. (1975). J. Virol. 21, 386.
6. Vigne, R., Brahic, M., Filippi, P., and Tamalet, J. (1977). J. Virol. 21, 386.
7. Brahic, M., Tamalet, J., Filippi, P., and Delbecchi, L. (1973). Biochimie 55, 885.
8. Haase, A. T., Garapin, A. C., Faras, A. J., Taylor, J. M., and Bishop, J. M. (1974). Virol. 57, 259.
9. Haase, A. T. and Baringer, J. R. (1974). Virol. 57, 238.
10. Weiss, M. J., Zeelon, E. P., Sweet, R. W., Harter, D. H., and Spiegelman, S. (1977). Virol. 76, 851.
11. Haase, A. T. and Varmus, H. E. (1973). New Biol. 245, 237.
12. Haase, A. T., Traynor, B. L., Ventura, P. E., and Alling, D. W. (1976). Virol. 70, 65.
13. Petursson, G., Nathanson, N., Georgsson, G., Panitch, H., and Palsson, P. A. (1976). Lab. Investig. 35, 402.
14. Narayan, O., Silverstein, A. M., Price, D., and Johnson, R. T. (1974). Science 183, 1202.
15. Haase, A. T., Stowring, L., Narayan, O., Griffin, D., and Price, D. (1977). Science 195, 175.
16. Brahic, M., Filippi, P., Vigne, R., and Haase, A. T. (1977). J. Virol. 24, 74.
17. Narayan, O., Griffin, D. W., and Silverstein, A. M. (1977). J. Infect. Dis. 135, 800.

18. Nathanson, N., Panitch, H., Palsson, P. A., Petursson, G., and Georgsson, G. (1976). Lab. Investig. 35, 444.
19. Overall, J. C., Jr., and Glasgow, L. A. (1970). Science 167, 1139.
20. Friedman, R., and Ramseur, J. (1974). Proc. Nat. Acad. Sci. USA 71, 3542.
21. Narayan, O., Griffin, D. E., and Chase, J. (1977). Science 197, 376.

IMMUNE RESPONSES IN VISNA VIRUS INFECTION OF SHEEP[1]

Diane E. Griffin[2], Opendra Narayan, and Robert J. Adams

Departments of Medicine, Neurology and Comparative Medicine, Johns Hopkins University, Baltimore, Maryland 21205

ABSTRACT The infection and primary immune responses of sheep inoculated intracerebrally with visna virus were studied. Virus was detectable early in the cerebrospinal fluid (CSF) and later in CSF cells and peripheral blood leukocytes (PBL). Virus specific sensitized cells, as measured by stimulation of ^{3}H-thymidine incorporation, were present in the CSF and PBL within one week, peaked by 2 weeks, and were no longer detectable by 6 weeks. Histopathology showed an initial acute meningoencephalitis which gradually changed to a more chronic inflammatory process with germinal center formation. CSF showed a marked mononuclear pleocytosis and elevated protein followed by a relative increase in CSF IgG and the appearance of virus neutralizing antibody. It is concluded that the initial immune responses of sheep to infection with visna virus are prompt, vigorous, and similar to those seen in acute viral infections, except that the virus is not eliminated eventually resulting in the slowly evolving neurologic disease visna.

INTRODUCTION

Visna is a slowly progressive neurologic disease of sheep which occurred naturally in Iceland from 1935 to 1951 (1). The etiologic agent of this disease has subsequently been identified as a exogenous retrovirus (2) which replicates by means of a DNA intermediate which probably integrates in host cell DNA (3). Susceptible sheep inoculated intracerebrally with visna virus develop persistent infections of the brain, lung and lymphoid tissues (4, 5), but do not develop clinical evidence of disease until months to years after infection (6). Contributing to virus persistence are the ability of the virus to sequester proviral DNA in infected cells (7) and the abili-

[1]This work supported in part by grant #NS-12127 from the National Institutes of Health.
[2]Investigator, Howard Hughes Medical Institute.

ISBN 0-12-668350-6

ty of the virus to undergo mutation resulting in antigenically distinct virus strains (8). Both of these factors appear to operate within an individual infected animal and may allow the virus to escape clearance by the immune system. For this reason the nature and time course of the initial cellular and humoral immune responses within the CNS and in the periphery of sheep infected with visna virus were studied.

MATERIALS AND METHODS

Virus - Strain 1514 visna virus was plaque purified, grown and assayed in primary sheep choroid plexus cells (5). Stock virus assayed 10^6 TCD_{50}/ml. Concentrated virus (9) was suspended in NET buffer and had 10^8 TCD_{50}/ml.

Sheep and Inoculations - Twelve 6-12 month old female Border Leicester sheep were inoculated into both cerebral hemispheres with 10^8 TCD_{50} concentrated visna virus in one ml. Two control animals were sham inoculated with 1 ml NET buffer and two with MEM. Prior to inoculation mammary tissue biopsies were taken from each sheep and explant cultures grown (MTC). At regular intervals after inoculation sheep were bled and CSF was obtained by lumbar or cisternal tap. All procedures were performed under general anesthesia.

Antibody Assays - Neutralizing antibody was determined by titration of serum or CSF against 100 TCD_{50} visna virus strain 1514 (8).

Lymphocyte Stimulation - PBL, prepared as previously described (10), and CSF cells were suspended in RPMI 1640 (supplemented with 5% fetal calf serum, 4 μg/ml gentamicin and 25mM HEPES) at a concentration of 1-2 x 10^6/ml. Virus specific cell stimulation was assessed by placing 0.2ml of the leukocyte suspension into microtiter wells containing mitomycin C treated monolayers of uninfected or visna virus infected MTC. Stimulation was assessed by measuring incorporation of ^{3}H-thymidine into cells (1 μCi/well) 48-72 h later. Stimulation indices were calculated: $\frac{\text{cpm infected cells}}{\text{cpm uninfected cells}}$

Mitogen stimulations of PBL were similarly assessed using PHA 2.5 μl/ml. PWM, 10 μg/ml, Con A, 5 μg/ml and LPS, 10 μg/ml except that ^{3}H-thymidine was added from 24-48 h.

CSF - Cell counts and differentials were performed on fresh fluid, then the cells were removed for use in the cell assay and the cell-free fluid frozen at -20°C. CSF IgG was quantitated using a radial immunodiffusion assay (11). Total protein was measured by trichloroacetic acid precipitation (12).

<u>Pathology</u> - Sheep were killed for pathologic examination at 10 days, 4 wks and 3 mos after infection. Tissue was formalin fixed, paraffin imbedded, sectioned, and stained with hematoxylin and eosin.

RESULTS AND DISCUSSION

During the first months after inoculation host responses to the virus infection are actively evolving although there is no clinical evidence of disease. CSF cell counts and differentials show a mononuclear pleocytosis which peaks at 2-3 weeks and recedes to low but elevated values (Fig. 1). Virus can be recovered from cell free CSF for the first 2 weeks, but only from the CSF cells, as well as PBL, thereafter. Pathologic examination of central nervous system (CNS) tissue reveals an acute meningoencephalitis at 10 days which is beginning to resolve and develop more chronic changes by 4 weeks.

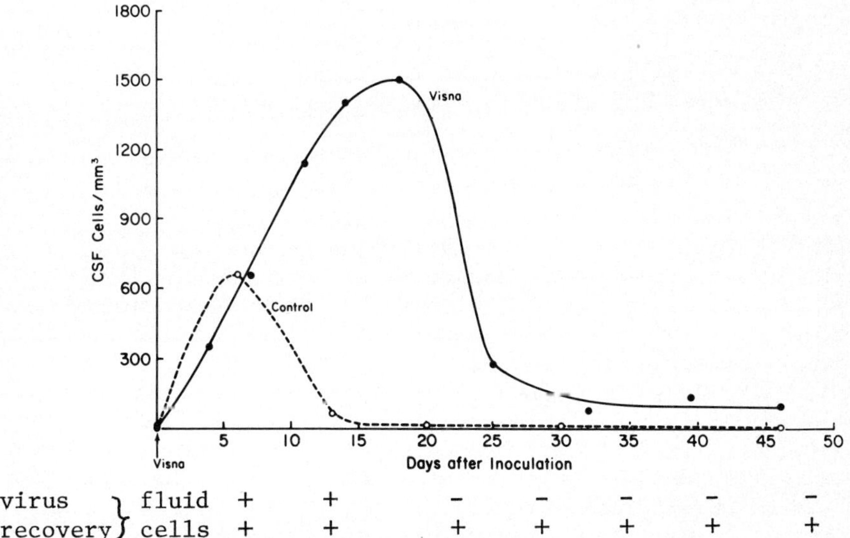

Figure 1 - Cerebrospinal fluid cell counts of sheep inoculated with visna virus (5) or tissue culture media (2, control). Virus recovery from the fluid and cells of the CSF of infected sheep during this period.

Cells sensitized to visna virus-specific antigen appeared within the CNS (CSF cells) and in the peripheral blood (PBL) within the first week after infection. These cells peaked at 2 weeks, but were detectable only transiently and by 6 weeks stimulation indices have returned to background levels (Fig. 2).

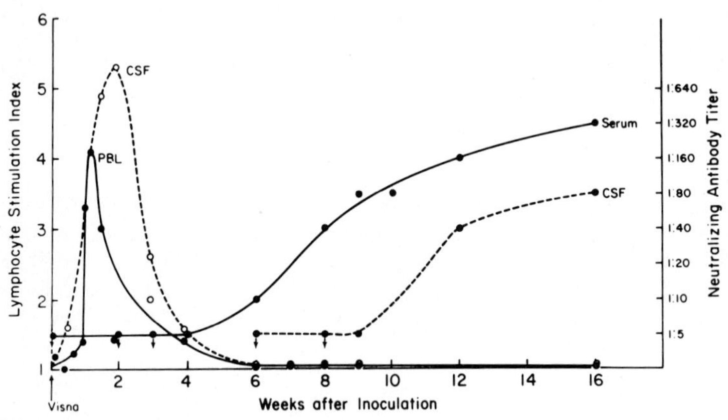

Figure 2 - Time course of virus specific cell stimulation and neutralizing antibody formation in blood and CSF after the intracerebral inoculation of sheep with visna virus.

The time course of detectable virus specific cell mediated immunity (CMI) is transient and similar to that seen in acute viral infections (13). The resolution of this CMI is temporally related to the disappearance of detectable virus within the CSF and it is possible either that virus is necessary for continued stimulation of the CMI and/or that the CMI contributes to clearing of cell free virus.

Thus, during the first 4 weeks after infection, there is evidence of a prompt virus-specific cell mediated immune response which is probably responsible for the acute meningoencephalitis which is in turn reflected by the mononuclear pleocytosis. The CSF shows evidence of impairment of the blood-brain barrier by the inflammatory process with greatly elevated CSF protein. The CSF IgG is elevated, but in proportion to the total protein.

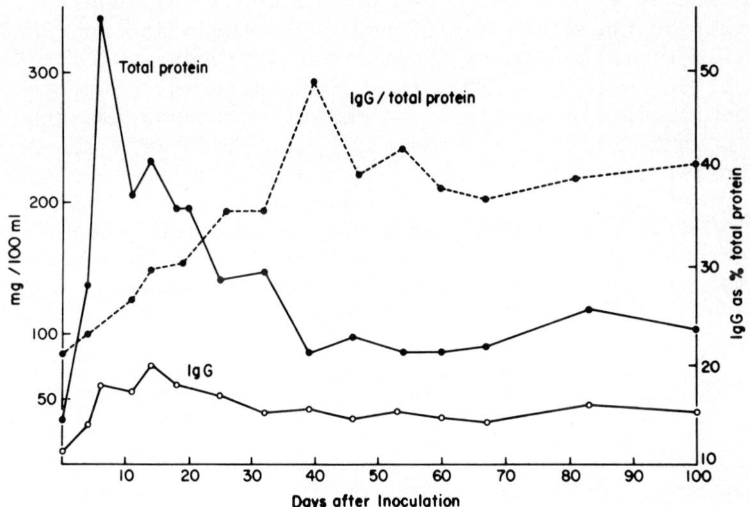

Figure 3 - Time course of the changes in CSF protein composition of sheep inoculated intracerebrally with visna virus.

During the 2nd and 3rd month after infection a more chronic inflammatory process is established with dense perivascular cuffs and germinal center formation. CSF cells and protein are only moderately increased over normal, but the percent of total protein which is IgG has increased (Fig. 3). Virus specific neutralizing antibody appears within the serum and then in the CSF (Fig. 2). By 4 mos individual sheep may have amounts of antibody in the CSF which approaches or exceeds that in the serum (14). These facts suggest local production of antibody within the CNS.

Since some persistent virus infections are associated with immunosuppression (15, 15), the responses of PBL to known B and T cell mitogens were assessed. No significant differences in mitogen stimulation betewen virus- and sham-inoculated sheep were detected during the first 45 days after inoculation (10). There is therefore, no evidence for a generalized immunosuppression by the virus. In addition there is evidence for a prompt virus specific cellular immune response and vigorous humoral immune response to the initial infection. This immune response may contribute to clearance of cell-free virus but does not eliminate persistently infected cells. It is possible that the majority of persistently infected cells may not be producing sufficient antigen for recognition. For instance the choroid plexus from one sheep studied showed 18% of cells had proviral DNA by in situ hybridization, 0.5% had detectable p30 antigen and none had detectable glycoprotein by fluorescent antibody staining (7).

Previous work has shown that much later in the course of infection antigenically distinct mutant viruses arise from the resevoir of persistently infected cells. These viruses are not neutralized by the antibody to the 1514 virus responsible for the initial infection. The appearance of these viruses may give rise to new rounds of virus replication, immune responses and CNS pathology which result in cumulative damage recognized clinically as a slowly evolving disease.

ACKNOWLEDGMENTS

We wish to thank Mr. Jack Bukowski and Ms. Diana Bobbie for technical assistance and Ms. Linda Kelly for manuscript preparation.

REFERENCES

1. Sigurdsson, B., Palsson, P.A. and Grimsson, H. (1957). J. Neuropath. Exp. Neurol. 16, 389.
2. Lin, F. H. and Thormar, H. (1970). J. Virol. 6, 702.
3. Haase, A. and Varmus, H. E. (1973). Nature (New Biol.) 245, 237.
4. Petersson, G., Nathanson, N., Georgsson, G., Panitch, H., and Palsson, P.A. (1976). Lab. Invest. 35, 402.
5. Narayan, O., Griffin, D. E. and Silverstein, A. M. (1977). J. Infect. Dis. 135, 800.
6. Gudnadottir, M. and Palsson, P.A. (1966). J. Immunol. 95, 1116.
7. Haase, A. T., Stowring, L., Narayan, O., Griffin, D. E., and Price, D. (1977). Science 195, 175.
8. Narayan, O., Griffin, D. E. and Chase, J. (1977). Science 197, 376.
9. Haase, A. T. and Baringer, J. R. (1974). Virol. 56, 238.
10. Griffin, D. E., Narayan, O., and Adams, R. J. (Manuscript submitted for publication).
11. Mancini, G., Carbonara, A. O. and Heremans, J. F. (1965). Immunochem. 2, 235.
12. Meulemans, O. (1960). Clin. Chim. Acta 5, 757.
13. Griffin, D.E. and Johnson, R. T. (1973). Cell. Immunol. 9, 426.

14. Griffin, D. E., Narayan, O., Bukowski, J. F., Adams, R. J. and Cohen, S. R. (Submitted for publication).
15. Friedman, H. (1975). In "Viruses and Immunity: Toward understanding Viral Immunology and Immunopathology" (C. Koprowski and H. Koprowski, eds.), pp. 17-47. Academic Press, New York.
16. Hotchin, J. (1971). N.Y. Acad. Sci. 181, 159.

PROGRESSIVE ANTIGENIC DRIFT OF VISNA VIRUS IN PERSISTENTLY INFECTED SHEEP[1]

Opendra Narayan, Diane E. Griffin[2], and Janice E. Clements

Department of Neurology, Johns Hopkins University
Baltimore, Maryland 21205

ABSTRACT Visna is a slow paralytic disease of sheep with a prolonged incubation period. The disease results from a persistent infection with an exogenous retrovirus. The virus replicates slowly in infected sheep and elicits normal immunological responses. These responses fail to clear the infection probably because of sequestered proviral DNA in tissue cells. Long term studies of infected sheep have shown that several months after the animals develop neutralizing antibody to the strain of virus inoculated, antigenic mutation of the virus occurs and this continues after development of antibody to the prior mutant virus. These mutants are virulent and, in the chronically infected animal, probably induce pathologic lesions which, cumulatively, may result in ultimate paralysis.

THE VISNA/MAEDI/PROGRESSIVE PNEUMONIA DISEASE COMPLEX

Visna is a slowly progressive paralytic disease of sheep preceded by a long incubation period (1). It is characterized histologically by intense infiltration and proliferation of mononuclear cells around the blood vessels in the brain and spinal cord, and diffuse leukomalacia (2). The disease was first recognized in Iceland in the 1930's as a rare occurrence among sheep afflicted with a more prevalent but also slowly evolving disease: maedi or chronic interstitial pneumonia (progressive pneumonia in the U.S.). The pneumonic disease occurs in sheep throughout the world, causing major agricultural problems in some countries. Small numbers of these animals also develop cerebral lesions indistinguishable from visna, indicating that the ratio of incidence between the two diseases may be common between pristine Icelandic and other more recently developed breeds of sheep (3).

The etiologic agents of this disease complex are a group

[1]Supported by grants #NS-12127 and #NS-10920.
[2]Investigator, Howard Hughes Medical Institute

ISBN 0-12-668350-6

of serologically related exogenous retroviruses which cause lytic productive infections in sheep cell cultures (4,5,6,7). Although strains of these viruses may have different tissue tropisms in the animal, thus causing primarily pneumonic or encephalitic disease, they elicit cross-reacting complement-fixing antibodies (6) and have an antigenically related P30 protein (8).

VISNA VIRUS REPLICATION IN SHEEP

Until recently, all studies of visna were performed in Iceland using local sheep. Studies reported here were done primarily with Hampshire sheep, an English breed common on the East Coast, and a strain of visna virus, 1514, obtained from an inoculated paralyzed Icelandic sheep (9,10). Fetal sheep were inoculated with virus and tissues assayed for infectivity at sequential intervals by inoculation of cell cultures with homogenates and by direct explantation. To test whether virus would replicate more extensively in an immunosuppressed host, some fetuses were thymectomized prior to inoculation with virus. As shown in Table 1, only minimal replication occurred in tissue and was not enhanced by immunosuppression.

Table 1

RECOVERY OF VISNA VIRUS[2] FROM TISSUE SUSPENSIONS AND EXPLANTS OF INTRACEREBRALLY INOCULATED FETAL LAMBS

Tissue	No. Tested		Suspensions		Explants	
Choroid Plexus	NT[3]	8	2	(25%)	8	(100%)
	T	13	2	(15%)	12	(92%)
Brain	NT	8	1	(12½%)	6	(75%)
	T	15	2	(11%)	12	(86%)
Lung	NT	8	2	(25%)	7	(90%)
	T	15	2	(12%)	9	(60%)
Spleen	NT	8	1	(14%)	6	(75%)
	T	14	0	(0%)	9	(64%)
Kidney	NT	7	0	(0%)	2	(30%)
	T	11	0	(0%)	3	(27%)
Buffy Coat	NT	7	0	(0%)	3	(43%)[4]
	T	15	---	---	4	(27%)

[1]Modified from Table 1, J. Inf. Dis. (1977) 135, 800 with permission from publishers.

[2]Virus recovery is based on 1 passage in culture

[3]NT = non-thymectomized T = thymectomized

[4]Virus isolation by cocultivation of buffy coat leukocytes with SCP cells

All tissue suspensions (10% wt/vol) failed to cause CPE if diluted at 1/10. Virus recovery was much more efficient from tissue explants, outgrowths of which produced virus and developed CPE. This manner of replication of the virus was noted in tissues of fetuses examined at 1, 2 and 9 weeks post inoculation, although the frequency of virus induction in individual explants decreased after the first 2 mos. No cell free virus was obtained from tissues of inoculated adult animals examined between 10 days and 1 year after inoculation although virus could still be recovered from tissue explants from these animals.

VIRUS PERSISTENCE IN INFECTED SHEEP

Despite the lack of exponential viral replication in tissues at any stage of infection, efficiency of virus recovery from explant cultures suggested a gradient of susceptibility of different tissues, the highest "virus yielders" being choroid plexus, brain, lung and lymphatic tissues. In addition, virus could be recovered routinely from peripheral blood leukocytes (PBL) of inoculated adult animals by cocultivation techniques (10). The great disparity between virus recoverability from tissue suspensions versus explants suggested that the virus genome was present in tissue but did not become fully expressed until cells were cultured *in vitro*. This was proven by in-depth examination of choroid plexus from a fetus infected for 3 weeks (Figure 1) (11).

FIGURE 1: STATE OF VISNA VIRUS IN TISSUE

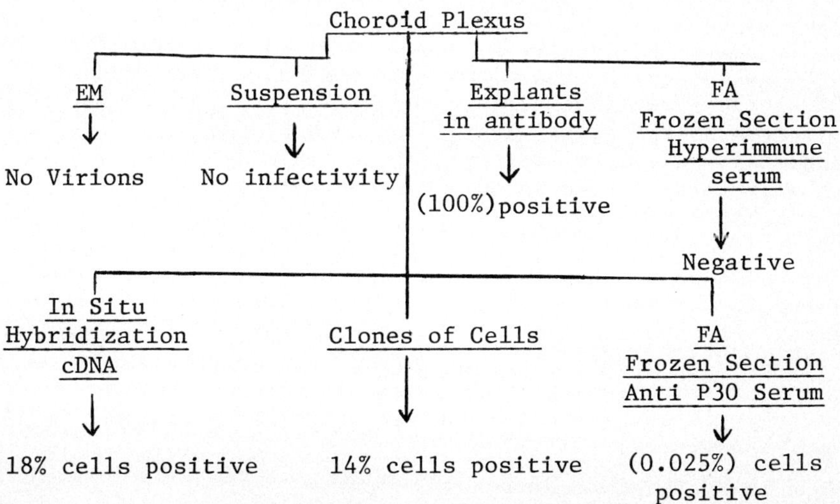

The number of cell clones from this tissue developing CPE correlated very closely with the number of cells in a frozen section of the tissue specifically labelled with cDNA by *in situ* hybridization. Although these data suggested a lysogenic state of infection, some virus replication, beyond the sensitivity of assay techniques, must have been in progress since between 6 and 12 weeks post inoculation nearly all animals produced neutralizing antibody (12). These antibodies are presumably directed to the glycoproteins in the virus envelope. However the sequestration of proviral DNA in tissue cells provided an effective means for persistent infection of the animal.

PERSISTENCE OF VIRAL GENOME IN PERIPHERAL BLOOD LEUKOCYTES

Virus recovery from PBL of infected sheep was regular but inefficient, requiring cocultivation of about 10^6 cells with indicator cells for induction of CPE. This suggested that only a few of these cells were infected (10, 13). Infected PBL provided a useful method for virus recovery and they may function *in vivo* as a source of infection for new cells. Cultured PBL inoculated with virus at high multiplicities did not support productive replication of the virus but virus could be recovered from the cells by cocultivation (Narayan, unpublished). PBL may harbor only proviral DNA, possibly in a lysogenic state, and reflect the replication occurring in other cell types in the infected animal.

ANTIGENIC DRIFT OF VIRUS IN PERSISTENTLY INFECTED SHEEP

Development of antibody in the animal had no effect on the recoverability of the virus from the PBL. Long-term studies have shown that viruses recovered from sheep prior to and several months after development of antibody were antigenically identical to the parental strain of virus used for inoculation (13). Subsequently however, viruses recovered from PBL of 3 of 4 sheep failed to be neutralized by the animals' sera (Figure 2) (14) suggesting mutation of the virus had occurred during the long infection of the animal and that antibody to the parental virus had provided a selective advantage for replication of the mutants (13). This hypothesis was proven in cell cultures which, when infected with plaque purified virus and maintained in medium with specific antiserum, yielded a virus which failed to be neutralized by the same serum (14).

FIGURE 2[1] VIRUS MUTATION IN FOUR PERSISTENTLY INFECTED SHEEP[2,3]

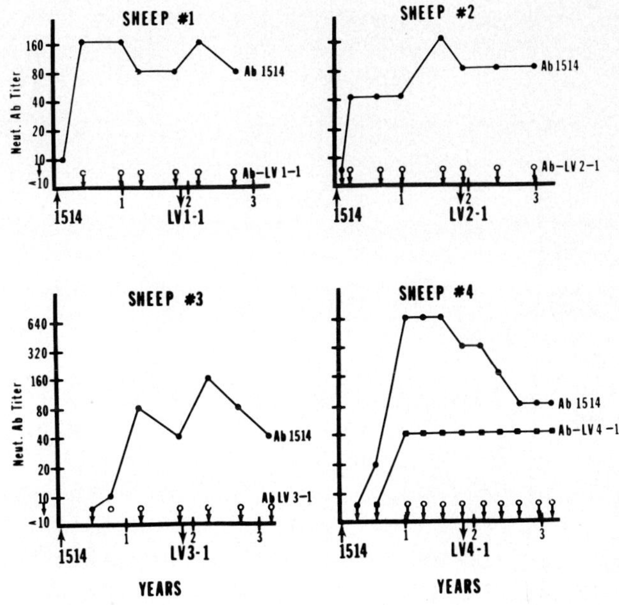

[1]Modified from Narayan, Griffin, Chase, (Science (1977) 197, 376 with permission of publishers.
[2]Leukocyte viruses (LV) recovered from sheep 1, 2, and 3 failed to be neutralized by the respective animal's sera. LV4-1 from #4 sheep was neutralized at a dilution of serum much lower than that neutralizing V1514.
[3]All 4 sheep were inoculated intracerebrally with plaque purified virus.

The demonstration of mutant viruses in chronically infected sheep raised two questions: Is antigenic mutation a continuous process in the face of neutralizing antibody, and does one antigenic population of a virus replace another after development of antibody to the first mutant? To answer these questions, sheep #1 shown in Figure 2, was selected for further study (13). Virus strain LV1-2 was isolated 6 months after the recovery of LV1-1. One year later 5 viruses, LV1-3 to LV1-7 were obtained simultaneously from PBL in 50ml of blood. The leukocyte viruses, plus parental virus 1514, were tested in batteries of neutralization tests using sequentially collected sera from the same animal. The neutralization patterns suggested that the viruses could be grouped into 3 antigenic strains: (a) parental 1514 virus and LV1-2; (b) LV1-3, LV1-4, LV1-7; (c) LV1-1, LV1-5, LV1-7. Antibody to each strain appeared to develop in sequential order (Figure 3). Thus although the viruses, including the parental strain, apparently coexisted in the animal, the antibody patterns suggested that the mutants may have evolved sequentially.

FIGURE 3 PROGRESSIVE MUTATION OF VISNA VIRUS IN SHEEP #1

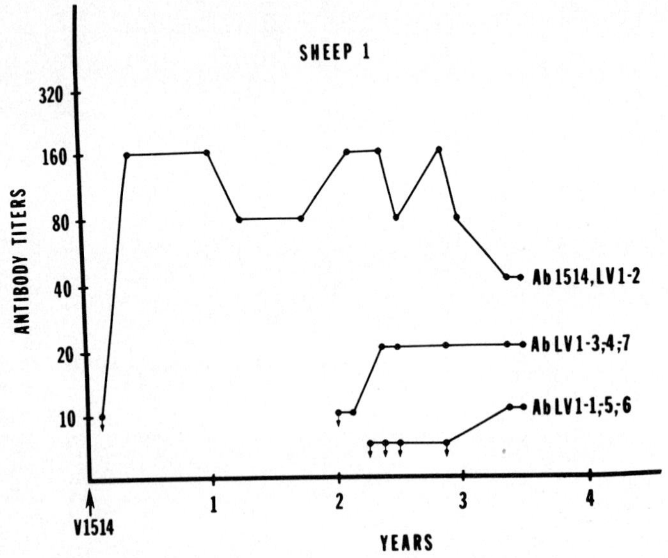

[1]Leukocyte viruses, LV1-1 to LV1-7 were recovered from the animal during years 2, 3, and 4 post inoculation.

Cultivation and plaque purification of the mutant viruses indicated that they were antigenically stable, did not revert to parental antigenicity and could replicate in cell cultures as efficiently as the parental virus (13). Further, they were also virulent for sheep, causing acute encephalitis after intracerebral inoculation (Narayan, unpublished).

These data show that in this persistent infection, the virus is not defective and the host is not immunologically compromised. Rather the slow disease appears dependent on a competent host immune system which exerts selective pressure for evolution of virulent mutants. Pathologic lesions induced by these mutants could cumulatively lead to clinical disease years after infection.

ACKNOWLEDGEMENTS

We thank Martha Somerville, Diana Bobbie and Linda Kelly for excellent technical assistance.

REFERENCES

1. Sigurdsson, B., Palsson, P., Grissom, H. (1958). J. Neuropath. Expl. Neurol. 16, 389.
2. Sigurdsson, B., Palsson, P., Van Bogaert, L. (1962). Acta Neuropath. 1, 343.
3. deBoer, G.F. (1975). Res. Vet. Sci. 18, 15.
4. Sigurdsson, B., Thormar, H., Palsson, P. (1960). Arch. ges Virusforsch. 10, 368.
5. Lin, F.H., Thormar, H. (1970). J. Virol. 6, 702.
6. Gudnadottir, M., Kristindottir (1967). J. Immunol. 98, 663.
7. Haase, A.T., Varmus, H.E. (1973). Nature New Biol. 245, 273.
8. Weiss, M.J., Zeelon, E.P., Swett, R.W., Harter, D.H., and Spiegelmann, S. (1971). Virology 76, 851.
9. Narayan, O., Silverstein, A.M., Price, D., Johnson, R.T. (1974). Science 183, 1202.
10. Narayan, O., Griffin, D.E., Silverstein, A.M. (1977). J. Inf. Dis. 135, 800.
11. Haase, A.T., Stowring, L., Narayan, O., Griffin, D.E., Price, D. (1977). Science 195, 175.
12. Griffin, D.E., Narayan, O., Adams, R.J. (1977). Submitted for publication.
13. Narayan, O., Griffin, D.E., Clements, J.E. (1978). Submitted for publication.
14. Narayan, O., Griffin, D.E., Chase, J. (1977). Science 197, 376.

DEMYELINATION PRODUCED BY WILD-TYPE AND TEMPERATURE-SENSITIVE MUTANTS OF MOUSE HEPATITIS VIRUS TYPE 4 (JHM)[1]

Martin V. Haspel, Peter W. Lampert and Michael B. A. Oldstone

Department of Immunopathology, Scripps Clinic and Research Foundation and Department of Pathology, University of California, San Diego
La Jolla, CA 92037

ABSTRACT Mouse hepatitis virus produces a fatal encephalomyelitis, but may under restrictive conditions produce primary demyelination. Early demyelination is produced by direct virus injury to the myelin synthesizing and maintaining oligodendrocytes. Temperature-sensitive mutants have been isolated which are associated with either a high or low incidence of demyelination. Using this new manipulative virus system, it will be possible to elucidate the molecular events leading to demyelination.

INTRODUCTION

Possible virus etiology has been proposed for many degenerative diseases of the central nervous system (1). The involvement of viruses in the pathogenesis of subacute sclerosing panencephalitis and progressive multifocal leukoencephalopathy is well accepted (2). The role of virus infection in multiple sclerosis, and other primary demyelinating diseases of man is, however, largely speculative. These conjectures are supported in part by evidence provided by comparative neurovirology. Virus induced demyelination has been demonstrated in dogs (canine distemper), goats (leukoencephalitis) and mice (mouse hepatitis virus, Theiler's encephalomyelitis virus) (3-6). Of these experimental models, only two utilize a small genetically characterized laboratory animal and are therefore, the most manipulative. This paper will briefly review the salient features currently known about

[1]This is publication No. 1488 from the Department of Immunopathology, Scripps Clinic and Research Foundation. This work was supported by USPHS grants AI 14209, NS 09053, NS 12428 and NS 14068.

ISBN 0-12-668350-6

mouse hepatitis virus induced demyelination. A discussion of demyelination produced by Theiler's encephalomyelitis virus is found elsewhere in this volume.

The JHM strain of mouse hepatitis virus was isolated in 1947 from the brains of two mice that spontaneously developed hind limb paralysis (4-5). The most striking feature of this encephalomyelitis was the widespread demyelination. Subsequent passage of the virus in the central nervous system resulted in a shortening of the incubation period with fatal encephalitis developing more frequently than paralytic disease. This virus, designated JHM, was subsequently shown to serologically crossreact with the other mouse hepatitis viruses and indeed be serologically related to human coronaviruses (7).

Variables Affecting the Pathogenesis of Wild Type Virus Induced Demyelination. One may conceptualize the involvement of the central nervous system by JHM as a disease continuum with acute fatal necrotizing encephalomyelitis and non-fatal demyelination at opposite ends of the spectrum. One might therefore predict that under restrictive conditions, the non-fatal demyelinating disease would predominate; the work of Weiner has shown this to be essentially correct (8). Inoculation of suckling mice resulted in the fatal disease, juvenile (4 week old) mice developed either demyelination or the fatal disease, while adult animals were resistant. The production of non-fatal demyelination was also dose dependent. Following a large infectious dose, all of the animals succumbed to the fatal disease, while few if any of the mice developed lesions after receiving a small dose.

Pathogenic Mechanism of Demyelination. Bailey et al. suggested in 1949 that the virus could directly affect the cells responsible for the maintenance of myelin (5). Direct cytotoxic effect by the virus upon the myelin synthesizing and maintaining oligodendrocytes has been reported (9). Coronavirus particles are readily observed in oligodendrocytes (Fig. #1). The sequence of the events leading to demyelination appears to begin with the infection inducing proliferative changes in the oligodendrocytes (10). These hypertrophic oligodendrocytes exhibit multiple connections of their plasma membrane with myelin lamellae, as well as connections to redundant loops of myelin (Fig. #1). These proliferative changes are followed by degeneration of the oligodendroglia leading to demyelination. Macrophages are then observed to strip the degenerated myelin from otherwise intact axons (Fig. #2).

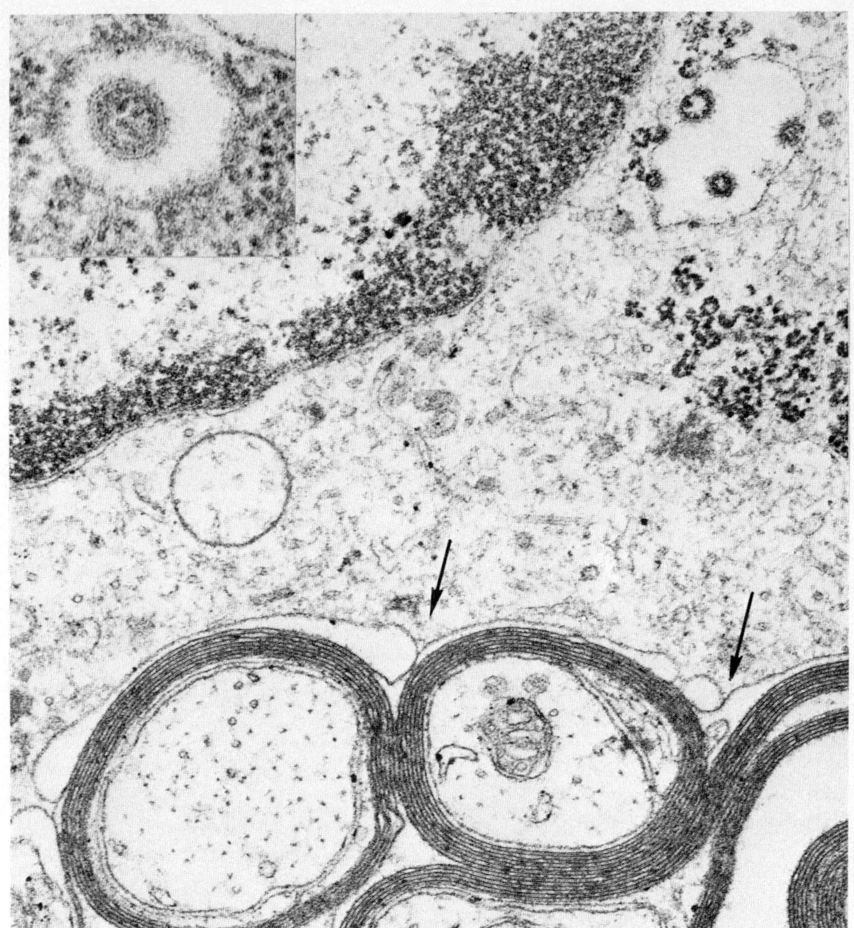

Figure 1. Oligodendrocytes, identified by its plasma membrane connections with myelin lamellae (arrows) shows coronavirus (JHM, wild type) budding into a cistern of endoplasmic reticulum (upper right) x55,000. The inset shows an intracisternal virus particle (x120,000). [Reproduced with permission, Lab. Invest. 33:440, 1975.]

Immunosuppression by cyclophosphamide or inoculation of thymus-less nude mice results in an increase in the severity of the disease (8, Haspel et al., unpublished observation). Furthermore, it has been reported that immunoglobulins were not detected in demyelinating lesions (8). In addition, only a scanty infiltrate of lymphocytes is observed in early lesions (10). Thus, it appears that early demyelination by JHM may not involve an immunopathologic mechanism but rather result from a direct virus induced injury to oligodendrocytes.

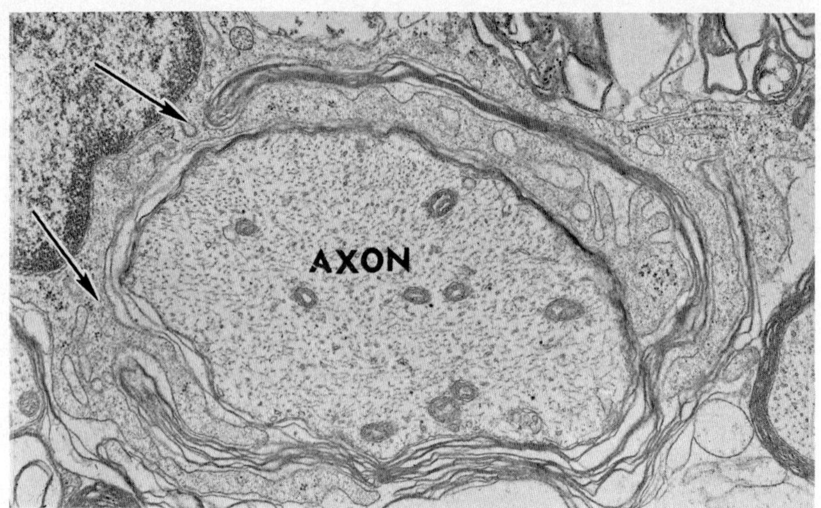

Figure 2. Macrophage invading (arrows) a myelin sheath stripping the myelin lamellae off the axon. Spinal cord 7 days after infection with wild type JHM virus. x10,000. [Reproduced with permission, Acta Neuropath. 27:74, 1973.]

Demyelination by Temperature-Sensitive Mutants. Since wild type virus infection produces demyelination inconsistently, we attempted to attenuate the wild type virus and thereby decrease the frequency of the fatal disease while maintaining the ability of the virus to produce demyelination. Temperature-sensitive (ts) mutants were isolated using techniques previously described (11). Selection for temperature-sensitive mutants resulted in a dramatic decrease in the incidence of the acute fatal encephalomyelitis (Fig.#3). The LD_{50} for 4 week old BALB/c st mice was 2 plaque forming units (PFU) of the wt virus, whereas intracerebral inoculation of 10,000 PFU of any of the ts mutants failed to produce 50% mortality. This attenuation resulted from the ts mutation as ts^+ revertant virus regained full virulence (3 PFU produced 60% mortality). Despite the attenuation, most of the mutants retained the ability to produce demyelination. The lesions in the spinal cord consisted of circumscribed patchy areas of demyelination (Fig. #4).

The mutants differed in their ability to produce demyelination (Fig. #3). There was no apparent relationship, unlike the wt virus, between the lethality of the virus dose and the incidence of demyelination. Ts mutant 8 was able to produce demyelination in 58% of the inoculated mice, while only 3% of the animals (1/38) developed the acute fatal encephalomyelitis. Thus, ts mutant 8 will permit the analysis of the early events of demyelination in the absence of the

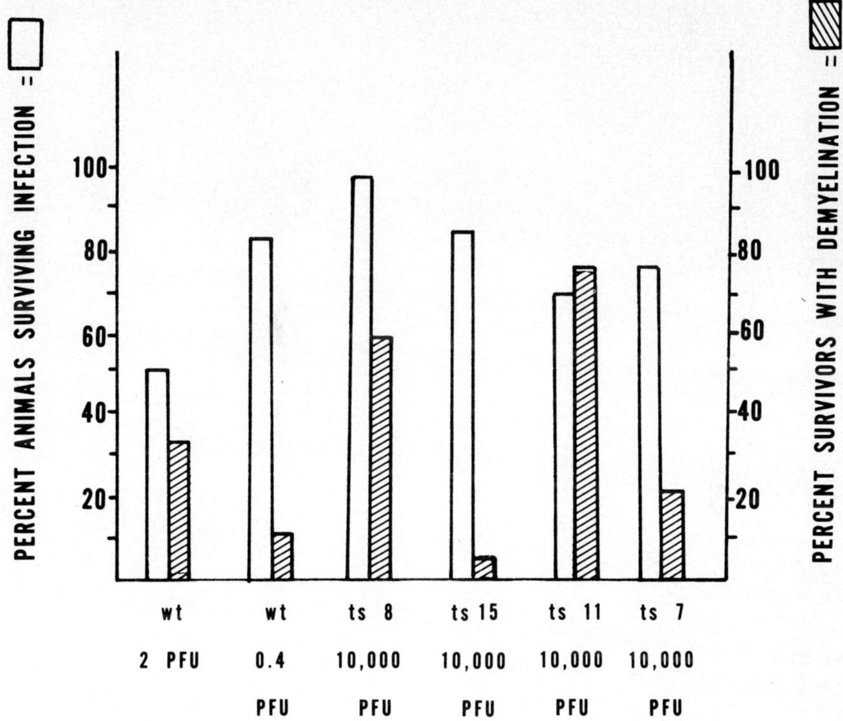

Figure 3. The neurovirulence and induction of demyelination by mouse hepatitis virus.

acute fatal disease. The differences observed among the mutants can not be attributed to partial activity "leakiness" of the mutants *in vitro*. Ts mutants 8 (58% demyelination) and 15 (5% demyelination) are equally "leaky" *in vitro* whereas ts mutants 7 and 11 are completely turned off at the non-permissive temperature (Table 1). Ts mutant 15 does replicate *in vivo*, thus the low frequency of demyelination is not simply due to an inability to replicate in the central nervous system. Reversion of the mutants *in vivo* cannot, likewise, explain these differences among the mutants, as all virus isolates retained the temperature-sensitive mutation. Thus, the ability of a mutant to produce demyelination cannot be attributed to simplistic variables and may in fact be due to the inherent molecular biology of the mutation. The availability of ts mutants which produce demyelination with either high or low frequency may theoretically permit the determination of the virus gene product(s) required for oligodendroglial cytolysis.

Table 1. Properties of Representative ts Mutants

	Virus				
Property	wt	ts7	ts8	ts11	ts15
EOP 39.5°/34°C[1]	1.3×10^{0}	1.3×10^{-5}	4.0×10^{-4}	1.6×10^{-5}	4.5×10^{-4}
Titer 34°/39.5°C[2]	5.0×10^{-1}	$>5.6 \times 10^{3}$	3.5×10^{1}	$>1.7 \times 10^{3}$	2.5×10^{1}
"Leakiness"[3]		0	+	0	+
Virus antigen synthesis at 39.5°C[4]	+	0	+	0	+

1. Efficiency of plating at 34° and 39.5°C. Wild type plaques were counted 39.5°C.
2. Titer 12 hr after infection (moi <0.1) of NCTC-1469 cells.
3. Some functional activity at 39.5°C resulting in mutant progeny virus.
4. Both cytoplasmic and surface antigens as detected by immunofluorescence.

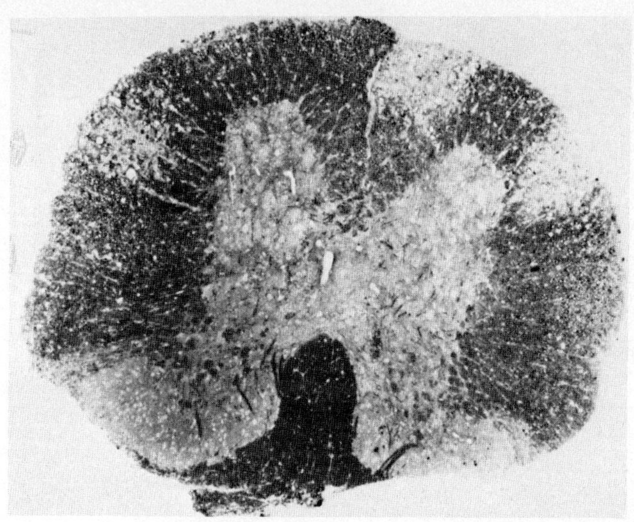

Figure 4. Patchy areas of demyelination in the white area of the spinal cord of a BALB/c st mouse 11 days after intracerebral inoculation of 10,000 PFU of ts mutant 11 (x50).

ACKNOWLEDGEMENTS

We thank Peggy Farness and Bob Garrett for excellent technical assistance and Cathy Doria for assistance in preparing this manuscript.

REFERENCES

1. Gajdusek, C. D. (1972). Adv. Geront. Res. 4,201.
2. Weiner, L. P., Johnson, R. T., and Herndon, R. M. (1973). New Engl. J. Med. 288,1103.
3. Cork, L. C. and Davis, W. C. (1972). Lab. Invest. 32,359.
4. Cheever, F. S., Daniels, J. B., Pappenheimer, A. M. and Bailey, O. T. (1949). J. Exp. Med. 90,181.
5. Bailey, O. T., Pappenheimer, A. M., Cheever, F. S., and Daniels, J. B. (1949). J. Exp. Med. 90,195.
6. Dal Canto, M. C. and Lipton, H. L. (1977). Amer. J. Path. 88,497.
7. McIntosh, K. (1974). Curr. Top. Microbiol. Immunol. 63,85.
8. Weiner, L. P. (1973). Arch. Neurol. 28,298.
9. Lampert, P. W., Sims, J. K., and Kniazeff, A. J. (1973). Acta Neuropath. (Berl.), 24,76.
10. Powell, H. C., and Lampert, P. W. (1975). Lab. Invest. 33,440.
11. Haspel, M. V., Duff, R., and Rapp, F. (1975). J. Virol. 16,1000.

THE RELATIONSHIP OF THEILER'S MOUSE ENCEPHALOMYELITIS VIRUS PLAQUE SIZE WITH PERSISTENT INFECTION[1]

Howard L. Lipton

Department of Neurology
Northwestern University Medical School
Chicago, Illinois 60611

ABSTRACT The present report characterizes the patterns of disease which result after intracerebral inoculation of mice with cell culture adapted Theiler's mouse encephalomyelitis viruses. GDVII virus, a large plaque variant, produced a rapidly fatal encephalitis, and was at least a thousand-fold more virulent than 4 small plaque variants which resemble Theiler's original strains. While it was demonstrated that these small plaque viruses persist in the host's spinal cord but not brain, surviving animals that had received approximately one 50 per cent lethal dose of GDVII virus did not appear to chronically harbor large plaque virus. However, the evidence presented does not entirely exclude this possibility. The persistent CNS infection with the small plaque viruses resulted in chronic demyelinating disease which appears to be similar in most respects to late disease following infection with brain-derived DA virus. Depending on the virus dose demyelinating disease may occur after a prolonged incubation.

INTRODUCTION

Over 40 years have elapsed since Max Theiler discovered a new group of murine enteroviruses and showed that they cause chronic central nervous system (CNS) infection in mice (1,2). Historically, this was the first report of a chronic virus infection in an animal host. These viruses are often referred to as the Theiler's mouse encephalomyelitis viruses (TMEV). In the original studies infectious virus was isolated from the CNSs of animals surviving acute flaccid paralysis (by titra-

[1]This work was supported by National Institutes of Health Grant No. 1 R01 AI 14139-01. H.L.L. is the recipient of USPHS Career Development Award 1 K04 AI00228-01.

ISBN 0-12-668350-6

tion back into mice) for as long as one year after intracerebral (IC) inoculation, and for 5 months in infected but asymptomatic mice. While Theiler (2) and others (3) had initially characterized the acute pathological changes showing the resemblance to human poliomyelitis, the occurrence of histological changes during the chronic phase of infection was not recognized. Subsequently, extensive lesions of myelin destruction in the spinal cords of chronically infected mice were described by Daniels, et al (4).

In the early 1940's this infection gained some prominence because it served as an animal model for human poliomyelitis. But with the advent of successful polio vaccines, the importance of this model declined. The fact that these viruses caused persistent infection in animals had initially raised no eyebrows. It is also surprising that this information was overlooked in the 1960's when there was a tremendous increase in interest in slow and chronic virus infections. A number of factors probably contributed to the lack of interest: (a) ordinary enteroviruses appeared to be less promising candidates for causing chronic CNS disease in humans than did measles virus and the unconventional-transmissable agents, (b) the TMEVs may have been regarded as atypical of other enteroviruses in their ability to cause persistent infections (this point still remains to be determined), and (c) a convenient virus assay system was not available for those isolates resembling Theiler's original (TO) strains.

I was originally attracted to the TMEVs as much from the prospect of studying a virus-induced demyelinating disease as a chronic infection. It happens that very few virus infections cause CNS demyelination, an important pathological reaction in human disease.

Since the brain-derived TO strains did not replicate and produce a cytopathic effect in cell culture systems without blind subpassage, mouse brain stocks of virus were used in our initial pathogenesis studies (5,6,7). At that time it was felt that the adaptation process might attenuate the TO viruses. The principal features of the pathogenesis of TMEV infection using brain-derived virus are illustrated in Figure 1. It was found that young adult mice developed flaccid paralysis during the first month after IC inoculation, from which most animals recovered (early disease). Subsequently, almost all surviving mice then developed a spastic paralytic gait disorder between one to 3 months (late disease). The histopathologic changes were markedly different. In early disease there was predominantly involvement of gray matter structures in the CNS which resembled human poliomyelitis, while the spinal cord leptomeninges and white matter were the sites of involvement during late disease. Collections of mononuclear cells were

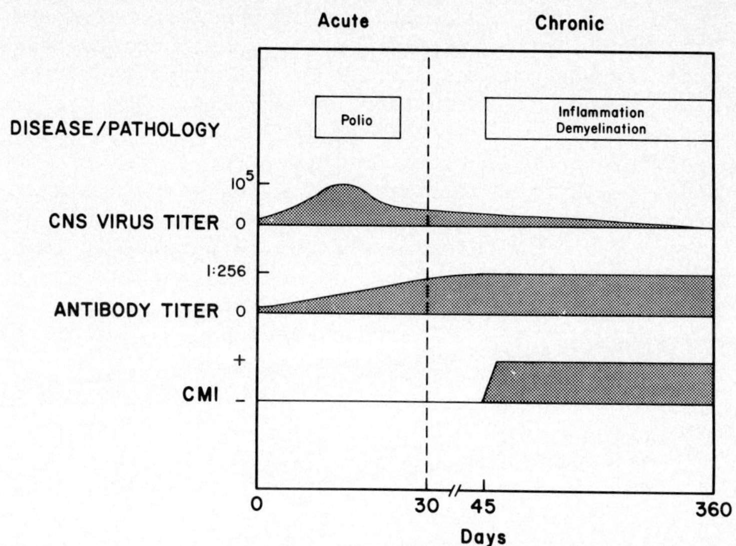

Figure 1. The pathogenesis of TMEV infection in young adult mice after IC inoculation of brain-derived DA virus. This figure summarizes the temporal relationships of virus replication and host immunity in the acute (early disease) and chronic (late disease) phases of the infection. CMI stands for cell-mediated immunity.

found in the leptomeninges, in perivascular sites and extending into the white matter where single as well as large numbers of demyelinated axons were found amidst inflammatory cells. These latter changes were observed throughout the first year.

CNS virus replication tended to parallel the phases of clinical disease just described. During the first 3 weeks there was exponential virus growth in the CNS reaching a maximum at 2 weeks with lower levels of infectious virus then persisting for as long as 14 months. On the basis of the histopathology and fluorescent antibody staining for virus antigen, TMEV initially appeared to replicate in neurons, particularly motor neurons in the brainstem and spinal cord. The cell type(s) supporting virus growth during the chronic phase of the infection still remains to be elucidated.

Finally, the temporal development of humoral and cellular immunity to TMEV was found to be slowed or delayed. Although neutralizing antibodies were detected by one week, there was an unusually gradual rise in titers over the first 2 months of infection. In addition, cellular immune responsiveness to virus was not detectable until 2 months, but once established its activity was present throughout the first year.

The present investigation characterizes the patterns of disease which result after infection of mice with cell culture adapted TMEVs. GDVII virus, a large plaque variant, produced a rapidly fatal encephalitis in mice and was at least a thousand-fold more virulent than 4 TO strains, all small plaque variants. The TO viruses were found to persist in the host's spinal cord but not brain, and depending on the virus dose, produce demyelinating disease after a prolonged incubation period. However, the evidence presented does not exclude the possibility that GDVII virus may also cause persistent CNS infection in surviving animals.

RESULTS

Correlation of Plaque Size and Virulence. Four TO strains of TMEV - DA, WW, TO4 and Yale viruses - were recently adapted by blind subpassage to produce cytopathic effect in primary baby mouse kidney or L929 cells (8). In contrast to GDVII virus which forms large plaques, all of the adapted TO viruses produced small plaques in L929 (8) and BHK21 cells. Another TMEV isolate, FA virus, which is reported to be similar in pathogenicity to GDVII virus, was obtained from the American Type Culture Collection, Rockville, MD. It has also produced large plaques in BHK21 cells. Purified stocks of these TMEV strains except FA virus were prepared in BHK21 cells, and the neurovirulence determined by IC inoculation of 4 to 6-week-old outbred mice (CD-1 male mice from a barrier colony at Charles Rivers Breeding Lab, Portage, MI) with ten-fold dilutions of virus. GDVII virus regularly produced a fatal acute encephalitis; the 50 per cent mean lethal dose (LD_{50}) was approximately 10 plaque forming units (PFU) (for 4 replicate experiments the mean $LD_{50} \pm$ the standard error was 8.7 ± 2.47 PFU). Depending on the virus dose affected mice died from 2 to 10 days after inoculation. While the LD_{50} for FA virus has not been determined, it is reported to cause acute encephalitis after a short incubation period.

In contrast, DA, WW, TO4 and Yale viruses generally were not lethal for mice during the first month of observation with administration of almost 10^5 PFU. The LD_{50} for these small plaque variants then exceeded 10^4 FPU. The correlation of disease state in mice and plaque size is shown in Table I. Therefore, in comparison to GDVII virus the TO strains are at least a thousand-fold less virulent. While only a limited number of isolates have been examined so far, it would appear that TMEV

that died mostly of trauma at ages ranging from 17 to 83.[3] Included in this series were one patient who died of lymphoma and one who died of MS. In the MS case, a large area of extensive demyelination was found in the pons at the trigeminal nerve root entry zone.[3]

Since another MS case processed some time ago had lesions of similar character at the vagus nerve entry zone into the medulla oblongata, we extended the search for HSV to vagus and other nerve ganglia. Studying a series of 9 cadavers, the virus was isolated from the jugular portion of the vagus ganglion from one cadaver and from the superior cervical ganglia of two others.[4] Virus was isolated by cultivation of the ganglionic tissue in culture either alone or in contact with indicator cells of either human or green monkey origin. It is possible that in some of the tissues in which HSV could not be reactivated, the viral genome was still present and could have been detected either by complementation or recombination performed after superinfection of the ganglionic tissue with properly selected temperature-sensitive mutants of HSV.[5] This might have raised the percentage of positive trigeminal ganglia to 80%, a figure more in line with the 75% of the American adult population showing neutralizing antibodies against HSV, and with the very high incidence of reactivation of facial HSV after trauma caused by surgery involving facial nerves.

The HSV isolated from an MS case does not seem to differ in its biological properties from the HSV isolated from any other cadaver. Since antigenic determinants of HSV are very poorly identified, there has been no attempt to use serological techniques to differentiate various isolates of HSV from nerve ganglia. Lonsdale and his associates[6], however, tried to identify various HSV isolates from nerve ganglia by restriction endonuclease analysis. In these assays the profile of HSV isolated from an MS case could be distinguished from other isolates; however, every isolate tested by this method showed a distinct profile different from that of all other isolates. The only identical restriction enzyme profiles were those of two viruses isolated from different ganglia of the same cadaver.

A parallel can be established between recurrent attacks of herpes labialis and the alternating periods of clinical exacerbation and remission in MS. Therefore we were curious to determine whether in Japan where MS is a very rare disease (1-4 cases/100,000), the prevalence of herpes labialis and of latent HSV in trigeminal ganglia is significantly lower than in America where incidence of MS is quite high. The results of survey conducted among 245 young adult Japanese of both sexes indicated that herpes labialis occurs at a significantly lower rate than it does in American populations (Table 1).[7]

TABLE I
HSV IN JAPAN AND U.S.A.

Country	Herpes Labialis*	HSV in Trigeminal Ganglia
Japan	59/245 (24%)	7/29** (7/52)***
U.S.A.	52/146 (35%)	22/44 (30/85)

* Based on response to questionnaire
** Positive/cases tested
*** Positive/ganglia tested

Latent HSV was less frequently observed in trigeminal ganglia of Japanese cadavers than in American cadavers. Does this mean that infection with HSV is less common in Japan than in the U.S.A.? Not at all if we consider that approximately the same percentage of the Japanese population has HSV antibody as does the American population. We can conclude that in Japan either the virus is more effectively eliminated from the organism (including nerve ganglia) or that it persists in latent form at a site(s) other than facial nerve ganglia.

No data were obtained on the incidence of oropharyngeal infection with HSV in Japanese children as compared to American children, but since the age at which HSV antibodies first appear seem to be similar[7] in the two countries, we can assume that primary infection with HSV occurs at the same age in both countries. After oropharyngeal infection, the virus presumably migrates to the trigeminal ganglia where it persists during the individual's lifetime. Since the prevalence of HSV encephalitis is very low one might ask what factors keep the potentially lethal virus from invading the brain. Or, to rephrase the question, is there ever a time after primary infection when HSV migrates to the brain only to be "driven back" to the trigeminal ganglion where it resides in a latent form?[8] Quite obviously it is impossible to solve this problem through the study of human subjects. In experiments with animals, my associate, Dr. Hamada, has injected mice in the lip with HSV then followed the spread of the infection into the nervous system.[9] As shown in Table 2, 3 days after infection the virus could only be recovered from ipsolateral trigeminal ganglia. On the sixth day after

TABLE II
CENTRIPETAL MIGRATION OF HSV AFTER INITIAL INFECTION OF TG

DAYS AFTER INFECTION	ISOLATION OF HSV FROM: TG	P	M	CE	TH	CR	ANTIBODY IN MOUSE SERA** (CPM IN RIA***)
3	2/3	0/3	0/3	0/3	0/3	0/3	411
6	5/5	5/5	3/3	3/5	2/3	3/5	542
9	1/3	1/3	1/3	1/3	1/3	1/3	1009
12	3/3	0/3	0/3	0/3	0/3	0/3	1633
>12	4/6	0/3	0/3	0/3	0/3	0/3	>2200

* by lip inoculation with an overall mortality of 8/60
TG = Trigeminal ganglion; P = Pons; M = Medulla;
CE = Cerebellum; TH = Thalmus; CR = Cerebrum.
** Diluted 1:500.
*** Normal mouse serum 364 cpm.

infection, however, the virus spread throughout the entire brain and HSV could still be recovered from all parts of the brain on the 9th day after infection. From the 12th day on, however, the virus could only be detected in the trigeminal ganglion, where it persists during the lifetime of the animal. Mortality in these experiments did not exceed 15 percent of the innoculated animals, yet HSV was recovered from the pons of at least 50 percent of the infected mice. Thus, we have to assume that within 1 week of primary exposure to HSV asymptomatic infection of brain takes place and that within 2 weeks of exposure the virus is again confined to the trigeminal ganglion (or possibly other nerve ganglia) where it resides in latent form during the lifetime of the animal. If a parallel situation exists in humans infected with HSV, it means that cases of herpetic encephalitis of infancy and early childhood may occur as the result of invasion of brain tissue by HSV from the trigeminal ganglion after primary oropharyngeal exposure. On the same analogy, it is also quite possible that in some human cases in which HSV establishes itself in the trigeminal ganglion an asymptomatic infection of the brain tissue occurs shortly after primary exposure.

Since the serum antibody levels of mice infected with HSV rose dramatically between the 6th and 9th days after infection (Table II), it is possible that antibody may be instrumental in the elimination of HSV from brain tissue. The virus is then driven back to the trigeminal ganglion which apparently shelters it from further onslaught by the antibody. Whether HSV is eliminated from the brain tissue of all infected individuals or whether it may remain in latent form in the brain cells of some "predisposed" subjects is impossible to ascertain at present. It is certain, however, that at some point in adult life the virus can become activated and in addition to causing herpes labialis it may also cause herpes encephalitis since this disease does not occur as the result of primary encounter with the virus. Trauma, infection, fever, irradiation or emotional stress precipitate attacks of herpes labialis,[8] but it is doubtful that these events also precipitate centripetal migration of the virus and development of CNS disease. One should not forget, however, that it has been known since 1897[8] that trauma may be a precipitating event of the first attack of MS.

When the trigeminal ganglion is severed from its axon by surgical procedure or removed from the body and maintained in culture, the latent function of HSV are expressed and infected cells in tissue culture are lysed. Trigeminal sensory-root section[10] in man almost invariably precipitates development of facial herpetic lesions, but, again, there is no evidence that such events play a role in precipitating a CNS disease. Thus, the factors that may activate HSV in the ganglia _in situ_ and precipitate centripetal migration of HSV, a sequence that certainly occurs in case of herpes encephalitis of adults, remain largely unknown. However, the presence of this virus at such a short distance from the brain certainly raises the question of whether in addition to being involved in the massive destruction of brain cells characteristic of herpetic encephalitis the virus may not play a role in the more subtle damage inflicted on cells in chronic CNS diseases.

In these cases it is doubtful that cells are killed as the result of viral replication. It is much more probable that the interaction between cells of the immune system and virus-infected cells indirectly inflicts cellular damage. We have no data for such a mechanism in HSV infection in vivo, but the results obtained with parainfluenza 1 virus may shed light on this problem. Some time ago[11] we showed that infection of young adult mice with the 6/94 strain of parainfluenza 1 virus causes chronic, degenerative, mostly asymptomatic lesions of white matter. These lesions can be induced by either live or inactivated virus.[12] As shown in Table 3, lesions cannot be induced in nude mice unless they receive thymus grafts from their normal littermates.[12] Since treatment

TABLE III
WHITE MATTER LESIONS INDUCED IN MICE BY INACTIVATED PARAINFLUENZA 1 VIRUS

Mice	Treatment	Pathological Lesions	Serum Antibody*
Nude (nu/nu)	None	0	23 ± 5.0
	Thymus grafted	+	82 ± 22.0
Balb/c	None	+	202 ± 53
	ATS **	0	14.3 ± 7.5

* μg/ml of serum
** Anti-thymocyte serum

of normal mice with anti-thymocyte serum also prevented development of white matter destruction it became apparent that the T cell played a role in mediating cell damage. Proof of the damaging role of T cells was obtained in experiments in which athymic nude mice received an adoptive transfer of lymphocytes obtained from spleens of Balb/c mice immunized intraperitoneally 4 days before with inactivated parainfluenza virus. Three hours after intraperitoneal transfer of spleen cells, nude mice were intracerebrally injected with inactivated parainfluenza 1 virus. Seven days after adoptive transfer mice showed white matter degeneration accompanied by infiltration of lymphocytic cells in subarachnoid spaces and of mononuclear cells in cerebral white matter.[12] These changes were comparable to those observed in immuno-competent mice inoculated with inactivated virus. Lymphocyte suspensions depleted of B cells were still able to induce lesions after adoptive transfer (Table IV). No lesions

TABLE IV
WHITE MATTER LESIONS PRODUCED BY ADOPTIVE TRANSFER OF LYMPHOCYTES FROM MICE IMMUNIZED WITH PARAINFLUENZA 1 VIRUS

Donor Mice	Cells for Adoptive Transfer	Ratio of Recipient Mice Showing White Matter Lesions	Serum Antibody*
Immunized	Spleen Lymphocytes	2/3	31.3 $\pm$ 12
	B Cell-Depleted Spleen Lymphocytes	5/7	13.3 $\pm$ 4
Normal	B Cell-Depleted Spleen Lymphocytes	0/4	None

* μg/ml of serum

were observed in mice that received inactivated virus only or in those that were injected with primed lymphocytes but not exposed to intracerebral inoculation of the virus. These experiments suggest that viral antigen deposited in the brain may generate cells that cause degenerative lesions of the brain cells present at the site of the viral antigen. Since replication of the virus in tissue is not required to mediate destruction of cerebral white matter and since viral antigen is present in the brain at such a low concentration that it cannot be detected by immunofluorescent staining,[12] it appears that activation of

HSV infection is not a prerequisite for damage of cells in chronic CNS disease but that consistent presence of viral antigen even at undetectable levels may be essential in order for T cells to attack brain cells exposed to this antigen. The constant presence of HSV in the close vicinity of the brain, or possibly in brain tissue itself, suggests such a hypothesis. One must only await the proof.

REFERENCES

1. Bastian, F. O., Rabson, A. O. and Yee, C. L., et al. (1972). Science 178: 306.
2. Baringer, J. R., and Swoveland, P. (1973). New England Journal of Medicine 288: 648 - 650.
3. Warren, K.G., Gilden, D. H., Brown, S. M., Devlin, M., Wroblewska, Z., Subak-Sharpe, J. and Koprowski, H. (1977). Lancet Sept. 24, 1977: 637 - 640.
4. Warren, K. G., Brown, S. M., Wroblewska, Z., Gilden, D., Koprowski, H. and Subak-Sharpe, J. (1978). New England Journal of Medicine (In press).
5. Brown, M., Jamieson, A. T., D. M. Lonsdale, Subak-Sharpe, J. H., Warren, K. G., and Koprowski, H. (1978). Fourth International Congress for Virology, The Hague, The Netherlands, August 30 - September 6, 1978.
6. Lonsdale, D. M., Brown, S. M., Subak-Sharpe, J. H., Warren, K.G., and Koprowski, H. (1978). Fourth International Congress for Virology, The Hague, The Netherlands, August 30 - September 6, 1978.
7. Warren, K. G., Wroblewska, Z., Okabe, H., Yonezawa, T., Rorke, L. B., Gilden, D. H. and Koprowski, H. (1978). American Journal of Epidemiology (submitted for publication).
8. Koprowski, H. and Warren, K. G. (1978). Perspectives in Biology and Medicine. In press.
9. Hamada, Takeshi, et al. Personal communication, (1978).
10. Carton, C. A., and Kilbourne. (1952). New England Journal of Medicine 246 (5): 172 - 179.
11. Zgorniak-Nowosielska, I., Iwasaki, Y., Tachovsky, T., Tanaka, R., and Koprowski, H. (1976). Archives in Neurology 33: 55 - 62.
12. Iwasaki, Y., Gerhard, W., and Koprowski, H. (1978). American Journal of Pathology (submitted for publication).

LATENT HERPETIC INFECTIONS IN THE CENTRAL NERVOUS SYSTEM OF EXPERIMENTAL ANIMALS[1]

Jack G. Stevens

Department of Microbiology and Immunology
UCLA School of Medicine, Los Angeles, California 90024

ABSTRACT Latent infections with Herpes simplex virus can be established in 1) rabbit brains following corneal inoculation, 2) the spinal cords of mice after footpad inoculation, 3) the central nervous system of mice following intravenous inoculation and 4) mouse brains after intracerebral inoculation of temperature-sensitive viral mutants. These systems can be exploited for studies of the possible role of such infections in the genesis of both acute herpetic encephalitis and chronic degenerative nervous system diseases.

INTRODUCTION

Following our initial demonstration that latent herpetic infections could be established experimentally in sensory ganglia of laboratory animals (1), others quickly showed that a similar infection occurred naturally in man (Reviewed in 2, 3, 4). These findings are significant since they are basic to the acceptance of one now generally accepted hypothesis concerning the pathogenesis of recurrent herpetic disease. According to that hypothesis [(which has been discussed in detail elsewhere (2)] virus harbored in ganglia is reactivated, travels in associated nerves, and produces disease on surface areas served by these nerves.

More recently, we have shown that latent infection can be established in the central nervous system of experimental animals (5, 6, 7). These studies are important for at least two reasons. First, since at least some cases of acute herpetic-encephalitis in man appear to result from reactivation of latent infections (8, 9), the brain itself offers the most simple explanation for the site at which

[1]The work from my laboratory presented here was supported by U. S. Public Health Service Grants AI-06246 and NS-08711.

ISBN 0-12-668350-6

latent agent is harbored. Secondly, the etiologies of several important chronic diseases of the central nervous system are unknown, but have long been postulated to include persistent viruses. If such agents are indeed involved, Herpes simplex virus has several properties which make it a very attractive candidate. Obviously, as with the concept of nervous system involvement in cutaneous disease, these theories would be made more substantive if virus were known to persist in the central nervous system.

This paper summarizes the results obtained from four different systems, all showing that latent herpetic infections can be established in the central nervous system of rabbits and mice.

RESULTS

Latent Infections in Rabbit Brains after Corneal Inoculation. New Zealand White rabbits, six weeks old were used for corneal inoculation of virus. Here one cornea was anesthetized, scarified and then infected with two drops of undiluted stock virus ($\sim 10^7$ plaque forming units). Several days later, each of 21 rabbits inoculated developed an acute encephalitis from which 14 recovered. These latter animals were used for studies of viral latency. They were killed at various intervals, and the trigeminal ganglia and half the brain ipsilateral to the eye inoculated were removed. The removed portions of the brain were then divided into four anatomic regions: a) brain stem (that portion from the inferior coliculus to 1 cm. caudal to the obex, including the pons), b) cerebellum, c) cerebrum (the region outside the lateral ventricles), and d) diencephalon (the region from the superior coliculus forward and within the bounds of the lateral ventricles). Portions of each tissue were frozen for later assays; the remainder was co-cultivated with RK_{13} cells. The results of these experiments are presented in Table 1. As can be seen, 5 of the 14 cultures of brain stems demonstrated cytopathic effects after 7 to 19 days of co-cultivation, and infectious virus was recovered. Four of these 5 brain stems were assayed directly for virus; none could be isolated. In addition, each of the 5 rabbits harbored latent virus in the ipsilateral trigeminal ganglion. It is also important to note that virus could not be detected in other portions of the brain, and spinal cords were not studied. Finally, all isolates were identified as Herpes simplex virus by neutralization tests involving specific hyperimmune rabbit sera.

TABLE 1

RECOVERY OF INFECTIOUS HERPES SIMPLEX VIRUS AFTER IN VITRO CULTIVATION OF LATENTLY INFECTED BRAIN STEM FROM RABBITS

Rabbit number	Months post infection	HSV latent in trigeminal ganglia	HSV latent in brain stem
1	3	+	+
2	3	+	-
3	3	+	-
4	1	+	-
5	1	+	-
6	1	+	+
7	1	+	+
8	2	-	-
9	2	-	-
10	12	-	-
11	5	+	-
12	12	+	+
13	9	+	-
14	9	+	+
Totals		11/14	4/14

Rabbits were inoculated on one scarified cornea with HSV. Animals were then killed at intervals, and tissues were removed, divided, and cocultivated with RK_{13} cells. Viral induced cytopathic effects in the RK_{13} cells were scored at daily intervals and isolates were identified immunologically as HSV.

For further details, see Reference 5.

Latent Infections in the Spinal Cords of Mice after Footpad Inoculation. In these and subsequent experiments 4-8 week old outbred Swiss-Webster mice were employed as the experimental animal. Here, about 40% of the mice inoculated with ~ 4,000 PFU in one scarified rear footpad (1) became paralyzed in both hind legs within 10 days after infection with HSV. Two-thirds of these died with acute encephalitis, while the remainder underwent a complete clinical recovery or were left with varying degrees of posterior paralysis. Since the leg contralateral to the foot infected became paralyzed, we reasoned that acute viral infection of the spinal cord was a certainty.

To determine whether latent HSV was present in the spinal cords of the mice which had recovered from the acute disease but still presented paralytic sequelae, the animals were killed, and their spinal cords were removed. In addition, 4-7 lumbosacral spinal ganglia contralateral to the side of inoculation were also collected. Both tissues were co-cultivated with RK_{13} cells. The results of several such experiments involving a total of 53 mice are presented in Table 2. As can be seen, 4 mice were found to harbor latent HSV in their spinal cords; viral specific cytopathic effects were produced in the RK_{13} cells after 6-18 days of co-cultivation, and infectious virus was recovered. Two of these 4 spinal cords were ground in a TenBroek homogenizer and assayed for infectious virus directly; none was found. It is also of importance to note that latent virus was recovered from the corresponding contralateral lumbosacral spinal ganglia in each of these mice. Again, the isolates were identified as HSV by neutralization studies.

Latent Infections in Mouse Central Nervous System following Intravenous Inoculations. In this study, mice were inoculated in a tail vein with 3×10^3 to 3×10^4 plaque forming units of Herpes simplex virus; all demonstrated neurological disease by 7 days after infection. After an interval of 35 to 109 days, the mice were sacrificed and the brain was removed and co-cultivated with RK_{13} cells. As can be noted from Table 3, 3 of 12 mice harbored latent Herpes simplex virus in the central nervous system (1 in the brain and 2 in the spinal cord). It should also be noted that no virus could be detected by direct assay methods from the DNA of 6 additional mice inoculated in the same manner, and serum neutralization tests established that the identity of those agents isolated as Herpes simplex virus.

Although the virus was injected intravenously in these experiments, it could reasonably be argued that leakage at the injection site resulted in infection of local nerve endings, centripetal passage of virus through nerves and subsequent establishment of latent infections in the central nervous

TABLE 2

RECOVERY OF INFECTIOUS HERPES SIMPLEX VIRUS AFTER IN VITRO CULTIVATION OF LATENTLY INFECTED LUMBOSACRAL SPINAL CORDS FROM MICE

Experiment number	Number of mice	Mice with right ganglia positive	Mice with spinal cords positive
		Mice tested	Mice tested
1	5	5/5	0/5
2	7	6/7	0/7
3	5	1/5	1/5
4	6	4/6	1/6
5	6	4/6	1/6
6	7	4/7	0/7
7	4	3/4	0/4
8	3	3/3	0/3
9	4	3/4	1/4
10	6	3/6	0/6
Totals	53	36/53	4/53

Mice were infected with HSV by inoculation of a rear foot-pad. After 1-8 mos., animals which had recovered from the acute neurologic disease were killed. Tissues were removed, chopped, and cocultivated with RK_{13} cells. Viral induced cytopathic effects in the RK_{13} cells were scored at daily intervals, and isolates were identified as HSV by immunologic methods.

For further details, see Reference 5.

TABLE 3

RECOVERY OF INFECTIOUS HERPES SIMPLEX VIRUS AFTER IN VITRO CULTIVATION OF LATENTLY INFECTED CENTRAL NERVOUS SYSTEM FROM MICE

Animal number	Virus inoculated (PFU)	Time between infection and sacrifice (Days)	Tissue processed for detection of latent infections and results	
			Brain	Spinal cord
1	3,000	109	.	.
2	3,000	109	.	.
3	3,000	109	.	.
4	8,000	56	.	+
5	8,000	56	.	.
6	8,000	56	.	.
7	30,000	27	.	+
8	30,000	27	.	.
9	30,000	27	.	.
10	30,000	35	+	.
11	30,000	35	.	.
12	30,000	35	.	.

Mice were inoculated intravenously with HSV. Animals were then killed at intervals, and tissues were removed, divided, and co-cultivated with RK_{13} cells. Viral induced cytopathic effects in the RK_{13} cells were scored at daily intervals and isolates were identified immunologically as HSV.

For further details, see Reference 6.

system. This possibility was ruled out when it was shown that similar results obtained when the tails of injected mice were amputated 2.5 cm. proximal to the injection site immediately following inoculation (6). Of course, these experiments do not prove that the central nervous system was seeded directly by virus in the bloodstream; it could have become infected indirectly by virus traveling in nerves associated with vessels or other organs originally seeded by the viremia. However, it can be concluded with certainty that the initial infection was blood borne.

Latent Infections in Mouse Brains after Intracerebral Inoculation of Temperature-Sensitive Mutants. Recently, this laboratory has been intensively involved with investigations concerning biochemical phenomena associated with the establishment and maintenance of latent herpetic infections in the nervous system. A somewhat indirect approach, but one which has proved to be illuminating, has been inoculation of phenotypically defined temperature-sensitive mutants of the virus into the brains of mice (which manifest the restrictive temperature). As has been discussed elsewhere (4, 7), given some basic assumptions, it can be predicted that those mutants unable to establish latent infections possess lesions in genes whose products are critical for establishment of such infections. This work is still in progress, and some 15 Type 1 mutants obtained from two laboratories (10, 11), have been shown to fall into "latency positive" or "latency negative" groups (7, Lofgren & Stevens, in preparation). A summary of findings with the first 5 mutants studied is presented in Table 4.

The results relevant to this communication are that "latency positive" mutants which persist in the brain do exist. Functionally, these agents fit into the broad category of replication defective viruses. Since replication defective variants of one virus (measles) are associated with at least some cases of the classic chronic central nervous system disease subacute sclerosing panencephalitis,(12) one may wonder if defective Herpes simplex virus is responsible for any of the others.

DISCUSSION

The experiments presented here show directly and unequivocally that Herpes simplex virus establishes latent infections in the central nervous system of both mice and rabbits. In addition, they suggest that the latent infection can be established by spread from both bloodstream and peripheral nerves. Additional experiments supporting these latter conclusions are presented in detail elsewhere (6, 13, 14). The major results also extend the observations of Good and Campbell (15, 16) and Schmidt and Rasmussen (17).

TABLE 4

RECOVERY OF INFECTIOUS VIRUS FOLLOWING IN VITRO CULTIVATION OF BRAINS FROM MICE PREVIOUSLY INOCULATED WITH TEMPERATURE-SENSITIVE MUTANTS OF HERPES SIMPLEX VIRUS

Mutant Inoculated	Dose (PFU)	Mice with brains positive/mice tested	Temperature sensitivity of virus recovered from the brain
ts D	10^5	4/12	4/4 Mutant
	10^6	6/8	4/4 Mutant
ts G	10^5	3/12	3/3 Mutant
	2.5×10^5	3/7	3/3 Mutant
ts S	10^5	1/11	1/11 Mutant
	10^6	2/5	2/2 Mutant
ts J	10^5	3/10	3/3 Mutant
	10^6	2/7	2/2 Mutant
ts I	10^5	0/12	
	10^6	0/8	
Wild-type[a]	$10\text{-}10^3$	1/20	

Mice had been inoculated intracerebrally 3 weeks previously with viral mutants at dosages shown. Brain tissue surrounding the needle tract was cocultivated with monolayers of BHK cells at 31°C (the nonrestrictive temperature) and viral specific cytopathic effects were scored. Several reisolates taken at random were identified as Herpes simplex virus by immunologic methods.

[a]These mice were the surviving groups in which 50% of mice had died with acute encephalitis over the range of dosages noted. At higher doses, all mice died.

For further details, see Reference 7.

By pharmacologically inducing encephalitis in rabbits previously infected with Herpes simplex virus, they suggested that the disease resulted from reactivation of latent infection in the brain.

In a more general sense, our findings in experimental animals satisfy a critical, basic event required for support of the concept that the central nervous system itself is the source of virus for at least some cases of herpetic encephalitis in man. Thus, it seems quite possible that the virus establishes latent infections in the brain and is later reactivated to produce the disease.

Finally, the possible role of latent viral infection in the genesis of chronic degenerative diseases of both neurons and supporting cells can be seriously investigated in these experimental systems. As mentioned above, we have recently shown that certain conditionally lethal (temperature-sensitive) viruses of defined phenotypes restricted at mouse body temperature can establish latent infections in the brains of these animals. Using this system the potential role of defective Herpes simplex virus in the genesis of chronic central nervous system diseases can also be systematically investigated.

REFERENCES

1. Stevens, J. G., and Cook, M. L. (1971). Science 173, 843.
2. Baringer, J. R., (1975). Progr. Med. Virol. 20, 1.
3. Stevens, J. G. (1975). Curr. Top. Microbiol. Immunol. 70, 31.
4. Stevens, J. G. (1978). Adv. Cancer Res. 26, 227.
5. Knotts, F. B., Cook, M. L. and Stevens, J. G. (1973). J. Exp. Med. 138, 740.
6. Cook, M. L. and Stevens, J. G. (1976). J. Gen. Virol. 31, 75.
7. Lofgren, K. W., Stevens, J. G., Marsden, H. S. and Subak-Sharpe, J. H. (1977). Virology 76, 440.
8. Leider, W., Megoffin, R. L., Lennette, E. H., and Leonards, L. M. R. (1965). New Engl. J. Med. 273, 341.
9. Kibrick, S. and Godding, G. W. (1965). In D. C. Godjisek et al (eds.) NINDB, Monogr. No. 2. U. S. Government Printing Office.
10. Marsden, H. S., Crombie, I. K. and Subak-Sharpe, J. H. (1977). J. Gen. Virol. 31, 347.
11. Courtney, R. J., Schaffer, P. A. and Powell, K. L. (1976). Virology 75, 306.
12. Ter Meulen, V., Katz, M. and Muller, D. (1972). Curr. Top. Microbiol. Immunol. 57, 1.
13. Cook, M. L. and Stevens, J. G. (1973). Infect. Immun. 7, 272.

14. Knotts, F. B., Cook, M. L. and Stevens, J. G. (1974). J. Inf. Dis. 130, 16.
15. Good, R. A. and Campbell, B. (1945). Proc. Soc. Exptl. Biol. Med. 59, 305.
16. Good, R. A. (1947). Proc. Soc. Exptl. Biol. Med. 64, 360.
17. Schmidt, J. R. and Rasmussen, A. F. Jr. (1960). J. Infect. Dis. 106, 154.

ALEUTIAN DISEASE AND ALEUTIAN DISEASE VIRUS OF MINK[1]

H.J. Cho, and David D. Porter

Animal Diseases Research Institute (Western), Agriculture Canada, Lethbridge, Alberta, TIJ 3Z4, Canada, and the Department of Pathology, UCLA School of Medicine, Los Angeles, California, 90024 U.S.A.

ABSTRACT Aleutian disease, a chronic or persistent viral infection of mink, is caused by a naturally temperature-sensitive 23 to 25 nm icosahedral virus, probably a parvovirus, which contains single-stranded DNA. Mink often respond to infection with enormous increases in immunoglobulin levels and specific antiviral antibody. Serious lesions including glomerulonephritis and arteritis result from tissue deposition of circulating immune complexes, and disease can be decreased by immunosuppression or infection in utero, or increased by inactivated virus vaccine. Control of the disease may be possible by identifying infected mink by specific serologic tests and removing infected animals from mink ranches.

INTRODUCTION

Aleutian disease (AD) was described in ranch-raised mink in 1956 by Hartsough and Gorham (1). The disease received this name because it was first found in mink homozygous for the recessive Aleutian coat color gene. Subsequently, the disease has been found in all types of mink, however, marked differences in the average severity of disease have been noted with low virulence virus stocks.

Mink are members of the order Carnivora and family Mustelidae, which includes the wolverine, marten, badger, fisher, ferret, weasel, skunk and otter (2). Mink (Mustela vison) mate once a year in the early spring and an average of 4 young are born after a 50 day gestation period (3). The young are weaned at 6 to 8 weeks of age and reach adult size at 7 months of age, at which time many are sacrificed for pelts on commercial ranches.

[1]This work was supported by Agriculture Canada, Grant AI-09476 from the National Insitute of Allergy and Infectious Diseases, National Institute of Health, and the Mink Farmer's Research Foundation.

ISBN 0-12-668350-6

AD is the major health problem of commercial mink ranching and we estimate that the disease causes losses of $10 million annually in North America. The prevalence of AD on mink ranches varies from none to 75% (4), and large numbers of mink of the Aleutian genotype may die of the disease before pelting season, while early deaths of other color types are relatively infrequent.

We (5,6) and others (7,8) have reviewed AD relatively recently with emphasis upon the pathology and pathogenesis of the lesions and the immunologic disturbances caused by infection. In this review we will emphasize new findings about the virus, the host response to infection, and means of controlling or eradicating the infection.

THE VIRUS

Viral Etiology of AD. The transmissible nature and probable viral etiology of AD were first described in 1962 (9,10, 11). The disease was experimentally reproduced by inoculating cell-free filtrates of diseased tissues. Based on this indirect evidence, it was generally assumed AD was caused by a virus. During the 1960's and early 1970's, a number of investigators attempted to isolate and visualize the virus and to propagate it in cell culture. This task proved frustratingly difficult to achieve. However, the AD virion was eventually visualized and purified by Cho and Ingram (12) and the virus has been propagated in cell culture by Porter and coworkers (13). Although Kenyon et al. (14) have also reported isolation of ADV by affinity chromatography, their isolate has characteristics of a picornavirus as it contained RNA as determined by the orcinol method and by ^{3}H-uridine labelling in mouse L cell culture (15,16). A recent biochemical study of ADV (17) and the behavior of the virus in cell culture (13) reveals that ADV is an autonomous parvovirus which is naturally temperature-sensitive in its replication in vitro.

Multiplication and Distribution of the Virus In Vivo. Although ADV was regarded as a "slow virus" the virus replicates rapidly in vivo. Maximal titers of 10^8 to 10^9 ID_{50}/gm were found in spleen, liver, and lymph node 10 days after experimental infection (18). After 10 days the virus titers in the tissues slowly fall so that two or more months after infection spleen titers of 10^5 ID_{50}/gm and serum titers of 10^4 ID_{50}/ml are commonly found (6). The virus is readily recovered from whole blood, serum, feces, urine, saliva, and various mink tissues (19,20,21). Using the Pullman strain of ADV, Eklund et al. (19) have shown markedly higher ADV titers in mink homozygous for the Aleutian gene than in non-Aleutian mink. With the Utah strain of ADV, Porter et al.(18) did not find a difference in the magnitude of virus replication among mink of different genotype. With the Guelph strain of ADV, it has been shown that similar amounts of ADV antigen were produced in both Aleutian and non-Aleutian mink (22). These differences could be explained by the fact that the Pullman strain is less virulent than the Utah or Guelph strains of ADV (23).

Using immunofluorescence techniques, Porter and coworkers (18) examined ADV infected mink tissues and found viral antigen mainly in the cytoplasm of macrophages in spleen and lymph node and of Kupffer cells in the liver. These findings have been confirmed by others (8) with small amounts of ADV antigen also being found in the nucleus of some cells.

Visualization of the Virus In Vivo. Electron microscopic examination of mink tissues, collected 10 to 13 days after experimental ADV infection, revealed numerous virus-like particles (20 to 22 nm diameter) in macrophages of the spleen, mesenteric lymph node, and liver (Kupffer cells) (24). The virus-like particles were principally present in vacuoles within the cytoplasm of these cells, and similar particles were occasionally observed inside the nucleus. The particles were always present in scattered form and no crystalline aggregation was observed. Treatment of thin sections of liver, spleen, and lymph nodes from ADV-infected mink with ADV antibody-conjugated ferritin resulted in ferritin labelling of areas in the cytoplasm of some cells. These ferritin-labelled vacuoles resembled the vacuoles containing virus-like particles in epon-embedded tissue. The immunoferritin also labelled areas outside the vacuoles in the cytoplasm and occasionally the nucleus of some cells. Since ADV antigen was present in the cytoplasm and occasionally also in the nucleus, it appears that ADV replicates in these cells. The virus-like particles described here seem different from those observed by Tsai and coworkers (25), who identified virus-like structures in the form of crystalline arrays within the cytoplasm of endothelial cells in the kidney of ADV-infected mink. However, others (6) have failed to find similar particles in renal endothelial cells at any stage of infection.

Multiplication of the Virus in Cell Culture. After many attempts to grow ADV in cells of various species, it was found that the agent would replicate in primary or a continuous line of feline renal cells if the original isolations were made at temperatures markedly lower than the body temperature (39°C) of mink. The original isolations of ADV were done at 31.8°C, but the highly virulent Utah-1 strain of virus rapidly acquired a higher temperature optimum for growth; by passage 10 at 31.8°C, the temperature optimum was found to be 37°C. The cultured virus produced AD in mink and could be reisolated from mink, and various preparations had equal infectivity for cell cultures and mink. The culture grown virus had properties identical to those of ADV purified from mink, and viral replication could be inhibited by inhibitors of DNA synthesis. The virus was found to be markedly cell-associated, and some infectivity was bound to membranes with a density in CsCl of 1.29 gm/ml (13). Recent studies have shown that less virulent strains of ADV may be even more temperature-sensitive and require even lower temperatures for growth in cell culture. Antigen expression without the production of infectivity by the Connecticut strain of ADV in feline cells (26) has been observed.

Purification and Isolation of the Virus. To facilitate virus extraction from infected mink tissue, a suspension containing as high a virus titer as possible was sought to infect mink. The infectivity titer of the ADV inoculum could be increased 10^3 to 10^4 times over that found in chronically infected mink by rapid serial passage of the virus in mink (18).

ADV virion antigen as measured by the counterimmunoelectrophoresis (CIEP) test can be obtained by homogenizing mink tissue from animals infected for 9-13 days, near the peak of viral proliferation. The antigen can be substantially purified by repeated fluorocarbon extraction and concentrated by sedimentation at 105,000g for 2 hours (22). The partially purified ADV antigen is in the form of complexes with antibody. Such complexes can be dissociated by acid treatment, heat, or additional repeated fluorocarbon extractions. The antigen can be further purified by molecular sieving chromotography. When applied to columns of Sephadex G-200 (Pharmacia) the ADV antigen elutes in the void volume, but on columns of Sepharose 6B (Pharmacia) it elutes with an average partition coefficient of 0.13. ADV antigen purified through the Sepharose 6B step contains 23-25 nm icosahedral virus particles and such preparations will produce AD when inoculated into mink (12). Examination of precipitin lines formed in CIEP by the reaction of ADV antigen with mink antibody shows virions cross-linked by antibody. These findings have been reproduced in other laboratories (27).

Physico-Chemical Properties of the Virus. Table 1 summarizes the known properties of ADV. Several of the characterization studies have been performed using crude tissue

TABLE 1
PHYSICO-CHEMICAL PROPERTIES OF ALEUTIAN DISEASE VIRUS

Description	Properties	Reference
Size	23-25 nm	12,13,27
Morphology	Icosahedral	12,13,27
Buoyant density in CsCl	1.41-1.43 gm/ml	13,31
Sedimentation constant	125 S	a
Particle molecular wt.	5.4-6.9 x 10^6 daltons	Calculated
Replication site	Nuclear	13,26
Nucleic acid		
Strandedness	Single stranded	17,22
Type	DNA	13,17,22
Density in CsCl	1.733 gm/ml	17
Molecular weight	1.2 x 10^6 daltons	17
Structural protein	3 or 4 polypeptides	17
Classification	Parvovirus	13,17,22,27
Dialysis	Not dialyzable	32
Ether	Stable	13,19
Fluorocarbon	Stable	33
pH3	Stable	34
Ultraviolet light	Inactivated	35
Heat		
56°C - 30 min	Stable	13,19
60°C - 30 min	Partially inactivated	13,19
80°C - 30 min	Inactivated	13,19
Deoxycholate	Stable	13,33
Protease enzymes	Stable	33
Nuclease enzymes	Stable	33
2.5 M KI, pH7.0	Stable	13
2.0 M KSCN, pH7.0	Partially inactivated	13
0.05 N NaOH	Inactivated	33
0.5 N HCl	Inactivated	33
0.5% iodine	Inactivated	33
0.5% NaOCl	Inactivated	b

[a] Cho, Pan and Trautman, unpublished
[b] Porter, unpublished

extracts of virus prepared from persistently infected mink, thus, the reported properties are almost certain to be those of virus-antibody complexes. In general, results have been similar when antibody-free and complexed virus were directly compared. ADV is infectious despite being complexed with antibody (13,28) and virus extracted from tissue collected as early as 9 to 13 days after infection is already bound to antibody (29,30).

Due to recent developments in the purification of ADV and the success of virus multiplication in cell culture, characterization of antibody-free ADV is now possible.

The morphology of ADV is the same as for described parvoviruses, and the virion probably has 32 capsomeres. Where optimal resolution is achieved, the capsomeres can be seen to be tubular structures 4.5 nm in diameter with a 1 to 2 nm central hole (13,31).

ADV infectivity of frozen, thawed, and sonicated cultures in CsCl density gradients was broadly distributed over a range of 1.26 to 1.43 gm/ml. A similar preparation extracted with Freon 113 showed a major infectivity peak at 1.41 to 1.43 and a minor peak at 1.29 gm/ml (13). It is probable that infectivity of cultured ADV found at 1.29 gm/ml is associated with cell membranes, which can be visualized. Equilibrium centrifugation of a highly concentrated ADV preparation (20 ml of purified ADV preparation from 20 kg of early infected mink tissues) yielded three visible bands at buoyant densities of 1.295, 1.332, and 1.405 to 1.416 gm/ml in CsCl (31). Electron microscopic examination of these three bands (Fig. 1) has shown that the lightest band contains mainly empty particles and the middle band contains mainly complete and a few empty particles. In the heavy band both complete and empty particles were observed and, in addition, many smaller ring-shaped particles measuring about 10 to 12 nm, which are believed to be mink ferritin, were present. ADV antibody did not clump these ferritin particles but did aggregate the AD virions. ADV antibody produced in mink by the middle and heavy bands reacted with virus particles of all three bands and only a single precipitin band was observed. Furthermore, the polypeptide compositions of the heavy and middle band particles are identical (17), thus the different density populations of AD virions show immunological specificity (31). Standard particles (ρ=1.405-1.416 had a CIEP antigen titer 16 times higher and an infectivity titer 10^4 ID_{50} higher than the middle band ADV (ρ=1.332). The middle band ADV has a particle to CIEP antigen ratio comparable with standard ADV but possessed much lower infectivity, and thus is obviously defective in terms of infec-

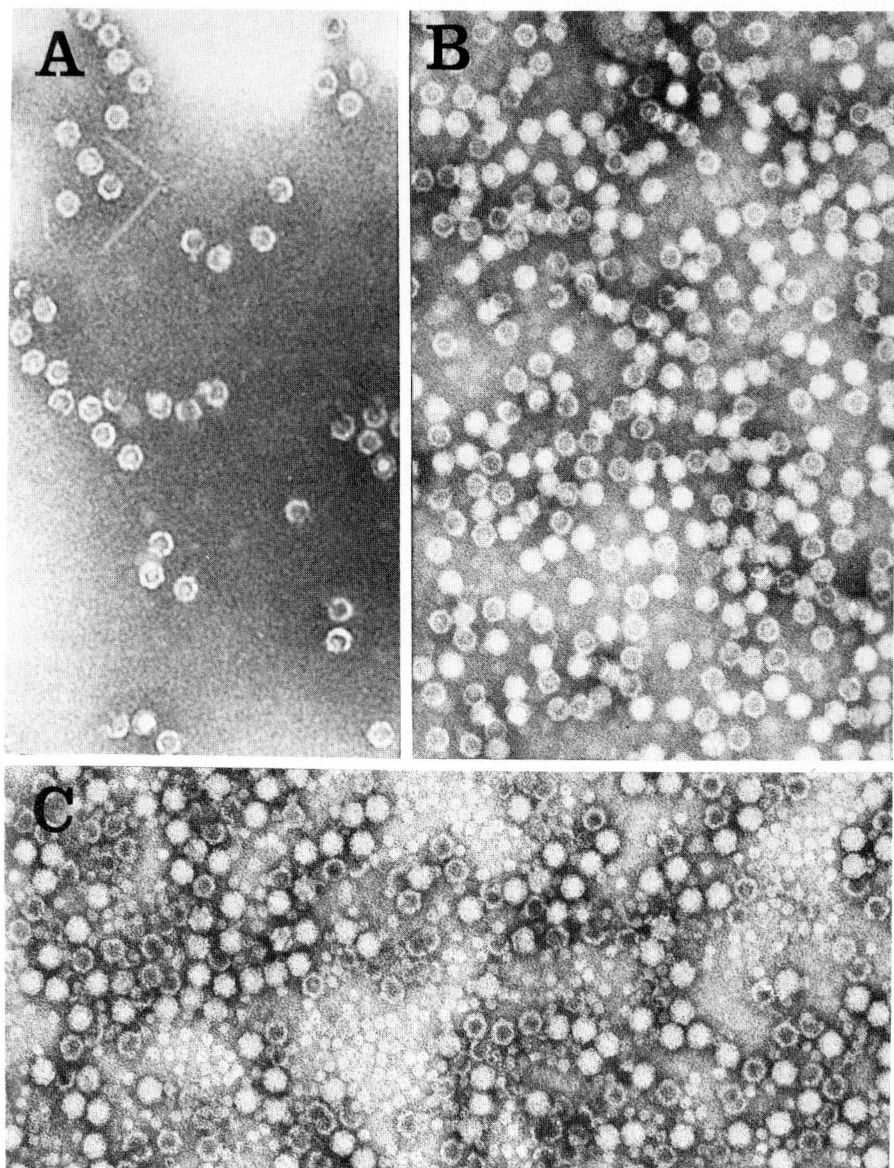

Figure 1. Aleutian disease virus. A. Empty particles ρ= 1.295. B. Full and empty particles ρ=1.332. C. Full and empty particles and mink ferritin ρ=1.405-1.416. Infectivity is principally in this band. X150,000. Reprinted from (31) by permission of the Canadian Journal of Comparative Medicine.

tivity. At present we do not know whether these middle band particles represent defective interfering (DI) particles of ADV. However, homologous interference has been observed with certain defective parvovirus particles (36). The defective particles may play a significant role in relation to the persistence of ADV in vivo.

In addition to the particles of three densities described above, small amounts of infectivity are often, but not always, noted at a density of 1.48 (13). These very heavy infectious particles are often seen in other parvovirus preparations (36).

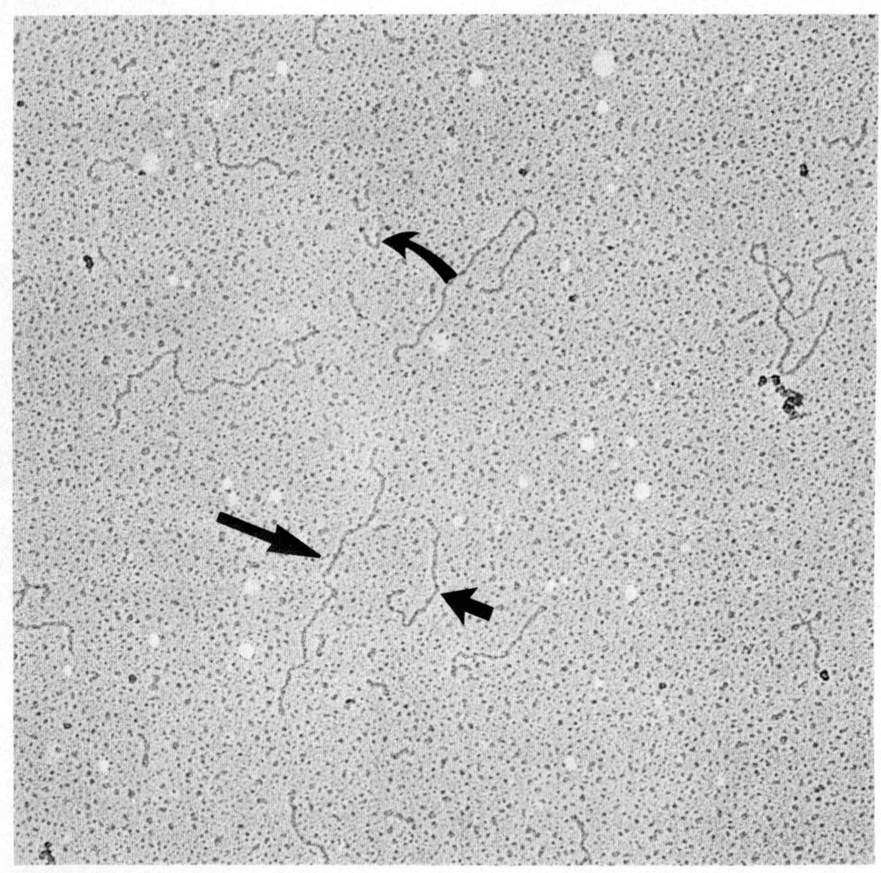

Figure 2. Aleutian disease viral DNA. Strands of 1.2μ (long arrow), 0.55μ (short arrow), and 0.25μ (curved arrow) are present. X50,000. Reprinted from (17) by permission of the Journal of Virology.

STRUCTURAL VIRAL COMPONENTS

Nucleic acid. The single stranded nature of ADV nucleic acid was suggested initially by observing a flame red fluorescent reaction under UV light after purified ADV or ADV-antibody complexes formed in the CIEP immune precipitates were stained with 0.01% acridine orange (22). Figure 2 shows nucleic acid species from ADV. Three different sizes of molecules, with approximate lengths of 1.2, 0.55, and 0.25 μm, were observed in proportions of 42%, 33%, and 25%, respectively. The molecular weights were estimated to be 1.2×10^6, 0.5×10^6, and 0.23×10^6 for the large, intermediate, and small molecules, respectively. ADV particles from buoyant densities of 1.33 and 1.41 gm/cm^3 contained morphologically similar-appearing nucleic acid. After analytical centrifugation in CsCl, ADV nucleic acid banded at a buoyant density of 1.733 gm/cm^3. However, in Cs_2SO_4, ADV nucleic acid had a lower buoyant density than that of double stranded RNA. In addition, when ADV nucleic acid was treated before centrifugation with pancreatic DNase, no visible band was observed after equilibrium centrifugation in CsCl. The reaction of formaldehyde (1.8%) with intact ADV caused an increase in absorption and a shift of the absorption maximum to a longer wavelength, from 264 nm to 268 nm. The thermal denaturation curve of ADV DNA showed no sharp rise when the sample was heated to 100°C. These results indicate that ADV contains a single stranded DNA (17).

Viral proteins. Polyacrylamide gel electrophoresis of ADV (17) revealed the presence of four polypeptides, with molecular weights of 30,000, 27,000, 20,500, and 14,000. These polypeptides were present in a ratio of 10:3:10:1 respectively. There was no difference between the polypeptide patterns of ADV particles with a buoyant density of 1.41 gm/cm^3 and those with a buoyant density of 1.33 gm/cm^3.

CLASSIFICATION OF ADV

Considering the known properties of ADV such as its size and morphology, the buoyant density of its virions, its intranuclear replication in vivo, and in vitro, its stability to heat and to many chemicals and low pH, the more recent direct evidence that it contains single stranded DNA, and the fact that it does not require helper virus in its replication, ADV can be tentatively classified as a member of the autonomous parvovirus group (13,17,22). The proteins of ADV have a molecular weight of approximately one-third those of previously studied parvoviruses, and single-stranded DNA of three discrete size classes is atypical of parvoviruses (36). No immunologic relationship of ADV to a number of parvoviruses could be shown (13).

THE DISEASE PROCESS

Gross and Microscopic Lesions. The most consistent lesion of AD is a widespread plasmacytosis which involves the bone marrow, lymph nodes, spleen, liver and kidney. Plasmacytosis is most easily evaluated in the kidney, since normally no plasma cells are present. The degree of plasma cell infiltration of the organs is directly proportional to the serum gamma globulin levels and to other lesions such as glomerulonephritis and arteritis (4). It should be noted that no osteolytic bone lesions similar to those of multiple myeloma have ever been seen.

Gross lesions include marked splenomegaly and lymphadenopathy, which become evident 1½ or 2 months after experimental infection. The kidneys become enlarged and pale, often with some petechial hemorrhages, a similar time after infection, and subsequently may develop many tiny cysts. Cystic changes of the liver are seen infrequently. The major cause of death is renal failure, and small numbers of animals die from rupture of inflamed arteries or an enlarged spleen. Mink of the Aleutian genotype typically die 80-150 days after infection with the highly virulent Utah-1 strain of ADV, while half of non-Aleutian mink will live for a year, and some will survive for their full lifespan of approximately 8 years after infection.

Immunoglobulin Abnormalities. ADV infection causes a profound hypergammaglobulinemia, with levels as high as 11 g/100 ml. We found 570 normal mink to have a gamma globulin level of 0.74 g/100 ml and 683 naturally infected mink had 3.5 g/100 ml (4). The increased gamma globulin is predominantly IgG (37), although some increases in IgM and IgA levels can be shown (38). The elevation of gamma globulin is due to overproduction, since the half-life of IgG is decreased in chronically infected mink (37). The IgG may show some restriction of electrophoretic mobility by 40 days after infection (39), and 10% of mink living for a year after infection develop a monoclonal gammopathy, sometimes with Bence-Jones proteinuria (40).

Very high levels of ADV antibody can be shown in chronically infected mink by immunofluorescence, complement fixation or CIEP techniques; the latter technique is best suited for routine use (41). Antibody can usually be detected by 7 to 10 days after experimental infection and by 30 days titers exceed those seen in other viral infections using similar techniques. Although the antibody can be shown to combine with the virion surface (12, 13, 27), the virus is not neutralized in vivo or in vitro (13, 18, 28). Numerous experiments (cited in 6) have shown that the mink do not make excessive responses to other antigens presented before, simultaneously or following ADV infection. The proportion of the increased immunoglobulin which is viral specific is not presently known.

Immune Complexes. Serum from chronically infected mink contains ADV infectivity bound to IgG which can be removed by treatment with anti-IgG (28). In addition, immune complexes of various smaller sizes can be demonstrated in the serum of chronically infected mink by analytical ultracentrifugation (37). The complexes contain IgG, but the nature of the antigens which are smaller than virions has not been studied. Severe immune complex arteritis (42) and glomerulonephritis (18, 43, 44, 45) are produced by tissue deposition of immune reactants, and IgG, IgM, C3 and after acid elution of Ig, ADV antigens are readily shown in the lesions. Viral-specific antibody has been shown in eluates of the diseased glomeruli (18).

Studies of Host Responsiveness to ADV. In most circumstances, highly virulent ADV strains will cause persistent infection and progressive disease in any genetic type of mink at any age. Recently, non-persistent infections in non-Aleutian mink have been described (46), and persistent but non-progressive infections in similar mink have also been noted (47). These studies employed the highly virulent Utah-1 and field strains of ADV, and limited analysis indicates that a single host gene is not responsible for the lack of disease progression (46). Less virulent ADV strains, exemplified by the Pullman strain, cause severe disease in Aleutian mink and mild or no disease in non-Aleutian mink (19, 23, 45).

Since AD appears to be the result of maximal antigenic stimulation and tissue deposition of immune complexes, it is not surprising that immunosuppresive therapy prevents lesion formation (48), and that lesions are minimal or absent during the peak of viral replication, prior to the development of substantial ADV antibody titers (18). Mink which are infected in utero have higher ADV titers and markedly depressed ADV antibody titers and lesions as compared with mink infected as young adults (49). This observation was interpreted as being due to immunologic hyporesponsiveness of the fetal mink, and this hyporesponsiveness disappeared by six months of age. Passive transfer of ADV antibody to mink at the time of peak viral replication produces acute necrotizing inflammatory lesions, and immunization of mink with a killed virus vaccine causes accelerated AD when the mink are challenged with live virus (50).

CONTROL AND ERADICATION OF AD

As described in previous sections, the pathogenesis of AD is believed to have an immunological basis and immunization of mink with formalin inactivated virus vaccine followed by live virus challenge resulted in a marked increase of the severity of lesions (50). At the present time, effective therapeutic or prophylactic measures are not available, therefore, attempts to control this disease on a ranch basis require identification and removal of carrier animals.

Diagnosis of ADV Infection by Nonspecific Tests. For the diagnosis of AD, the iodine agglutination test (IAT), which measures increased gamma globulin, has been widely used in the field (51). The limitation of the test is that it is nonspecific and gives negative results in the early stage of the disease before serum gamma globulin levels have increased. On the other hand, the IAT will react positively in the presence of high levels of gamma globulin irrespective of the cause of these elevated levels.

In Denmark, over a 10-year period, about 250,000 of one million breeding mink have been tested annually by IAT, however, the prevalence of ADV infection is still very high, (M. Hansen, personal communication). IAT detects only 16 to 65% of CIEP reactors (5,52,53).

Serum protein electrophoresis has also been employed to detect hypergammaglobulinemia. Larsen (54) considered a gamma globulin of up to 14% of total serum protein as normal. An and Ingram (47) considered mink serum gamma globulin in concentrations above 21%, (three standard deviations above the normal mean) to be hypergammaglobulinemic. Although serum electrophoresis is more accurate than IAT, the test is more difficult and expensive to perform and it has not been possible to utilize it on a sufficiently large scale to control AD on infected mink ranches.

Specific Serological Tests. Five specific serological tests are available for measuring ADV antibody. The sensitivity, reproducibility, and specificity of immunofluorescence, complement fixation, and two CIEP tests have been recently compared (41). All four methods were reliable and specific for ADV antibody. Although CIEP is the least sensitive test, it is easy to perform, very reliable, reasonably inexpensive, and suitable for large scale testing.

Eradication of AD Using CIEP Test. Recently Cho and Greenfield (55) have shown that the CIEP test can be utilized as a tool to detect ADV infection on a large scale on commercial mink ranches under field conditions. Using this test, it has been shown that ADV antibody is detected consistently in the serum of mink 7 days after experimental ADV infection (56). To eradicate AD, the CIEP test was applied to three ADV infected commerical mink ranches. All mink on the ranches were tested during the pelting season and before the breeding season for 4 consecutive years. By culling all mink that were positive for ADV antibody, it was demonstrated that the disease could be eliminated.

ECOLOGY OF AD

The host range of ADV is not well defined, but ADV antibody can be demonstrated in domestic ferrets (Porter, unpublished) and feral skunks, racoons, and foxes (5). On mink ranches, both horizontal and vertical transmission of ADV can be observed. The relative importance of the two routes of infection is not clear at the present time (7).

CONCLUSIONS AND OUTLOOK

The means by which ADV avoids host responses and causes persistent infection in mink is not clear. The temperature sensitive nature of the virus (13) is an attractive candidate for the mechanism of viral persistence (57). The failure of ADV antibody to neutralize infectivity (13,18) could also be the cause of persistent infections, either alone, or in conjunction with the temperature sensitivity of viral replication. Of course, other causes for viral persistence could be equally or more important than those which have been experimentally noted. A great deal of work needs to be done on the characterization of the virus. Measurement of specific cellular immunity to ADV and interferon responses of the host should be studied. Systematic studies of the host range of ADV and the route of viral transmission under field conditions would provide useful information. Practical control of the disease may be possible using specific serologic tests and eliminating antibody positive mink.

REFERENCES

1. Hartsough, G.R., and Gorham, J.R. (1956). Nat. Fur News 28, 10.
2. Ewer, R.F. (1973). "The Carnivores" Cornell University Press, Ithaca, New York.
3. Bowness, E.R. (1968). Canad. Vet. J. 9, 103.
4. Porter, D.D., and Larsen, A.E. (1964). Am. J. Vet. Res. 25, 1226.
5. Ingram, D.G., and Cho, H.J. (1974). J. Rheumatology 1, 74.
6. Porter, D.D., and Larsen, A.E. (1974). Prog. Med. Virol. 18, 32.
7. Gorham, J.R., Henson, J.B., Crawford, T.B., and Padgett, G.A. (1976). In "Slow Virus Diseases of Animals and Man" (R.H. Kimberlin, ed.) pp. 135-158. North-Holland Publishing Co., Amsterdam.
8. Henson, J.B., Gorham, J.R., McGuire, T.C., and Crawford, T.B. (1976) In "Slow Virus Diseases of Animals and Man" (R.H. Kimberlin, ed.) pp. 175-205. North-Holland Publishing Co., Amsterdam.
9. Henson, J.B., Gorham, J.R., Leader, R.W., and Wagner, B.M. (1962). J. Exp. Med. 116, 357.
10. Karstad, L., and Pridham, T.J. (1962). Canad. J. Comp. Med. 26, 97.
11. Trautwein, G.W., and Helmboldt, C.F. (1962). Am. J. Vet. Res. 23, 1280.
12. Cho, H.J., and Ingram, D.G. (1973). Nature N. Biol. 243, 174.
13. Porter, D.D., Larsen, A.E., Cox, N.A., Porter, H.G., and Suffin, S.C. (1977). Intervirology 8, 129.
14. Kenyon, A.J., Gander, J.E., Lopez, C., and Good, R.A. (1973). Science 179, 187.
15. Yoon, J.-W., Kenyon, A.J., and Good, R.A. (1973). Nature N. Biol. 245, 205.
16. Yoon, J.-W., Dunker, A.K., and Kenyon, A.J. (1975). Virology 64, 575.
17. Shahrabadi, M.S., Cho, H.J., and Marusyk, R.G. (1977). J. Virol. 23, 353.
18. Porter, D.D., Larsen, A.E., and Porter, H.G. (1969). J. Exp. Med. 130, 575.

19. Eklund, C.M., Hadlow, W.J., Kennedy, R.C., Boyle, C.C., and Jackson, T.A. (1968). J. Infect. Dis. 118, 510.
20. Gorham, J.R., Leader, R.W., and Henson, J.B. (1964). J. Infect. Dis. 114, 341.
21. Kenyon, A.J., Helmboldt, C.F., and Nielsen, S.W. (1963). Am. J. Vet. Res. 24, 1066.
22. Cho, H.J., and Ingram., D.G. (1974). J. Immunol. Methods 4, 217.
23. Bloom, M.E., Race, R.E., Hadlow, W.J., and Chesebro, B. (1975). J. Immunol. 115, 1034.
24. Shahrabadi, M.S., and Cho, H.J. (1977). Canad. J. Comp. Med. 41, 435.
25. Tsai, K.S., Grinyer, I., Pan, I.C., and Karstad, L. (1969). Canad. J. Microbiol. 15, 138.
26. Hahn, E.C., Ramos, L., and Kenyon, A.J. (1977). Infect. Immun. 15, 204.
27. Chesebro, B., Bloom, M., Hadlow, W., and Race, R. (1975). Nature 254, 456.
28. Porter, D.D., and Larsen, A.E. (1967). Proc. Soc. Exp. Biol. Med. 126, 680.
29. Cho, H.J., and Ingram, D.G. (1972). J. Immunol. 108,555.
30. Crawford, T.B. (1973). Fed. Proc. 32, 842. Abst.
31. Cho, H.J. (1977). Canad. J. Comp. Med. 41, 215.
32. Tabel, H., and Ingram, D.G. (1970). Arch. ges. Virusforsch. 32, 53.
33. Burger, D., Gorham, J.R., and Leader, R.W. (1965). In "Slow, Latent, and Temperate Virus Infections" (D.C. Gajdusek, C.J. Gibbs, and M. Alpers, eds.) pp. 307-313. U.S. Government Printing Office, Washington, D.C.
34. Haagsma, J. (1969). Tijdschr. Diergeneesk. 94, 824.
35. Goudas, P., Karstad, L., and Tabel, H. (1970). Canad. J. Comp. Med. 34, 118.
36. Rose, J.A. (1974). In "Comprehensive Virology" (H. Fraenkel-Conrat and R.R. Wagner, eds.) Vol. 3, pp. 1-61. Plenum Press, New York.
37. Porter, D.D., Dixon, F.J., and Larsen, A.E. (1965). J. Exp. Med. 121, 889.
38. Porter, D.D., Porter, H.G., and Larsen, A.E. (1977). Fed. Proc. 36, 1268. Abst.
39. Tabel, H., and Ingram, D.G. (1970). Canad. J. Comp. Med. 34, 329.
40. Porter, D.D., Dixon, F.J., and Larsen, A.E. (1965). Blood 25, 736.
41. Crawford, T.B., McGuire, T.C., Porter, D.D., and Cho, H.J. (1977). J. Immunol. 118, 1249.
42. Porter, D.D., Larsen, A.E., and Porter, H.G. (1973). Am. J. Pathol. 71, 331.

43. Henson, J.B., Gorham, J.R., Padgett, G.A., and Davis, W.C. (1969). Arch. Pathol. 87,21.
44. Pan, I.C., Tsai, K.S., and Karstad, L. (1970). J. Pathol. 101, 119.
45. Johnson, M.I., Henson, J.B., and Gorham, J.R. (1975). Am. J. Pathol. 81, 321.
46. Larsen, A.E., and Porter, D.D. (1975). Infect. Immun. 11, 92.
47. An, S.H., and Ingram, D.G. (1977). Am. J. Vet. Res. 38, 1619.
48. Cheema, A., Henson, J.B., and Gorham, J.R. (1972). Am. J. Pathol. 66, 543.
49. Porter, D.D., Larsen, A.E., and Porter, H.G. (1977) J. Immunol. 119, 872.
50. Porter, D.D., Larsen, A.E., and Porter, H.G. (1972). J. Immunol. 109, 1.
51. Henson, J.B., Gorham, J.R., and Leader, R.W. (1962). Nat. Fur News 34, 8.
52. Gierloff, B., and Thordal-Christensen, A. (1975). Dansk Vet. Tidsskr. 58, 589.
53. Greenfield, J., Walton, R., and MacDonald, K.R. (1973). Res. Vet. Sci. 15, 381.
54. Larsen, A.E. (1965). Nat. Fur News 37, 18.
55. Cho, H.J., and Greenfield, J. (1978). J. Clin. Microbiol. 7, 18.
56. Cho, H.J., and Ingram, D.G. (1973). Canad. J. Comp. Med. 37, 217.
57. Preble, O.T., and Youngner, J.S. (1975). J. Infect. Dis. 131,467.

EQUINE INFECTIOUS ANEMIA: VIRION CHARACTERISTICS, VIRUS-CELL INTERACTION AND HOST RESPONSES[1]

Timothy B. Crawford, William P. Cheevers, Paula Klevjer-Anderson, and Travis C. McGuire

Department of Veterinary Microbiology and Pathology, Washington State University, Pullman, WA 99164

ABSTRACT EIA is a retravirus infection of horses that is expressed as an immunologically-mediated destruction of RBC's. Cycles of viral replication in macrophages, triggered by antigenic shifts of the virus, originate from an unknown site of persistence and lead to anemia and clinical disease. *In vitro*, infection of macrophages leads to cell destruction, whereas infection of fibroblasts leads to non-cytopathic persistent infection. It appears that, were it not for the ability of EIAV to lytically infect macrophages, the infection would have only trivial consequences. The virus is able to periodically exploit this ability due to its capacity to modulate antigenically and temporarily escape immunologic controls, which gives EIA its distinctive features. In this paper, the nature of the virus, the effects on the host cell and the host immune responses are examined and their role in pathogenesis and persistence discussed.

INTRODUCTION

Equine Infectious Anemia (EIA) has only recently been established as a new member of the group of non-oncogenic retravirus infections that lead to chronic diseases other than neoplasia (1,2). Although EIA has been recognized as a virus disease for almost 75 years, only recently has information of events at the cellular and molecular levels begun to emerge that promises to lead to a mechanistic explanation of the disease. This paper presents a concise review of EIA and a presentation of recent data from this laboratory, with emphasis on those aspects most relevant to pathogenesis at a cellular and molecular level. Clinical, diagnostic, epidemiologic and pathologic aspects are treated only briefly; they have been well covered in general reviews (3,4).

[1]This work was supported by NIH Grant AI07471; ARS, USDA Cooperative Agreement 12-14-100-9067(45); and Washington State University, ARC Project 0146.

ISBN 0-12-668350-6

General Features of EIA. The earmark that renders EIA unique among retravirus infections is its episodic character. Following infection, disease is not progressive, but occurs during unpredictable "cycles", between which the horse is clinically normal. These clinical episodes are associated with periods of rapid virus replication, and are characterized by 3 to 7 days of fever, depression and destruction of RBC's. An apparent breakdown of the control mechanisms which suppress virus replication between cycles allows a resurgent viremia that typically reaches about 10^4 /ml for 2 to 3 days, before subsiding (5). This outpouring of virus and associated antigens in the face of a vigorous immune response sets in motion a series of immunopathologic events that cause the clinical signs and lesions of EIA. The wide variation in frequency and severity of clinical cycles, even among horses infected with the same virus strain (6), is apparently due to differences between horses in effectiveness of the host defense mechanisms. The defenses seem to be immunologic and to become more effective with prolonged stimulation: most horses stop cycling altogether within a few months. Nevertheless, they harbor the virus for life. The usual mode of transmission is horizontal, via infected blood or secretions, although the possibility of genomic vertical transmission has not been eliminated.

CHARACTERISTICS OF EIA VIRUS

Ultrastructural Morphology. EIAV ultrastructure has been described by a number of workers. Its general features clearly place it with the retraviruses, though its definitive subclassification is still unclear. It has more features in common with visna virus and a bovine syncytia-producing virus than with prototype C-type viruses (7).

RNA-Dependent DNA Polymerase (RDDP) and its Reaction Product. The virion-associated RDDP of EIAV exhibits optimal endogenous activity in the presence of 0.02% triton X-100, 6×10^{-3}M Mg^{++} and pH 7.8 at 37°C. Optimal conditions are rather stringent in that minor deviations result in a rapid decline in reaction rate (1). The enzyme also catalyzes an exogenous oligo-dT-primed reaction using poly(A) but not poly(dT) as template (1,2). In isopycnic centrifugation, all RDDP activity co-bands at 1.15-1.16 gm/cm^3 with purified EIAV labeled in RNA with ^{14}C-UR (8). Under optimal conditions the EIAV-RDDP reaction proceeds linearly for many hours, catalyzing the synthesis of a DNA product which sediments at 4-10s (Fig. 1) and which is predominantly (≥ 90%) double stranded. Double-strandedness is shown by susceptibility of the 18-hr endogenous product to hydrolysis by S1 nuclease (Fig. 2).

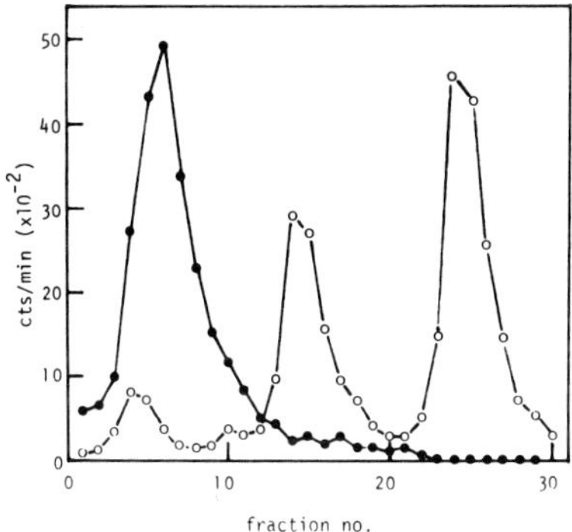

Fig. 1. Velocity sedimentation of EIAV RDDP reaction product. EIAV DNA synthesized *in vitro* (1) was purified by hydroxyapatite chromatography. 100 μl of DNA was mixed with 50 μl of ^{14}C-labeled marker RNA (4s, 18s, 28s) and analyzed by velocity sedimentation (8) (SW 41 rotor; 27.5 K; 15.6 hr; 23°C). (•–•) ^{3}H; (o–o) ^{14}C. Sedimentation from left to right.

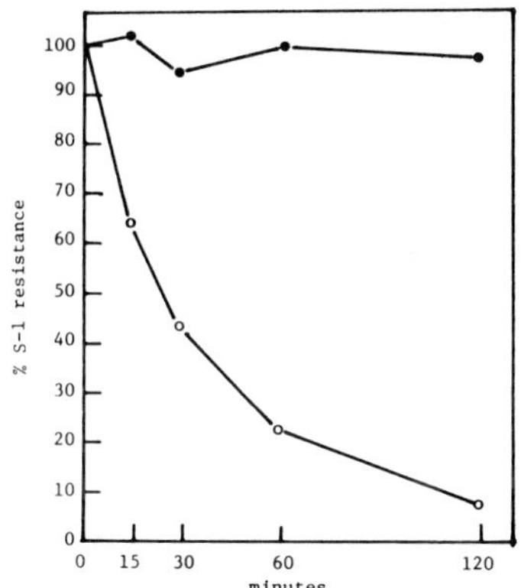

Fig. 2. Susceptibility of EIAV RDDP reaction product to hydrolysis by SI nuclease. Unheated (•–•) or heated (o–o) (100°C; 5 min) purified EIAV DNA was diluted 1:100 in 0.03 M Na acetate buffer, pH 4.5-0.3 M NaCl-0.005 M $ZnCl_2$-25 μg/ml denatured salmon sperm DNA and incubated at 37°C in the presence or absence of 500 units of nuclease. Plotted as residual TCA-insoluble radioactivity.

Note that the native DNA is resistant to Sl nuclease, whereas it is sensitive to this enzyme after heat-denaturation. Further, the chromatographic behavior of the native DNA on hydroxyapatite and benzoylated-naphthoylated DEAE-cellulose is consistent with a double-stranded configuration. Hybridization of heat-denatured DNA product to excess EIAV-RNA, followed by centrifugation of the hybrids in CS_2SO_4, shows that the product is comprised of (+) strand/(-) strand duplex molecules (Fig. 3). The representation of EIAV genome sequences in this product has not been determined; estimates from sedimentation rate and complexity analysis indicate a sequence representation of 35-50% based on a MW of 2.8 x 10^6 daltons for the genome subunit (8). In the presence of actinomycin D, the endogenous reaction rate is reduced by more than 50% (1), and the DNA product is predominantly single-stranded (∿80%) with (-) strand hybridization characteristics.

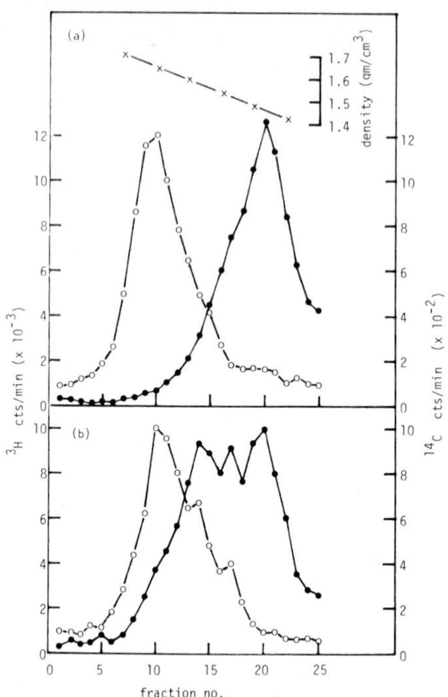

Fig. 3. Hybridization of EIAV RDDP reaction product to EIAV-RNA. [^{3}H] EIAV DNA was mixed with excess ^{14}C-labeled 62s EIAV RNA (8) in 0.02 M Tris buffer, pH 7.4-0.6 M NaCl-0.01 M EDTA, heated at 100°C for 5 min, rapidly cooled, and incubated for 72 hr at 68°C (a) or -20°C (b). The nucleic acids were then mixed with saturated Cs_2SO_4 to a final density of 1.6 gm/cm^3 and centrifuged at 35 K for 60 hr at 10°C. The gradients were fractionated and ^{3}H and ^{14}C radioactivity measured. (•-•) ^{3}H; (o-o) ^{14}C.

Viral RNA. The genome of EIAV has been characterized by velocity sedimentation, polyacrylamide gel electrophoresis, buoyant density in CS_2SO_4 and susceptibility to nuclease digestion (8). The nucleic acid of purified virus is resolved by sedimentation analysis into a fast-sedimenting genome component, which comprises about two-thirds of the virion RNA, and a slow-sedimenting component, which is probably comprised of host-derived tRNA and a trace amount of 5s rRNA. The fast-sedimenting RNA has a sedimentation coefficient of 62s and a MW of about 5.5 x 10^6 daltons, as determined by electrophoretic mobility. The 62s RNA is predominantly single-stranded, but contains double-stranded regions as indicated by partial resistance to RNase I and S1 nuclease and by a lower buoyant density in CS_2SO_4 than that of the single-stranded RNA derived from it by heat-denaturation. The products of heat denaturation of the 62s RNA include a small amount of 4s RNA and single-stranded 34s RNA with a MW of about 2.8 x 10^6 daltons. These data indicate that the EIAV genome consists of two 34s subunits of single-stranded RNA held in a high-MW complex with 4s RNA by a mechanism involving a small degree of base pairing. Such a structure conforms generally to the currently accepted models for retravirus genome structure.

Structural Polypeptides. Figure 4 shows a polyacrylamide gel stained with Coomassie blue after electrophoresis of SDS-disrupted EIAV purified by isopycnic centrifugation as previously described (8). Fourteen polypeptides, designated numerically from the origin, are resolved. By staining, the most prominent structural polypeptides appear to be 1, 2, 3, 8, 9 and 12. Polypeptides 4, 5, 10, 11, 13 and 14 are present in very low amounts and are difficult to demonstrate photographically. They are, however, readily apparent by absorbancy scanning and by visual examination of stained gels.

The MW of EIAV polypeptides was estimated from electrophoretic mobility relative to the migration of marker proteins of known MW (9). The results of 4 gels, presented in Table 1, indicate that the 14 structural polypeptides range in MW from about 10,000 to 79,000 daltons.

Some polypeptides, notably glycopeptides (10), do not conform to a linear relationship between mobility and the logarithm of MW in SDS-polyacrylamide gel electrophoresis, because of variation in the amount of SDS bound (11). Such anomalously migrating proteins may be detected by comparing mobilities at a variety of acrylamide concentrations, in which the relative mobilities at the different concentrations obey the relationship $\text{Log } R_F = \text{Log } Y_0 - K_R T$ (12) where R_F is the relative mobility of a protein in a gel of concentration T, Y_0 is the limiting relative mobility and K_R is the retardation coefficient. The mobilities of the prominent EIAV

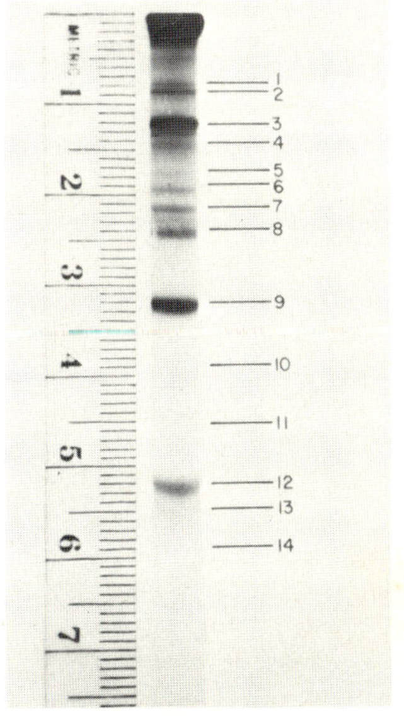

Fig. 4. Polyacrylamide gel electrophoresis of EIAV polypeptides. Purified virus was disrupted by heating at 100°C for 3 min in 100 µl of 1% SDS-1% 2-mercaptoethanol-10% glycerol. Electrophoresis of dissociated peptides was in phosphate-buffered continuous SDS gels containing 10% acrylamide at 4.5 mA/gel for 12 hr.

TABLE 1.

MOLECULAR WEIGHT OF EIAV PROTEINS

Band No.	Molecular Weight ($\times 10^{-3}$)				
	1	2	3	4	Average
1	75	78	78	84	78.8
2	74	77	77	82	77.5
3	63	65	63	65.5	64.1
4	57	58.5	58	58	57.9
5	50	53	50.5	51.5	51.3
6	47	49	46.5	48.5	47.8
7	43	43.5	42	43	42.9
8	39	40.5	38	40.5	39.5
9	30	30	27	28.5	28.9
10	23.5	23	-	-	23.3
11	18	18	17.5	-	17.8
12	14.5	13.5	10	12.5	12.6
13	13	12	-	-	12.5
14	9.5	9	-	-	9.3

TABLE 2.

QUANTITATION OF EIAV PROTEINS

Band No.	Percent of Radioactivity			
	1	2	3	Average
1,2	10.7	7.2	10.7	9.5
3	14.4	11.4	10.9	12.2
4-7	7.7	7.1	7.0	7.3
8	8.5	8.9	10.6	9.3
9	25.3	27.5	27.9	26.9
10	3.7	4.6	4.2	4.2
11	3.4	5.6	7.2	5.4
12	22.0	24.5	16.9	21.1
13	2.1	1.8	3.2	2.4
14	2.2	1.4	1.4	1.7

proteins (bands 1, 2, 3, 8, 9, and 12) were compared to monomers, dimers and trimers of lysozyme in split gels ranging from 5% to 15% in acrylamide concentration. These procedures indicated that EIAV proteins 1, 2 and 3 deviate markedly and protein 8 slightly from their expected rate of migration over a range of acrylamide concentrations. All of these bands are glycopeptides (see below), and the results indicate that their MW's as listed in Table 1 are inaccurate. Nevertheless, following the convention suggested by August *et al.* (13), we shall refer to these proteins according to their *apparent* MW.

The relative proportion of EIAV proteins was estimated by electrophoresis of disrupted virus labeled for at least 12 hours with [^{3}H] or [^{14}C]-amino acid mixtures. There was excellent correspondence between the migration pattern of labeled polypeptides and that obtained using stained gels (Fig. 5). Quantitation of the proportion of radioactivity associated with each peak of amino acid label from the gel in Fig. 5 and two other similar gels is presented in Table 2. Eighty percent of the virion protein is accounted for in polypeptides 1, 2, 3, 8, 9, and 12. Half is confined to peptides 9 and 12. This distribution of peptides is not markedly different from those of other mammalian type-C retraviruses (14), except for the prominent polypeptide 3, comprising ∿12% of the labeled protein. Peptides 9 and 12 are considered analogous to p30 and p15, the major non-glycosylated structural proteins of type-C mammalian retraviruses.

The structural glycopeptides of the EIAV virion were identified by labeling with [^{3}H]-glucosamine. The radioactivity profile of glucosamine-labeled polypeptides co-electrophoresed with [^{14}C]-amino acid-labeled peptides is shown in Fig. 5. In this gel, correspondence of [^{3}H] and [^{14}C] radioactivity peaks indicates that polypeptides 1, 2, 3, 8 and 14 are glycosylated. The percentage of glucosamine label in glycopeptides 1, 2 (42%) and 8 (21%) (gp77/79 and gp40) is similar to that usually obtained in the gp69/71 and gp45 components of murine type-C retraviruses (14). An analogue of glycopeptide 3 (gp64), comprising about 15% of the glucosamine label, has not been previously described in retraviruses. The minor glycoptide gp10 (band 14) has not been described in type-C viruses, although an analogous protein was noted in Mason-Pfizer monkey virus, a type-D mammalian retravirus (15). Preliminary assessment of surface labeling of EIAV with ^{125}I using lactoperoxidase (B.G. Archer, unpublished data) and diazodiiodosulfanilic acid (G.A. Robertson, unpublished data) is consistent with the conclusion that gp77/79 and gp10 are surface glycoproteins. Glycopeptides gp64 and gp40 are minimally labeled, if at all, using surface labeling procedures. The gp40 is possibly a degradation product of the major surface glycoprotein, as is the gp45 glycopeptide of murine viruses (16). The significance of gp64 is unknown.

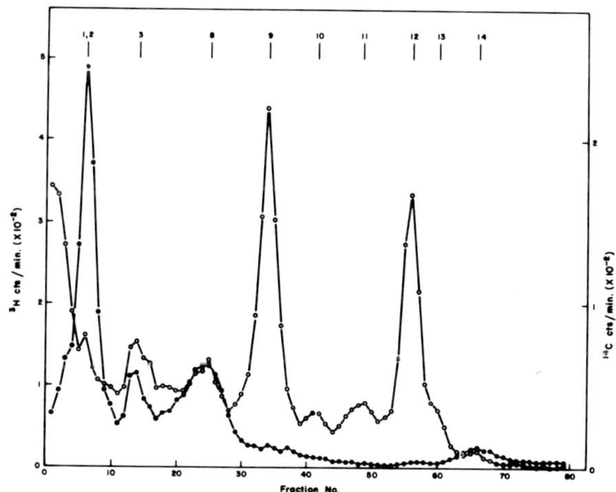

Fig. 5. Polyacrylamide gel electrophoresis of EIAV labeled with [^{14}C]amino acids and [^{3}H]glucosamine. Virus produced in a roller culture of EIAV-infected cells during a 24-hr period in the presence of L-[^{14}C]amino acid mixture and D-[6-^{3}H]glucosamine was purified and subjected to SDS-gel electrophoresis as described in Fig. 4. (•–•) ^{3}H; (o–o) ^{14}C.

EIA VIRUS-CELL INTERACTION

Integration of EIAV Provirus in Persistently Infected Cells. The method of acceleration of the reassociation of a double-stranded DNA "probe" synthesized *in vitro* by the endogenous EIAV RDDP was used to test for the presence of integrated provirus in persistent EIAV infection (17,18). Figure 6 shows the results of a typical experiment. In 0.26 M neutral phosphate buffer at 68°C, EIAV "probe" reassociates with second-order kinetics and a $C_0t\ \frac{1}{2}$ of 4×10^{-3} mol sec/L in the presence of salmon sperm DNA in vast excess. In the presence of excess DNA extracted from persistently-infected equine dermal fibroblasts, the EIAV "probe" reassociates with a $C_0t\ \frac{1}{2}$ of 2×10^{-3} mol sec/L. These results indicate the presence in the infected cell genome of integrated DNA sequences complementary to the EIAV "probe"; the acceleration factor of 2-fold at the concentration of infected cell DNA used is equivalent to the presence of 40-50 copies of the "probe" sequences per cell. Other experiments show that a 2×10^5-fold excess of normal horse spleen DNA has no effect on reassociation of the EIAV "probe" relative to salmon sperm DNA. Thus, establishment of EIAV persistence *in vitro* probably involves the acquisition by infected cells of integrated EIAV DNA sequences. It is reasonable, therefore, to assume that virus replication is by the provirus mode.

In Vitro Propagation Systems for EIAV. Cells from species other than Equidae have generally been refractory to infection by EIAV. Reports of replication in cells from other species (19,20) are either unconfirmed or have been subsequently contradicted by other workers (21). This uncertainty is partially a reflection of the cumbersome methods that formerly were the only available means of assaying viral infectivity: horses and more recently, macrophage cultures. The question should be reexamined using the more sensitive methods now available for detecting the genome and antigens of the virus.

Field isolates of wild-type (WT) EIAV replicate readily only in equine macrophages cultured from peripheral leukocytes (22). These cells do not replicate and are very fastidious, requiring precise culture conditions. Frequent spontaneous degeneration renders CPE unreliable as an indicator of viral replication, thus serologic detection of viral antigens is used to locate end points. A complete titration requires 2 to 3 weeks. Macrophage cultures inoculated with a large dose of virus express detectable antigen within 24 hours and release infectious virus by 20 hours (23). Viral particles bud from the cell membrane in profusion (24) and the cell begins to die within 36 hours. The CPE is not accompanied by cell fusion or other distinguishing features.

In contrast to macrophages, other equine cell types do not readily support replication of WT EIAV (21,25). Dividing equine cells apparently exert a degree of restriction over WT EIAV replication *in vitro*. Strains of EIAV that are able to bypass this restriction and grow in cultures of replicating fibroblasts have been developed by extensive passaging *in vitro*. One strain, developed by Malmquist et al. (26) is widely used as a source of virus for investigation and for serodiagnostic antigen. It retains its virulence for horses

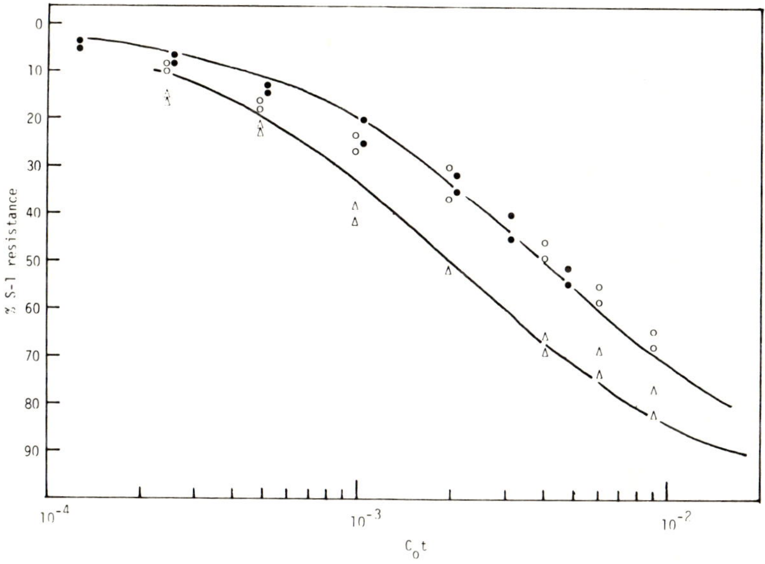

Fig. 6. Acceleration of reassociation of EIAV DNA by DNA of EIAV-infected fibroblasts. EIAV DNA was synthesized *in vitro* using the endogenous RDDP reaction in the presence of 1.23 x 10^{-5} M [^{3}H]TTP and 3.7 x 10^{-4} M dGTP, dATP and dCTP (1). The 18 hr double-stranded reaction product was purified by hydroxyapatite chromatography and mixed with a 25,000-fold excess of purified DNA from salmon sperm or EIAV-infected fibroblasts. The nucleic acids were sheared at 45,000 p.s.i., heat-denatured and reassociated at 68°C, either in 0.26 M or 0.55 M neutral phosphate buffer. Reassociation was monitored by development of resistance to SI nuclease (Fig. 2) and expressed as C_0t (18). (•-•) EIAV-DNA + salmon sperm DNA in 0.26 M PO_4 buffer. (o-o) EIAV-DNA + salmon sperm DNA in 0.55 M PO_4 buffer. (Δ-Δ) EIAV-DNA + EIAV-equine fibroblast DNA in 0.55 M PO_4 buffer.

after >50 passages in equine fibroblasts, and further, retains, through at least 3 clinical cycles in vivo, the ability to replicate in fibroblasts in vitro (Crawford and Klevjer-Anderson, unpublished). Another adapted strain was produced by extensive passaging of WT virus in the leukocyte culture system. After 43 passes, it was able to replicate in horse kidney cells (27), whereas non-passaged source virus would not. Interestingly, this strain had lost its pathogenicity for horses (28).

There are fundamental differences in the in vitro virus-cell relationship between the macrophages and the replicating cells such as the fibroblast, that stimulate speculation as to whether there might be in vivo correlates of pathogenetic significance. In macrophages, replication is apparently uncontrolled, leading to a very large number of budding particles per cell (24), disruption of cellular machinery and rapid CPE. In contrast, replicating cells such as fibroblasts exert much tighter controls on viral expression; indeed, with WT virus, replication is almost completely restricted. However, the level of restriction (attachment, integration, transcription, translation or assembly) has not yet been determined. Adapted strains are able to replicate in fibroblasts, but do so in relative harmony with the cell. Fewer particles are produced per cell (7) and deleterious effects of replication are not obvious (26). One is tempted to wonder whether the inability of the macrophage to control viral expression is related to its lack of mitotic activity. It is possible that the lack of DNA synthetic activity leads to a block or reduction of provirus integration. Autonomous replication of the episomal provirus could thus allow escape from the transcriptional controls normally exerted over integrated provirus. Alternative explanations include differences at the translational level, or in the processing of transcribed RNA that determines whether it becomes mRNA or 60s-70s genomic RNA. To study these possibilities, experiments were initiated to define the parameters of viral genome function at the molecular level.

Virus-Cell Interactions During Persistent Infection in Vitro. The first objective was to characterize the nature of the persistent infection of fibroblasts by the fibroblast-adapted strain of EIAV (26). Persistently infected fibroblast cultures, maintained as described previously (1) were examined, with non-infected cells as controls. No significant differences were found in the growth rates of the two (Fig. 7A). Similar saturation densities and plating efficiencies were also found between infected ($1.8 \times 10^5/cm^2$; 80%) and non-infected (2×10^5; 81%) cells. No morphologic alterations or disorientation of growth patterns suggestive of transformation were found, and the infected cells were not able to grow in reduced serum medium (2% FCS).

To examine the possibility that a physiological effect on the cell might be evident soon after primary infection but disappear after a few passages, cells were infected at a MOI of 40 fibroblast infectious units per cell (FIU, defined below) and the subsequent growth rate compared to mock infected cultures. There was no observable effect on the cell growth kinetics or saturation density (Fig. 7B).

EIAV has heretofore been titrated only in horses and macrophage cultures; no titration in fibroblast cultures has been reported. For studying the kinetics of viral replication in fibroblasts, the amount of virus necessary to infect a fibroblast culture (FIU) had to be defined. A concentrated stock of fibroblast-grown virus was titrated in the conventional macrophage system and in fibroblast cultures by the following method. Replicate cultures of fibroblasts in 60 mm petri plates were infected with 0.2 ml of each $\log_{10}$ virus dilution by allowing it to adsorb onto 1 day-old cultures for 2 hours. Cultures were washed, maintained with weekly subculturing for 10 weeks, and monitored weekly by IF for appearance of viral antigen. The development of fluorescent cells in cultures infected with the highest positive dilution required 5 weeks. The stock titered 5×10^7 $TCID_{50}$ in fibroblasts and 5×10^9 $TCID_{50}$ in macrophages. Thus one FIU was equivalent to approximately 10^2 macrophage infectious units, a value similar to one suggested by Kono et al., using a different strain of adapted virus (27).

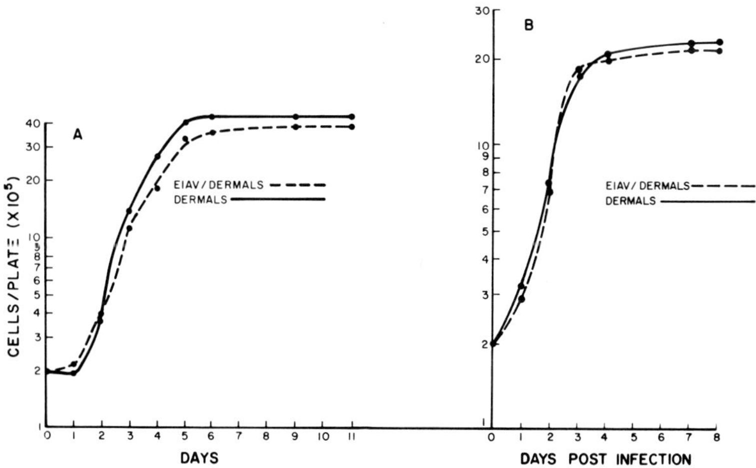

Fig. 7. Growth rates of non-infected (—) and infected (--) fibroblasts. (A) long-term persistent infection. (B) Immediately after primary infection.

Due partially to this relative insensitivity, attempts to define the kinetics of viral replication in fibroblasts during primary infection have been unsuccessful. When using IF as the criterion of infection, the relationship between virus dilution and number of infected cells at 3 days post infection was not linear. Experiments described below indicate that at least 2 other factors contribute to this nonlinearity and therefore render a fluorescent focus assay unusable. They relate to fluctuations in viral protein synthesis and sensitivity of IF.

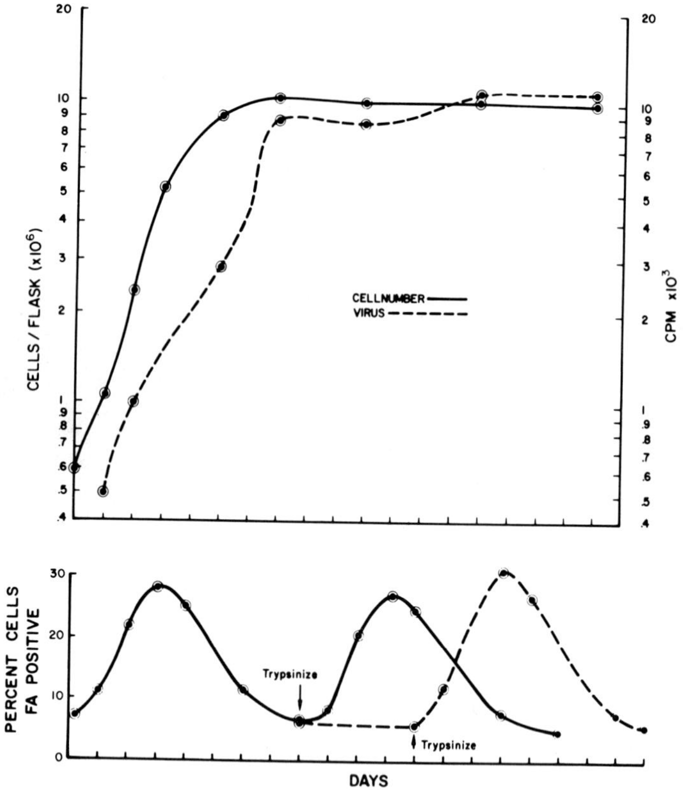

Fig. 8. Viral antigen synthesis and virion production in persistently infected cultures. Virion production (upper panel) was monitored by release of [^{3}H]-UR labeled virus into the medium over a 24 hr period. Percentage of cells IF positive for intracytoplasmic viral antigen in relation to the growth phase of the culture is shown in the bottom panel. (—) cultures passaged on reaching confluence, (--) cultures left at confluence 5 days before passaging.

Viral Replication Kinetics in Persistent Infection in Vitro. To study the nature of the persistent infection in fibroblasts, the number of cells synthesizing IF-detectable viral antigen and the release of labeled virus into the medium was monitored daily, while the cultures were passaged at varying intervals. Replicate cultures were labeled with 10 μC/ml ^{3}H-UR for 24 hours. Virus released into the medium was purified and analyzed by TCA-insoluble radioactivity.

The percentage of IF-positive cells increased to a maximum of about 30% as the culture entered logarithmic growth, then declined to about 5% as the culture reached stationary phase, remaining there as long as the cultures were stationary. Upon passage, this pattern was repeated (Fig. 8). In contrast, the release of virions was not related to the growth phase of the culture. The amount of virus released increased concomitantly with cell number, reached a plateau and remained constant for up to 30 days in stationary cultures.

These same parameters were examined in persistently infected cultures synchronized by isoleucine deprivation (29). Isoleucine-deficient MEM with 10% dialyzed FCS was added to log phase cultures for 26 hours to arrest cells in the G_1 phase. Release by restoration of complete medium initiated synchronous DNA synthesis and cell division (Fig. 9). Viral protein synthesis was biphasic, reaching maxima of about 46% IF-positive cells during the late G_1 and $S+G_2$ phases of the cell cycle. In contrast, no relationship to the cell cycle was seen in release of virions; the amount released was proportional to the number of cells in the culture. The inability to demonstrate 100% IF-positive cells during synchronization could be the result of (a) lack of total synchrony (b) inadequate sensitivity of IF and/or (c) refractory cells within the population.

In this system, as with RSV (30), the rate of virus production is uniform and independent of events related to the cell cycle. Even though sufficient to supply the needs of viral assembly, the amount of viral protein in productively-infected resting cells is largely below the detection level of routine IF. Cell-cycle dependent processes do, however, markedly affect the rate of protein synthesis. This rate increases at certain points in the cycle, leading to excessive accumulations that become detectable by IF during the G_1 and $S+G_2$ phases. Cell-cycle dependent variation in the expression of retravirus proteins is not uncommon (31). The fact that fluctuations in the amount of viral proteins in the cell are not reflected in virus release is support for the concept that the rate-limiting step in virus production is the amount of available genomic RNA. Further studies are needed to deter-

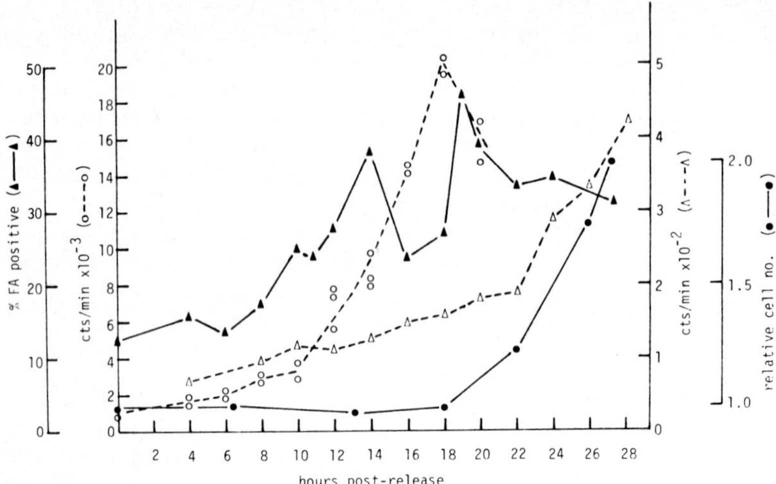

Fig. 9. Viral antigen synthesis and virion production in synchronized cells, both monitored as in Fig. 8. DNA synthesis measured by uptake of [^{3}H]-thymidine.

mine whether the cell-cycle dependence of protein synthesis exists *in vivo* and what effect this has on the usefulness of IF as an investigative tool.

In Vivo Studies. Clinical cycles are associated with periods of rapid viral growth in mononuclear cells that are apparently monocytes and tissue macrophages (3). Whether there are other cells infected but not detectable by IF is not known. Titers are high in blood and in most tissues and organs during acute disease. Most of the infectivity in the plasma is in the form of infectious virus-antibody complexes (3). Between cycles, infectivity declines in both serum and tissues, the liver and spleen retaining higher titers than other organs. Some animals that experience a severe form of the disease, chronic EIA, exhibit prolonged viremia and persistence of high titers of virus in most organs. This form is often fatal. In the majority of cases however, the frequency and severity of clinical episodes declines with time, about 90% occurring within 1 year of infection (5). These features and other data suggest a control mechanism that is more efficient in some horses than others, and that improves with time after infection. The ability of immunosuppressive drugs to induce recrudescence (5) suggests the controls are immunologic in nature. Other data (see below) implicates the cellular, rather than the humoral, immune system.

The cell type in which the virus persists between cycles is not known. IF detects few or no infected cells during stable asymptomatic infection (3). It isn't known if this reflects insensitivity of the technique, absence of infected cells, or latent infection. The *in vitro* studies reported here suggest the first of these 3 possibilities. The site of persistence may be a cell other than the macrophage. The existence of an *in vivo* version of the persistent infection in fibroblasts *in vitro* is an attractive possibility.

THE ROLE OF THE IMMUNE RESPONSE IN DISEASE AND VIRAL PERSISTENCE

Adequate consideration of the pathogenesis of persistent infections by conventional viruses requires proper attention to the host immune response. This section is concerned mainly with immune function as it relates to virus elimination, but immunologic factors in lesion production are first briefly summarized; for a fuller treatment see Henson and McGuire (3).

The anemia of EIA is the result of immune-mediated RBC destruction (3) presumably initiated by binding of EIAV hemagglutinin subunits (32) to the RBC in the presence of circulating antibody. It is exacerbated by bone marrow depression and perhaps by a hyperactive phagocytic system (33). Depressed circulating C3 levels are consistent with the presence of circulating infectious virus-antibody complexes and a subclinical, non-progressive, immune-complex mediated glomerulitis (3). Although during acute disease lymphoid necrosis is prominent, no immunosuppression has been reported in either acute or asymptomatic EIA, though selective depression of one major immunoglobulin class - IgG(T) - follows long-term infection (34). The cause of lymphoproliferation and mononuclear infiltration of liver and kidney during active disease are most likely related to the cellular immune response to virus infected cells.

The Immune Response to Internal Virion Antigens. Horses make antibody to a number of proteins not accessible on the intact virion. At least 2 of the proteins yield precipitin lines with serum from infected horses: the major core protein and a smaller protein, designated previously as p15 (35,36) which is probably the same as our p12 herein. The widely used diagnostic test (37) is based on detection of antibody against group-specific determinants on the major core protein, previous designations for which have ranged from p25 to p28 (2,38,39), and herein designated p29. This protein, which is the most likely candidate to carry interspecies reactivity, is antigenically unrelated to the analogous protein of several other mammalian retraviruses (2, Crawford and McGuire, unpublished).

IgG from EIAV infected horses fixes complement with viral antigen, but it was found that some sera contain high levels of an unusual, non-complement fixing (CF) immunoglobulin, IgG(T), which competitively inhibits CF by IgG (3). Because of the possibility that IgG(T), a major subclass of IgG in the horse, might have other important roles in the biology of this infection, an effort to examine its function was undertaken. The ability to precipitate with antigen was studied because of the unusual nature of the precipitin curve of IgG(T). Ovalbumin, for example, which has several multiple--but antigenically unique--determinants, precipitates with IgG(T) in a narrow range at equivalence, but the complexes are soluble not only in antigen excess (as expected), but also in antibody excess ("flocculating" curve).

The following experiments were suggested by the findings of Archer and Krakauer that some unusual structural property of IgG(T) rendered intramolecular binding thermodynamically more favorable than intermolecular binding, relative to IgG (40). The ability of IgG(T) to precipitate with antigen was examined with systems in which the number and spacing of haptenic determinants on the carriers could be controlled. Affinity chromatography-purified IgG(T) would not precipitate with DNP-substituted proteins at any DNP:carrier ratio, whereas IgG would precipitate at suitable ratios (Table 3). In passive hemagglutination experiments, IgG(T) would cross-link (hemagglutinate) RBC's coated with ovalbumin, DNP_4-, or DNP_9-ovalbumin, but would not cross-link RBC's coated with DNP, whereas IgG would (Table 4). The thermodynamic explanations for these observations will be elaborated more fully elsewhere (McGuire et al., in preparation), but they center around the fact that, where antigenic determinants are sufficiently close together, IgG(T) prefers to occupy both its

Table 3
Precipitation by IgG and IgG(T)

Antigen	Precipitation	
	IgG	IgG(T)
DNP_5BSA	-	-
DNP_8BSA	-	-
$DNP_{12}BSA$	+	-
DNP_{20}ovalbumin	+	-
Ovalbumin	+	+

Table 4
Passive HA by IgG and IgG(T)

Antigen on RBC	HA	
	IgG	IgG(T)
Ovalbumin	+	+
DNP_4ovalbumin	+	+
DNP_9ovalbumin	+	+
DNP[a]	+	-

[a] DNFB-treated RBC's

binding sites on the same molecule. Where this type of binding is sufficiently favored, precipitation and agglutination reactions are inhibited (41). This phenomenon accounts for the observation that horse seras with sufficient IgG(T) against the EIAV p29 will neither fix complement nor precipitate in gel diffusion. Further, the p29-IgG(T) inhibition is dependent on the antibody-antigen ratio: reduction of the IgG(T) excess by simply diluting the serum can bring about precipitation (39). Thus, this antibody class is functionally important in modulating 2 immunologic parameters associated with EIAV antigens: CF and precipitation. It is not known whether it affects viral neutralization or the biologic activity of antigen-antibody complexes. We present evidence below that it also interfaces with the cell mediated immune (CMI) system.

The Immune Response to Antigens of the Viral Envelope. Horses make neutralizing antibody to EIAV (3,5) but its effectiveness is periodically circumvented by the ability of the virus to change antigenically. Kono, et al. (42) have shown that a new antigenic variant becomes predominant in the blood in concert with each new clinical cycle, thus permitting a temporary escape and rapid replication by the virus. The appearance of antibody to the new variant in a few days apparently reestablishes immunologic control, and viremia subsides. The envelope proteins involved in the variation have not been identified, but results presented herein indicate the gp77/79 and gp10 are likely candidates, since they appear to be accessible on the envelope. The extreme unwieldiness of the available assay system for WT EIAV greatly hampers studies based on viral neutralization; very few labs are willing to attempt them. To date, data in this area have originated almost exclusively from one laboratory in Japan, and currently stands without confirmation. It is not clear how much antigenic relatedness exists between viral isolates. If present though, it is apparently minor (5,42) and inadequate to stimulate cross-protection by antibody. As suggested by Kono (5) it is likely that some mechanism other than neutralizing antibody is responsible for the eventual cessation of cycles seen in most horses.

Antigenic variation also occurs in visna of sheep (43), driven by the selective pressure of antibody (44), where it may play a role in the maintenance of persistent infection as well as lesion progression. Another interesting parallel is the antigenic drift of African trypanosomes. Changes in the 64,000 dalton surface glycoprotein of the protozoan are responsible for antigenic variation (45). Though this molecule also carries cross-reacting determinants, only the variant-specific determinants are accessible on the surface (46).

While there is sufficient genomic capacity in trypanosomes to permit several genes coding for this surface glycoprotein, the limited size of the EIA genome (8) restricts genetic mechanisms of antigenic variation to mutation and recombination. Study of the envelope glycoproteins of EIAV should allow testing of these hypotheses and determine if exploitable homology exists between variants.

T-lymphocytes in infected horses are responsive to purified inactivated EIAV (47). The degree of sensitivity rises in association with each disease cycle, and the response is not strain specific (48). The stimulating antigens have not been identified, but are most likely in the viral envelope since the antigen used was intact virus. Though studies with virion antigens reflect the general state of the cellular immune system, the cellular response to cell membrane antigens (see below) is more relevant to mechanisms of viral persistence.

The Immune Response to Viral Associated Cell Membrane Antigens. There are 2 lines of evidence suggesting the presence of EIAV-specified determinants on the membrane of persistently infected cells. Using a radioimmune binding assay, we have shown that EIAV-infected horses produce antibody that binds specifically to the membrane of persistently infected cells (49). The fact that the antibody levels were low may be because only cross-reacting antibody was detected: the serum source and the target cells were not antigenically homologous. Second, we have recently found low-level direct cytotoxic killing of antigenically heterologous persistently infected cells by lymphocytes from infected horses (Fujimiya et al., in preparation). Thus infected cells *in vitro* express recognizable membrane antigens that are not strain-specific.

We have recently examined whether horse leukocytes can kill cells through the antibody-dependent cell mediated cytotoxicity (ADCC) mechanism. Using the model system of DNP-coated erythrocytes, anti-DNP sera, and IgG and IgG(T) purified from these sera by affinity chromotography, we have shown that horse leukocytes will lyse RBC's sensitized with IgG but not RBC's sensitized with IgG(T). The lysis can be mediated by lymphocyte-monocyte mixtures, neutrophils and macrophages. Further, IgG(T) efficiently inhibits the reaction, apparently by competing with IgG for the antigenic determinants, even when present at low IgG(T): IgG ratios (Fujimiya et al., in preparation). These data are consistent with the previous observation that horse monocytes and neutrophils lack a membrane receptor for IgG(T) (50).

To date, we have not been able to demonstrate ADCC against persistently-infected cell in EIA. Further experiments are needed to confirm this observation and to define

TABLE 1
HI ANTIBODIES IN HUMAN SERA

Age	JCV[a] +/Total	%+	Age	BKV[b] +/Total	%+
0-4	21/133	16	1-5	17/46	37
5-9	19/69	27.5	6-10	43/52	83
10-14	13/20	65			
> 50	120/157	76	> 50	40/53	75

[a]Data from Padgett and Walker, J. Inf. Dis. 127:467, 1973; Padgett and Walker, unpublished.
[b]Data from Gardner, Brit. Med. J. 1:77, 1973.

olds had antibodies to BKV and 32% to JCV. In 249 5-9 year-olds these percentages had risen to 82% and 69% respectively (Table 2). By 9 years of age more children had been infected with BKV than with rubella or measles viruses, and infection with JCV was only slightly lower than with these other viruses. In marked contrast only 12% of the 5-9 year-olds had antibodies to cytomegalovirus, a virus excreted in urine. Besides demonstrating a high incidence of infection by BKV and JCV, this study raises the possibility that children may excrete these agents by routes additional to the urinary.

Third, serologic surveys of animal sera for HI antibodies against JCV have given negative results (1). Animals screened included many commonly owned as household pets and several species of primates. These results suggest that there is no animal reservoir for JCV.

TABLE 2
PREVALENCE OF ANTIBODIES TO VIRAL AGENTS IN 5-9 YEAR-OLD MEXICAN CHILDREN[a]

Virus	% With Antibody
BK	82
Rubella	77
Measles	73
JC	69
Cytomegalovirus	12

[a]Golubjaknikov, *et al.*, unpublished

WHAT HAS BEEN LEARNED ABOUT THE ROUTE(S) OF EXCRETION?

The only known route of excretion for either BKV or JCV is in urine. BKV has been found in urine of persons with reduced immune competence due to congenital defects, therapy following organ transplantation, and therapy for malignancies (9,10,11). JCV has also been found in urine of immunosuppressed persons and in urine of a healthy pregnant woman (10,12,13). Most of what is known about urinary excretion of these viruses comes from studies of renal transplant patients, especially those of Gardner, Coleman, and colleagues in England (9,10,14), and Lecatsas and colleagues in South Africa (15,16). In long-term studies, between 30 and 50% of renal transplant patients have been found to excrete papovavirus virions at one time or another. Usually excretion is of short duration, but some patients excrete large quantities of virus for long periods of time. Seventy-eight renal transplant recipients have been studied for over one year by Hogan and colleagues at the University of Wisconsin. Twenty-one % have shed virus during this period of observation. Virus identification was accomplished by immunofluorescent staining of urinary sediments containing inclusion-bearing cells. In addition to individuals excreting only BKV or JCV, three patients were found to excrete BKV on one occasion and JCV on another (Hogan, et al., unpublished). In none of these studies was excretion of the papovavirus accompanied by signs or symptoms of illness.

Excretion of virus by children undergoing a primary infection has not been studied, and, in fact, little is known about the primary infection. Mäntyjärvi, et al. (6) found significant antibody rises to BKV in sera of 11 of 66 children participating in a study of respiratory infections. Ten of the 11 underwent a seroconversion. Seven children had concommitent signs of mild upper respiratory illness and fever, in 3 cases this was attributable to another virus infection. The other 4 children were asymptomatic. Since infection is a common occurrence during childhood, it is highly unlikely that it is regularly associated with a major illness. Probably most infections are subclinical or trivial, but this area requires further investigation.

IS THERE A PERSISTENT INFECTION WITH REACTIVATION?

One could argue, just by analogy with what is known about polyoma virus in mice and SV40 virus in rhesus monkeys, that infection of man by BKV and JCV leads to persistence of the viruses; however, there are some data to support this contention. It comes from serologic studies of renal transplant recipients.

Three groups (14,17, and Hogan, et al., unpublished) have measured antibody levels in patients before and after transplantation. Two of them also examined urine specimens for presence of virus. The results of the serologic tests are shown in Table 3. Patients who seroconverted after transplantation may have been experiencing a primary infection. However, many patients had antibody at the time of transplantation, and they still exhibited a significant increase in titer later. In some cases, the rise in antibody level was correlated with excretion of the corresponding virus in the urine.

Virus activity occurring in immunosuppressed persons who had measurable levels of serum antibody (evidence of previous exposure) prior to immunosuppression can best be explained by reactivation.

IS REACTIVATION OR THE POSSIBILITY THEREOF OF ANY CONCERN?

Since the primary infection is probably trivial and most adults excreting virus have no illness attributable to the virus, is reactivation of any medical concern? Our answer is yes, for three reasons.

First, JCV is clearly associated with PML a progressive, lethal brain disease that usually occurs in persons with impaired immunity, including renal transplant recipients (18). PML was once considered to be a rare disease that occurred mainly in older people. However, just since 1971, JCV has been identified in lesions of 37 cases of PML (12,19,20 and Padgett, unpublished). During 1977 we identified JCV in 6 cases of PML. The patients' ages were 5, 11, 27, 39, 47 and 51 (Padgett, et al., unpublished).

TABLE 3

HI ANTIBODY RISES IN RENAL TRANSPLANT PATIENTS

Source of Data	Virus Used	#/Total	Rises	Sero-Convert
Coleman, et al.[a]	BK	17/36	13	4
Krech, et al.[b]	BK	10/33	8	2
Hogan, et al.[c]	BK	10/48	5	3
	JC		2	5

[a]Coleman, et al., Brit. Med. J., 3:371-375, 1973.
[b]Krech, et al., Z. Immun.-Forsch., 148:341-355, 1975.
[c]Hogan, et al., unpublished.

TABLE 4
TYPES OF TUMORS INDUCED BY BK VIRUS IN HAMSTERS

Fibrosarcomas
Undifferentiated and poorly differentiated sarcomas
Ependymomas
Choroid plexus papillomas
Reticulum cell sarcoma

Second, Coleman, et al. (13) have found papovavirus virions in urine of 5 pregnant women (sample size unstated) and have identified the virus from one of them as JCV. In Japan, Taguchi, et al. (21) found anti-BKV antibodies in the IgM fractions of cord blood from three babies. The mothers had had significant rises in antibody titers to BKV during their pregnancy. Infection or reactivation during pregnancy might lead to infection of the fetus with unknown results.

Third, both of these human papovaviruses are oncogenic in some experimental animals. In newborn hamsters BKV (Table 4) induces mainly undifferentiated or fibrosarcomas and ependymomas (22,23,24,25,26,27,28). In contrast, in the same animal JCV induces a wide variety of malignant tumors (Table 5), particularly in the nervous system (29,30,31, ZuRhein, et al., unpublished). Also, both viruses can alter human cells in vitro so that they acquire many of the characteristics of transformed cells (32,33, Padgett and Walker, unpublished). Because of these facts and because large numbers of people are infected by these viruses, which are persistent and may become active whenever immunity is depressed,

TABLE 5
TYPES OF TUMORS INDUCED BY JC VIRUS IN HAMSTERS

Medulloblastoma
Undifferentiated neuroectodermal
Glioblastoma
Ependymoma
Pineocytoma
Meningioma
Neuroblastoma
Leiomyosarcoma
Hemangioma

consideration must be given to the possibility that they are involved in the etiology of human cancer. This area of investigation is receiving increasing attention. Published results (34,35,36,37) have generally been negative but they are not within the scope of this presentation. Hopefully, a definitive answer will be obtained within the next few years.

REFERENCES

1. Padgett, B. L., Rogers, C. M., and Walker, D. L. (1977). Infect. Immun. 15, 656.
2. Brown, P., Tsai, T., and Gajdusek, D. C. (1975). Am. J. Epidem. 102, 331.
3. Brown, P., and Morris, J. A. (1976). Proc. Soc. Exp. Biol. Med. 152, 130.
4. Gardner, S. D. (1973). Br. Med. J. 1, 77.
5. Padgett, B. L., and Walker, D. L. (1973). J. Infect. Dis. 127, 467.
6. Mäntyjärvi, R. A., Meurman, O. H., Vihma, L., and Berglund, B. (1973). Ann. Clin. Res. 5, 283.
7. Shah, K. V., Daniel, R. W., and Warszawski, R. (1973). J. Infect. Dis. 128, 784.
8. Portolani, M., Marzocchi, A., Barbanti-Brodano, G. and LaPlaca, M. (1974). J. Med. Microbiol. 7, 543.
9. Gardner, S. D., Field, A. M., Coleman, D. V., and Hulme, B. (1971). Lancet i, 1253.
10. Gardner, S. D. (1977). In "Recent Advances in Clinical Virology" (A. P. Waterson, ed.), Vol. I. pp. 93-115. Churchill Livingstone, Edinburgh.
11. Henry, K., Bird, R., Watson, G., and Hugh-Jones, K. (1977). Lancet i, 695.
12. Rand, K. H., Johnson, K. P., Rubinstein, L. J., Wolinsky, J.S., Penney, J. B., Walker, D. L., Padgett, B. L.,and Merigan, T. C. (1977). Ann. Neurol. 1, 458.
13. Coleman, D. V., Daniel, R. A., Gardner, S. D., Field, A.M. and Gibson, P. E. (1977). Lancet ii, 709.
14. Coleman, D. V., Gardner, S. D., and Field, A. M. (1973). Br. Med. J. 3, 371.
15. Lecatsas, G., Prozesky, O. W., Van Wyk, J., and Els, H. J. (1973). Nature, Lond. 241, 343.
16. Lecatsas, G., Schoub, B. D., Rabson, A. R., and Joffe, M. (1976). Lancet ii, 907.
17. Krech, U., Jung, M., Price, P. C., Thiel, G., Sege, D., and Reutter F. (1975). Z. Immun.-Forsch. 148, 341.
18. Walker, D. L. (1978). In "Handbook of Clinical Neurology" (P. J. Vinken and G. W. Bruyn, eds.), Vol. 34. North-Holland Publ. Co., Amsterdam, in press.
19. Padgett, B. L., and Walker, D. L. (1976). Prog. Med. Virol. 22, 1.

20. Padgett, B. L., Walker, D. L., ZuRhein, G. M., Hodach, A. E., and Chou, S. M. (1976). J. Infect. Dis. 133, 686.
21. Taguchi, F., Nagaki, D., Saito, M., Haruyama, C., Iwasaki, K., and Suzuki, T. (1975). Japan. J. Microbiol. 19, 395.
22. Shah, K. V., Daniel, R. W., and Strandberg, J. D. (1975). J. Natl. Cancer Inst. 54, 945.
23. Näse, L. M., Kärkkäinen, M., and Mäntyjärvi, R. A. (1975). Acta Path. Microbiol. Scand. 83, 347.
24. Van der Noordaa, J. (1976). J. Gen. Virol. 30, 371.
25. Costa, J., Yee, C., Tralka, T. S., and Rabson, A. S. (1976). J. Natl. Cancer Inst. 56, 863.
26. Dougherty, R. M. (1976). J. Natl. Cancer Inst. 57, 395.
27. Corallini, A., Barbanti-Brodano, G., Bortoloni, W., Nenci, I., Cassai, E., Tampieri, M., Portolani, M., and Borgatti, M. (1977). J. Natl. Cancer Inst. 59, 1561.
28. Greenlee, J. E., Narayan, O., Johnson, R. T., and Herndon, R. M. (1977). Lab. Invest. 36, 636.
29. Walker, D. L., Padgett, B. L., ZuRhein, G. M., Albert, A. E., and Marsh, R. F. (1973). Science 181, 674.
30. Varakis, J. N., and ZuRhein, G. M. (1976). Acta Neuropathol. 35, 243.
31. Varakis, J. N., ZuRhein, G. M., Padgett, B. L., and Walker, D. L. (1976). J. Neuropath. Exp. Neurol. 35, 314.
32. Shah, K. V., Hudson, C., Valis, J., and Strandberg, J. D. (1976). Proc. Soc. Exp. Biol. Med. 153, 180.
33. Fareed, G. C., Takemoto, K. K., and Gimbrone, M. A., Jr. (1978). In "Microbiology - 1978" (D. Schlessinger, ed.) American Society for Microbiology, Washington.
34. Corallini, A., Barbanti-Brodano, G., Portolani, M., Balboni, P. G., Grossi, M. P., Possati, L., Honorati, C., LaPlaca, M., Mazzoni, A., Caputo, A., Veronesi, U., Orefice, S., and Cardinali, G. (1976). Infect. Immun. 13, 1684.
35. Costa, J., Yee, C., and Rabson, A. S. (1977). Lancet ii, 709.
36. Becker, L. E., Narayan, O., and Johnson, R. T. (1976). J. Canad. Sci. Neurol. 3, 105.
37. Fiori, M., and Di Mayorca, G. (1976). Proc. Natl. Acad. Sci. 73, 1662.

MEASLES VIRUS PERSISTENT INFECTIONS- BIOLOGICAL AND BIOCHEMICAL STUDIES[1]

Edward C. Hayes and Hans J. Zweerink

Department of Microbiology and Immunology
Duke University Medical Center, Durham, N. C. 27710

ABSTRACT A HEp-2 cell line persistently infected with measles virus was established. All cells in the culture contained viral antigens and produced temperature-sensitive infectious virus with a thousand-fold reduction in yield compared to lytically infected HEp-2 cells. Total particle production of persistently infected cells was approximately 50% of that of lytically infected cells. The former particles contained an excess of the presumed viral matrix polypeptide and reduced amounts of polypeptide P4. The synthesis of viral polypeptides in infected cells was also determined. Lytically infected Vero cells contained high levels of two non-structural polypeptides. Persistently infected HEp-2 cells accumulated more nucleocapsids than lytically infected HEp-2 cells.

INTRODUCTION

Over the last few years it has been recognized that measles virus may be involved in prolonged degenerative diseases of the central nervous system such as subacute sclerosing panencephalitis and multiple sclerosis (for a review see Ref. 1). The protracted disease pattern reflects a chronic infection, similar to persistent infections that can be established *in vitro* with measles virus. Studies of the mechanisms by which measles virus persistent infections are established and maintained *in vitro* are of obvious importance for an understanding of the possible role of this virus in chronic diseases.

This report presents a comparison between cells persistently and lytically infected with measles virus; in particular the nature of the viral particles released by these cells, and the viral polypeptides present in their nucleus and cytoplasm.

[1]This work was supported by PHS Grant No. A1-10132, T32-CA-0911 and K04-NS-0086.

ISBN 0-12-668350-6

RESULTS

The Establishment and General Properties of HEp-2 Cell Cultures Persistently Infected With Measles Virus. Monolayer cultures of HEp-2 cells were infected with the plaque-purified Edmonston vaccine strain of measles virus. A few cells survived and readily established a culture (HEp-2-PI) that has been passaged more than 250 times at three-day intervals. Measles specific antigens were detected by fluorescent antibody staining in more than 90% of these cells; approximately 10^3 plaque-forming units (PFU) per 10^6 cells were released in the medium over a 48 hour period, compared to more than 10^6 PFU per 10^6 lytically infected HEp-2 cells (HEp-2-Lytic). By infectious center assay, it was shown that all cells in the HEp-2-PI culture were capable of producing infectious viral particles.

Temperature-Sensitive and Interfering Particles. At 39.5^o lytically produced measles virus plaqued with 65% efficiency (compared to 37°) whereas virus produced by HEp-2-PI cultures plaqued only with 3% efficiency. Neither virus plaqued at 41°. Incubation of HEp-2-Lytic and HEp-2-PI cultures at 39.5^o reduced the viral yield by 66% and 88% respectively. After two subsequent passages of HEp-2-PI cells at 39.5^o infectious virus and viral antigens had disappeared completely. On return to 37° they remained free of infectious virus and viral antigens for at least 5 passages.

The presence of interfering viral particles in HEp-2-PI cultures was strongly suggested by the finding that supernatants inhibited the replication of prototype measles virus in HEp-2 or Vero cells by 60% or more. Treatment at pH 2 eliminated interfering activity, and centrifugation at 100,000 x g for 1 hour pelleted the interfering activity. Thus interference is not caused by soluble substances such as interferon.

Particles Released by Persistently and Lytically Infected Cultures. Infected cells were exposed to ^{35}S-methionine for 24 hours and viral particles were purified from the supernatants and polypeptides were analyzed by SDS-PAGE (Figure 1).

Each viral preparation contained one major host polypeptide (actin). Extensive cytopathology in lytically infected HEp-2 cultures resulted in viral preparations containing many other host polypeptides. Six major viral polypeptides were identified. One was associated with the surface glycoprotein (P1 or G, mol. wt. 74-76,000), and one with another surface component, the fusion protein (P5 or F_1, mol. wt. 45,500); one with the nucleocapsid (P3 or NC, mol. wt. 59,000) and one presumably with the matrix protein (P6 or M, mol. wt. 35,000).

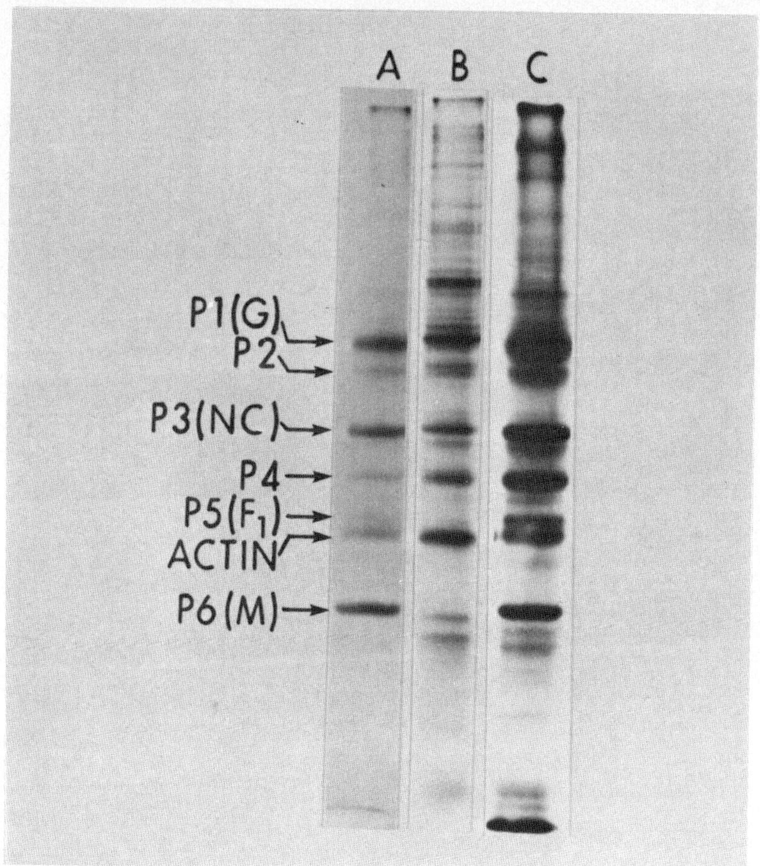

FIGURE 1. Virion Polypeptides. HEp-2 and Vero cells were infected at a multiplicity of 0.01 and labeled between 24 and 48 after infection with ^{35}S-methionine (25 μCi/ml) in MEM containing 0.45 mg per liter methionine and 5% dialyzed fetal calf serum. An actively growing culture of HEp-2-PI cells was labeled under the same conditions. The culture supernatants were harvested, cell debris was removed (1000 x g for 5 min.) and virions pelleted through 40% sucrose [w/v in TE buffer (0.01 M Tris pH 7.2; 0.002 M EDTA)] onto a 70% sucrose cushion. They were centrifuged to equilibrium in a 15-40% (w/w) potassium tartrate gradient. Virions, banding at approximately 24% tartrate were analyzed on 10% discontinuous SDS-polyacrylamide gels (2) and polypeptides were visualized by fluorography (3). A, B and C: particles derived from HEp-2-PI, HEp-2-Lytic and Vero-Lytic cultures respectively.

The function of the remaining two polypeptides (P2 and P4, mol. wt. 70,000 and 51,000 respectively) is not known. These results agree with those of others (4-6). There were no qualitative differences among particles released from HEp-2-PI, HEp-2-Lytic and Vero-Lytic cells in regard to the number of viral polypeptides and their molecular weights. However, differences in the relative amounts of certain polypeptides were observed (Table 1). HEp-2-PI particles contained much more P6 but much less P4 than HEp-2-Lytic and Vero-Lytic particles.

If one assumes that the amounts of label incorporated into virions is an estimate of the relative number of particles, then one can calculate that HEp-2-PI cells produce approximately one-half the number of particles compared to lytically infected HEp-2 cells. However the number of infectious viral particles released from HEp-2-PI cells is approximately 1000-fold less than from HEp-2-lytic cultures. The large number of non-infectious HEp-2-PI particles may represent interfering particles that were discussed above.

Intracellular Virus-Specific Polypeptides. Measles virus specific serum (human convalescent) was used to precipitate ^{35}S-methionine labeled viral polypeptides from the cytoplasm and nuclei of HEp-2-PI, HEp-2-Lytic and Vero-Lytic cells, as well as uninfected HEp-2 and Vero cells. The immunoprecipitates were analyzed by SDS-PAGE.

TABLE 1

RELATIVE AMOUNTS OF VIRION POLYPEPTIDES IN PARTICLES PRODUCED BY THREE TYPES OF INFECTED CELLS [a]

Polypeptides	HEp-2-PI	HEp-2-Lytic	Vero Lytic
P1 (G)	30	34	36
P2	5	8	10
P3 (NC)	28	21	23
P4	3	20	15
P5 (F_1)	3	10	7
P6 (M)	30	7	9

[a]The relative amounts of the polypeptides in the virions was determined by scanning the gels in Figure 1 and by integrating the areas under the peaks with a PDP-11 computer. They were expressed as a percentage or the total viral polypeptides in each gel.

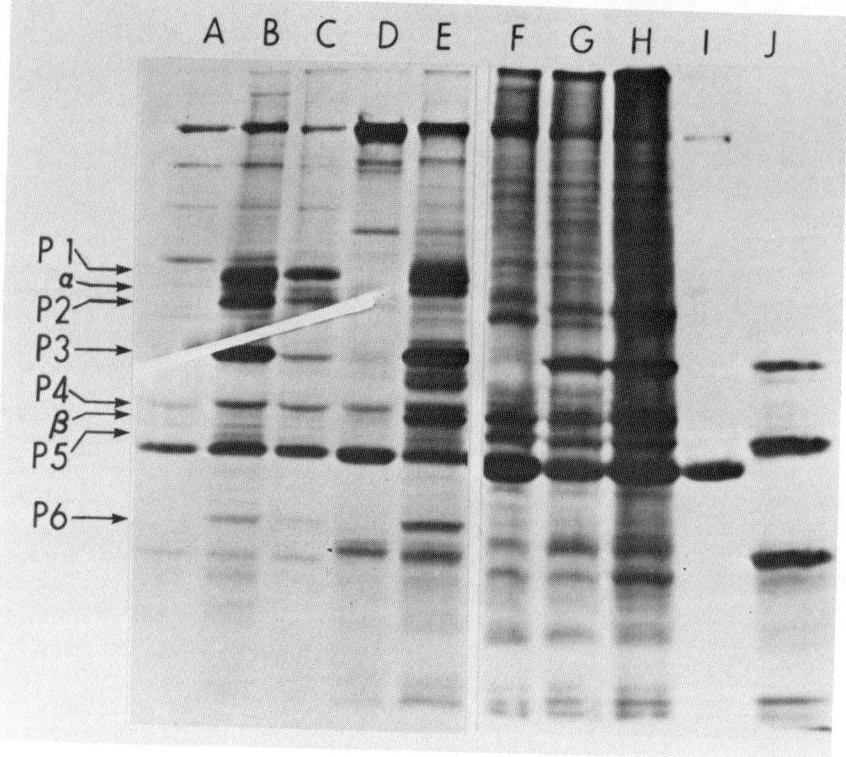

FIGURE 2. Measles virus polypeptides in infected cells. ^{35}S-Methionine-labeled cells (see Figure 1) were removed from the surface with 0.02% EDTA and washed with NET buffer (0.15 M NaCl; 0.005 M EDTA; 0.05 M Tris pH 7.2). Samples of 10^6 cells were incubated for 15 minutes at 22° in 200 μl NET buffer and 1% Triton-X100. Nuclei were separated from the cytoplasmic extract by centrifugation, washed once in NET buffer and re-suspended in 20 μl NET buffer and 0.1% SDS. After sonication they were diluted to 200 μl with NET buffer. Samples (10 μl) of these extracts were incubated at 22° for 1 hour with an excess of measles virus specific human convalescent serum (1 μl) in the presence of an extract from 2 x 10^5 unlabeled uninfected homologous cells (in 20 μl). After adsorption of the antigen-antibody complexes to a 20 μl column of protein A-Sepharose beads and extensive washing, they were eluted with SDS buffer and analyzed on gels (see Figure 1). A-E, cytoplasmic extracts; F-J nuclear extracts. A and F, uninfected HEp-2; B and G, HEp-2-PI; C and H, HEp-2-Lytic; D and I, uninfected Vero; E and J, Vero-Lytic.

Figure 2 shows three classes of polypeptides in cellular extracts. 1) Viral structural polypeptides G,P2,NC,P4,F_1 and M (see Figure 1) are present in all extracts from infected cells. 2) Host-specific polypeptides were identified by comparing precipitates from uninfected and infected cells; actin is one of the major host-contaminants, the other one a polypeptide of high molecular weight (approx. 200,000). 3) Non-structural viral polypeptides. A few polypeptide species with migration rates distinctly different from the viral structural polypeptides were present in infected cell extracts but not in uninfected cell extracts. They are labeled α and β. In the nuclear extracts the background of host-specific polypeptides was more pronounced and it was difficult to discern non-structural viral polypeptides.

Major differences were observed between cytoplasmic extracts of the HEp-2 (PI and Lytic) and Vero cells: polypeptide P2 was absent in Vero cells, and significant amounts of the non-structural polypeptides α and β were present. Because of these differences a quantitative comparison of Vero and HEp-2 cells is difficult. The most striking difference between cytoplasmic extracts of HEp-2-PI and HEp-2-Lytic cells (see Table 2) is the low level of nucleocapsid polypeptide in

TABLE 2

RELATIVE AMOUNTS OF VIRAL POLYPEPTIDES IN IMMUNE PRECIPITATES OF CYLOPLASMIC EXTRACTS OF HEp-2-PI, HEp-2-LYTIC AND VERO-LYTIC CELLS [a]

Polypeptides	HEp-2-PI	HEp-2-Lytic	Vero Lytic
P1 (G)	37	43	32
α	<1	<1	9
P2	19	27	<1
P3 (NP)	32	13	25
P4	7	13	10
β	<1	<1	15
P5 (F_1)	1-2	1-2	1-2
P6 (M)	4	2	6

[a]The relative amounts of polypeptides in the virions was determined as discussed in the legend to Table 1.

HEp-2-Lytic cells (13% versus 32% in HEp-2-PI). HEp-2-PI and HEp-2-Lytic nuclear extracts gave nearly identical polypeptide patterns: polypeptides P2, NC and P4 were present in relatively similar amounts. (Thus nucleocapsids did not accumulate in nuclei of HEp-2-Lytic cells). The nuclear extract from Vero-Lytic cells was different in that P2 was absent (as in the cytoplasmic extract) and NC and P5 were present. In addition, a polypeptide that migrated slightly faster than the M-polypeptide was also present. This nuclear polypeptide pattern needs to be viewed with caution because actin was present in the immune precipitate of the nuclei of uninfected Vero cells but not in that of infected cells. An artifact such as proteolytic activity during the preparation of extracts can not be ruled out.

DISCUSSION

Lytically and persistently infected HEp-2 cells show little difference in regard to viral polypeptides that accumulate during a 24-hour labeling period. On the other hand there is a difference in the polypeptide composition of the particles released by the two cultures. HEp-2-PI particles contain significantly more of the matrix polypeptide (P6) but less of P4. This would suggest a possible defect at the level of virion assembly in HEp-2-PI cells. However, the validity of such conclusion depends on factors that have not been investigated in sufficient detail. The ratio of infectious to non-infectious viral particles (complete or incomplete) that are produced during lytic infections is not known. Furthermore, the purity of the viral preparations is difficult to ascertain.

REFERENCES

1. Morgan, E. M. and Rapp, F. (1977). Bact. Revs. 41, 636.
2. Maizel, J. V. (1971). In "Methods in Virology", (K. Maramorosh and H. Koprowski: eds.) Academic Press, New York, Vol. 5, pp. 179-246.
3. Bonner, W. M., and Laskey, W. A. (1974). Eur. J. Biochem. 46, 83.
4. Mountcastle, W. E., and Choppin, P. W. (1977). Virology 73, 463.
5. Wechsler, S. L., and Fields, B. N. (1978). J. Virol. 25, 285.
6. Hardwick, J. M., and Bussell, R. H. (1978). J. Virol. 25, 687.

PERSISTENCE AND TRANSFORMATION BY HUMAN CYTOMEGALOVIRUS[1]

Fred Rapp and Laszlo Geder

Department of Microbiology and
Specialized Cancer Research Center
The Milton S. Hershey Medical Center
The Pennsylvania State University College of Medicine
Hershey, Pennsylvania 17033

ABSTRACT Infection of human embryo lung fibroblasts with a freshly isolated human cytomegalovirus resulted in persistent infection and subsequent transformants. Cytomegalovirus-specific antigens were detected in these cells by indirect and anti-complement immunofluorescence techniques. The transformed cells formed colonies in soft agarose and induced tumors in athymic nude mice receiving subcutaneous transplants. Local invasion of adjacent structures was rare. Cytomegalovirus-related antigens were again detected by indirect and anti-complement immunofluorescence techniques in cells cultured in vitro from the tumors and karyotypic analysis confirmed their human origin. Neither virus DNA nor virus proteins were detected in transformed cells superinfected with cytomegalovirus. Herpes simplex virus production in cytomegalovirus-transformed human embryo lung cells was delayed when compared to virus production in normal human embryo lung cells and the spread of cytopathology was slower in the transformed cells. After prolonged in vitro cultivation, the transformed human embryo lung cells yielded a double-enveloped herpesvirus designated HMCV. The growth characteristics of HMCV were consistent with the cell specificity of herpes simplex virus but the cytopathology induced was similar to herpes simplex virus or cytomegalovirus, depending on the host cell. Neutralization and immunofluorescence tests revealed relatedness to cytomegalovirus, while molecular studies of virus DNA (molecular weight, buoyant density, and RNA-DNA hybridization) revealed limited relatedness to herpes simplex viruses.

[1]This work was supported by Contract #NO1-CP-5-3516 within the Virus Cancer Program of the National Cancer Institute and by Grants CA 18450 and CA 16365 awarded by the National Cancer Institute.

ISBN 0-12-668350-6

INTRODUCTION

Previous studies (1, 2) have shown that some properties of human cytomegalovirus (CMV) are commonly associated with certain properties of known oncogenic viruses (i.e., stimulation of host cell DNA and RNA synthesis). Oncogenic transformation of hamster embryo fibroblasts (HEF) by ultraviolet (UV)-inactivated CMV has also been achieved (3). We also previously noted that normal human prostate cells which had been infected with a human CMV *in vivo*, developed long-term persistent and then latent CMV-infection *in vitro*, and grew to passage levels well above those routinely attained by normal cells (4). During the latent period, virus was no longer rescuable, although CMV-specific antigens and nucleic acids continued to be detected in the cells. This observation suggested that persistent infection of human cells with an adequate "live" CMV preparation might result in transformation. Therefore, further studies of the transforming ability of human cells by non-inactivated CMV strains were undertaken.

METHODS

Cells. Human embryo lung (HEL) cells were transformed with a genital strain (Mj) of human CMV (5, 6). The tumor cell cultures were prepared and maintained as published elsewhere (6) and HEL cells were obtained from HEM Research, Inc., Rockville, Md. and maintained as before (4). Primary human embryo kidney (HEK), primary human amnion (HA), the RK-13 established rabbit kidney cell line, and Flow 5000 human embryo cell cultures were obtained from Flow Laboratories and maintained in Ham's medium with fetal calf serum (10%), sodium bicarbonate (0.075%), penicillin (100 IU/ml), and streptomycin (100 μg/ml). Normal human epithelioid kidney (HK) and human epithelioid kidney cancer (HKC) cell cultures were prepared in our laboratory as described earlier (7) and were in passages 1-4 at the time of the experiments. Mouse embryo fibroblast (MEF), HEF, and rabbit kidney (RK) cell cultures were prepared in our laboratory as primary cell cultures and Vero monkey and 3T3 mouse cell lines were maintained using Dulbecco's medium with 5% calf serum, 0.075% sodium bicarbonate, and antibiotics.

Viruses. CMV laboratory strains AD169 and C87 (3, 8), CMV-Mj (4) and CMV-K9V (9) were used. CMV cervical isolate "Walsh" was kindly supplied by T. Weller.

Animals. Weanling athymic homozygous (nu/nu) nude mice (NIH Swiss Webster, sixth backcross generation) and NIH Swiss heterozygous mice were obtained from Life Sciences, St. Petersburg, Fl. Inbred LSH hamsters were obtained from Lakeview Hamster Colony, New Field, N.J. and Dla:(NZW) female rabbits were obtained from Dutchland Lab Animals, Inc., Denver, Pa. Transplantation to athymic weanling nude mice, autopsy, and histological examination were carried out as described previously (6).

Immune Sera and Serologic Tests. Human antisera to CMV and herpes simplex virus (HSV) were obtained from hospital patients, and CMV- and HSV-hyperimmune sera were obtained from R. Lausch (6, 10). Anti-complement and indirect immunofluorescence (IF), virus neutralization and microcytotoxicity tests were carried out as published elsewhere (3, 6, 11). Virus rescue experiments have been described (5, 6). Band velocity sedimentation, buoyant density centrifugation, restriction enzyme cleavage, and RNA-DNA hybridization were carried out as described in a series of papers (12-20).

RESULTS

Persistence and Transformation by Cytomegalovirus Isolates. HEL, normal human prostatic fibroblasts (HPF), HEK, and primary HA cell cultures were inoculated at a low multiplicity of infection (m.o.i. of 0.3-0.03 plaque-forming units [PFU]/cell) with non-inactivated laboratory CMV strains AD169 and C87. Short-term persistent infection of these cells was achieved. The virus-yielding cells could not be maintained longer than 5 *in vitro* passages. Long-term persistent infection of HEL cell cultures was established with a new cervical CMV isolate (Walsh) and with the K9V strain isolated from a patient with Kaposi sarcoma. Cell transformants were not observed within a period of 26 weeks (Table 1).

Long-term persistent infection (6 and 26 weeks) of HEL cells with the prostatic CMV strain Mj resulted in subsequent transformation of these cells (Fig. 1). Two cell lines (CMV-Mj-HEL-1 and CMV-Mj-HEL-2) have been developed which appear as short fibroblastoid and epithelioid cells. CMV-specific membrane antigens have been detected in these cells in early *in vitro* passages by indirect IF techniques, and by lymphocyte microcytotoxicity assays (5).

TABLE 1

TRANSFORMATION EXPERIMENTS WITH HUMAN CYTOMEGALOVIRUS ISOLATES

CMV isolates	Target cells	Method	Results
AD169	HEL[a], HPF, HEK, HA	live virus	Persistent infection (4)[b]
C87	HEL, HPF HEK, HA	live virus	Persistent infection (5)
Walsh	HEL	live virus	Persistent infection (15-24)
K9V	HEL	live virus	Persistent infection (15-24)
Mj	HEL	live virus	Persistent infection
	HEF	DEAE dextran + live virus	followed by transformation

[a]HEL, human embryo lung fibroblasts; HPF, human prostatic fibroblasts; HEK, human embryo kidney; HA, human amnion; HEF, hamster embryo fibroblasts.

[b]Number of *in vitro* passages with persistently infected cells.

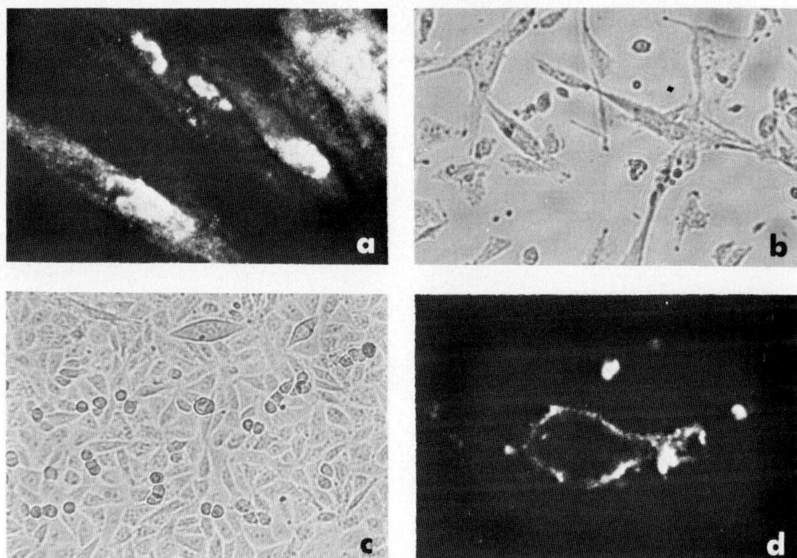

FIGURE 1. a) CMV-antigens in persistently infected HEL cells detected by indirect IF test (X1250). b) Early-phase growth of transformed HEL cell population (X500). c) Unstained photomicrograph of CMV-Mj-HEL-2-transformed cells (X500). d) CMV-related membrane antigens in CMV-transformed HEL cells detected by indirect IF test (X1250).

Properties of CMV-Transformed HEL Cells: Oncogenicity in Athymic Nude Mice and Resistance to Superinfection with Herpesviruses. Cells in early passages (2.5×10^7 CMV-Mj-HEL-2) induced tumors in 47% of the athymic nude mice receiving subcutaneous transplants (Fig. 2). The tumors were composed of poorly differentiated, small, polygonal cells with large nuclei and scanty cytoplasm embedded in an abundant collagenous matrix. Local invasion of adjacent structures was rare. No tumors were observed in nude mice inoculated with control HEL cells. The tumor cells cultured *in vitro* were small and epithelioid. Concentration densities of approximately 10^6 cells/cm^2 were reached with transformed cells, while normal HEL cells grew to densities of 4×10^5 cells/cm^2. CMV-related antigens were detected by indirect IF and lymphocyte microcytotoxicity assays in cells cultured from the tumors *in vitro*. No infectious virus could be rescued from either transformed or tumor cells in early passages by cocultivation with human embryo fibroblasts, treatment with iododeoxyuridine (IUdR), or by UV irradiation (6).

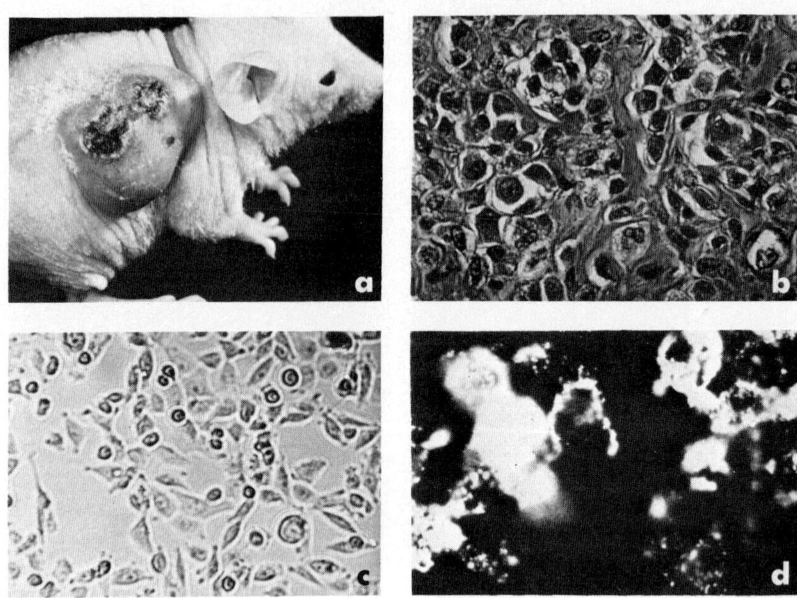

FIGURE 2. a) Athymic nude mouse with tumor induced by CMV-transformed HEL cells. b) Histology of tumor; stained with hematoxylin and eosin (X500). c) *In vitro* culture of CMV-Mj-HEL-2,T-1 tumor cells (X500). d) CMV-related membrane antigens in tumor cells detected by indirect IF test (X1250).

The CMV-transformed HEL cell lines were completely resistant to superinfection with CMV-strains. Both control HEL and CMV-Mj-HEL-1 cells were infected at an m.o.i. of 5. The lack of growth of AD169 in the transformed cells is shown in Fig. 3.

Coverslip cultures of CMV-Mj-HEL-1 cells demonstrated no virus-specific antigens by indirect IF tests 48 hr after infection with strain AD169, although CMV antigens were detected in the virus-infected control HEL cells. Virus DNA was not synthesized in the transformed cells (Fig. 4). However, stimulation of cell DNA synthesis in the transformed cultures infected with CMV was observed (Fig. 4).

Growth of HSV-2 (m.o.i. of 5) in control HEL cells and CMV-transformed cell lines was delayed by 4 hr in the transformed cell lines; otherwise, virus production appeared to be normal when compared to control cells. Using a low m.o.i. (0.03 PFU/cell) the delay in the yield of infectious virus was more pronounced. At 24 hr after inoculation, virus titers in the transformed cells were 1-3 logs lower than in HEL cells.

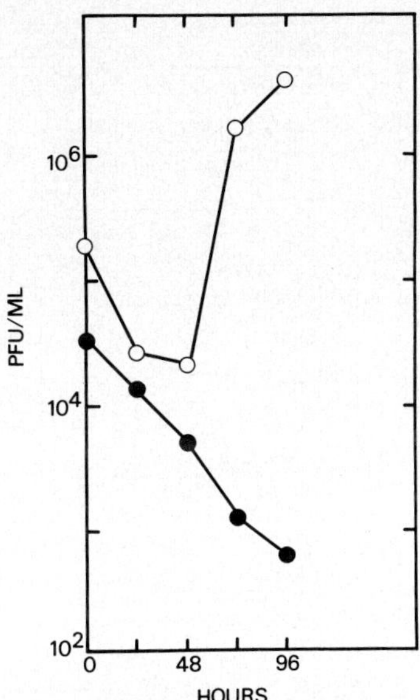

FIGURE 3. Replication of CMV AD169 in HEL ○--○, and in CMV-Mj-HEL-1 cells ●--● inoculated at an m.o.i. of 5.

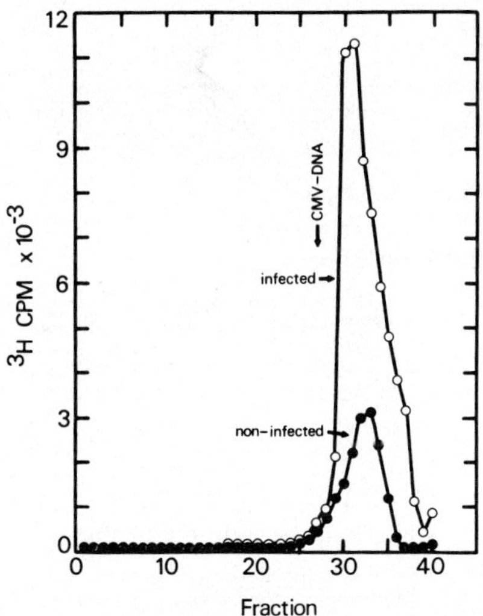

FIGURE 4. Buoyant density in CsCl of DNA extracted from CMV-Mj-HEL-2,T-1 cells infected with AD169 at an m.o.i. of 5 ○--○, DNA from noninfected cells ●--●. An internal marker (K. pneumoniae DNA, density 1.717) was included in the gradient to indicate the region where CMV DNA bands.

Alterations in Properties of CMV-Transformed Cell Lines During Prolonged In Vitro Cultivation. One of the original transformed cell lines (CMV-Mj-HEL-2) and one of the tumor lines derived from a neoplasm of an athymic nude mouse inoculated with CMV-Mj-HEL-2 cells, have shown diverse alterations during one year of *in vitro* cultivation. CMV-Mj-HEL-2 cells (2.5 x 10^7) in different passages were transplanted to 150 athymic nude mice 2-4 weeks after birth. Cells in early passages (1-20) produced tumors in 47% of the mice. At passages >20, tumorigenicity increased to 100%. The latent period for tumor development was an average of 15 days for cells in low passages, and gradually decreased to 6 days by passages 193 and higher. Inoculation with 2.5 x 10^7 tumor cells (CMV-Mj-HEL-2,T-1) grown *in vitro* between passages 1-20 was only moderately oncogenic; only 2 of 12 nude mice developed tumors after a latent period of 28 days. The tumorigenicity then decreased to undetectable levels after 20 passages. At passage 146, however, oncogenicity was again observed; 66% of the transplanted nude mice developed tumors after a latent period of 40 days. The average latent period was 22 days (10).

With increasing passage levels, the percentage of CMV-Mj-HEL-2 cells expressing CMV-related antigens gradually decreased from 40-20% by passages 28-40, and to zero by passage 60. The CMV-Mj-HEL-2,T-1 cells maintained a relatively constant 50% antigenic expression level. The percentage decreased, however, to 0-7% at passage numbers 146 and above and a concomitant increase in tumorigenicity was observed. Although the tumor cell line regained its tumor inducing potential by high *in vitro* passages, the oncogenic potential of these cells was less than that of the parental cell line. The lowest cell count of the parental cell line to induce tumor development in 100% of the inoculated mice was 10^7 cells and the length of the latent period was an average of 25 days. Sixty days and 2.5 x 10^7 cells were necessary to achieve 100% tumor development when the tumor cells were used. Oncogenicity was increased by higher cell doses, but transplantation of even 4 x 10^7 cells gave inconsistent results with varying lengths of latent period and frequency of tumor development.

No karyotypic change was observed in the parental cell line during one year of *in vitro* cultivation. The tumor cells, which had karyotypic markers identical to the parent line at low passages, consisted of a mixed population of cells by passage 146. One-half of the cell population had an average chromosome number of 47-48. Others developed polyploidy with an average chromosome number of 90-93.

Induction of Immune Response in Athymic Nude Mice. Three of six pooled serum samples taken 3-4 weeks after the first sign of tumor development in nude mice inoculated with CMV-Mj-HEL-2 cells had IF antibodies to CMV-infected and CMV-transformed HEL cells. One of the six pooled sera also demonstrated neutralizing antibodies to CMV. One of two pooled sera from nude mice inoculated with CMV-Mj-HEL-2,T-1 cells showed IF antibodies to antigens in CMV-infected cells and low titers of neutralizing antibodies to CMV.

Analogous cells were inoculated into nude mice bearing CMV-Mj-HEL-2 tumors 2 weeks after the first signs of tumor development. Fewer animals developed secondary tumors from CMV-Mj-HEL-2 and CMV-Mj-HEL-2,T-1 cells compared to control mice infected with normal HEL cells prior to the challenge doses (10). This phenomenon is similar to the concomitant immunity observed in immunologically competent inbred animals.

Virus Induction. Infectious virus was not released and could not be induced from the parental cell lines during 228 *in vitro* passages. The CMV-Mj-HEL-2,T-1 tumor line, however, spontaneously released an infectious virus on three separate occasions at passages 90, 94, and 117. The affinity of the virus to different cell cultures was as broad as that of HSV. Restricted growth has been observed, however, in adult HK, adult HKC, adult HE, and human embryo fibroblast cell cultures. In HEL cells, the growth pattern and cytopathic effects were similar to those seen with CMV (Table 2).

The virus (HMCV) is a double-enveloped herpesvirus that is sensitive to ether and inhibited by IUdR. No antigenic relationship to HSV was detected using HSV-immune sera in neutralization (Table 3) and IF tests although these tests revealed CMV-related antigenicity. The virus did not induce clinically manifested disease in newborn and weanling heterozygous mice, athymic nude mice, newborn and weanling hamsters, or rabbits. Pooled normal rabbit, hamster, mouse, guinea pig, African green monkey, and Rhesus monkey sera had no neutralizing activity against the isolate.

In neutral sucrose gradients (Fig. 5a), HMCV DNA sedimented just behind T4 DNA. Using the Burgi-Hershey equation the molecular weight of HMCV DNA was calculated to be 96 x 10^6 daltons. In alkaline sucrose gradients (Fig. 5b), HMCV DNA sedimented very heterogeneously; the largest size sedimented just behind T4 DNA. Measured by equilibrium centrifugation in CsCl using HSV-2 DNA as a standard, HMCV DNA had a higher buoyant density than HSV-2 DNA (1.729 g/cc) and was far removed from the buoyant position of CMV DNA (1.717

TABLE 2

CYTOPATHOLOGY AND MAXIMUM VIRUS YIELD IN CELL CULTURES OF HUMAN AND ANIMAL ORIGIN INOCULATED WITH HMCV

Cells	Type of CPE	Maximum virus yield PFU/ml
HEL	CMV-like	5.0×10^4
HEK	HSV-like, syncytial	7.0×10^5
HKC	HSV-like, syncytial	5.0×10^4
HE	HSV-like, rounded cells	5.0×10^4
HA	HSV-like, rounded cells	1.1×10^6
Vero	HSV-like, rounded cells	1.3×10^6
MEF	CMV-like	1.2×10^4
3T3	No CPE	1.0×10^0
HEF	HSV-like, rounded cells	3.0×10^6
RK	HSV-like, rounded cells	2.7×10^7
RK-13	HSV-like, rounded cells	5.0×10^6
CMV-Mj-HEL-2	Restricted, rounded cells	6.0×10^3
CMV-Mj-HEL-2,T-1	Restricted, rounded cells	1.3×10^3

TABLE 3

NEUTRALIZATION TESTS WITH HMCV FROM RK CELLS

	% plaque reduction with serum compared to control without serum							
Virus	Pre-CMV immune rabbit	Anti-CMV AD169 rabbit	Anti-CMV-Mj rabbit	Pre-HMCV immune rabbit	Anti-HMCV rabbit	Anti-HSV-1 rabbit	Anti-HSV-2 hamster	Anti-HSV human
HMCV	0	53	56	0	87	0	13	0
CMV-Mj	0	55	99	0	0	0	0	0
HSV-2	0	ND[a]	ND	ND	0	99	100	100

[a]ND, not done.

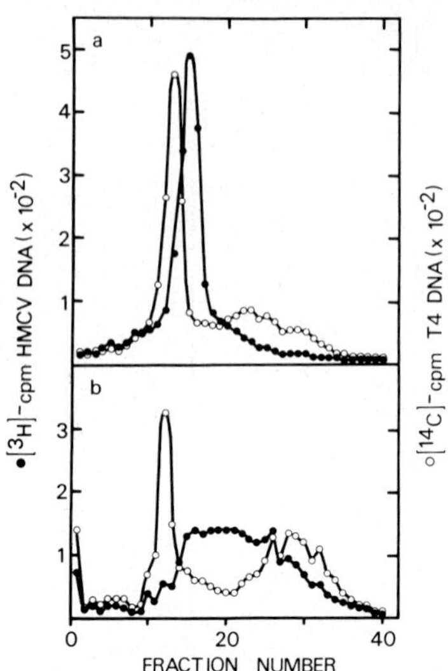

FIGURE 5. Sucrose gradient sedimentation of HMCV DNA. [^{3}H]TdR-labeled HMCV DNA was isolated from the Hirt supernatant and purified by glycerol gradient sedimentation. The HMCV DNA was dialyzed and concentrated. An aliquot of [^{3}H]HMCV DNA (●--●) was mixed with [^{14}C]T4 DNA (○--○). Half the mixture was sedimented through a neutral sucrose gradient (a) and the other half through an alkaline sucrose gradient (b). Sedimentation is from right to left.

g/cc) (Fig. 6). To measure the buoyant density of HMCV DNA, ^{14}C-DNA was mixed with ^{3}H-labeled HSV-2 DNA (1.729 g/cc), *K. pneumoniae* DNA (1.717 g/cc), and cellular DNA (1.695 g/cc). The three DNAs of known buoyant density were used to plot a line of density versus distance (data not shown). From this standard curve, the buoyant density of HMCV DNA was 1.730 g/cc and using the equation of Schildkraut *et al.* (21) the guanine plus cytosine content of HMCV DNA was calculated to be 71.5%.

The restriction enzyme cleavage pattern of HMCV DNA was determined and directly compared to the patterns of CMV, HSV-1, and HSV-2 DNAs. Each of the four ^{32}P-labeled DNAs was cleaved by endonuclease *EcoR$_I$* and placed in a separate well in an agarose slab gel. Following electrophoresis, the gel was dried, placed on X-ray film, and autoradiographs were developed after appropriate time intervals. Figure 7 shows an example of an autoradiograph of the *EcoR$_I$*-cleaved DNA; the cleavage patterns for CMV, HSV-1, and HSV-2 are

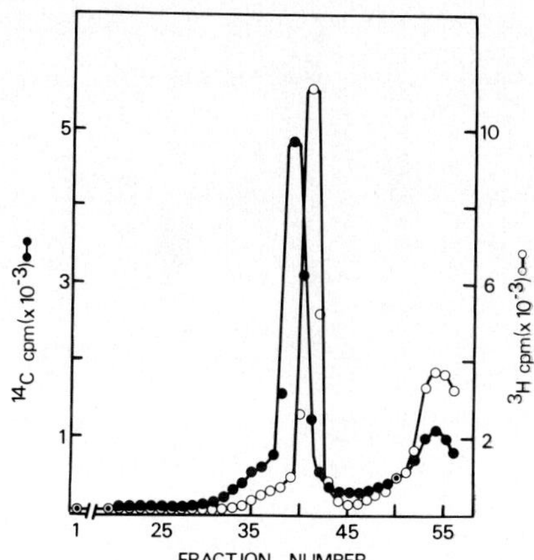

FIGURE 6. Buoyant density centrifugation of HMCV DNA. ^{14}C-labeled HMCV DNA (●--●) was isolated from the Hirt supernatant as was ^{3}H-labeled HSV-2 DNA (○--○). The Hirt supernatants also contained a small amount of host cell DNA. The DNAs were mixed and sedimented isopycnically in a CsCl gradient. At equilibrium, fractions were collected from the bottom. The meniscus is on the right. Density increases to the left. For convenience, the bottom twenty fractions were not graphed.

indistinguishable from previously published patterns (15, 22-28). As clearly seen in figure 7, the $\underline{EcoR}_I$ cleavage pattern of HMCV DNA is unique and totally different from the $\underline{EcoR}_I$ cleavage patterns of the other three DNAs.

The extent of base sequence homology among HMCV, HSV-1, HSV-2, and CMV DNAs was measured as a first approximation using RNA-DNA hybridization. ^{32}P-labeled HMCV RNA, isolated as total RNA in the HMCV-infected cell half-way through the lytic cycle, was hybridized to each of the four virus DNAs, denatured, and immobilized on nitrocellulose filters. Calf thymus DNA and blank filters served as the negative controls; HMCV DNA served as the positive control. Filters were used in triplicate. The virus DNAs were labeled with either [^{3}H] or [^{14}C] thymidine so that loss (if any) of virus DNA from the filter during hybridization was measured and appropriate corrections made. The hybridization experiment was repeated three times with similar results. The results of one such experiment are shown in Table 4. If the filters containing

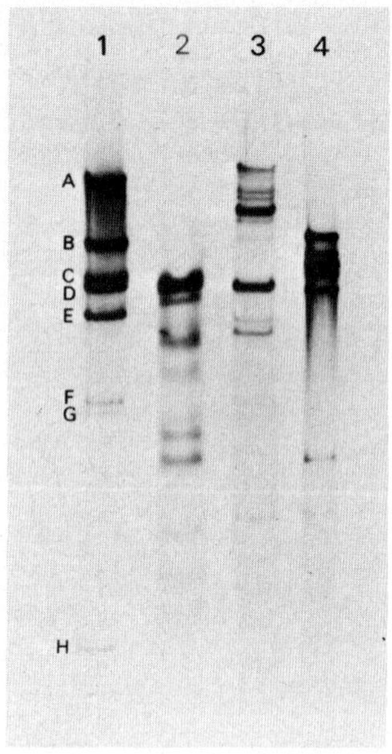

FIGURE 7. $\underline{EcoR_I}$-digestion products of HMCV DNA. ^{32}P-labeled HMCV DNA, along with the standards, ^{32}P-labeled CMV (AD169), HSV-2 (333), and HSV-1 DNAs, were individually digested with restriction enzyme $\underline{EcoR_I}$. The digestion products of each DNA were separated by electrophoresis through a 0.5% agarose slab gel. The gel was dried and placed on Kodak medical X-ray film RP/R2. After an appropriate time, the film was developed. Track 1: HMCV DNA. Track 2: CMV (AD169) DNA. Track 3: HSV-3 (333) DNA. Track 4: HSV-1 DNA.

denatured calf thymus DNA are taken as zero specific hybridization and the filters containing denatured HMCV DNA are taken as 100% hybridization, then extensive hybridization was detected between HMCV RNA and HSV-2 DNA. Less, but still measurable, hybridization was detected between HMCV RNA and HSV-1 DNA, and no hybridization was detected between HMCV RNA and CMV DNA.

DISCUSSION

Previous studies have shown that UV-irradiated HSV and CMV retain their ability to transform HEF cells and that the transformed cells are oncogenic in hamsters (3, 29, 30).

TABLE 4

HYBRIDIZATION OF HMCV RNA WITH HERPESVIRUS DNAs

DNA on filter	^{32}P cpm on filter	^{32}P cpm, background subtracted[a]	% DNA remaining on filter	% hybridization[b]
HMCV	640 ± 158[c]	372	44	100
CMV	243 ± 12	0	52	0
HSV-1	341 ± 35	73	100	20
HSV-2	539 ± 87	271	100	73
CT[d]	268 ± 68	0	-	0
Blank	194 ± 43	0	-	0

[a]The filters containing calf thymus DNA are taken as nonspecific background and set equal to zero cpm.

[b]The hybridization conditions are presumably those of large DNA excess, so that a modest loss of denatured DNA from a few filters would not affect the measurements. If, on the other hand, the hybridization conditions were those of RNA excess, then the cpm on the filters would be corrected for DNA loss by assuring the RNA cpm on the filters are directly proportional to the DNA on the filters. In this case, the final percent hybridization for HSV-1 and HSV-2 DNAs would be 9% and 32% respectively.

[c]The average value plus-and-minus one standard deviation for triplicate filters.

[d]CT, calf thymus DNA.

However, no human cells in primary culture have been successfully transformed with human herpesviruses and then established in long-term culture. We have reported the first evidence that nonestablished human cells can be transformed *in vitro* by a non-inactivated genital CMV isolate, that they obtain an indefinite *in vitro* lifetime, and that such cells have all the characteristics of virus-transformed cells (5, 6, 7). Transformation is a rare event and not all the CMV strains seem to have detectable transforming potential. The expression of easily detectable CMV-related antigens in the transformed cells seems to be a property of cell cultures in early passages. It is more difficult to detect virus-specific antigens in high *in vitro* passages. In fact, the oncogenicity of the transformed cells seems to be inversely proportional to the rate of expression of CMV-related antigens in the cell population (10). A similar increase in oncogenicity in later passages of virally transformed cells has been observed with HSV-2-transformed hamster cells (30). "Fading" of virus-specific antigens has also been observed in CMV-transformed-HEF cells by Lausch (personal communication). Most probably, some antigens are expressed but the indirect IF technique is not sensitive enough to detect minute amounts of CMV-related antigens in the transformed cells at high passages. This hypothesis is supported by the finding that sera of CMV-Mj-HEL-2 tumor-bearing nude mice react with CMV-infected cells in the IF test but display no reaction with uninfected human embryo fibroblasts. Significant neutralization of CMV-Mj was also observed with one of the pooled sera from the transplanted mice. Even though athymic nude mice are immunologically deficient, this study appeared feasible since normal B cell functions are present (31) and antibody responses to virus antigens have been demonstrated.

The resistance to superinfection with CMV exhibited by CMV-transformed human cells is another indication of the presence of the CMV genome in these cells. Approximately 0.4 genome equivalents per cell (unpublished results) are present in these lines. Transformation is not always followed by complete resistance to superinfection with the homologous virus. Simian virus 40-transformed lines are usually resistant (32) but permissive clones of transformed cells can be isolated. HSV-2-transformed cells can be partially resistant (33) or completely resistant (34) to HSV.

The growth characteristics of a newly isolated virus (HMCV) obtained from the CMV-transformed tumor cell line were consistent with the cell specificity of HSV but the cytopathology was similar to HSV or CMV, depending on the host cell. Neutralization and IF tests revealed relatedness to CMV, while molecular studies of the virus (molecular weight, buoyant position and RNA-DNA hybridization) showed limited relatedness to HSV.

Infection of primary HK cell cultures with HMCV resulted in persistent infection, subsequent viral transformation, and the establishment of epithelioid cell lines (unpublished data). Virus-related antigens were detected in two of four cloned sublines by indirect IF techniques after treatment with IUdR. Two of four cloned sublines induced poorly differentiated malignancies in 66% of athymic nude mice that received subcutaneous transplants. Infectious virus was rescued from one of the nontumorigenic clones after temperature shock treatment at passage 26. The virus isolate was confirmed to be identical to the original HMCV strain by neutralization tests. Virus-related antigens were detected by IF in IUdR-treated tumor cells cultured *in vitro* and karyotypic analysis confirmed the human origin of the transformed cells.

Thus, a number of cytomegalovirus isolates appear capable of inducing persistent infection of cells *in vitro* with minimal (and limited) cytopathology. This relationship changes when a transformant appears, probably because the transformed cell is resistant to superinfection and outgrows the "normal" and "persistently infected" cells. These results constitute further evidence of the transforming capacity of non-inactivated herpesviruses and raise the possibility that similar events may occur under natural conditions (i.e., in the human host).

REFERENCES

1. St. Jeor, S. C., Albrecht, T. B., Funk, F. D., and Rapp, F. (1974). J. Virol. 13, 353.
2. Tanaka, S., Furukawa, T., and Plotkin, S. A. (1975). J. Virol. 15, 297.
3. Albrecht, T., and Rapp, F. (1973). Virology 55, 53.
4. Rapp, F., Geder, L., Murasko, D., Lausch, R., Ladda, R., Huang, E.-S., and Webber, M. M. (1975). J. Virol. 16, 982.
5. Geder, L., Lausch, R., O'Neill, F., and Rapp, F. (1976). Science 192, 1134.
6. Geder, L., Kreider, J., and Rapp, F. (1977). J. Natl. Cancer Inst. 58, 1003.
7. Geder, L., Sanford, E. J., Rohner, T. J., and Rapp, F. (1977). Cancer Treatment Reports 61, 139.

8. Rowe, W. P., Hartley, J. W., Waterman, S., Turner, H. C., and Huebner, R. J. (1956). Proc. Soc. Exp. Biol. Med. 92, 418.
9. Glaser, R., Geder, L., St. Jeor, S., Michelson-Fiske, S., and Haguenau, F. (1977). J. Natl. Cancer Inst. 59, 55.
10. Geder, L., Laychock, A., Gorodecki, J., and Rapp, F. (1977). The Third International Symposium on Oncogenesis and Herpesviruses, Cambridge, MA, pp. 84 (abstr.).
11. Sanford, E. J. Dagen, J. E., Geder, L., Rohner, T. J., and Rapp, F. (1977). J. Urol. 118, 809.
12. Hirt, B. (1967). J. Mol. Biol. 26, 365.
13. Pater, M. M., Hyman, R. W., and Rapp, F. (1976). Virology 75, 481.
14. Iltis, J. P., Oakes, J. E., Hyman, R. W., and Rapp, F. (1977). Virology 82, 345.
15. Oakes, J. E., Hyman, R. W., and Rapp, F. (1976). Virology 75, 145.
16. Oakes, J. E., Iltis, J. P., Hyman, R. W., and Rapp, F. (1977). Virology 82, 353.
17. Sharp, P. A., Snyder, B., and Sambrook, J. (1973). Biochemistry 12, 3055.
18. Sheldrick, P., and Berthelot, N. (1974). Cold Spring Harbor Symp. Quant. Biol. 39, 667.
19. Wadsworth, S., Jacob, R. J., and Roizman, B. (1975). J. Virol. 15, 1487.
20. Raskas, H., and Green, M. (1971). In "Methods in Virology" (K. Maramorosch and H. Koprowski, eds.), Vol. 5, pp. 247. Academic Press, Inc., New York.
21. Schildkraut, C. L., Marmur, J., and Doty, P. (1962). J. Mol. Biol. 4, 430.
22. Clements, J. B., Cortini, R., and Wilkie, N. M. (1976). J. Gen. Virol. 30, 243.
23. Hayward, G. S., Frenkel, N., and Roizman, B. (1975). Proc. Natl. Acad. Sci. USA 72, 1768.
24. Hayward, G. S., Jacob, R. J., Wadsworth, S. C., and Roizman, B. (1975). Proc. Natl. Acad. Sci. USA 72, 4243.
25. Huang, E.-S., Kilpatrick, B. A., Huang, Y. T., and Pagano, J. (1976). Yale J. Biol. Med. 49, 29.
26. Kilpatrick, B. A., Huang, E.-S., and Pagano, J. S. (1976). J. Virol. 18, 1095.
27. Skare, J., and Summers, W. C. (1977). Virology 76, 581.
28. Skare, J., Summers, W., and Summers W. (1975). J. Virol. 15, 726.
29. Duff, R., and Rapp, F. (1971). Nature (New Biol.) 233, 48.
30. Duff, R., and Rapp, F. (1973). J. Virol. 12, 209.

31. Burns, W. H., and Allison, A. C. (1971). Antigens 3, 479.
32. Sambrook, J. (1972). Adv. Cancer Res. 16, 141.
33. Doller, E., Duff, R., and Rapp, F. (1973). Intervirology 1, 154.
34. Darai, G., and Munk, K. (1976). Int. J. Cancer 18, 469.

EPSTEIN-BARR VIRUS INFECTION OF EPITHELIAL CELLS AND LYMPHOCYTES

Joseph S. Pagano

Cancer Research Center and The Departments of Medicine and Bacteriology and Immunology,
The University of North Carolina School of Medicine,
Chapel Hill, North Carolina 27514

ABSTRACT. EBV is now believed to infect and replicate initially in epithelial cells in the oropharynx. The secondary target cell of the virus is thought to be the B-lymphocyte. The EBV genome is carried in the form of supercoiled DNA as a plasmid or episome in such cells. A consideration of the primary and secondary cell type infected by EBV together with the existence of the episomal form of the viral genome provide the basis for a unifying hypothesis that may account for many features of the pathogenesis of infectious mononucleosis, Burkitt's lymphoma and nasopharyngeal carcinoma.

INTRODUCTION

Characteristically herpes-group viruses establish persistent infections after the initial acute episode. These infections sometimes take the form of chronic active infection with replication of virus for months or even years. However, the hallmark of herpes-group infections is their complete latency and the episodic reactivation of the latent virus or its genome. The symptoms accompanying reactivation are sometimes quite different from those of the initial infection; reactivation may also be asymptomatic. The Epstein-Barr Virus (EBV) is still unique among the human herpes-group viruses in providing an *in vitro* cellular model of latency that seems to correspond in several ways on the cellular level to latent EBV infection in man.

In any consideration of latency and reactivation of virus infection the cell type in which the virus replicates initially and the cells in which the viral genetic information persists are of crucial importance. Not only for herpes-group viruses but for many viruses generally we in all likelihood have a distorted view of viral replication. These imperceptions stem from the fact that the viruses are as a rule studied *in vitro* in fibroblastic cells, whereas the primary cell type involved in the host is rarely of mesenchymal origin. In the case of EBV *in vitro* studies have been conducted

ISBN 0-12-668350-6

exclusively in cells of lymphoid origin. As we shall see such experimental work while it has adduced useful data does not offer a complete picture of virus-cell interactions in the intact organism.

The Epstein-Barr Virus is linked to three quite different diseases: Burkitt's lymphoma (BL), nasopharyngeal carcinoma (NPC) and infectious mononucleosis (IM). Any hypothesis that purports to establish a causal relation between EBV and, for example, Burkitt's lymphoma must also take into account the association of the virus with the other two conditions. The interaction of EBV with its natural target cells furnishes, I believe, many of the elements needed for a unifying hypothesis that links the virus and the two malignant and one benign conditions associated with it (1). The key elements of this hypothesis are the primary and secondary target cells of EBV and the episomal form of the viral genome. In this paper I shall review some of the features of EBV-cell interactions that provide the basis for a cohesive view of the pathogenesis of these three diseases in which the infected cell types play a central role.

Target Cell: Earlier Data. Virtually all of what we know about Epstein-Barr virus has come from studies of the virus in relation to lymphoid tissue.

(a) Burkitt's lymphoma cell lines. The virus was first isolated by Epstein, *et al*. (2) from explanted lymphoblastoid cells originating from a Burkitt's lymphoma. Such cells and later the fresh tissue from which the cell line had originated were shown to contain the Epstein-Barr virus genome (3,4, 5,30). Cell lines established from BL are of two general types: virus-producing and non-virus-producing; there are also Burkitt's cell lines of intermediate type that appear to make virus structural antigens but not infectious virus.

(b) Infectious mononucleosis cell lines. These are of two general types, those established from the peripheral blood of patients with acute infectious mononucleosis or who were exposed to virus infection at some time in the past. Some such cell lines shed transforming virus in variable amounts *in vitro* (6). Examples of two cell lines are shown in Figure 1. The other type of cell line that can be established with materials from infectious mononucleosis is produced by infection *in vitro* of cord-blood lymphocytes with fresh infectious virus recovered from the throats of patients with infectious mononucleosis. Cord-blood lines transformed with EBV are characteristically non-virus-producing and carry only a small number of viral genomes (7,8).

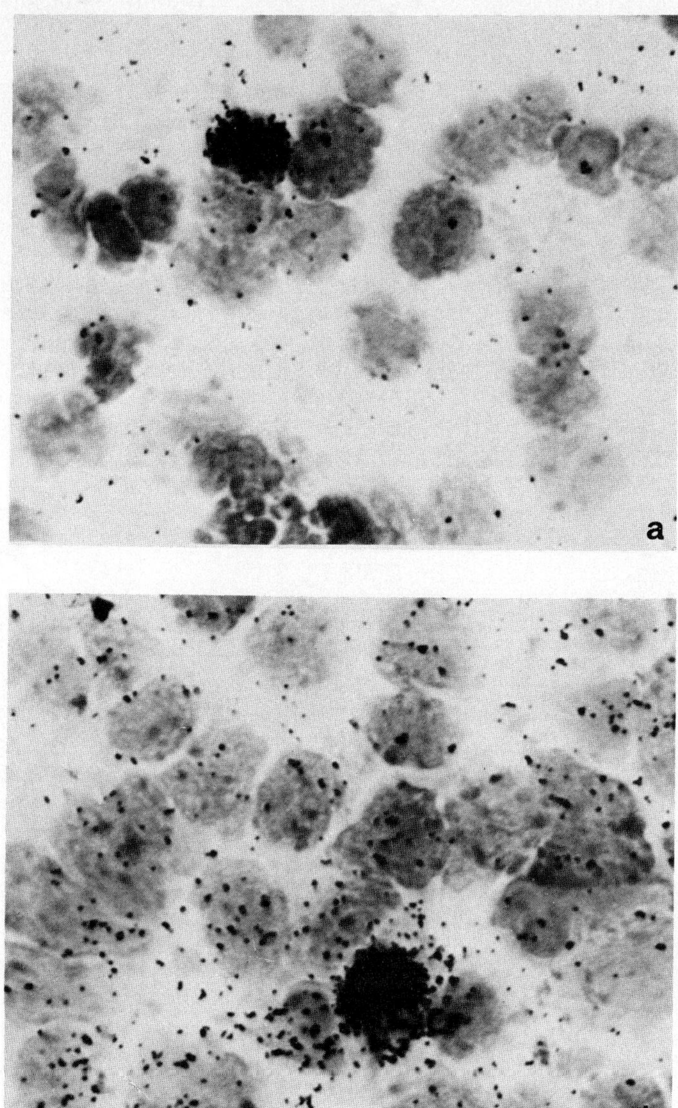

Figure 1. Cytohybridization with Epstein-Barr Virus ^{3}H cRNA (5 x 10^{5} CPM per slide) to lymphocytic lines derived from the peripheral blood of 2 patients with infectious mononucleosis. A, cell line IM 6; B, cell line IM 8. Exposure time, 3 weeks. (From Huang & Pagano (1974), reference No. 22.)

(c) B-Lymphocytes. It has become clear in the last few years that the cell type in which EBV is found is the B-lymphocyte (9,10,5). B-lymphocytes have receptors for EBV, can replicate the virus *in vitro* and carry the viral genome in a latent form, whereas T-lymphocytes do not carry receptors for EBV and cannot be infected. T-lymphocyte lines established from patients with acute lymphocytic leukemia do not bear the EBV genome.

(d) T-lymphocytes . The atypical lymphocyte response that is the hallmark of the blood-cell response in patients with acute infectious mononucleosis consists predominantly of T-lymphocytes (11, 12). This T-lymphocyte response is prominent in the pathogenesis of IM, but it appears to represent a reactive phenomenon that corresponds to the cytotoxic effects demonstrable *in vitro* on EBV-bearing B-lymphocytic target cells (13, 14, 15).

RESULTS

Primary Target Cell. EBV has long been assumed to replicate in lymphoid cells *in vivo* as well as *in vitro*, with the primary site of replication of the virus being perhaps tonsillar tissue followed by spread to circulating B-lymphocytes as suggested in Figure 1. However, this evidence poses a paradox in the absence of concrete data to show that EBV replicates in such B-lymphocytes while the cells are still in the circulation. The negative evidence comes from cRNA-DNA hybridization, both on membrane filters and by *in situ* methods, of pellets of white blood cells taken from the circulation of a number of patients with acute infectious mononucleosis (8). Additional negative evidence comes from the failure of Klein *et al*. to find any cells in the circulation of patients with infectious mononucleosis that contain EBV viral capsid antigen (VCA) or early antigen (EA), concomitants of viral replication. A few circulating cells that contain the Epstein-Barr virus nuclear antigen (EBNA), a concomitant of the repressed viral genome, have, however, been found (16).

The primary cell type in which EBV replicates in man has been identified only recently. The epithelial nature of the primary cell was anticipated by a scattering of other observations. The first was the discovery that EBV apparently replicates in the parotid gland (17). Niederman and his colleagues came to this conclusion by sampling saliva from patients with IM, and we have confirmed this observation by cannulation of Stenson's duct (18). The parotid gland is predominantly an epithelial not a lymphoid organ. Additionally Glaser _et al_. (19) have shown that nasopharyngeal carcinoma, the basic cell type of which is epithelial in origin, has receptors for EBV. Trumper _et al_.showed that treatment of NPC with IUdR not only induces the appearance of EA, but that passage of NPC in nude mice leads to the appearance of EBV VCA and even virions in the tumor cells (20). NPC contains the EBV genome in a latent form (4, 21). Therefore it seems clear that nasopharyngeal carcinomatous tissue can support replication of EBV.

Lemon _et al_. then showed that epithelial cells recovered from the throats of patients with acute infectious mononucleosis harbor large amounts of EBV DNA, as visualized by _in situ_ cytohybridization (see Figure 2). Not all epithelial cells contain the viral DNA, and some cells contain more than others, an observation consistent with the asynchronous nature of the infection. In some infected cells there were heavy accumulations of viral DNA over the nuclei. Additionally, smaller epithelial cells obtained by cannulation of Stenson's duct and presumed to be cells from the parotid gland or its ducts also contained large amounts of viral DNA (Figure 3). Such large accumulations of hybridizable DNA in the _in situ_ cytohybridization technique are virtually tantamount to active replication (22, 23).

Since transforming virus could be recovered from the throats of all of these patients in appreciable titers (up to 10^5 TU/ml) there was the possibility that the viral DNA seen in the epithelial cells was merely virus absorbed to the cell surface even though the absorption of such large amounts of virus was unlikely. Figure 4 shows that when epithelial cells from an uninfected person were exposed to 10^5 TU/ml of fresh EBV-containing secretions little absorbed virus was detected. Additional control experiments have been carried out on some of the same specimens with SV40 and CMV cRNA with negative results. It thus appears that the primary cells infected during acute exposure to EBV are epithelial cells lining the oropharynx, perhaps the anterior and midpharynx and probably the salivary ductal cells as well. Presumably replication of virus in such cells ordinarily results in cell death, as with replication of herpes-group viruses generally, and this feature may be responsible for the pharyngitis of the acute phase of infectious mononucleosis.

Secondary Target Cell: B-lymphocytes. Receptors on the B-lymphocytes for the Epstein-Barr virus are closely identified with if not identical to receptors for complement. EBV

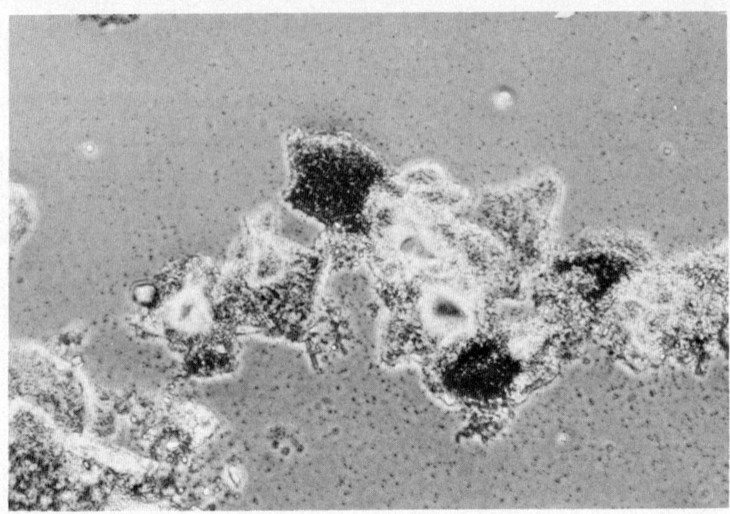

Figure 2. Cytohybridization with EBV-specific cRNA to oropharyngeal epithelial cells from a patient with infectious mononucleosis. 8 x 10^5 CPM ^{3}H cRNA (specific activity 1x 10^7 CPM μg) was applied to the coverslip preparation. Exposure time, 6 weeks. (Unstained, phase contrast.)

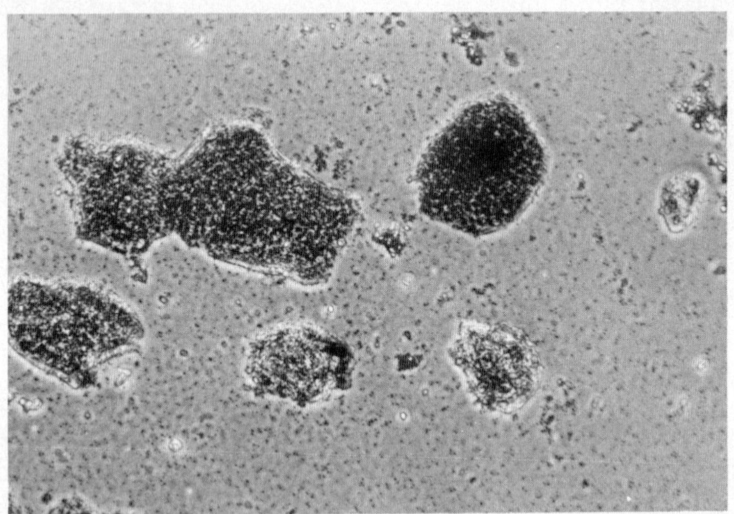

Figure 3. Cytohybridization with EBV cRNA to epithelial cells recovered from Stensen's duct secretions of a patient with acute infectious mononucleosis. Conditions as in Figure 2. Transforming EBV was recovered from the throat of this patient and the patient that yielded the cells shown in Figure 2.

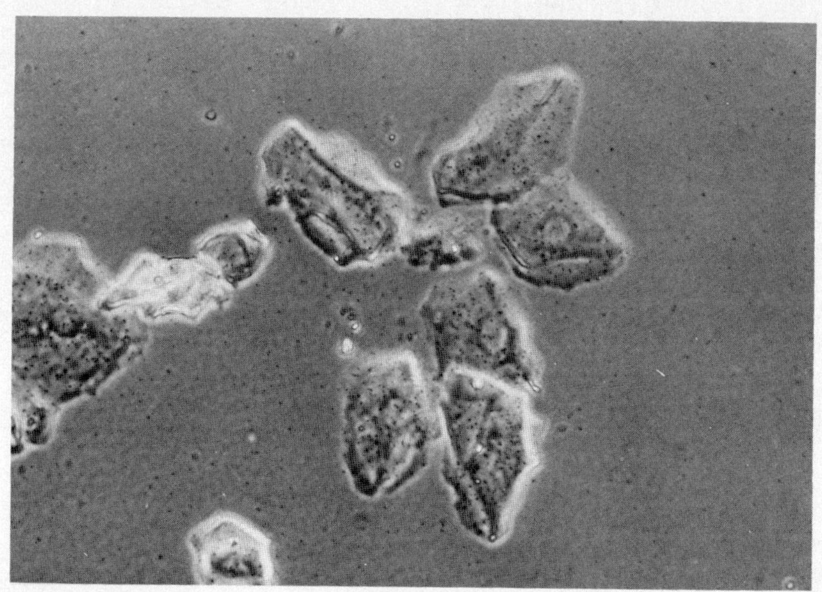

FIGURE 4. Cytohybridization with EBV cRNA to epithelial cells from the throat of an EBV-antibody-negative patient after adsorption of the cells with about 10^5 units of transforming virus recovered from the patient whose cells are shown in Figure 3. Conditions as in Figure 2. (From Lemon *et al*. (1977), reference No. 18.)

alone of all the herpes-group viruses is known to use specific or relatively specific receptors for absorption to the cell surface. Capping of B-lymphocytes with C_3D removes the receptors for the virus, and conversely exposure of B-lymphocytes to EBV causes capping of the C_3D receptors. Cells that have been exposed to the virus no longer exhibit the capping phenomenon after exposure to complement (9, 10). This is the chief evidence for the identity or covalent linkage of C_3D and EBV receptors. However, the receptors for EBV have not yet been isolated, and this curious and biologically significant similarity has not yet been established beyond doubt as an identity.

The rare established B-lymphocyte lines that do not contain EBV genome (24) fall into two classes: those that can and those that cannot be infected with EBV. The 698 cell line which cannot be infected with EBV does not possess C_3D receptors (25).

T-lymphocytes. T-lymphocytes cannot be infected with EBV, nor do they possess receptors for either C_3D or EBV (10,8). The MOLT cell line which was formerly considered a T-cell line (26) presents a special case inasmuch as these cells have some T-cell characters, i.e., ability to rosette sheep-red cells, but MOLT cells also have complement receptors--in other words MOLT cells have both B- and T-cell character. EBV attaches to these cells, but does not penetrate them (27). The MOLT cell line does not contain detectable EBV DNA (5, 28).

Epithelial Cell Receptors. The question of whether epithelial cells possess complement receptors inasmuch as they must possess EBV receptors is unanswered, in part because of the lack of bona fide epithelial cell lines. At least some human epithelial cells such as kidney tubules do contain C_3D receptors (29). Furthermore, as mentioned earlier, Glaser and de Thé have presented evidence that nasopharyngeal carcinomas have receptors for EBV inasmuch as they can be infected with the virus. A definition of any differences among the complement and virus receptors of the various cell types should prove instructive.

Classes of B-lymphocytes. There are essentially two classes of B-lymphocytes with respect to Epstein-Barr virus; permissive of virus replication and nonpermissive (3,30). Permissive cells exhibit a spectrum of capability to replicate the virus and virus products. Some cell lines contain or release virus particles, both infectious and noninfectious, whereas others seem only to make various EBV antigens associated with replication of the virus (28). Nevertheless in such cases there is relatively prolific viral DNA replication as shown by *in situ* cytohybridization (See Figure 1) which implies that the virus-induced DNA polymerase (31) is active. Cell lines that produce virus, virus DNA or structural products in complete or abortive fashion originate both from explanted Burkitt's lymphomas (e.g., the P_3J subline of HR1 cells) and from the peripheral blood of patients with past or active infectious mononucleosis. The prototype of cells that are restrictive for replication of the virus is produced by the infection of umbilical cord lymphocytes with IM virus from the throat. EBV-genome-bearing cell lines established from cord cells in this way do not produce virus (6,7,32); such cells do contain EBNA(16) and LYDMA(33).

It is noteworthy that such IM cell lines, which contain the episomal form of the EBV genome in the form of molecules with a contour length slightly less than that of the EBV genome *et al.*(34), contain only a few copies of the genome--in the vicinity of 10. This limited copy number raises the question of whether the virus-induced DNA polymerase ever becomes functional in such cells (1). The alternative is induction of the enzyme for a brief period of functioning, then virus DNA synthesis without cell killing, followed by repression of the virus polymerase; this sequence of events seems rather unlikely.

A related question concerns whether or not a round of viral DNA replication is needed to set in train the events leading to transformation of cord-blood lymphocytes by EBV. There is no decisive evidence on this point. Indeed two conflicting arguments have been presented (35,36), but neither interpretation is as yet decisive. From what has been said before about the limited EBV genome copy number in cord-blood cells in the absence of other concomitants of EBV replication it is attractive to predict that transformation of cord cells may not require the induction of virus-induced DNA polymerase (36). Experiments with a specific inhibitor of EBV DNA replication such as acycloguanosine (37) rather than an inhibitor such as phosphonoacetic acid which inhibits cellular DNA synthesis to some extent as well as viral DNA replication--and hence may interfere with cell transformation by the former effect--need to be done.

The EBV Episome. Explanted Burkitt's lymphomas also occasionally yield a non-permissive cell line. The prototype of such a BL-cell line is Raji (See Figure 5B) which contain 55-60 copies of the EBV genome mostly in a non-integrated form (38) in the chromosomes and recognized as a novel episomal or plasmid form of eukaryotic DNA (39). Adams and Lindahl (40) showed that the nonintegrated viral DNA in Raji cells existed in a supercoiled form with a contour length equivalent to approximately 100 x 10^6 daltons (Figures 6, 7). (This genome form has been variously designated as plasmid and episome because of the uncertainty as to its relation to possible homologous integrated viral sequences.) This molecular form was predicted to be replicated not by the virus-induced polymerase, but because of its strictly regulated copy number, by the host DNA polymerase. Strong indirect evidence supporting this notion comes from the lack of effect of phosphonoacetic acid on the number of copies of the EBV episome (41,42) and more recently by acycloguanosine (Shaw and Colby, unpublished data).

So far only data on the contour length of supercoils

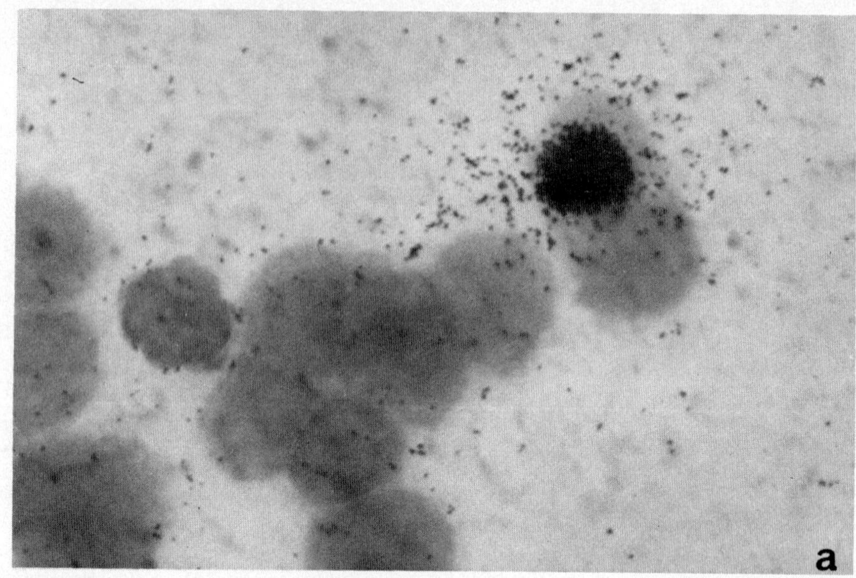

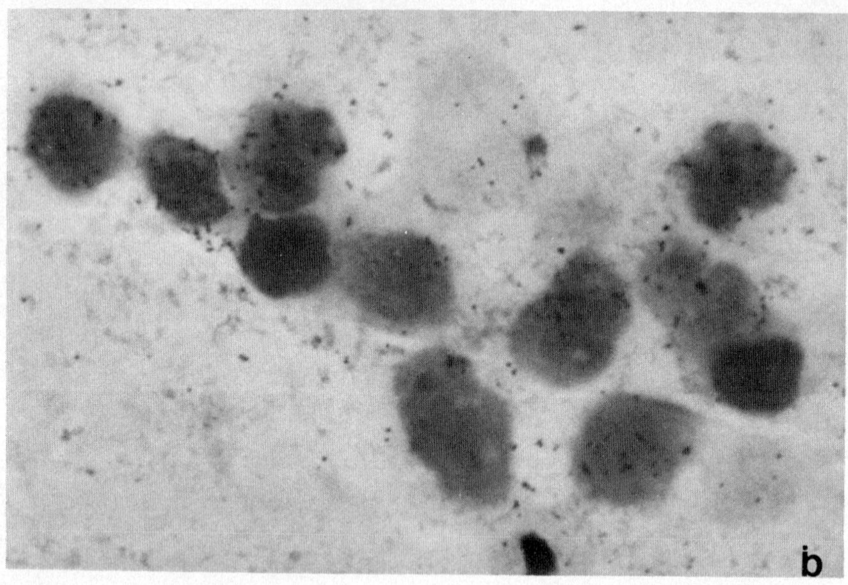

Figure 5. Cytohybridization with EBV cRNA to virus-producing and non-virus-producing cell lines. A. The virus-producing line, P_3HR1 (exposure time, 2 weeks); B. The non-virus-producing line, Raji (exposure time, 4 weeks). (From Huang & Pagano (1977), reference No. 23.)

from various cell lines have been published. As mentioned before some IM lines contain a molecule about 95% the length of the HR1 DNA molecule. The Namalva cell line contains a supercoiled molecule of about 50 million daltons (Okasinski and Pagano, unpublished data). Shaw and Colby have very recently been able to recover supercoiled EBV DNA from the HR1 virus-producing cell line; presumably these molecules come from the approximately 90% of the cell population that is not actively replicating viral DNA (See Figure 5A). The contour lengths are being compared to Raji supercoils and authentic virion DNA (J. Shaw, J. Griffith, & C. Moore). Better characterization of the supercoils by restriction endonuclease analysis is just beginning and is expected to provide fresh insights into the relation of the supercoils to virion DNA and to each other (Shaw, unpublished data).

In addition to the episomal forms, lymphoblastoid cell lines such as Raji are believed to contain integrated viral sequences as well (43). The evidence which is largely based on the existence of viral DNA sequences in DNA that is significantly lighter than pure viral DNA is suggestive but not decisive because of the experimental problem of the intrusion of remnants of episomal DNA during attempts to purify cellular DNA with integrated viral sequences. For this reason it has not yet been possible to determine with certainty the number of copies, nor whether fractional copies of the EBV genome are present in the integrated state. Recently Lindahl et al. have analyzed the Ramos cell line which originated from a Burkitt's lymphoma. This line which contains only 2 EBV copies per cell is believed to contain only integrated viral DNA (44).

The functions of this unique molecular form, the episome, must be of unique importance. We have from the time of the first work with this form of EBV DNA considered that the episome may be the molecular basis of latency for herpes-group viruses generally. The episomal form of the viral genome can clearly serve as the repository not only of viral information, but of the intact genome which can be reactivated, if we presume that it is the episomal form that is being reactivated rather than any integrated form of the viral genome. In Raji cells there is ordinarily no escape into replication of the latent viral genome, but reactivation of the genome can be artificially accomplished by treatment with IUdR. (45). In the case of HR1 cells the transition from the latent episome to replicating viral DNA presumably occurs continually and spontaneously. The supercoiled form can be recovered from HR1 cells (Shaw and Colby) after treatment with acycloguanosine. Episomes are found not only in lymphoid cells, but also in epithelial

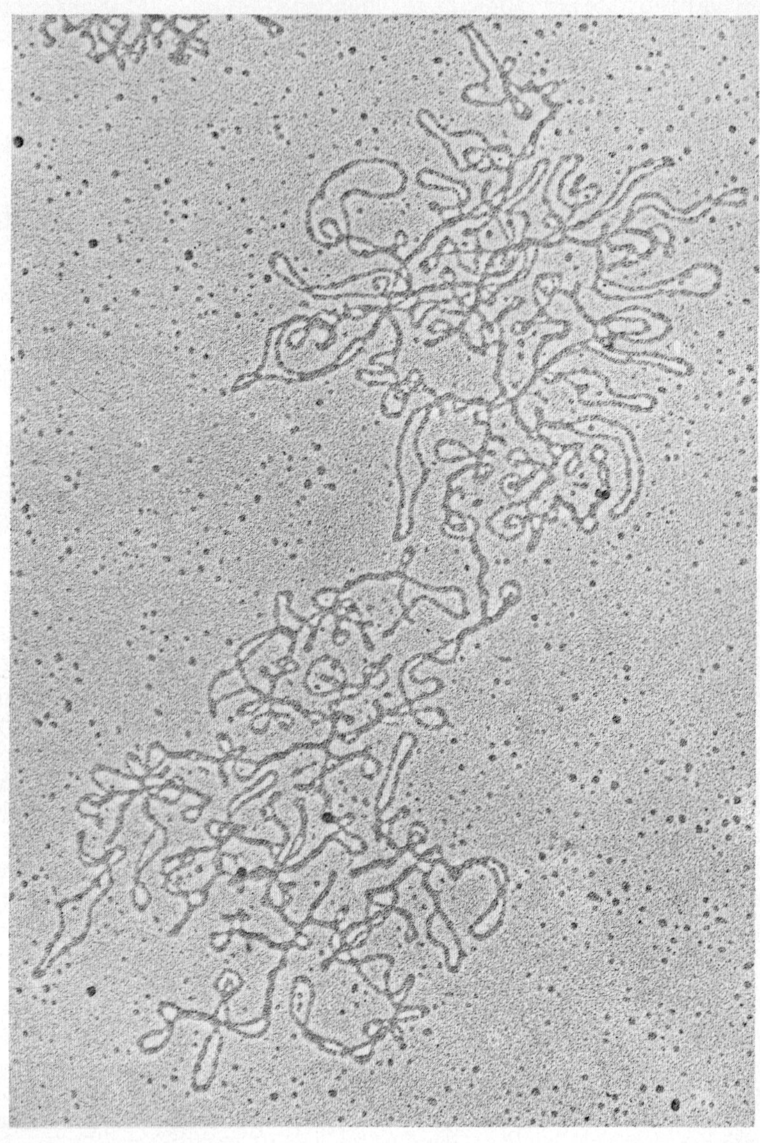

Figure 6. The EBV episome. Supercoiled EBV DNA recovered from Raji cells. The DNA was purified by ethidium bromide-cesium chloride density gradient centrifugation and identified as EBV DNA by cRNA-DNA hybridization. (Unpublished data of Dr. James Shaw. Photograph courtesy of Claire Moore.)

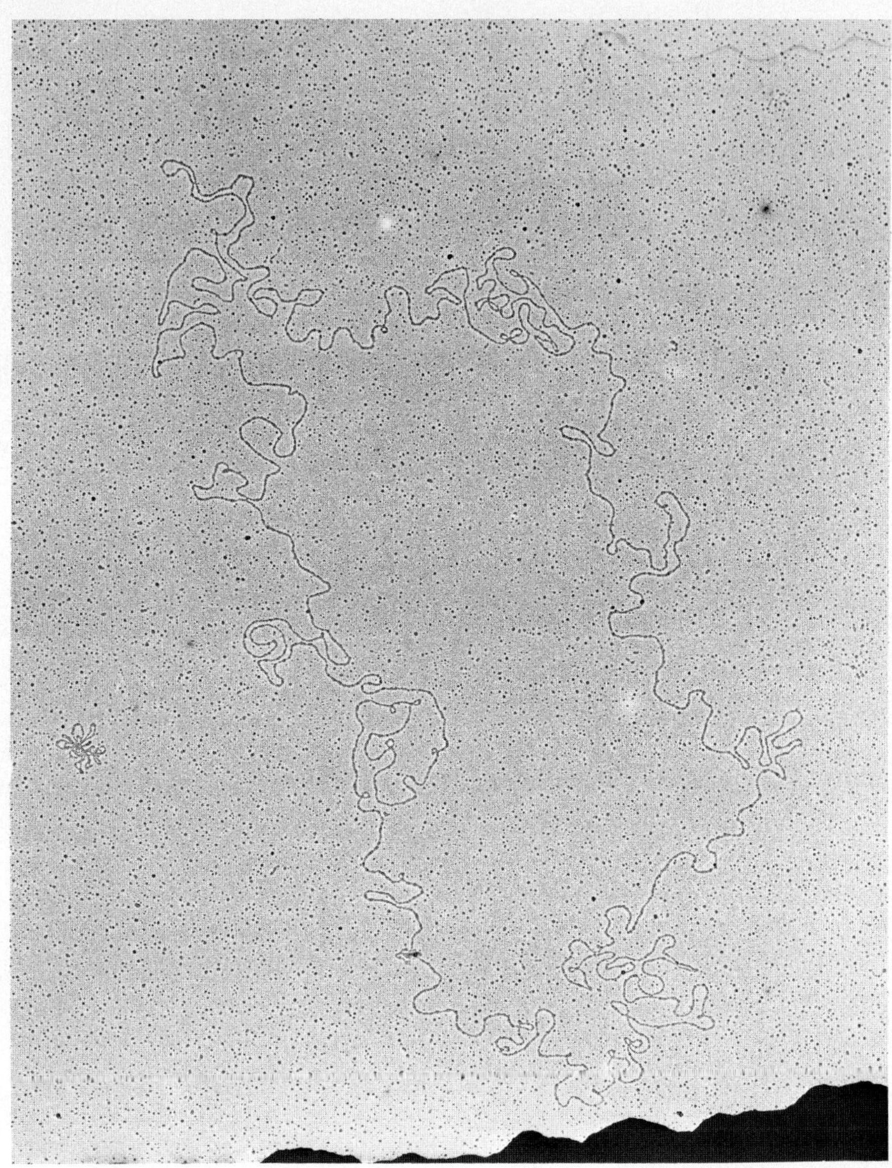

Figure 7. Open circular form of EBV DNA recovered from Raji cells. (Unpublished data of Dr. James Shaw. Photograph courtesy of Claire Moore.)

i.e., in nasopharyngeal carcinomatous cells (46).

As to the mechanism of reactivation of the latent episome one possibility is that the episomal form found in Raji cells lacks DNA sequences involved with replication. This seems unlikely in light of the evidence that the episome can be induced to replicate. Another possibility is that replication of the episome in Raji cells is repressed by a DNA-binding protein such as EBNA (47). In HR1 cells repression of replication of the viral genome found in the majority of the cell population presumably undergoes a cyclic release of the repressed state so that viral DNA replication can ensue.

An intriguing question is whether the episome is expressed in infectious mononucleosis; IM cells proliferate in culture in a manner resembling malignant cell lines, but they are not actually malignant. IM cell lines are polyclonal in contrast to Burkitt's cell lines which are monoclonal (48,49,50). Could it be that the plasmid can direct lymphoproliferation, but that integration of specific viral transforming sequences is needed before a truly malignant clone of cells can emerge? Cell lines bearing episomes also express other viral functions including EBNA, LYDMA and, most recently, EBV-specific thymidine kinase (51). The EBV thymidine kinase, first identified in Raji cells that have been superinfected with EBV (51), can also be detected at low levels in non-superinfected Raji cells (Chen, unpublished data). Furthermore, the virus-induced enzyme has been obtained from virus-producing HR1 cells by affinity chromatography (Fife, Chen & Pagano, unpublished data), and uninfected Raji cells have been shown to be able to phosphorylate acycloguanosine, an effect associated with herpesvirus-specific thymidine kinase (Furman, Colby, Shaw & Pagano, unpublished data). Obviously all of these functions could be the result of integrated viral sequences, and it has not yet been possible to discriminate between the two possibilities.

DISCUSSION

Pathogenesis of Infectious Mononucleosis, Burkitt's Lymphoma and Nasopharyngeal Carcinoma. As I indicated in the introduction to this paper, any consideration of how EBV might cause three disease conditions must center on the target cells, both primary and secondary, for virus interactions and on the molecular state and function of the viral genetic material. A full understanding of these features of EBV infection should suffice to explain the genesis of the three diseases on a cellular level. In the living organism obviously many other factors such as genetic predisposition, existence of cofactors, and especially cell-mediated immunologic mechanisms obviously will come into play to determine whether or not transformed cell clones will arise and survive. I have touched upon some of these other considerations recently (1).

In any case, I would propose that a sequence of events something as follows occurs: primary infection with EBV either symptomatic and asymptomatic is initiated in the epithelial cells of the oropharynx. The genome circularizes in the process of replication as appears to be the case with herpes simplex virus replication. There are then either two outcomes: (a) replication of virus and death of the cell, or (b) arrest of the process with the circularized EBV genome becoming fixed in an epithelial cell which survives. Such an epithelial cell harboring the episomal form of EBV may be a potential progenitor cell for nasopharyngeal carcinoma. This cell would not be a malignant cell, and presumably a number of such events occur during ordinary primary infection. Virus shed from the epithelial cells then infects the secondary target cell, namely the B-lymphocyte. Here again, the entering genome circularizes, but in this case, it may never replicate while in the body. Such cells bearing the episome are, however, stimulated to proliferate into polyclonal lines. All of the above events are consistent with subclinical infection with EBV and infectious mononucleosis. We have proposed elsewhere that many of the symptoms of infectious mononucleosis, whether caused by EBV or CMV, may be caused by the cell-mediated immunologic responses stimulated by the B-lymphoproliferation induced by EBV infection (13, 52).

Against the background of such basic cellular interactions nasopharyngeal carcinoma and Burkitt's lymphoma may arise in analogous fashion. In both, the progenitor cells bearing the episomes, epithelial in one case and B-lymphocytic in the other case, become malignant under the direction of specific integrated viral sequences that have transforming effects on the respective cell types. The co-existence of the episomal form and integrated viral sequences is not incompatible. In fact, their co-existence is likely inasmuch as the episome is replicated by the host DNA polymerase, and it is in intimate apposition to the host-cell genomic material. In this perception the episome is a mediator of integration; the idea that there are frequent transpositions of viral genetic material from episome to integrated state follows. However, the emergence of a malignant cell line would depend upon the less frequent successful integration and functioning of specific viral transforming sequences. It follows from this scheme that if any other malignancies are eventually traced to EBV they are likely to be either epithelial at the site or sites where EBV initially replicates or B-lymphocytic in origin.

The importance of the primary cell type is underscored by the artificial character of the virus commonly used in the laboratory. This virus comes from P3HR1 cells, and it is no longer capable of transforming B-lymphocytes. Fresh virus newly released from epithelial cells, is most effective at acheiving this effect on B-lymphocytes. It is however, a generalization to assume that any virus shed by lymphocytes has lost the ability to transform lymphocytes inasmuch as we have isolated virus from a cell line established from the peripheral blood of a patient with infectious mononucleosis that can transform cord-blood lymphocytes (6). Moreover, virus from the throat washings from a patient with transfusion mononucleosis that has been passaged through marmoset lymphocytes also retains the ability to transform cord-blood lymphocytes; the latter cell line may be something of a special case. In short, a virus that is recovered from the primary target cell may have key biologic effects lost by laboratory strains of virus as pointed out by Klein elsewhere in this volume. In another system, namely, murine cytomegalovirus infection (53), Nedrud and Pagano (unpublished data) have data that suggest that virus that replicates in the natural primary cell type *in vitro*, namely, epithelial cells in tracheal organ cultures, retains virulence for mice that is lost by virus passaged *in vitro* in mouse fibroblasts.

REFERENCES

1. Pagano, J. S., and Okasinski, G. (1978). Proc. Third International Symposium on Oncogenesis and Herpesviruses, Boston, Mass., in press.
2. Epstein, M. A., Achong, B. G., and Barr, Y. M. (1964). Lancet 1, 702.
3. zur Hausen, H., and Schulte-Holthausen, H. (1970). Nature 227, 245.
4. Nonoyama, M., Huang, C.-H., Pagano, J. S., Klein, G., Singh, S. (1973). Proc. Natl. Acad. Sci. USA 70, 3265.
5. Pagano, J. S. (1975). Cold Spring Harbor Symposium on Quantitative Biology: Tumor Viruses 39, 797.
6. Pagano, J. S., Huang, C.-H., and Huang, Y.-T. (1976). Nature 263, 787.
7. Miller, G., Coope, D., Niederman, J., and Pagano, J.S. (1976). J. Virology 18, 1071.

8. Pagano, J. S. (1974). "Viruses, Evolution and Cancer," Chapter 4, Academic Press, New York.
9. Jondal, M., and Klein, G. (1973), J. Exp. Med. 138, 1365.
10. Jondal, M., Klein, G., Oldstone, M. B. A., Bokish, V., and Yefenof, E. (1976). Scand. J. Immunol. 5, 401.
11. Sheldon, P. J., Hemsted, E. H., Papamichail, M., et al. (1973). Lancet 1, 1153.
12. Virolainen, M., Andersson, L. C., Lalla, M., et al.(1973) Clin. Immunol. Immunopathol. 2, 114,
13. Hutt, L. M., Huang, Y.-T., Dascomb, H. E., and Pagano, J. S. (1975). J. Immunol. 115, 243.
14. Svedmyr, E., and Jondal, M. (1975), Proc. Natl. Acad. Sci. USA 72, 1622.
15. Royston, I., Sullivan, J. L., Periman, P. O., and Perlin, E. (1975). New Eng. J. Med. 293, 1159.
16. Reedman, B. M. and Klein, G. (1973). Int. J. Cancer 11, 499.
17. Niederman, J. C., Miller, G., Pearson, H. A., and Pagano, J.S. (1976). New England J. Med. 294, 1355.
18. Lemon, S. M., Hutt, L. M., Shaw, J. E., Li, J.-L. H., and Pagano, J. S. (1977). Nature 268, 268.
19. Glaser, R., de The, Lenoir, G., and Ho, J. H. C. (1976). Proc. Natl. Acad. Sci. USA 73, 960.
20. Trumper, P. A., Epstein, M. A., Giovanella B. C., and Finerty, S. (1977). Int. J. Cancer 20, 655.
21. zur Hausen, H., Wolf, H., and Werner, J. (1975). Cold Spring Harbor Symposium on Quantitative Biology 39,791.
22. Huang, E.-S. and Pagano, J. S., (1974). "Viral Immunodiagnosis," Chapter 16, pp. 279-299. (E. Kurstak and R. Morisset, eds.) Academic Press, New York.
23. Huang, E.-S. and Pagano, J. S., (1977). Methods In Virology, 6, pp. 457-497. Academic Press, New York.
24. Andersson, M. and Lindahl, T. (1976). Virology 73, 96.
25. Klein, G., Sugden, B., Leibold, W., and Menezes, J. (1974). Intervirology 3, 232.
26. Minowada, J., Ohnuma T., and Moore, G.E. (1972). Nature 251.
27. Menezes, J., Scigneurin, J. M., Patel, P., Bourkas, A. and Lenoir, G. (1977). J. Virology 22, 816.
28. Minowada, J., Nonoyama, M., Moore, G.E., Rauch, A.M., and Pagano, J. S. (1974). Cancer Res. 34, 1898.
29. Gelfand, M. C., Frank, M. M. and Green, I. (1975). J. of Experimental Med. 142, 1029.
30. Nonoyama, M., and Pagano, J.S. (1971). Nature New Biol. 233, 103.
31. Benz, W. C. and Strominger, J. L. (1976). Proc. Natl. Acad. Sci. USA 72, 2413.

32. Gerber, P., Nkrumah, F., Pritchett, R., and Kieff, E. (1976). Int. J. Cancer 17, 71.
33. Emberg, I., Masucci, G., and Klein, G. (1976). Int. J. Cancer 17, 197.
34. Adams, A., Bjursell, G., Kaschka-Dierich, C., and Lindahl, T. (1977). J. of Virology 22, 373.
35. Thorley-Lawson, D. and Strominger, J.L. (1976). Nature, 263, 332.
36. Lemon, S.M., Hutt, L.M., and Pagano, J.S. (1978). J, of Virology 25, 138.
37. Elion, G., Furman, P.A., Fyfe, J.A., deMiranda, P., Beauchamp, L., and Schaeffer, H.J. (1977). Proc. Nat. Acad. Sci. USA 74, 5716.
38. Nonoyama, M., and Pagano, J.S. (1972). Nature New Biol. 238, 169.
39. Pagano, J.S., Nonoyama, M., and Huang C.-H. (1973). "Possible Episomes in Eukaryotes," Proc. of the Fourth Lepetit Colloquium, North Holland Publishing Company, p. 318.
40. Adams, A. and Lindahl, T. (1975). Proc. Nat. Acad. Sci. USA 72, 1477.
41. Huang, E.-S., Huang, C.-H., Huong, S.-M., and Selgrade, M.J. (1976). Yale J. Biol. & Med. 49, 93.
42. Summers, W.C., and Klein, G. (1973). J. of Virology 18, 151.
43. Adams, A., Lindahl, T., and Klein, G. (1973). Proc. Nat. Acad. Sci. USA 70, 2888.
44. Anderson-Amvret, M. and Lindahl, T. (1978). J. of Virology 25, 710.
45. Hampar, B., Derbe, J.G., Martos, L.M., and Walker, J.L. (1972). Proc. Nat. Acad. Sci. USA 69, 78.
46. Kaschka-Dierich, C., Adams, A., Lindahl, T., Bornkamm, G., Bjursell, G., Klein, G., Giovanella, B.C., and Singh, S. (1976). Nature 260, 302.
47. Luka, J., Siegert, W., and Klein, G. (1977). J. of Virology 22, 1.
48. Bechet, J.M., Fialkow, P.J., Nilsson, K., Klein, G., and Singh, S. (1974). Exp. Cell Res. 89, 275.
49. Fialkow, P.J., Klein, G., Garther, S.M., and Clifford, K. (1970). Lancet 1, 384.
50. Nilsson, K. (1971). Int. J. Cancer 8, 432.
51. Chen, S.-T., Estes, J.E., Huang, E.-S., and Pagano, J.S. (1978). J. of Virology 26, 203.
52. Lemon, S.M, Hutt, L.M., and Huang, Y.-T. (1977). Clin. Immunol. and Immunopathol. 8, 513.
53. Mantyjarvi, R.A., Selgrade, M.J.K., Collier, A.M., Hu, S.-C., and Pagano, J.S. (1977). J. of Inf. Dis. 136, 444.

MURINE MODELS OF CYTOMEGALOVIRUS LATENCY

M. Colin Jordan

Department of Medicine, and Reed Neurological Research Center
UCLA, Los Angeles 90024

Although most individuals experience no illness as a consequence, human cytomegalovirus (CMV) infections are common throughout the world (1). Recently, however, the list of diseases associated with CMV infection has increased greatly, and in many instances, illness is thought to result from reactivation of virus previously latent in the human host. This is particularly true under conditions where the host immune response is impaired either by underlying disease or by immunosuppressive chemotherapy (2). Contrariwise, stimulation of the immune response, such as by pregnancy, transfusion of antigenically foreign blood components, or organ graft rejection is another potential mechanism of reactivation of latent CMV. Generally, the lung and liver have been the major organs sustaining damage under these various circumstances (1), (3).

Presently, the evidence that CMV causes a latent infection in man is indirect. It has been known for several years that infection often follows blood transfusion (4), suggesting that leukocytes from healthy donors may carry virus in a dormant state. The alternative possibility that transfusion of foreign blood components may activate virus previously latent in the recipient has been discussed by Lang (5). Nevertheless, there is no direct evidence for the existence of latent CMV in human leukocytes despite extensive studies involving over 800 blood donors (6). Other indirect evidence implicating CMV as a cause of latent infection in man includes: [1] the onset of CMV viruria in up to 40% of rheumatologic patients after initiation of immunosuppressive treatment (7), and [2] the gradual increase throughout gestation of cervical shedding of CMV in pregnant women (8). Here, reactivation of endogenous latent virus is thought to account for shedding of CMV in urine and cervical secretions respectively.

Because it is difficult to approach the question of latent CMV infection experimentally in man, investigators have turned to animal models. The mouse CMV (MCMV) system has been particularly useful because of the striking similarities between human and murine infection. The course of acute and chronic MCMV infection has been well characterized and, more recently, the first models of murine CMV latency have been developed with very interesting findings relevant to human disease. The data obtained with these models will be summarized here.

ISBN 0-12-668350-6

In 1975, Olding, et al (9) reported that mice inoculated *in utero* or at birth with tissue-culture passaged MCMV developed latent infection of splenic B-lymphocytes. The infection could be activated *in vitro* by co-cultivation of B-cells with allogeneic but not syngeneic mouse embryo indicator cells. Latent virus could also be activated by co-cultivation of splenic lymphocytes with bacterial lipopolysaccharide, a B-cell mitogen. In a subsequent paper, Olding, et al (10) noted that the majority of mice inoculated *in utero* or at birth did not have detectable active MCMV infection of other organs, thus indicating that recovery of virus from B-lymphocytes *in vitro* was probably not the result of spread of infectious virus from elsewhere in the animal. Using hybridization techniques, it was also shown that MCMV-DNA could be detected in amounts equivalent to 3 to 4 virus genomes per 100 spleen cells. Thus, in these experiments, latent infection of murine lymphocytes by MCMV was demonstrated, and allogenic stimulation was necessary to activate the infection *in vitro*. THese findings are compatible with human epidemiologic data indicating that CMV infections may be transmitted by blood transfusion and that allogenic stimulation (e.g., in renal transplantation or pregnancy) may be important in activation of latent infection.

In another model of latent MCMV infection, Mayo, et al (11) inoculated 5- to 6-week old DBA/2 mice with MCMV intraperitoneally and waited 20-24 weeks, at which time infectious virus was no longer recoverable from murine organs. Immunosuppressive treatment with cyclophosphamide consistently resulted in reactivation of MCMV in the salivary gland and, rarely, in the kidney or lung. As pointed out by the authors, this result does not necessarily indicate that salivary gland is the site harboring latent virus but rather it is the most permissive organ for virus replcation. Mayo, et al further demonstrated that MCMV infection could be transferred from latently infected to previously uninfected mice by transfusion of spleen cells. Pretreatment of recipient animals with cyclophosphamide enhanced the ability of donor spleen cells to transfer the infection. Thus, in this model, immunosuppression was demonstrated to be an important factor in activation of latent virus infection, both in the animal originally infected and in the recipient of latently infected spleen cells.

CHeung and Lang (12) have conducted studies on the possible development of latent MCMV infections in salivary gland and prostatic tissue of a small number of mice. Virus was recovered after prolonged explant culture or after establishment and passage of cell lines of these tissues from mice without previously detectable cytolytic infection. Although further verification is needed, these preliminary findings suggest the possibility of latent MCMV infection of secretory visceral organs and may have relevance to data implicating the transplanted kidney as a source of CMV infection in human patients (13).

In this laboratory (14), we have developed a model of MCMV latency in which virus becomes undetectable 16 to 20 weeks after subcutaneous inoculation in 95% of C_3H mice. The remaining 5% with persistent salivary gland infection are excluded from the experiments by surgical tissue biopsy and virus assay in cell culture. In the mice with latent infection, MCMV cannot be activated by prolonged co-cultivation of tissues (including spleen cells), explant culture techniques, or assay of tissue homogenates using centrifugal force to enhance sensitivity of virus detection. However, a two-week regimen of immunosuppression with cortisone acetate and antilymphocyte serum results in reactivation and dissemination of MCMV in virtually every mouse. Examination of the sequence of events during immunosuppression indicates that the liver is most often the site of initial detection of MCMV, although a viremia follows shortly thereafter with widespread dissemination of infection. Histologic studies performed during viral reactivation show severe involvement of numerous organs but, as in human patients, the most significant disease is found in the liver and lung. In the liver, diffusely scattered foci of inclusion-bearing hepatocytes surrounded by mononuclear cell infiltration are the dominant histologic features. Severe interstitial pneumonitis is also found consistently but these changes are more difficult to interpret since uninfected control animals may develop similar abnormalities. In the latter, however, the large numbers of MCMV intranuclear inclusions seen in the interstitial regions of infected mice are not found. Presumably, the immunosuppressive regimen predisposes control animals to the development of interstitial pneumonia caused by agents other than MCMV. Experiments are now underway using less potent immunosuppressive regimens to study more precisely the pathogenesis of interstitial pneumonia associated with reactivation of MCMV. Thus, the observations in this model are very similar to current understanding of latent CMV infection in man. The site of latent MCMV has not yet been defined, and virus cannot be activated in vitro by cocultivation of various tissues, including lymphocytes. However, reactivation and dissemination of virus with severe histopathologic damage in a variety of organs occurs regularly following immunosuppression.

From review of the murine models of CMV latency which have been developed and studied to date, it is apparent that there are similarities and differences among them. One general area of concern is the question of whether MCMV actually exists in a non-replicating and, therefore, latent state in the tissues, or whether extremely low-level virus replication continues but cannot be detected. The latter possibility cannot be excluded with currently available techniques and must always be considered in the interpretation of data derived from these systems. Nevertheless, most investigators have been careful to exclude the possibility of low-level cytolytic infection within the limitations of conventional techniques.

It is also clear that several factors effect the behavior of MCMV in the animal with regard to development and activation of latent infection (summarized in Table 1). The source of virus, the age and strain of mice, and the route of inoculation appear to be the major variables. For example, the MCMV strain used by Olding, et al (9) was attenuated by continuous passage in tissue culture, resulting in diminished virus replication in the liver and spleen during acute infection (16). When newborn mice are inoculated with this MCMV strain, recovery from lymphocytes is possible by cultivation with allogeneic but not syngeneic indicator cells. However, using virulent virus harvested from murine salivary glands, Mayo, et al (11) have been able to demonstrate latency in splenic lymphocytes of adult mice by transfusion techniques but activation of MCMV does not so clearly relate to allogeneic stimulation. This finding could be due to differences in virus or mouse strains, timing of infection, or a greater complexity of *in vivo* versus *in vitro* activation. In our system employing virulent virus given to adult C H mice by a different route (subcutaneously), latent MCMV cannot be activated *in vitro* by any of the techniques employed by other investigators and transfusion of latent virus in lymphocytes has not been demonstrated (14). Preliminary experiments using an intraperitoneal route of infection, however, indicate that virulent virus can be recovered from splenic lymphocytes in vitro, suggesting that different routes of inoculation may significantly alter the manner in which the viral infection is "processed" by the host.

In summary, several murine models of cytomegalovirus latency have been developed, and each has unique features which may be valuable in advancing our understanding of latent CMV infections in man. By combining the more relevant features of the models described to date, it should be possible to define the pathogenesis of latent human CMV infections, particularly those associated with blood transfusion, graft rejection, graft-versus-host disease, and immunosuppression.

Table 1. Summary of Major Features of Murine Models of Cytomegalovirus Latency

Investigators	Strain of Virus	Route of Inoculation	Age of Inoculation	Strains of Mice	Activation from Splenic Lymphocytes	Transfer of MCMV by Spleen Cells or Whole Blood	Dissemination by Immuno-suppression	Disease from Reactivation
Olding, et al.[9]	Attenuated	I.P.	Newborn, in utero	C_3H/HeJ Ha/ICR	Yes	N.D.*	N.D.	N.D.
Mayo, et al.[11]	Virulent	I.P.	5-6 weeks	DBA/2	N.D.	Yes	Yes	N.D.
Cheung, et al.[15]	Virulent	I.P.	7-8 weeks	C_3H Balb/c	N.D.	Yes	N.D.	N.D.
Jordan et al.[14]	Virulent	S.Q.	4 weeks	C_3H/st SJL/J	No	No	Yes	Yes

*N.D. = not done.

REFERENCES

1. Weller, T. H. (1971). N. Engl. J. Med. 285, 203, 267.
2. Craighead, J. E., Hanshaw, J. B., and Carpenter, C. F. (1967). JAMA 201, 725.
3. Meyers, J. D., Spencer, H. C., Watts, J. C., Gregg, M. B., Stewart, J. A., Troupin, R. H., and Thomas, E. D. (1975). Ann. Intern. Med. 82, 181.
4. Prince, A. M., Szmuness, W., Millian, S. J., and David, D. S. (1971). N. Engl. J. Med. 284, 1125.
5. Lang, D. J. (1972). Archiv fur die Gesamte Virusforschung 37, 385.
6. Bayer, W. L., Tegtmeier, G. E. (1976). Yale J. Biol. Med. 49, 5.
7. Dowling, J. N., Saslow, A. R., Armstrong, J. A., and Ho, M. (1976). Yale J. Biol. Med. 49, 77.
8. Numazaki, Y., Yano, N., Morizuka, T., Takai, S., and Ishida, N. (1970). Am. J. Epidemiol. 91, 410.
9. Olding, L. B., Jensen, F. C., and Oldstone, M. B. A. (1975). J. Exp. Med. 141, 561.
10. Olding, L. B., Kingsbury, D. T., and Oldstone, M. B. A. (1976). J. Gen. Virol. 33, 267.
11. Mayo, D. R., Armstrong, J. A., and Ho, M. (1977). Nature 267, 721.
12. Cheung, K. S., and Lang, D. J. (1977). Infect. Immun. 15, 568.
13. Ho, M., Suwausirikul, S., Dowling, J., Youngblood, L., and Armstrong, J. (1975). N. Engl. J. Med. 293, 1109.
14. Jordan, M. C., Shanley, J. D., and Stevens, J. G. (1977). J. Gen. Virol. 37, 419.
15. Cheung, K. S., and Lang, D. J. (1977). J. Infect. Dis. 135, 841.
16. Osborne, J. E., Walker D. L. (1970). Infect. Immun. 3, 228.

Author Index

Subject Index

R

S

T

A
B
C 8
D 9
E 0
F 1
G 2
H 3
I 4
J 5